High-Yielding Dairy Cows

High-Yielding Dairy Cows

Editor

João Carlos Caetano Simões

MDPI • Basel • Beijing • Wuhan • Barcelona • Belgrade • Manchester • Tokyo • Cluj • Tianjin

Editor
João Carlos Caetano Simões
University of Trás-os-Montes e
Alto Douro
Portugal

Editorial Office
MDPI
St. Alban-Anlage 66
4052 Basel, Switzerland

This is a reprint of articles from the Topical Collection published online in the open access journal *Animals* (ISSN 2076-2615) (available at: https://www.mdpi.com/journal/animals/special_issues/High_yielding_Dairy_Cows).

For citation purposes, cite each article independently as indicated on the article page online and as indicated below:

LastName, A.A.; LastName, B.B.; LastName, C.C. Article Title. *Journal Name* **Year**, *Volume Number*, Page Range.

ISBN 978-3-0365-3046-8 (Hbk)
ISBN 978-3-0365-3047-5 (PDF)

Contents

About the Editor

João Carlos Caetano Simões (DVM, PhD) is a Professor of Large Animal Medicine and Reproduction at the University of Trás-os-Montes e Alto Douro (UTAD). As a clinician and academic, he has dedicated 27 years of his professional career to working with veterinary students and producers, and has collaborated with national and European animal production associations. On the basis of this work experience and research, he has written numerous scientific and technical publications. In recent years, he has also edited several books and Special Issues on animal production and veterinary medicine for scientific journals: https://orcid.org/0000-0002-4997-3933. His main research interests include medicine of production, preventive veterinary medicine, large animal reproduction, animal production, and animal nutrition.

Preface to "High-Yielding Dairy Cows"

In recent decades, genetic, nutrition, and management improvements have resulted in high-yielding dairy cows. According to FAOSTAT data [1], 715,922,506 tons of fresh milk was globally produced in 2019. Currently, the total milk yield (reaching 100,000 kg [2]) and fat and protein contents reach unpreceded levels during the lifetime production of dairy cows, and further increases are expected in the next few years. These improvements pose threats and challenges to animal health and welfare due to metabolic stress faced.

Metabolic stress is characterized by excessive lipomobilization, inflammatory and immune dysregulations, and oxidative stress on body organic systems. It mainly occurs during the periparturient period. Cows are more susceptible to metabolic and infectious diseases such as ketosis, digestive alterations, metritis, mastitis, lameness, and different degrees of infertility.

The ultimate goal of this Special Issue was to aggregate new information striving for more efficient dairy cow production in confinement and pasture systems. A total of 20 papers (4 full reviews and 16 research studies) directly or indirectly assessing all aforementioned aspects in high-producing dairy cows were published from different regions worldwide, ensuring a global approach to this reality.

We hope that this Special Issue can effectively add some basic and applied scientific value to knowledge in the dairy industry to satisfy the global demand for milk protein and fat by an ever-expanding global population. This is one key pattern to ensure the accessibility of safe food to populations of distinct socio-economic statuses, mitigating the adverse environmental impacts of dairy production.

References

1. FAOSTAT (Food and Agriculture Organization of the United Nations Statistical Databases) 2021. Retrieved from http://faostat.fao.org/ (accessed on 4 December 2021).
2. Van Eetvelde, M.; Verdru, K.; de Jong, G.; van Pelt, M. L.; Meesters, M.; Opsomer, G. Researching 100 t cows: An innovative approach to identify intrinsic cows factors associated with a high lifetime milk production. *Prev. Vet. Med.* **2021**, *193*, 105392.

João Carlos Caetano Simões
Editor

animals

MDPI

Article

Metabolomics Approach Explore Diagnostic Biomarkers and Metabolic Changes in Heat-Stressed Dairy Cows

Shuangming Yue [1,2], Siyan Ding [1], Jia Zhou [1], Chao Yang [2], Xiaofei Hu [1], Xiaonan Zhao [1], Zhisheng Wang [1], Lizhi Wang [1], Quanhui Peng [1] and Bai Xue [1,*]

1 Animal Nutrition Institute, Sichuan Agricultural University, Chengdu 611130, China; yueshuangming@hotmail.com (S.Y.); dalanmaoding0906@sina.com (S.D.); zhoujia1@stu.sicau.edu.cn (J.Z.); hxf8982@126.com (X.H.); zxn780213675@126.com (X.Z.); wangzs@sicau.edu.cn (Z.W.); 12825@sicau.edu.cn (L.W.); 14101@sicau.edu.cn (Q.P.)

2 Department of Bioengineering, Sichuan Water Conservancy Vocation College, Chengdu 611845, China; mxp886@sicau.edu.cn

* Correspondence: xuebai@sicau.edu.cn

Received: 3 August 2020; Accepted: 16 September 2020; Published: 25 September 2020

Simple Summary: Heat stress results in a decline in the production performance of the dairy cows. This study explored the metabolomes of milk and blood plasma in heat-stressed cows by means of ^{1}H nuclear magnetic resonance (^{1}H NMR)-based metabolomics tools. Findings of the current experiment reveal that HS alters the metabolic composition of milk and blood plasma in lactating dairy cows. In brief, HS alters twelve metabolites in blood plasma and eight metabolites in milk, which are primarily involved in proteolysis, gluconeogenesis, and milk fatty acid synthesis, suggesting that these metabolites could be possible biomarkers for heat-stressed dairy cows.

Abstract: In the present experiment, we investigated the impact of heat stress (HS) on physiological parameters, dry matter intake, milk production, the metabolome of milk, and blood plasma in lactating Holstein dairy cows. For this purpose, 20 Holstein lactating cows were distributed in two groups in such a way that each group had 10 cows. A group of 10 cows was reared in HS conditions, while the other group of 10 cows was reared in the thermoneutral zone. The results of the experiment showed that cows subjected to HS had higher respiration rates ($p < 0.01$) and greater rectal temperature ($p < 0.01$). Results of milk production and composition explored that HS lowered milk production ($p < 0.01$) and milk protein percentage ($p < 0.05$) than cows raised in a thermoneutral place. Furthermore, HS increased the concentrations of N-acetyl glycoprotein, scyllo-inositol, choline, and pyridoxamine in milk, while HS decreased the concentrations of O-acetyl glycoprotein, glycerophosphorylcholine, citrate, and methyl phosphate in milk. Moreover, HS enhanced plasma concentrations of alanine, glucose, glutamate, urea, 1-methylhistidine, histidine, and formate in cows, while the plasma concentration of low-density lipoprotein, very-low-density lipoprotein, leucine, lipid, and 3-hydroxybutyrate decreased due to HS. Based on the findings of the current research, it is concluded that HS alters the milk and blood plasma metabolites of lactating Holstein dairy cows. Overall, in the current experiment, HS altered eight metabolites in milk and twelve metabolites in the plasma of lactating Holstein dairy cows. Furthermore, the current study explored that these metabolites were mainly involved in proteolysis, gluconeogenesis, and milk fatty acid synthesis and could be potential biomarkers for dairy cows undergoing HS.

Keywords: heat stress; dairy cow; milk production; metabolomics; biomarkers

1. Introduction

Heat stress (HS) is negatively influencing animal health, production, and reproduction performance in modern livestock production systems [1]. The impact of HS on animals' health, production, and reproduction performance is variable in different livestock species and breeds [2,3]. In beef cattle, HS is generally considered less crucial than dairy cattle because beef cattle have lower body size and metabolic rate, which results in lower body heat production [2,3]. Hence, dairy researchers are interested in conducting research on a topic related to the effect of HS on dairy animals' health, production, and reproduction performance. Holstein dairy cow is a popular dairy breed among dairy farmers due to its high milk production. However, the potential of Holstein cows to emit body heat by skin evaporation is reduced in hot and humid environments [4]. Therefore, it is assumed that Holstein cows are at a greater risk of facing HS [5]. Findings of previous experiments have explored that HS in dairy cows not only reduces milk production and milk quality but also results in metabolic disorders, rumen acidosis, negative energy balance, and even death [6–8]. These metabolic disorders and health problems are directly linked with economic losses in the dairy industry [2,9,10].

In dairy cow's production systems, the temperature and humidity index (THI) is considered as an index of the HS. Nevertheless, THI is only based on environmental humidity and temperature [1] and is not a direct biomarker of metabolic changes in dairy cows in response to HS. Therefore, in the dairy production system, vigorous metabolite biomarkers are required to identify the HS occurrence and to provide understandings related to physiological mechanisms in heat-stressed dairy cows.

Previous studies' findings have explained that HS in lactating dairy cows not only results in reduced milk production but also negatively influences the milk protein contents [11,12]. It is generally considered that lower dry matter intake (DMI) is the major reason for reduced milk production and milk protein contents in heat-stressed dairy cows [7,13]. However, the latest studies have reported that lower dry matter intake partially (about 35–50%) describes the reduction in milk production and milk composition in dairy cows [11,12]. Therefore, it could be assumed that change in post-absorptive metabolism may be responsible for a large portion of the reduction in productivity of dairy cows; however, the mechanisms triggering these changes still remain vague [14].

Metabolomics, a useful tool for analyzing the changes of metabolites in physiological fluids and tissues in response to internal and external stimulations, has been successfully applied for filtering out biomarkers for milk quality [15], energy metabolism [16], and rumen health in dairy cows [14]. However, only a few studies have attempted to identify biomarkers for the diagnosis of HS in dairy cows based on milk [15] and plasma [16] metabolomics. Previous studies have reported that blood and milk metabolites of lactating dairy cows change during HS [14,17–19]. However, to date, none of the studies have been carried out to evaluate the effect of HS on both milk and plasma metabolome and their association on the milk yield of mid-lactating dairy cows. In this study, we formulated the hypothesis that dairy cows suffering from HS may manifest signs of milk and blood metabolite changes, which can be used as biomarkers for evaluating the influence of HS on dairy cows. Furthermore, the other objective of the current experiment was to find changes in metabolic pathways due to HS and to identify the relationship of metabolic changes with milk yield reduction in heat-stressed dairy cows by comparing the metabolomes of milk and plasma between cows reared in HS and thermoneutral (TN) conditions.

2. Materials and Methods

The current experimental trial was conducted at Mei Jiadun Dairy Farms (Huang Gang, Hubei, China). All experimental procedure was in accordance with Chinese laws on Animal Experimentation (GB/T 35892-2018). Furthermore, all research protocols were permitted by the Institute Animal Care Committee of Sichuan Agricultural University, China (#SCAUAC201408-3).

2.1. Experimental Design and Animal Management

A total of twenty healthy Holstein cows from the Mei Jiadun Dairy Farms (Huang Gang, Hubei, China) herd, with healthy and symmetrical udders, were selected for the current experiment. All cows were similar in parity, days in milk, milk yield, and body weight (Table 1). Cows were maintained and reared under the same feeding and management regime before the start of the experiment. For these experiments, experimental animals were kept in a cowshed (closed type) to avoid environmental variation and photoperiods on cow's metabolism. Shed (150 m × 20 m) was equally divided into 10 parts with the help of temporary sidewalls of bamboos and steel pipes. Each part was designated to a single cow. Every part of the barn was bedded with geotextile mattresses and topped with a thin layer of sawdust. Each part of the barn contained a drinker and fan. To avoid light, shading nets were used around the cowshed.

Table 1. Characteristics of the experimental cows.

Parameter	TN [1] (Thermal Neutral)	HS [2] (Heat Stress)	*p*-Value
Number of cows	10	10	
Parity	3.0 ± 1.0	3.2 ± 1.1	0.78
Lactation days	146.2 ± 12	145 ± 25	0.95
Milk yield (kg)	26.8 ± 1.5	27.2 ± 2.1	0.95
Average bodyweight (kg)	604.4 ± 56.44	601.81 ± 78.97	0.95

[1] TN: Thermal neutral. [2] HS: heat stress.

Each experimental animal was reared in an individual pen. Fresh drinking water was ensured for each pen. All other managemental practices were also similar for all animals. Ten cows were considered for HS treatment, and similarly, ten cows were considered for TN treatment.

Samples from HS experimental animals were collected in the summer period (15 July to 30 August). In the summer season, THI was increased from 75 to 85 over one month and stabled at 80.50 for seven days. While the samples of TN treatments were gathered in the spring (15 march to 30 April). In the spring season, THI steadily increased from 58.8 to 62.3 over a one-month period. Before the onset of the experiment, all experimental animals were given an adaptation period of 15 days. A total of 30 days duration was given as an experimental period in the current study. All data were collected in 30 days of the experimental period.

Experimental animals were reared on total mixed ration (TMR). The frequency of feeding the animals was twice a day at 0500 and 1700. The TMR was supplied by the gate feeders, and 10% refusals were permitted. All experimental animals were milked twice a day (at 0500 and 1700). Animals were milked in a conventional milking parlor with a machine milking system (DeLaval, Tumba, Sweden). The milk production of each cow was recorded at each milking.

The diet used in this study was prepared to fulfill the requirements of a cow weighing 550 kg and producing 25 kg of milk (3.1% milkfat) daily, as suggested by National Research Council (NRC) 2001 [20]. Furthermore, the net energy for lactation (NEL) was 6.36 MJ/kg dry matter, which could meet the cow's needs for TN conditions. The TMR's ingredients and chemical composition are presented in Table 2.

Table 2. Composition (%. of dry matter (DM)) and nutritional value of the diet.

Item	%
Ingredient	
Silage (Corn silage)	19.63
Alfalfa hay	5.82
Hemarthriia altissima [1]	10.60
Sweet potato vine	13.95
Corn meal	23.50
Wheat bran	6.00
Soybean meal	9.70
Rapeseed meal	6.00
Calcium fatty acids	1.00
Limestone	1.20
Calcium superphosphate	1.00
Salt	0.60
Premix [2]	1.00
Total	100
Roughage-to-concentrate	50/50
Nutrient levels [3]	
NEL [4] (MJ/kg)	6.36
NEL (Mcal/kg)	1.51
CP [5]	16.18
NDF [6]	36.46
ADF [7]	22.96
Ca [8]	1.06
P [9]	0.56

[1] *Hemarthria altissima* and sweet potato vine fresh mowing. [2] Premix contained (per kilogram of premix): VA—5,00,000 IU, VD— 1,50,000 IU, VE—3000 IU, Fe (iron) —4.0 g, Cu (copper) —1.3 g, Mn—3.0 g, Zn (Zinc)—6.0 g, I (iodine)—80 mg, Se (selenium)—50 mg, and Co (cobalt)—80 mg. [3] Net energy for lactation was the calculated value; the others were measured values. [4] NEL: Net energy for lactation; [5] CP: Crude protein; [6] NDF: Neutral detergent fiber; [7] ADF: Acid detergent fiber; [8] Ca: Calcium; [9] P: Phosphorus.

2.2. Sampling and Analysis

The samples of fresh and refused TMR were collected on a weekly basis during the experimental duration (30 days). Collected samples of TMR were oven-dried at 65 °C for 24 h. Dried samples of TMR were ground through a 1 mm screen and were analyzed for dry matter, ash, calcium, phosphorus, Crude protein (CP), Neutral detergent fiber (NDF), and Acid detergent fiber (ADF). Proximate analysis was carried out following the protocol of Association of Official Analytical Chemists (AOAC) [21]. The contents of fiber in samples were determined following the protocol of Van Soest et al. [22].

In the last week of the summer period and spring period, morning milk samples (volume of 25 mL each) were collected at day 7 from the individual cow for the determination of milk components (fat, lactose, and protein) and somatic cells counts (SCC) (Arizona DHIA, Tempe, AZ, USA). In brief, a total of 25 mL of composite milk from each animal, with approximately equal volumes from each lactating udder quarter, was transferred to a sterile plastic bottle (Corning, Inc., Corning, NY, USA) and kept on ice until transport to the laboratory and storage at −80 °C for further analysis. Similarly, one-day milk samples (10 mL) from every cow were also collected on the last day of the experimental trial for ^{1}H nuclear magnetic resonance (^{1}H NMR) analysis. The collected sample of milk was defatted by centrifugation at 3000× *g* for fifteen min at four-degree Celsius. The skimmed milk samples were stored at −80 °C for H.NMR analysis.

Samples of blood were also collected from the coccygeal vein of the cows before morning feeding on the last day of the experimental trial. The blood samples were stored in evacuated tubes containing EDTA for anticoagulation. The collected samples were centrifuged, the same as described for the production of skimmed milk except for the centrifugation time. The centrifugation time was 10 min for blood samples. After the centrifugation procedure, blood plasma was aliquoted and stored at −80 °C until NMR analysis.

2.3. Measurements

Milk production and milk fat content were utilized to determine 4% fat-corrected milk. Rectal temperature (RTs) was determined 3 times per day by using a clinical thermometer (Nanjing, China). The respiratory rate was measured at 07:00, 14:00, and 22:00 twice a week by counting the total number of flank movements in one min and recorded as breaths per minute. DMI was recorded on a daily basis. DMI was measured by subtracting daily refusals from the daily offered feed. The following formula was used to calculate the temperature and humidity index (THI) following NRC 2001 [20].

$$THI = (Tdb + Twb) \times 0.72 + 40.6 \quad (1)$$

where Tdb. = dry bulb temperature (°C), Twb. = wet bulb temperature (°C).

The Tdb and Twb temperatures were recorded daily at 0700, 1400, and 2200. Cows were weighed for 3 consecutive days with an empty stomach in the morning, both at the beginning and the end of the experiment.

2.4. ^{1}H Nuclear Magnetic Resonance (^{1}H NMR) Spectroscopic Measurement

We followed the procedure of sample preparation and NMR experiments, as described in the literature [23]. In brief, samples of blood plasma were melted at room temperature. Thawed plasma was homogenized by using a vortex mixer. After that, 400 μL of plasma sample was placed in a 1.5 mL plastic tube and mixed with 170 μL of D_2O for detecting the signal. A 30 μL of Phosphate Buffer Saline (PBS) was added to reduce variations in the pH of the sample. After that, the samples of plasma were centrifuged at 12000 revolutions per minute (rpm) for ten minutes at four degrees Celsius. After centrifugation, 500 μL of supernatant was shifted into 5 Mm. nuclear magnetic resonance tubes. NMR tubes containing supernatant were stored at four degrees Celsius until further analysis. Nuclear magnetic resonance spectra of all samples were obtained at 298 K on a Bruker Avance 600 with a 599.91 Hz, the acquisition time of 0.9984 s, the spectral width of 8.01 kHz, and 2.1 s relaxation delay with 128 scans collected into 32K data points. One dimensional (1D) spectra were recorded using the Carr-Purcell-Meiboom-Gill (CPMG) experiment to suppress water signals and broad protein resonances.

All obtained ^{1}H.NMR spectra were phased manually, and the baseline was adjusted by the use of MesReNova 7.1software. Before Fourier transformation, Free Induction Decay Signal (FIDs) were multiplied by an exponential with a 0.3Hz line broadening factor. The nuclear magnetic resonance spectrum was a reference to the lactate resonance at 1.33 ppm. Each spectrum (0.5–9.0 ppm) was split into 0.002 ppm bins, removing the residue water region from 5.2 to 4.5 ppm. The left-over bins of each spectrum were normalized to a total spectral area of unity prior to pattern recognition.

2.5. ^{1}H NMR Data Handling

Free induction decays were multiplied by a factor of 1.0 Hz and phased manually. The chemical shift of each milk sample was referenced to the signal of the sodium trimethylsilyl (2, 2, 3, 3-2 H4) 1 propionate (TSP) (δ 0 ppm), and the chemical shift of each plasma sample was referenced to the signal of L-lactate (δ 1.33ppm). The NMR spectra of milk samples (δ 0.6–9.0 ppm) and plasma samples (δ 0.5–9.0 ppm) were further divided into 0.002 ppm integral region and integrated. The parts of δ 4.73–5.16 ppm and δ 3.33–3.34 ppm were eliminated to eradicate the baseline effect of imperfect water supersession and effects of ethanol on milk samples, and the region of δ 4.5–5.2 ppm was removed to eradicate baseline effects of inadequate water supersession on plasma samples.

Multivariate analyses, comprising Principal Component Analysis (PCA) and Orthogonal Partial Least Squares Discrimination Analysis (OPLS-DA), were carried out by the use of SIMCA-P11.0

software (UmetricsUmea, Sweden). PCA is known to show the core structure of datasets in an unbiased pattern and to reduce the dimensionality of experimental data. In the current experiment, PCA was used to explore the global clustering and variations in metabolic profiles between obtained samples. Furthermore, the data of mean center scaling was employed for PCA. For regression, partial least square discrimination analysis was used, and the R^2X, R^2Y, and Q^2 parameters were carried out to assess the model quality. The R^2Y and R^2X parameters, which characterize the fractions of X and Y variations, respectively, were utilized to assess the quality of the model. The predictive ability of the model was represented by the Q^2 parameter. Usually, the model is acceptable when the values of R^2Y, R^2X, and Q^2 are more significant than 0.5. After obtaining the initial overview of PCA and Partial Least Squares Discrimination Analysis (PLS-DA) analysis, a more sophisticated OPLS-DA model with the precise discriminant info was obtained from the HS and TN periods.

The difference of the metabolites in experimental treatments is represented as a coefficient of variation plots. The OPLS-DA model was utilized to increase the separation between samples by eliminating the variation in the X matrix unrelated to the Y matrix [24]. The multivariate models were validated by a 6-round cross-validation procedure to guard against overfitting and by permutation tests ($n = 200$). The color-coded coefficient of loading plots of the OPLS-DA model was utilized to determine the difference between the HS and TN periods.

Based on the number of samples in the current experiment, the correlation coefficients of $|r| > 0.63$ and $|r| > 0.60$ were used as the cutoff values for the significance of milk and plasma samples, respectively.

2.6. The Differential Metabolites Identification and the Analysis of the Metabolic Pathway

The OPLS-DA and PLS-DA were used to identify the difference in metabolites. SPSS 16.0 was used to calculate the receiver operating characteristic (ROC) curves, and the diagnostic value of these differentials was determined by the areas under the curves (AUCs). The Kyoto Encyclopedia of Genes and Genomes (KEGG) was used to find the metabolic pathways and enrichment analysis. After that, MetaboAnalyst3.0 (https://www.metaboanalyst.ca/) was used for the pathway analysis.

2.7. Calculation and Statistical Analysis

The effects of HS on physiological parameters and milk yield were analyzed by repeated-measures by using the PROC GLM procedure of SAS 9.0 software (SAS Institute Inc., Cary, NC, USA). Least square means were separated by using the Pare difference (PDIFF) procedure of SAS 9.0 software (SAS Institute Inc, Cary, NC, USA). The significance was declared at $p < 0.05$.

3. Results

3.1. Environmental Conditions

The THI values calculated during the experimental periods are shown in Figure 1. Results revealed that daily mean THI was between 58.5 and 62.3 during the spring season (avg. 60.8 in the whole spring season). The daily mean THI was ranged from 75.0 to 85.0 during the summer season (avg. 80.5 in the whole summer season). The mean daily THI in the entire summer period exceeded 72, indicating that dairy cows in the HS treatment suffered HS.

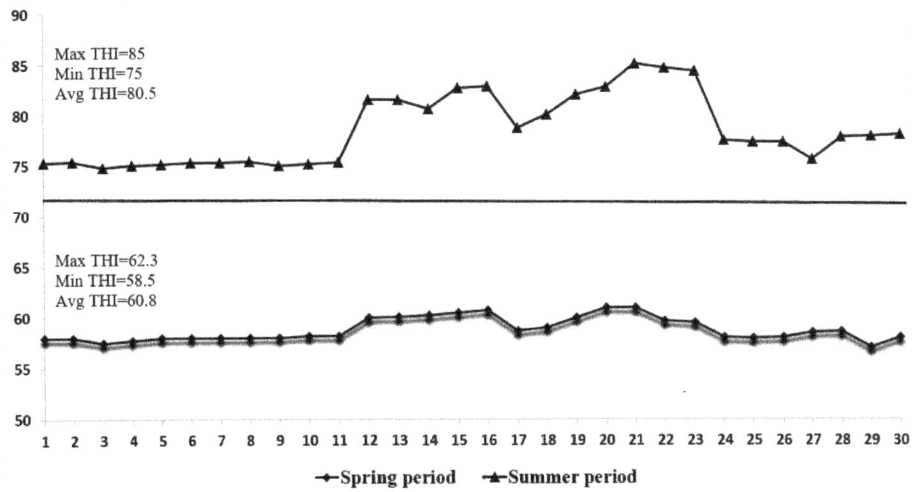

Figure 1. The temperature and humidity index (THI) values (mean/d) during the summer period ($n = 10$, triangle line) and spring period ($n = 10$, rectangle line).

3.2. Physiological Parameters and Animal Performance

The physiological parameters and performance of cows are illustrated in Table 3. Bodyweight did not differ both at the beginning and the end of the HS and TN periods. However, body weight change in the experimental treatments varied greatly ($p < 0.01$). The respiratory rates and RTs were higher ($p < 0.01$) in the HS experimental treatment than the TN treatment.

Table 3. Physiological parameters and performance of cows reared in heat stress and thermoneutral periods (mean ± SD).

Item	Thermoneutral Period	Heat Stress Period	*p*-Value
Respiration rate (breaths/min)	35.4 ± 4.7	82.3 ± 10.5	<0.01
Rectal temperature (°C)	38.6 ± 0.1	39.4 ± 0.3	<0.01
Initial [1] BW (kg)	604.4 ± 56.4	601.8 ± 79.0	0.91
Final BW (kg)	612.2 ± 55.6	565.5 ± 74.6	0.17
BW changes (kg)	7.8 ± 8.3	−36.4 ± 11.2	<0.01
[2] DMI (kg)	18.1 ± 2.5	15.5 ± 2.0	<0.05
Milk yield (kg/d)	28.6 ± 1.4	19.9 ± 2.1	<0.05
Protein (%)	3.2 ± 0.2	2.9 ± 0.9	<0.05
Protein yield (g/d)	914.5 ± 2.2	578.1 ± 1.8	<0.01
Fat (%)	3.6 ± 0.7	3.8 ± 0.6	0.12
Fat yield (g/d)	1026.6 ± 3.2	753.8 ± 2.5	<0.05
Lactose (%)	4.9 ± 0.2	4.7 ± 0.2	0.19
Lactose yield(g/d)	1407.2 ± 5.2	933.5 ± 3.8	<0.05
Somatic cells count 1000/mL	286± 8.2	312± 9.5	0.03

[1] BW = bodyweight, [2] DMI = dry matter intake.

Heat stress reduced milk production ($p < 0.05$), milk protein contents ($p < 0.05$), milk protein yield ($p < 0.01$), and DMI ($p < 0.05$). Heat stress did not alter milk lactose contents and milk fat contents ($p > 0.10$). Results showed that milk fat yield and lactose yield were lowered in the HS experimental cows as compared to the TN experimental group animals. Results showed that HS increased the SCC ($p = 0.03$) of cows.

3.3. ^{1}H NMR Spectra of Milk and Blood Plasma Samples

The representative 600 MHz of Varian 600 (Agilent, Santa Clara, CA, USA) spectra gave an overview of the metabolic profiles from milk and plasma (Supplementary Materials Figures S1–S3) of cows in the HS and TN groups. By visual inspection of the ^{1}H NMR spectra, different metabolite patterns were observed between the experimental groups. Milk concentrations of pyridoxamine and scyllo-inositol and plasma concentrations of glucose and formate appeared to be higher in cows of the HS group. To get a more intensive analysis of the difference in metabolites in the experimental treatments, NMR data was further analyzed by using multivariate analysis, including PLS-DA, PCA, and OPLS-DA, to obtain the significant differences of obtained metabolites between experimental treatments.

3.4. Identification of Different Metabolites

The PCA and PLS-DA of the ^{1}H.NMR data from milk and plasma represented a separation in the experimental treatments (Figure 2, Supplementary Materials Figures S4 and S5). OPLS-DA analysis revealed that concentrations of eight metabolites in the samples of milk and twelve metabolites in the samples of plasma were significantly different between the HS and TN experimental groups (Figures 3 and 4). The cows in the HS groups had a higher level of choline, N-acetyl glycoprotein (NAG), scyllo-inositol, and pyridoxamine and lower levels of citrate, O-acetyl glycoprotein (OAG), glycerol phosphorylcholine (GPC), and methyl phosphate in milk as compared to cows in the TN groups. The cows from the HS group had greater levels of glutamate, alanine, glucose, histidine, urea, 1-methylhistidin, and formate and lower levels of Very low density lipoprotein (VLDL), low density lipoprotein (LDL), lipid, leucine, and 3-hydroxybutyrate in plasma as compared to the cows in the TN experimental group (Table 4). We further employed ROC analysis to find the diagnostic values of obtained different metabolites for distinguishing the HS experimental cows from the TN experimental cows (Figure 5A,B). Most of the metabolites showed excellent diagnostic potential with the AUC values of 0.70–1.0.

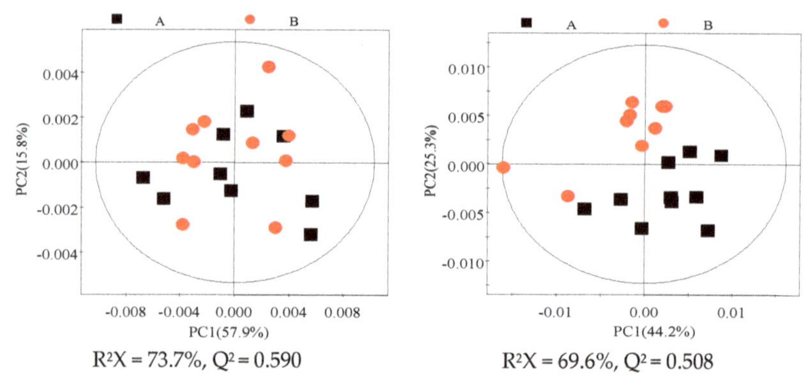

Figure 2. Principal Component Analysis (PCA) scores plot based on ^{1}H.NMR spectrum of aqueous phase acquired from the HS (heat stress) and TN treatments (**left**). PCA scores plot based on ^{1}H NMR spectra of plasma acquired from the HS and TN treatments (**right**). A = heat stress (HS) group; B = thermal neutral (TN) group.

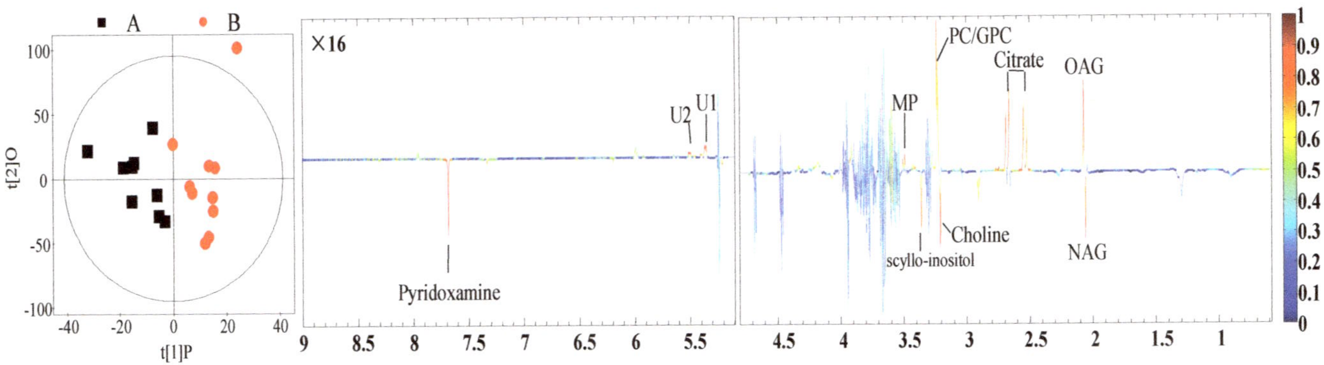

Figure 3. Orthogonal Partial Least Squares Discrimination Analysis (OPLS-DA) scores plots (**left panel**) and the corresponding coefficient loading plots (**right panel**) of milk. A = heat stress treatment; B = thermal neutral treatment.

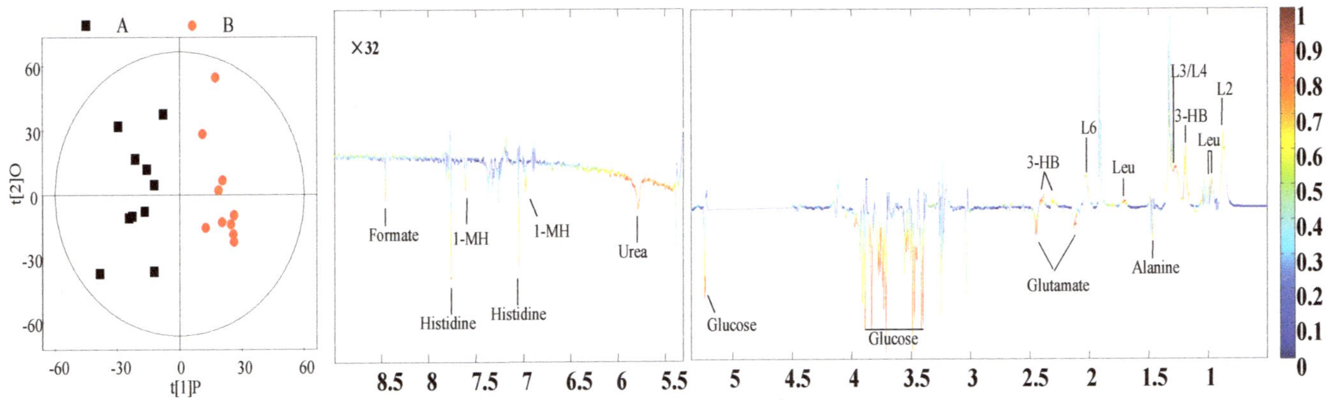

Figure 4. OPLS-DA scores plots (**left panel**) derived from the ^{1}H. NMR spectra of plasma and the corresponding coefficient loading plots (**right panel**) acquired from the different pairwise groups. A = heat stress group; B = thermal neutral group.

Table 4. OPLS-DA coefficients derived from the ^{1}H NMR data of metabolites in the milk and plasma of cows reared in the heat stress and thermoneutral zone.

	Metabolites [1]	Identification (ppm) and Multiplicity [2]	Correlation Coefficients (r) [3]	*p*-Value
Milk	N-Acetyl glycoprotein	2.06 (s) [4]	−0.73	0.02
	O-Acetyl glycoprotein	2.07 (s)	0.68	0.01
	Citrate	2.53 (d), 2.67 (d) [5]	0.74	<0.01
	Choline	3.20 (s)	−0.64	0.03
	Glycerophosphorylcholine	3.23 (s)	0.63	0.01
	Methyl phosphate	3.49 (d)	0.68	0.1
	scyllo-Inositol	3.36 (s)	−0.66	0.02
	Pyridoxamine	7.67 (s)	−0.68	0.02
Plasma	L2, L4: VLDL	0.89 (br), 1.29 (br) [6]	0.81	<0.01
	L3: LDL	1.27 (br)	0.66	<0.01
	L6: Lipid	2.01 (br)	0.64	<0.01
	Leucine	0.96 (t) [7], 1.69 (m) [8]	0.75	<0.01
	3-Hydroxybutyrate	1.20 (d)	0.74	<0.01
	Alanine	1.48 (d)	−0.64	<0.01
	Glutamate	2.13 (m), 2.46 (m)	−0.65	<0.01
	Glucose	3.42 (t), 3.54 (dd) 3.71 (t), 3.73 (m) 3.84 (m), 5.23 (d)	−0.71	<0.01
	Urea	5.78 (br)	−0.66	0.02
	Histidine	7.05 (s), 7.76(s)	−0.63	<0.01
	1-Methylhistidine	6.96 (s), 7.61(s)	−7.64	<0.01
	Formate	8.45 (s)	−0.68	<0.01

[1] VLDL: very-low-density lipoprotein; LDL, low-density lipoprotein. [2] Multiplicity: s represents singlet, d represents doublet, t represents triplet, q represents quartet, dd represents doublet of doublets, m represents multiplet, br represents broad resonance. [3] Correlation coefficients: positive and negative signs show a positive and negative correlation in the concentrations of milk metabolites in dairy cows in the TN (thermoneutral) group relative to cows in the HS (heat stress) group. [4] s: singlet; [5] d: doublet; [6] br: broadlet; [7] t: triplet; [8] m: multiplet.

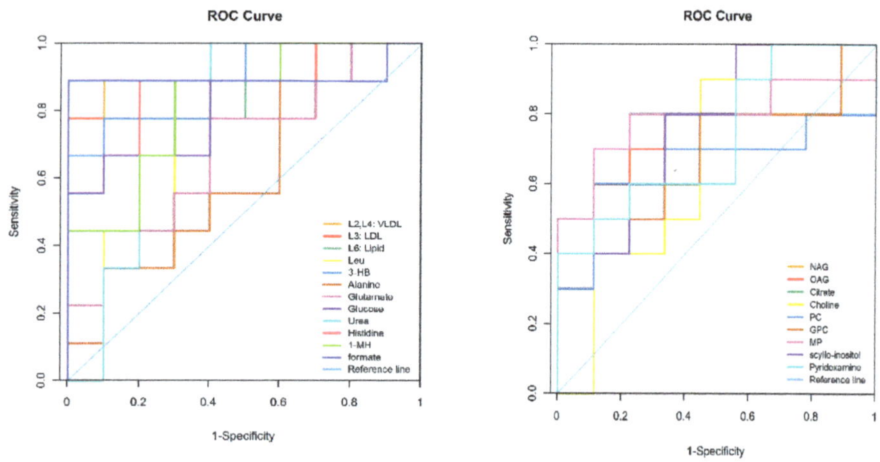

(**A**) (**B**)

Figure 5. ROC (receiver operating characteristics) analysis of biomarkers from the milk (**A**) and plasma (**B**) for cows from the HS (heat stress) and TN (thermoneutral) periods.

3.5. Integration of Key Different Metabolic Pathways

In the current experiment, a total of twelve pathways were obtained. Those metabolites were involved in various biochemical pathways, such as proteolysis, glycolysis, lipolysis, gluconeogenesis, galactose, inositol phosphate metabolism, muscle protein catabolism, purine metabolism, and alanine metabolism (Figure 6A,B). The obtained results were combined to draw a metabolic network map, as shown in Figure 7, to represent the biological significance of the variation of plasma and milk concentrations of metabolites as a result of HS.

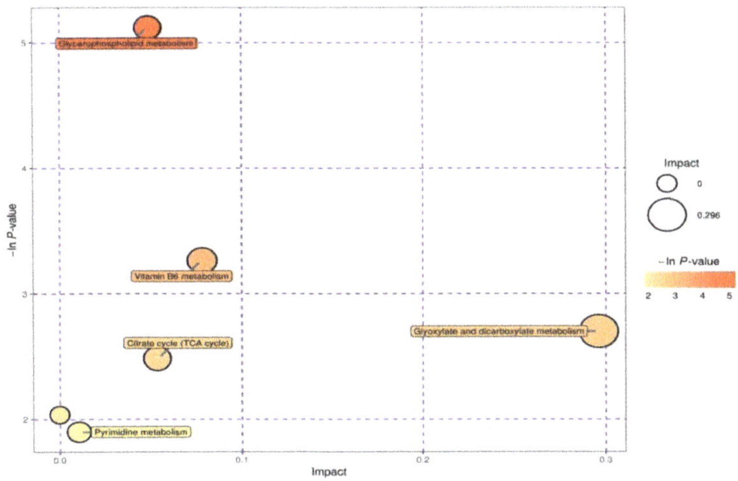

(A)

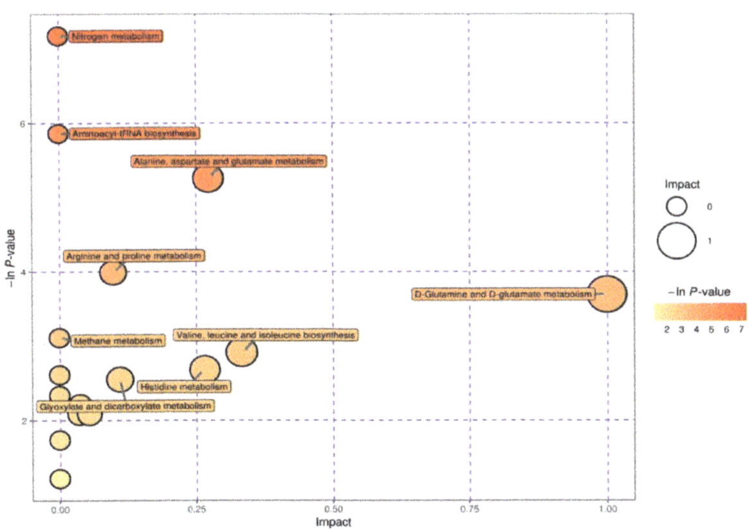

(B)

Figure 6. The metabolome view map of significant metabolic pathways characterized in milk (**A**) and plasma (**B**) for cows from the HS (heat stress) and TN (thermoneutral) periods. This figure tries to find pathways that were significantly changed based on enrichment and topology analysis.

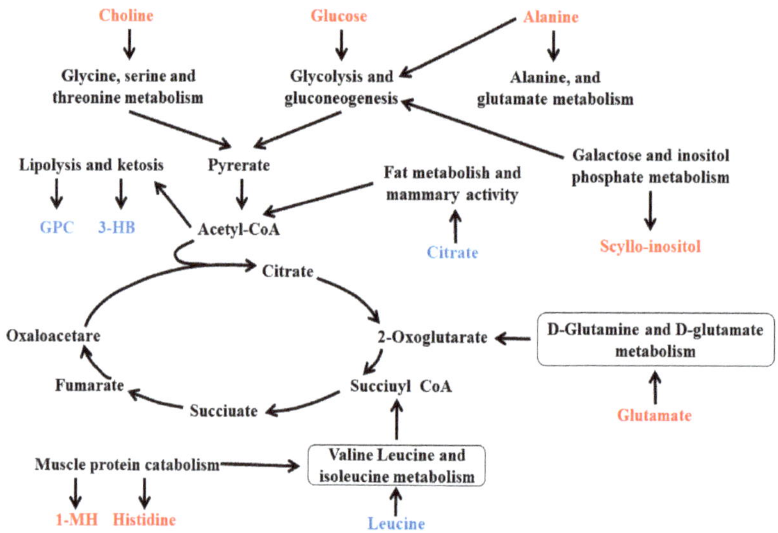

Figure 7. Metabolic network of the potential biomarkers in plasma and milk, which were different between the HS (heat stress) and TN (thermoneutral) periods. The increased metabolites in the HS period are represented in red. The decreased metabolites in the HS period are represented in blue. GPC: glycerol phosphoryl choline; 3-HB: 3-hydroxybutyrate; 1-MH: 1-Methylhistidine.

4. Discussion

In dairy production systems, THI has been used as a vibrant indicator to evaluate HS in dairy cows. It is normally believed that THI above 68 results in HS in dairy cows [25]. In the present experiment, THI results showed that average THI was above 80.5 in the HS experimental group (Figure 1), indicating HS in cows of this group. Moreover, previous studies have reported that higher RTs and respiration rates are prominent indicators of heat-stressed dairy cows [26–29]. In the present experiment, higher RTs and respiration rates in cows of the HS experimental treatments further clarified that cows in the HS experimental treatment suffered from HS.

In this study, HS reduced DMI and milk production, which is similar to the findings of previous studies [1,7,30]. It has already been reported that HS reduces the DMI, milk yield, and milk protein contents [15,19]. However, recent studies have explored that in heat-stressed dairy cows, low DMI only partially describes the dairy cow milk productivity and changes in milk composition [17,18]. In the current study, HS reduced DMI dramatically in cows, and thus cows were unable to meet energy demands for maintenance and lactation. Therefore, it could be assumed that HS cows were in negative nutrient balance, as supported by bodyweight changes in HS cows as compared to cows in the thermoneutral zone (Table 3). In negative energy balance, processes of both lipolysis and proteolysis started in cows [31].

3-Methylhistidine (3-MH) is a product of the posttranslational methylation of histidine in both myosin and actin, and the plasma concentration of 3-MH is a good indicator of protein mobilization in cows because it is not further metabolized in the body [32]. Interestingly, we did not find the changes in the plasma concentration of 3-MH. Instead, we found the upregulation of 3-Methylhistidine(1-MH)in the plasma of heat-stressed dairy cows. 1-MH is the degradation product of anserine (β-alany1-1-methy1-histidine) that is a dipeptide found in the muscles of many animals, including cows [33]. 1-MH is used as a biomarker for meat intake in humans because anserine is not found in human muscle [34]. This methylation product of histidine cannot be synthesized in the human body unless anserine-containing meat is consumed. In the current study, it could be assumed that anserine was degraded in cow muscle and resulted in the production of 1-MH during HS, which ultimately enhanced the plasma concentration of 1-MH. Therefore, the plasma concentration of 1-MH could be a possible biomarker for monitoring protein mobilization in HS dairy cows.

In normal and pathological conditions, the rate of protein synthesis is correlated with the intracellular amino acid pool as compared to the blood plasma concentration of amino acid [35,36]. Results showed that HS increased concentrations of glutamate, alanine, and histidine in the plasma of cows. It has been reported that glutamate, alanine, and histidine are the main precursors of glucose production via gluconeogenesis [12,37]. It has also been reported that during periods of food deprivation, gluconeogenesis and glycolysis are regulated by alanine to ensure glucose production [37,38]. Furthermore, it has also been reported that under HS condition, alanine synthesis increases, and alanine accumulates as the major amino acid [16,39]. In the current study, higher levels of alanine were found in the HS experimental group, representing that during HS, gluconeogenesis was started in cows. Therefore, high plasma alanine concentration is also a sign of inadequate cellular energy substrates of dairy cows in HS periods.

Glutamate is essential for carbohydrate and amino acid metabolism because it is a chief precursor of glucose synthesis during the process of gluconeogenesis [40]. It has been reported that in many amino acids' metabolism, proline, ornithine, glutamine, arginine, and histidine converge on glutamate itself [41]. Moreover, under stressful cell conditions, glutamate can be synthesized for neurotransmitters and glutathione to protect cells from damage from the redox crisis [42]. It has been reported that HS increases the synthesis of glutamate, which is often part of an adaptive strategy for the body in response to stressors [12,16,37,39]. A higher level of glutamate was found in the heat-stressed cows, highlighting that glutamate was involved in multiple metabolic processes under HS. The accumulation of excess amino acids in the plasma indicated that amino acid catabolism was probably started in the liver [43]. These amino acids can further degrade to produce 2-oxoglutarate. Through the tricarboxylic acid cycle (TCA), 2-oxoglutarate is converted to oxaloacetic acid, which is used to synthesize glucose to supplement the energy demand. Based on findings, it could be concluded that dairy cows subjected to HS suffer from protein catabolism and lack of energy substrates.

Formate is an essential endogenous one-carbon metabolite in animals, who participate in a vital one-carbon pool of intermediary metabolism, and used as a valuable biomarker for impaired one-carbon metabolism [44]. In dairy production, the endogenous source of formate is rumen methanogens, which use microbial fermentation to produce products, such as formic acid and hydrogen, which are reduced to produce methane. Therefore, there is normally little hydrogen and formic acid in the rumen [45]. Our results showed that plasma concentrations of formate in heat-stressed cows were increased, suggesting that HS led to impaired methanogenesis, which led to high levels of formate entering the blood. Formate is observed to be toxic to animals as it has been reported to be involved in the toxicity observed with methanol poisoning [46]. Metabolic acidosis and optic nerve damage are connected with the toxicity of formate [47]. Hovda et al. [48] demonstrated that metabolic acidosis happened only at higher concentrations of plasma formate. Our results showed impaired one-carbon metabolism and possible signs of a metabolic acidosis of the cows during the HS period.

Current study findings explored that plasma concentrations of glucose were increased in the animals of the HS group, similar to the research results of Srikandakumar and Johnson in Holstein cows [49]. The higher plasma concentration of glucose could be explained by the theory of higher uptake of glucose from the kidney and the intestine and the higher production of hepatic glucose during HS [50–52].

Significantly, decreased levels of 3-hydroxybutyrate (3-HB) were observed in the HS animals. The 3-HB, the main ketone body, is generated from lipolysis in the liver mitochondria and could be utilized as an energy source as the supply of blood glucose is insufficient [53]. Therefore, the decreased 3-hydroxybutyrate plasma levels in heat-stressed cows might indicate that the inhibition of β-oxidation of fatty acid occurred in cows suffering HS [54]. The upregulation of plasma concentrations of glucose in the HS period in this study supported the theory that β-oxidation of fatty acid is inhibited in heat-stressed cows. Previous studies have also reported that HS limits fat mobilization to limit heat generation [55,56]; therefore, it could be assumed that lipolysis may not be an energy-generating process for dairy cows during HS. However, further evidence is required to confirm this hypothesis. In addition, plasma 3-HB is derived not only from liver fatty acid metabolism but also from rumen butyrate [57]. A decrease of DMI in animals of the HS experimental group may be related to a decrease in plasma 3-hydroxybutyrate. In any case, the decreased plasma 3-HB in this study might suggest the inhibitory effect of HS on lipolysis in dairy cows. Plasma 3-HB level may also be used as a biomarker for evaluating the susceptibility to HS.

A significantly decreased level of leucine was seen in cows of the HS experimental groups in the current study. The essential amino acid leucine can stimulate muscle protein synthesis [56]. Fried et al. [58] stated that the increased concentration of leucine in the blood led to an increase in protein turnover in a mouse model. A decreased leucine concentration in blood plasma causes physical and mental fatigue [59]. Leucine is also concerned with different disease conditions; for example, the disease of maple syrup urine disease [60]. It has also been reported that the metabolism of leucine is essential for heart disease [61]. Taegtmeyer et al. [62] observed that the supplementation of leucine aided in repair of the heart during ischemic injury. Therefore, it could be hypothesized that the decreased level of leucine observed in our study might lead to mental fatigue and ischemic injury of the heart of dairy cows during HS.

VLDL is likely an essential source of fats for extrahepatic tissues [63]. Our results showed that the plasma concentration of VLDL decreased due to HS. The result of the lower plasma level of VLDL is similar to the research results of Tian et al. [16], who observed that HS negatively influenced the level of plasma VLDL in the animals suffered from HS. It is known that hepatocytes are involved in the assembly and secretion of VLDL particles [64]. The lower level of VLDL could be justified with the lesser production of the essential components of VLDL like cholesterol, apolipoproteins, and phospholipids (especially GPC) [65]. In the current study, the level of GPC decreased, but the concentration of choline was upgraded in the milk of HS cows. This result might indicate that HS causes abnormalities in liver lipid metabolism. It has been reported that a reduced ratio of GPC to choline leads to a metabolic change of phosphatidylcholine and GPC to choline during the HS period [66]. It has been further reported that decreased GPC levels in the milk are advantageous for choline release and adaptation to hot weather conditions. Previous studies have also reported that choline is positively associated with good coagulation parameters [67]. The high choline content, together with the decreased citrate content in the dairy cow, during the HS period, could improve milk coagulation properties [67].

As a common protein modification, protein glycosylation is linked to protein functions, such as biological recognition, enzymatic protection, and protein folding [68,69]. The N-acetyllactosamine units are typically present in glycoprotein, which is constituted of an O- or N-linked core structure. It has been reported that N-acetyllactosamine core structures are carriers of terminal carbohydrate epitopes, which confer a specific biological property to the glycoprotein [69]. Furthermore, it is believed that approximately all of the important molecules concerned with the adaptive and innate responses are glycoproteins, and the majority of circulating glycoproteins are acute phase reactants and immunologic proteins. Many studies have reported that HS impairs cow immune function [70,71] and results in higher disease incidence, especially in the period after postpartum [72]. Moreover, it has also been reported that HS also has a negative impact on the offspring [73]. The study of Nardone et al. [74] explained that HS in dairy cows during the late pregnancy reduced the concentration of IgA and IgG and the percentage of total protein of the colostrum from the first milking of primiparous dairy cows. In another study, higher mortality in heat-stressed neonatal calf was observed due to impaired colostral antibody absorption [75]. So far, no study has reported any glycan-dependent functions for the glycoproteins present in the cow's milk. Therefore, in the current experiment, the increased milk level of NAG and the decreased milk level of OAG in animals of the HS experimental groups might have experienced an impaired immune function and incorrect glycosylation of immunoglobulins by mammary gland epithelial cells. Thus, both NAG and OAG may be useful milk biomarkers for heat-stressed cows.

Scyllo-inositol is an isomer of myoinositol, which can be produced in vivo from glucose-six-phosphate [76]. Inositol is a galactose metabolism product, and it has been reported that a part of serum inositol is derived from the degradation of dietary phytate by some specific species of intestinal microbes [77]. Milk scyllo-inositol may either be up taken from blood or synthesized by mammary gland epithelial cells. It has been reported that in transgenic mice with an Alzheimer's phenotype, scyllo-Inositol reverses the memory deficit [78]. In the current study, the content of scyllo-inositol was upregulated by HS, which indicated that HS leads to inositol metabolism disorder and cellular toxicity of mammary gland epithelial cells.

Pyridoxamine is one form of vitamin B6, serving as a coenzyme in various enzymatic reactions, such as transamination and decarboxylation reactions [79]. The deficiency of vitamin B6 can cause various processes of the body, such as nephrotic syndrome and inflammation [80]. Milk pyridoxamine, therefore, is an essential functional nutrient beneficial to animal health. However, the biological significance and the mechanism of the elevated milk concentration of pyridoxamine during the HS period is not known.

Citrate is not the main constituent of milk, but it plays as a buffer role in the udder of dairy cows. It has been reported that citrate regulates $Ca^{2+}$and H^{+} ions homeostasis and maintain the fluidity of milk by effecting casein micelles [81]. In the case of citrate deficiency in the udder, the clumping of Ca^{2+} appears, which causes injury to the parenchymatous tissues of the alveoli of the udder and leads to the damage of barriers between blood and milk and inflammatory reactions in the alveolar tissue of the udder [82]. In the present experiment, the level of citrate in milk decreased significantly during the HS period, indicating that HS may seriously affect udder health. Furthermore, citrate is known to be involved in the TCA, and in dairy cows, it is known as a biomarker of energy balance [83]. Moreover, citrate is correlated with ketone bodies in milk and de novo fatty acids synthesis [84]. There is strong evidence that citrate in milk comes from citrate produced within the mammary secretory cell rather than from that in the blood plasma [85]. The previous study of Tian et al. [16] reported that the HS decreased the milk citrate concentration due to changes in mammary gland function rather than a disturbance in the blood citrate metabolism. Therefore, the lower level of citrate in heat-stressed cow's milk could reflect the decreased mammary activity rather than imbalanced energy metabolism. In the current experiment, HS increased the SCC, similar to the findings of the recent study [86]. It has been reported that higher milk SCC due to HS indicates a mammary immune response to simulated infection [87]; therefore, SCC results of the current study suggest that HS may seriously affect udder health.

5. Conclusions

In this study, results explored that HS reduced the DMI and influenced the metabolites in the milk and plasma of dairy cows. Results showed that a total of eight metabolites in milk and 12 metabolites in plasma were altered due to HS. These metabolites were mainly involved in gluconeogenesis, protein degradation and synthesis, and milk fat synthesis. These metabolites in the milk and plasma could be potential biomarkers for HS. In addition, this work found several metabolites (especially in milk) that have rarely been studied.

Supplementary Materials: The following supplementary material is available online at http://www.mdpi.com/2076-2615/10/10/1741/s1. Figure S1: A representative 600 MHz ^{1}H.NMR spectrum (δ0.5–5.0 and δ5.0–9.5) of milk from the HS period and TN period, Figure S2: A representative 600 MHz ^{1}H.NMR spectrum (δ0.5–5.5 and δ5.5–9.0) of plasma from the HS period and TN period, Figure S3: least squares discrimination analysis (PLS-DA) scores plots (**a**) derived from ^{1}H.NMR spectra of the aqueous phase of milk extracts. A = heat stress group; B = thermal neutral group, Figure S4: PLS-DA scores plots (**a**) derived from ^{1}H.NMR spectra of serum obtained from different groups, and cross-validation (**b**) by permutation test (n = 300). A = heat stress group; B = thermal neutral group. Figure S5: Partial least squares discrimination analysis PLS-DA score plots (left panel) derived from ^{1}H NMR spectra of serum obtained from different groups and cross validation (left panel) by permutation test (n = 300). A = Heat Stress group; B = thermal neutral group.

Author Contributions: S.Y., S.D. and B.X. planned the experiment and finalized the draft of the manuscript with consent from all contributing authors. S.D., J.Z., C.Y., X.H. and X.Z., carried out the experiment, finalized, and analyzed the collected data. Z.W., L.W. and Q.P. improved and revised the final draft of the manuscript. All authors have read and agreed to the published version of the manuscript.

Funding: This experiment was conducted with the help of a project funded by the Science and Technology Support Program of the National Science and Technology Ministry, Grant number 2012BAD12B02.

Acknowledgments: The authors would like to express their thanks to the Mei Jiaidun Dairy Cow Farms (Hubei, China) for providing research trial facilities. The authors also acknowledge the support of Mei Jiaidun Dairy Cow Farms for experimental animals' care, management, feeding, and handling.

Conflicts of Interest: The authors declare no conflict of interest.

References

1. Bernabucci, U.; Biffani, S.; Buggiotti, L.; Vitali, A.; Lacetera, N.; Nardone, A. The effects of heat stress in Italian Holstein dairy cattle. *J. Dairy Sci.* **2014**, *97*, 471–486. [CrossRef] [PubMed]
2. St-Pierre, N.; Cobanov, B.; Schnitkey, G. Economic losses from heat stress by US livestock industries. *J. Dairy Sci.* **2003**, *86*, E52–E77. [CrossRef]
3. Nardone, A.; Ronchi, B.; Lacetera, N.; Ranieri, M.; Bernabucci, U. Effects of climate changes on animal production and sustainability of livestock systems. *Livest. Sci.* **2010**, *130*, 57–69. [CrossRef]
4. Rhoads, M.L.; Kim, J.W.; Collier, R.J.; Crooker, B.A.; Boisclair, Y.R.; Baumgard, L.H.; Rhoads, R.P. Effects of heat stress and nutrition on lactating Holstein cows: II. Aspects of hepatic growth hormone responsiveness. *J. Dairy Sci.* **2010**, *93*, 170–179. [CrossRef]
5. Liu, J.; Li, L.; Chen, X.; Lu, Y.; Wang, D. Effects of heat stress on body temperature, milk production, and reproduction in dairy cows: A novel idea for monitoring and evaluation of heat stress—A review. *Asian Australas. J. Anim. Sci.* **2019**, *32*, 1332–1339. [CrossRef]
6. Collier, R.J.; Doelger, S.G.; Head, H.H.; Thatcher, W.W.; Wilcox, C.J. Effects of heat stress during pregnancy on maternal hormone concentrations, calf birth weight and postpartum milk yield of holstein cows. *J. Anim. Sci.* **1982**, *54*, 309–319. [CrossRef]
7. West, J. Effects of heat-stress on production in dairy cattle. *J. Dairy Sci.* **2003**, *86*, 2131–2144. [CrossRef]
8. Shwartz, G.; Rhoads, M.L.; Vanbaale, M.J.; Rhoads, R.P.; Baumgard, L.H. Effects of a supplemental yeast culture on heat-stressed lactating Holstein cows. *J. Dairy Sci.* **2009**, *92*, 935–942. [CrossRef]
9. Wang, J.P.; Bu, D.P.; Wang, J.Q.; Huo, X.K.; Guo, T.J.; Wei, H.Y.; Zhou, L.L.; Rastani, R.R.; Baumgatd, L.H.; Li, F.D. Effect of saturated fatty acid supplementation on production and metabolism indices in heat-stressed mid-lactation dairy cows. *J. Dairy Sci.* **2010**, *93*, 4121–4127. [CrossRef] [PubMed]
10. Cheng, J.B.; Wang, W.Y.; Zheng, N.; Li, S.L.; Zhang, Y.D.; Zhang, X.K. Natural Period Change of Heat Stress Reveals Unique "Heat-stressed Milk Protein Decrease Syndrome" in Mid-lactation Dairy Cows. *China Anim. Husb. Vet. Med.* **2014**, *41*, 73–84. (In Chinese)
11. Shahid, R.; Qureshi, Z.I.; Rahman, Z.U.; Ahmad, N. Effect of recombinant bovine somatotropin (rbST) and oxytocin on health biomarkers, reproductive performance and milk composition of Nili-Ravi buffaloes (*Bubalus bubalis*). *Pak. Vet. J.* **2019**, *39*, 553–557. [CrossRef]
12. Cowley, F.C.; Barber, D.G.; Houlihan, A.V.; Poppi, D.P. Immediate and residual effects of heat stress and restricted intake on milk protein and casein composition and energy metabolism. *J. Dairy Sci.* **2015**, *98*, 2356–2368. [CrossRef] [PubMed]
13. Fuquay, J.W. Heat stress as it affects animal production. *J. Anim. Sci.* **1981**, *52*, 164–174. [CrossRef] [PubMed]
14. Tian, H.; Zheng, N.; Wang, W.; Cheng, J.; Li, S.; Zhang, Y.; Wang, J.Q. Integrated Metabolomics Study of the Milk of Heat-stressed Lactating Dairy Cows. *Sci. Rep.* **2016**, *6*, 24208. [CrossRef] [PubMed]
15. Sun, H.Z.; Wang, D.M.; Wang, B.; Wang, J.K.; Liu, H.Y.; Guan, L.L.; Liu, J.X. Metabolomics of Four Biofluids from Dairy Cows: Potential Biomarkers for Milk Production and Quality. *J. Proteome Res.* **2015**, *14*, 1287–1298. [CrossRef] [PubMed]
16. Tian, H.; Zheng, N.; Wang, W.; Cheng, J.; Li, S.; Zhang, Y.; Wang, J.Q. Identification of diagnostic biomarkers and metabolic pathway shifts of heat-stressed lactating dairy cows. *J. Proteom.* **2015**, *125*, 17–28. [CrossRef] [PubMed]

17. Abeni, F.; Calamari, L.; Stefanini, L. Metabolic conditions of lactating Friesian cows during the hot season in the Po valley. 1. Blood indicators of heat stress. *Int. J. Biometeorol.* **2007**, *52*, 87–96. [CrossRef]
18. Calamari, L.; Abeni, F.; Calegari, F.; Stefanini, L. Metabolic conditions of lactating Friesian cows during the hot season in the Po valley. 2. Blood minerals and acid-base chemistry. *Int. J. Biometeorol.* **2007**, *52*, 97–107. [CrossRef]
19. Calamari, L.; Petrera, F.; Stefanini, L.; Abeni, F. Effects of different feeding time and frequency on metabolic conditions and milk production in heat-stressed dairy cows. *Int. J. Biometeorol.* **2013**, *57*, 785–796. [CrossRef]
20. National Research Council. *Nutrient Requirements of Dairy Cattle*, 7th ed.; National Academy Press: Washington, DC, USA, 2001.
21. AOAC International. *Official Methods of Analysis*, 18th ed.; AOAC international: Gaithersburg, MD, USA, 2005.
22. Van Soest, P.J.; Robertson, J.B.; Lewis, B.A. Methods for dietary fiber, neutral detergent fiber, and nonstarch polysaccharides in relation to animal nutrition. *J. Dairy. Sci.* **1991**, *74*, 3583–3597. [CrossRef]
23. Cloarec, O.; Dumas, M.E.; Trygg, J.; Craig, A.; Barton, R.H.; Lindon, J.C.; Holmes, E. Evaluation of the Orthogonal Projection on Latent Structure Model Limitations Caused by Chemical Shift Variability and Improved Visualization of Biomarker Changes in ^{1}H NMR Spectroscopic Metabonomic Studies. *Anal. Chem.* **2005**, *77*, 517–526. [CrossRef] [PubMed]
24. Trygg, J.; Wold, S. Orthogonal projections to latent structures (O-PLS). *J. Chemom.* **2002**, *16*, 119–128. [CrossRef]
25. Armstrong, D. Heat stress interaction with shade and cooling. *J. Dairy Sci.* **2002**, *77*, 2044–2050. [CrossRef]
26. Dikmen, S.; Alava, E.; Pontes, E.; Fear, J.M.; Dikmen, B.Y.; Olson, T.A.; Hansen, P.J. Differences in thermoregulatory ability between slick-haired and wild-type lactating holstein cows in response to acute heat stress. *J. Dairy Sci.* **2008**, *91*, 3395–3402. [CrossRef] [PubMed]
27. Dikmen, S.; Hansen, P. Is the temperature-humidity index the best indicator of heat stress in lactating dairy cows in a subtropical environment? *J. Dairy Sci.* **2009**, *92*, 109–116. [CrossRef] [PubMed]
28. Zimbelman, R.; Rhoads, R.; Rhoads, M.; Duff, G.; Baumgard, L.; Collier, R. A re-evaluation of the impact of temperature humidity index (THI) and black globe humidity index (BGHI) on milk production in high producing dairy cows. In Proceedings of the 24th Southwest Nutrition and Management Conference, 1111 N. Dunlap Ave, Savoy, IL, USA, 26–27 February 2009; pp. 158–169.
29. Kadzere, C.; Murphy, M.; Silanikove, N.; Maltz, E. Heat stress in lactating dairy cows: A review. *Livest. Prod. Sci.* **2002**, *77*, 59–91. [CrossRef]
30. Rhoads, M.; Rhoads, R.; VanBaale, M.; Collier, R.J.; Sanders, S.; Weber, W.J.; Crooker, B.A.; Baumgard, L. Effects of heat stress and plane of nutrition on lactating Holstein cows: I. Production, metabolism, and aspects of circulating somatotropin. *J. Dairy Sci.* **2009**, *92*, 1986–1997. [CrossRef]
31. Belić, B.; Cincović, M.; Stevančević, M.; Toholj, B. Parameters of the negative energy balance (NEFA, BHB and glucose) and neutrophil to lymphocyte ratio in cows after calving- comparison of two indicators of stress. *Vet. Stanica* **2011**, *42*, 281–284.
32. Harris, C.I.; Milne, G. The urinary excretion of N-methyl histidine by cattle: Validation as an index of muscle protein breakdown. *Br. J. Nutr.* **1981**, *45*, 411–422. [CrossRef]
33. Hussain, R.; Ali, F.; Rafique, A.; Ghaffar, A.; Jabeen, G.; Rafay, M.; Liaqat, S.; Khan, I.; Malik, R.; Khan, M.K.; et al. Exposure to sub-acute concentrations of glyphosate induce clinico-hematological, serum biochemical and genotoxic damage in adult cockerels. *Pak. Vet. J.* **2019**, *39*, 181–186. [CrossRef]
34. Tůma, P.; Samcova, E.; Balinova, P. Determination of 3-methylhistidine and 1-methylhistidine in untreated urine samples by capillary electrophoresis. *J. Chromatogr. B* **2005**, *821*, 53–59.
35. Wang, C.; Guo, F. Branched chain amino acids and metabolic regulation. *Sci. Bull.* **2013**, *58*, 1228–1235. [CrossRef]
36. Gao, S.T.; Guo, J.; Quan, S.Y.; Nan, X.M.; Fernandez, M.S.; Baumgard, L.H.; Bu, D.P. The effects of heat stress on protein metabolism in lactating Holstein cows. *J. Dairy Sci.* **2017**, *100*, 5040–5049. [CrossRef] [PubMed]
37. Guo, J.; Gao, S.; Quan, S.; Zhang, Y.; Bu, D.; Wang, J. Blood amino acids profile responding to heat stress in dairy cows. *Asian Australas. J. Anim. Sci.* **2018**, *31*, 47–53. [CrossRef] [PubMed]
38. Meijer, A.J. Amino Acids as Regulators and Components of Nonproteinogenic Pathways. *J. Nutr.* **2003**, *133*, 2057–2062. [CrossRef]

39. Ai, Y.; Cao, Y.; Xie., Z.L.; Zhang., Y.S.; Shen., X.Z. Relationship between Free Amino Acids in Cow's Blood and Decreasing Milk Protein under Heat Stress (In Chinese). *Food Sci.* **2015**, *36*, 38–41. [CrossRef]
40. Newsholme, P.; Lima, M.R.; Procopio, J.; Pithoncuri, T.C.; Doi, S.Q.; Bazotte, R.B.; Curi, R. Glutamine and glutamate as vital metabolites. *Braz. J. Med. Biol. Res.* **2003**, *36*, 153–163. [CrossRef]
41. Brosnan, J.T. Glutamate, at the Interface between Amino Acid and Carbohydrate Metabolism. *J. Nutr.* **2000**, *130*, 988. [CrossRef]
42. Salgado, M.C.; Meton, I.; Anemaet, I.G.; Baanante, I.V. Activating transcription factor 4 mediates up-regulation of alanine aminotransferase 2 gene expression under metabolic stress. *Biochim. Biophys. Acta* **2014**, *1839*, 288–296. [CrossRef]
43. Stahel, P.; Purdie, N.G.; Cant, J.P. Use of dietary feather meal to induce histidine deficiency or imbalance in dairy cows and effects on milk composition. *J. Dairy Sci.* **2014**, *97*, 439–445. [CrossRef]
44. Lamarre, S.G.; Morrow, G.P.; Macmillan, L.; Brosnan, M.E.; Brosnan, J.T. Formate: An essential metabolite, a biomarker, or more? *Clin. Chem. Lab. Med.* **2013**, *51*, 571–578. [CrossRef]
45. Morgavi, D.P.; Forano, E.; Martin, C.; Newbold, C.J. Microbial ecosystem and methanogenesis in ruminants. *Animal* **2010**, *4*, 1024–1036. [CrossRef] [PubMed]
46. Barceloux, D.G.; Bond, G.R.; Krenzelok, E.P.; Cooper, H.; Vale, J.A. American academy of clinical toxicology practice guidelines on the treatment of methanol poisoning. *J. Toxicol. Clin. Toxicol.* **2002**, *40*, 415–446. [PubMed]
47. Medinsky, M.A.; Dorman, D.C.; Bond, J.A.; Moss, O.R.; Janszen, D.B. Pharmacokinetics of methanol and formate in female cynomolgus monkeys exposed to methanol vapors. *Res. Rep. Health Eff. Inst.* **1997**, *77*, 1–30.
48. Hovda, K.E.; Andersson, K.S.; Urdal, P.; Jacobsen, D. Methanol and Formate Kinetics during Treatment with Fomepizole. *Clin. Toxicol.* **2005**, *43*, 221–227. [CrossRef]
49. Srikandakumar, A.; Johnson, E.H. Effect of heat stress on milk production, rectal temperature, respiratory rate and blood chemistry in Holstein, Jersey and Australian Milking Zebu cows. *Trop. Anim. Health Prod.* **2004**, *36*, 685–692. [CrossRef] [PubMed]
50. Garriga, C.; Hunter, R.R.; Amat, C.; Planas, J.M.; Mitchell, M.A.; Moreto, M. Heat stress increases apical glucose transport in the chicken jejunum. *Am. J. Physiol.-Regul. I* **2006**, *290*, 195–201. [CrossRef] [PubMed]
51. Ikari, A.; Nakano, M.; Suketa, Y.; Harada, H.; Takagi, K. Reorganization of ZO-1 by sodium-dependent glucose transporter activation after heat stress in LLC-PK1 cells. *J. Cell. Physiol.* **2005**, *203*, 471–478. [CrossRef] [PubMed]
52. Febbraio, M.A. Alterations in Energy Metabolism during Exercise and Heat Stress. *Sports Med.* **2001**, *31*, 47–59. [CrossRef]
53. Akram, M. A Focused Review of the Role of Ketone Bodies in Health and Disease. *J. Med. Food* **2013**, *16*, 965–967. [CrossRef]
54. Zammit, V.A. Ketogenesis in the liver of ruminants—Adaptations to a challenge. *J. Agric Sci.* **1990**, *115*, 155–162. [CrossRef]
55. Belhadj Slimen, I.; Najar, T.; Ghram, A.; Abdrrabba, M. Heat stress effects on livestock: Molecular, cellular and metabolic aspects, a review. *J. Anim. Physiol. Anim. Nutr.* **2016**, *100*, 401–412. [CrossRef]
56. Qu, M.; Wei, S.; Chen, Z.; Wang, G.; Zheng, Y.; Yan, P. Differences of hormones involved in adipose metabolism and lactation between high and low producing Holstein cows during heat stress. *Anim. Nutr.* **2015**, *1*, 339–343. [CrossRef] [PubMed]
57. DeFrain, J.; Hippen, A.; Kalscheur, K.; Schingoethe, D. Feeding lactose increases ruminal butyrate and plasma β-hydroxybutyrate in lactating dairy cows. *J. Dairy Sci.* **2004**, *87*, 2486–2494. [CrossRef]
58. Fried, S.K.; Watford, M. Leucing weight with a futile cycle. *Cell Metab.* **2007**, *6*, 155–156. [CrossRef] [PubMed]
59. Blomstrand, E.; Celsing, F.; Newsholme, E. Changes in plasma concentrations of aromatic and branched-chain amino acids during sustained exercise in man and their possible role in fatigue. *Acta Physiol. Scand.* **1988**, *133*, 115–121. [CrossRef]
60. Harris, R.A.; Zhang, B.; Goodwin, G.W.; Kuntz, M.J.; Shimomura, Y.; Rougraff, P.; Dexter, P.; Zhao, Y.; Gibson, R.; Crabb, D.W. Regulation of the branched-chain α-ketoacid dehydrogenase and elucidation of a molecular basis for maple syrup urine disease. *Adv. Enzym. Regul.* **1990**, *30*, 245–256. [CrossRef]
61. Huang, Y.; Zhou, M.; Sun, H.; Wang, Y. Branched-chain amino acid metabolism in heart disease: An epiphenomenon or a real culprit? *Cardiovasc. Res.* **2011**, *90*, 220–223. [CrossRef]

62. Taegtmeyer, H.; Harinstein, M.E.; Gheorghiade, M. More than bricks and mortar: Comments on protein and amino acid metabolism in the heart. *Am. J. Cardiol.* **2008**, *101*, S3–S7. [CrossRef]
63. Bell, A.W. Lipid metabolism in liver and selected tissues and in the whole body of ruminant animals. *Prog. Lipid Res.* **1979**, *18*, 117–164. [CrossRef]
64. Gruffat, D.; Durand, D.; Graulet, B.; Bauchart, D. Regulation of VLDL synthesis and secretion in the liver. *Reprod. Nutr. Dev.* **1996**, *36*, 375–389. [CrossRef] [PubMed]
65. Fast, D.G.; Vance, D.E. Nascent VLDL phospholipid composition is altered when phosphatidylcholine biosynthesis is inhibited: Evidence for a novel mechanism that regulates VLDL secretion. *Biocheim. Biophys. Acta.* **1995**, *1258*, 159–168. [CrossRef]
66. Cheng, X.; Luming, W.; Yongxue, L.; Jianzhong, Z.; Baozhang, L.Q. Effect of acute heat stress on the contents of phospholipids and the fluidity of cell membrane in rat lung. *Chin. J. Appl. Physiol.* **1992**, *8*, 321–324.
67. Sundekilde, U.K.; Frederiksen, P.D.; Clausen, M.R.; Larsen, L.B.; Bertram, H.C. Relationship between the metabolite profile and technological properties of bovine milk from two dairy breeds elucidated by NMR-based metabolomics. *J. Agric. Food. Chem.* **2011**, *59*, 7360–7367. [CrossRef]
68. Moremen, K.W.; Tiemeyer, M.; Nairn, A.V. Vertebrate protein glycosylation: Diversity, synthesis and function. *Nat. Rev. Mol. Cell Biol.* **2012**, *13*, 448–462. [CrossRef] [PubMed]
69. Fiete, D.; Srivastava, V.; Hindsgaul, O.; Baenziger, J.U. A hepatic reticuloendothelial cell receptor specific for SO_4-4GalNAcβ1, 4GlcNAcβ1, 2Manα that mediates rapid clearance of lutropin. *Cell* **1991**, *67*, 1103–1110. [CrossRef]
70. Do Amaral, B.; Connor, E.; Tao, S.; Hayen, M.; Bubolz, J.; Dahl, G. Heat stress abatement during the dry period influences metabolic gene expression and improves immune status in the transition period of dairy cows. *J. Dairy Sci.* **2011**, *94*, 86–96. [CrossRef] [PubMed]
71. Tao, S.; Dahl, G. Invited review: Heat stress effects during late gestation on dry cows and their calves. *J. Dairy Sci.* **2013**, *96*, 4079–4093. [CrossRef]
72. Thompson, I.; Dahl, G. Dry-period seasonal effects on the subsequent lactation. *Prof. Anim. Sci.* **2012**, *28*, 628–631. [CrossRef]
73. Tao, S.; Monteiro, A.; Thompson, I.; Hayen, M.; Dahl, G. Effect of late-gestation maternal heat stress on growth and immune function of dairy calves. *J. Dairy Sci.* **2012**, *95*, 7128–7136. [CrossRef]
74. Nardone, A.; Lacetera, N.; Bernabucci, U.; Ronchi, B. Composition of colostrum from dairy heifers exposed to high air temperatures during late pregnancy and the early postpartum period. *J. Dairy Sci.* **1997**, *80*, 838–844. [CrossRef]
75. Stott, G.; Wiersma, F.; Menefee, B.; Radwanski, F. Influence of environment on passive immunity in calves. *J. Dairy Sci.* **1976**, *59*, 1306–1311. [CrossRef]
76. Holub, B.J. *Advances in Nutritional Research*; Springer: Boston, MA, USA, 1982.
77. Steer, T.E.; Gee, J.N.; Johnson, I.T.; Gibson, G.R. Biodiversity of human faecal bacteria isolated from phytic acid enriched chemostat fermenters. *Curr. Issues Intest. Microbiol.* **2004**, *5*, 23–39. [PubMed]
78. Fenili, D.; Brown, M.; Rappaport, R.; McLaurin, J. Properties of scyllo–inositol as a therapeutic treatment of AD-like pathology. *J. Mol. Med.* **2007**, *85*, 603–611. [CrossRef] [PubMed]
79. Snell, E.E. Chemical Structure in Relation to Biological Activities of Vitamin B6. In *Vitamins & Hormones*; Academic Press: Cambridge, MA, USA, 1958; pp. 77–125.
80. Depeint, F.; Bruce, W.R.; Shangari, N.; Mehta, R.; O'Brien, P.J. Mitochondrial function and toxicity: Role of B vitamins on the one-carbon transfer pathways. *Chem. Biol. Interact.* **2006**, *163*, 113–132. [CrossRef]
81. Shennan, D.B.; Peaker, M. Transport of Milk Constituents by the Mammary Gland. *Physiol. Rev.* **2000**, *80*, 925–951. [CrossRef]
82. Firyal, S.; Tanveer, S.; Awan, A.R.; Tayyab, M.; Wasim, M.; Nawaz, M.; Sattar, H.; Nawaz, S.; Saeed, S.; Ali, M.M. Association of bovine tumor necrosis factor alpha gene polymorphism with mastitis in Nili Ravi buffaloes. *Pak. Vet. J.* **2019**, *39*, 128–131. [CrossRef]
83. Dhillon, K.S.; Singh, J. A new horizon in the pathobiology, aetiology and novel treatment of mastitis in buffalo. *Buffalo Bull.* **2013**, *32*, 26–34.
84. Banks, W.; Clapperton, J.L.; Girdler, A.K. Effect of dietary unsaturated fatty acids in various forms on the de novo synthesis of fatty acids in the bovine mammary gland. *J. Dairy Res.* **1990**, *57*, 179–185. [CrossRef]

85. Linzell, J.; Mepham, T.; Peaker, M. The secretion of citrate into milk. *J. Physiol.* **1976**, *260*, 739–750. [CrossRef]
86. Yue, S.; Wang, Z.; Wang, L.; Peng, Q.; Xue, B. Transcriptome Functional Analysis of Mammary Gland of Cows in Heat Stress and Thermoneutral Condition. *Animals* **2020**, *10*, 1015. [CrossRef] [PubMed]
87. Salama, A.A.K.; Contreras-Jodar, A.; Love, S.; Mehaba, N.; Such, X.; Caja, G. Milk yield, milk composition, and milk metabolomics of dairy goats intramammary-challenged with lipopolysaccharide under heat stress conditions. *Sci Rep.* **2020**, *10*, 5055. [CrossRef] [PubMed]

animals

MDPI

Article

Novel Facets of the Liver Transcriptome Are Associated with the Susceptibility and Resistance to Lipid-Related Metabolic Disorders in Periparturient Holstein Cows

Ryan S. Pralle [1,2], Wenli Li [3,*], Brianna N. Murphy [3], Henry T. Holdorf [1] and Heather M. White [1,*]

1 Department of Animal and Dairy Sciences, University of Wisconsin-Madison, Madison, WI 53706, USA; praller@uwplatt.edu (R.S.P.); hholdorf@wisc.edu (H.T.H.)
2 School of Agriculture, University of Wisconsin-Platteville, Platteville, WI 53818, USA
3 Dairy Forage Research Center, USDA-Agricultural Research Service, Madison, WI 53706, USA; bmurphy8@wisc.edu
* Correspondence: wenli.li@usda.gov (W.L.); heather.white@wisc.edu (H.M.W.); Tel.: +1-608-890-0056 (W.L.); +1-608-263-7783 (H.M.W.)

Simple Summary: Energy and nutrient demands of the early lactation period can result in the development of metabolic disorders, such as ketosis and fatty liver, in dairy cows. Variability in the incidence of these disorders suggests that some cows have an ability to adapt. The objective of this study was to discover differences in liver gene expression that are associated with a cow's susceptibility (disposition to disorder during typical conditions) or resistance (disposition to disorder onset and severity when presented a challenge) to metabolic disorders. Cows in a control treatment and a ketosis induction protocol treatment were retrospectively grouped into susceptibility and resistance groups, respectively, by a machine learning algorithm using lipid biomarker concentrations. Whole-transcriptome RNA sequencing was performed on liver samples from these cows. More susceptible cows had lower expression of glutathione metabolism genes, while less resistant cows had greater expression of eicosanoid-metabolism-related genes. Additionally, differentially expressed genes suggested a role for immune-response-related genes in conferring susceptibility and resistance to metabolic disorders. The overall inferred metabolism suggests that responses to oxidative stress may determine susceptibility and resistance to metabolic disorders, with novel implications for immunometabolism.

Citation: Pralle, R.S.; Li, W.; Murphy, B.N.; Holdorf, H.T.; White, H.M. Novel Facets of the Liver Transcriptome Are Associated with the Susceptibility and Resistance to Lipid-Related Metabolic Disorders in Periparturient Holstein Cows. *Animals* **2021**, *11*, 2558. https://doi.org/10.3390/ani11092558

Academic Editors: João Simões and Michael Hässig

Received: 14 July 2021
Accepted: 25 August 2021
Published: 31 August 2021

Abstract: Lipid-related metabolic disorders (LRMD) are prevalent in early lactation dairy cows, and have detrimental effects on productivity and health. Our objectives were to identify cows resistant or susceptible to LRMD using a ketosis induction protocol (KIP) to discover differentially expressed liver genes and metabolic pathways associated with disposition. Clustering cows based on postpartum lipid metabolite concentrations within dietary treatments identified cows more or less susceptible (MS vs. LS) to LRMD within the control treatment, and more or less resistant (MR vs. LR) within the KIP treatment. Whole-transcriptome RNA sequencing was performed on liver samples (−28, +1, and +14 days relative to calving) to assess differential gene and pathway expression (LS vs. MS; MR vs. LR; $n = 3$ cows per cluster). Cows within the MS and LR clusters had evidence of greater blood serum β-hydroxybutyrate concentration and liver triglyceride content than the LS and MR clusters, respectively. The inferred metabolism of differentially expressed genes suggested a role of immune response (i.e., interferon-inducible proteins and major histocompatibility complex molecules). Additionally, unique roles for glutathione metabolism and eicosanoid metabolism in modulating susceptibility and resistance, respectively, were implicated. Overall, this research provides novel insight into the role of immunometabolism in LRMD pathology, and suggests the potential for unique control points for LRMD progression and severity.

Keywords: dairy cow; transition period; ketosis; fatty liver; RNA-Seq; clustering; liver metabolism

1. Introduction

The physiological adaptations necessary to make the transition from a gravid, non-lactating state through parturition and into a lactating state represent several metabolic challenges for periparturient dairy cows. The principal challenges include onset of negative net energy and nutrient balance spurred by a reduction in voluntary feed intake and increasing lactation energy requirements, insulin resistance, immunosuppression, and mineral imbalance [1–4]. Responses to these challenges, such as the degree of body nutrient mobilization and hepatic nutrient metabolism, can vary [5–8]. Maladaptation to these challenges can result in numerous pathologies, including the lipid-related metabolic disorders (LRMD) hyperketonemia (HYK) and fatty liver syndrome (FLS). These comorbid disorders are prevalent in early lactation dairy cows, with an incidence of 43–53% and 50% for HYK and FLS, respectively [9–12]. Hyperketonemia—defined as elevated concentrations of blood β-hydroxybutyrate (BHB)—and bovine FLS—characterized by a substantial accumulation of liver triglyceride (TG)—have been associated with unfavorable performance outcomes, including greater risk of comorbidities, decreased reproductive efficiency, productive losses, and premature culling [7,9,13–15]. The deterministic estimate for total cost per case of hyperketonemia is USD 375 and USD 256 for primiparous and multiparous cows, respectively [16]. The financial impact of fatty liver on dairy production is difficult to determine, but was previously expected to cost the industry at least USD 60 million annually [14]. Therefore, these LRMD represent a concern not only for the health and wellness of dairy cows, but also for the cost of dairy production.

Our understanding of the pathology of these disorders has historically focused on understanding the liver metabolism of long-chain fatty acids (FA) mobilized from adipose tissue TG [14,17]. An apparent excess of liver uptake of endogenous FA surpasses the oxidative and secretory capacity of hepatocytes, promoting ketogenesis and storage of re-esterified FA [14,17,18]. Thus, most research regarding the prevention and treatment of HYK and FLS has focused on limiting FA substrates by managing dairy cows' prepartum obesity or excessive mobilization of FA during the postpartum period [19–21]. Alternative approaches have been focused on nutritional interventions to support oxidation via TCA [22–24] or to improve liver TG secretion [25–27].

The advent of genome-wide gene expression profiling using microarray or RNA-Seq techniques has expanded our capability to identify and understand the underpinning regulatory mechanisms of physiological states. With respect to the liver transcriptome, research concerning LRMD has focused on comparing prepartum dietary treatments (lower vs. greater dietary energy) [28–31] or feed restriction models [32,33] to either increase the risk of metabolic disorders or induce negative energy balance. In addition, the liver transcriptome has been compared between cows overfed dietary energy achieving subclinical HYK and those not progressing to HYK [34]. These papers have given insight into the pathogenesis of LRMD, such as highlighting the role of key transcription factors (i.e., peroxisome proliferator-activated receptor) and the contributions of steroid biosynthesis [28–32]. However, the liver transcriptome of periparturient dairy cows experiencing LRMD absent a challenge protocol has not been compared to apparently healthy controls. Additionally, there has been no evaluation of whether the differential biology observed during these experiments is representative of any differential biology in LRMD that occur without a challenge protocol.

We hypothesize that periparturient cows have divergent physiological adaptations to lactation that affect their susceptibility or resistance to LRMD. Furthermore, we hypothesize that there are unique genes and metabolic pathways responsible for LRMD susceptibility and resistance. Our presupposition is that dairy cows developing LRMD pathologies absent of a challenge represent an LRMD-susceptible population of cows, while cows with delayed onset or severity of LRMD when challenged represent a population of LRMD-resistant cows. To evaluate our hypotheses, we clustered periparturient cows based on postpartum lipid metabolite concentrations within a control treatment (CTL; no dietary challenge) and a ketosis induction protocol (KIP; dietary challenge with oversupply of energy prepartum and postpartum feed restriction). Within the CTL treatment, we identified a cluster of cows apparently less susceptible (LS) to LRMD and a cluster more susceptible (MS) to LRMD; additionally, we determined a cluster apparently more resistant (MR) to LRMD and a cluster less resistant (LR) to LRMD in the KIP treatment. Our objectives were to evaluate the liver transcriptome of these clusters via RNA sequencing within the original treatment for differentially expressed genes (DEGs) and enriched metabolic pathways (EMPs), as well as to evaluate the common and unique liver transcriptome features of the susceptibility and resistance models.

2. Materials and Methods

2.1. Animal Experimental Design

This research was part of a previously detailed experiment that enrolled multiparous Holstein cows (n = 25) at the University of Wisconsin–Madison Dairy Cattle Instruction and Research Center [35]. All animal use and handling protocols were approved by the University of Wisconsin–Madison College of Agricultural and Life Sciences' Animal Care and Use Committee (protocol A005467-R01). Blocked by expected calving date, cows were randomly assigned to a CTL (n =13) or a KIP (n = 12) treatment. The experimental period began at −28 days relative to calving (DRTC) and ended at +56 DRTC. Control cows were allowed ad libitum intake of diets formulated to meet the needs of dry or lactating cows, respectively (Supplementary Table S1). The KIP cows were offered a daily top-dress of dry, cracked corn (6 kg) in addition to ad libitum access to the dry cow ration [35]. Post-calving, KIP cows were offered ad libitum access to the lactating cow ration until +14 DRTC, at which time feed intake was restricted to 80% of ad libitum intake, based on the average voluntary intake from the 3 days preceding restriction. Blood BHB was monitored daily for all postpartum cows with a BHBCheck meter (PortaCheck, Moorestown, NJ); cows that achieved a blood BHB $\geq$ 3.0 mmol/L were treated for clinical ketosis, re-alimented to feed, and allowed ad libitum intake for the remainder of the experiment. All KIP cows and 2 CTL cows achieved blood BHB $\geq$ 3.0 mmol/L [35]; treatment for clinical ketosis included intravenous dextrose (250 mL; Phoenix Scientific Inc., St. Joseph, MO, USA; 50% dextrose), orally administered Propylene Advantage (300 mL/d for 3 to 5 days; TechMix LLC, Stewart, MN), and a B-vitamin complex injected intramuscularly (20 mL; Sparhawk Laboratories, Inc., Lenexa, KS, USA). Additional cow illnesses monitored by the herd veterinarian included displaced abomasum (n = 1 LR cow), clinical hypocalcemia (n = 1 LR cow), mastitis (n = 1 LS cow at +53 DRTC), retained placenta (n = 0 cows), and metritis (n = 0 cows).

Methods regarding sample collection and analysis are detailed in depth in the companion manuscript [35]. Briefly, feed intake and milk yield data were recorded daily, and composition analysis was performed on monthly composites of feed samples and on milk samples collected weekly. Sampling occurred at −28, −14, +1, +14, +28, +42, and +56 DRTC for body weight (BW), body condition score (BCS), blood samples, and blind percutaneous liver biopsies (Supplementary Methods) [35]. Sample collection at +14 DRTC preceded feed restriction for the KIP treatment. Additional body weights and BCSs were evaluated at −7 and +7 DRTC, and additional blood samples were collected at −7, −5, −3, +3, +5, and +7 DRTC. Blood fractions (serum or plasma) were quantified for BHB and glucose, respectively, using Catachem VETSPEC reagents on the Catachem Well-T AutoAnalyzer (Catachem, Oxford, CT, USA). Plasma FA concentration and liver TG concentration were determined using colorimetric, enzymatic assays [22,35,36].

2.2. *k*-Means Clustering and Retrospective Selection

Cows assigned to the CTL treatment were absent of an additional dietary challenge, and were leveraged to identify cows LS or MS to LRMD. Meanwhile, cows assigned to the KIP treatment were dietarily challenged to induce LRMD and leveraged to identify cows MR or LR to LRMD. To avoid subjectively choosing cows representing differential LRMD susceptibility or resistance, a k-means clustering algorithm (R, version 3.5.2) was used to empirically group cows within their original dietary (CTL or KIP) treatment based on metabolic characteristics. Variables supplied to the algorithm included concentrations of plasma FA, serum BHB, and liver TG from +1, +14, and +28 DRTC, as well as the maximum postpartum concentration of each lipid metabolite from any DRTC timepoint. This allowed for a longitudinal assessment of LRMD status, and the maximum concentrations served as a proxy for LRMD severity. All variables were tested for normality ($p \leq 0.05$, Shapiro–Wilk test) and transformed ($\log_{10}$ or reciprocal) to an empirically Gaussian distribution ($p > 0.05$, Shapiro–Wilk test). For each original treatment, we evaluated algorithms allowing for 2–5 clusters, 1000 iterations, and 1000 random starts. Based on silhouette plot evaluation (R package: cluster, version 2.1.0) [37] and the number of cows within the largest 2 clusters ($n \geq 4$), the optimal number of clusters was 2 and 4 for CTL and KIP cows, respectively. Two of the clusters from within the KIP treatment had too few cows to be considered for RNA-Seq ($n \leq 2$) and were excluded. From the clusters within CTL and KIP treatments, cows ($n = 4$ per cluster) were randomly selected to represent the cluster, and proceeded to RNA isolation and library preparation for RNA-Seq. Poor RNA integrity, detailed subsequently, resulted in the elimination of one cow per cluster; therefore, each cluster was represented by 3 randomly selected cows ($n = 6$ per original treatment) with adequate RNA integrity. Assignment of clusters as LS or MS in the CTL treatment and LR or MR in the KIP treatment was based on the statistical evaluation of the metabolites supplied for clustering (see Statistical Analysis). Greater concentrations of LRMD biomarkers suggest progressed pathology and increased risk of negative performance outcomes [14,17,38,39]; therefore, the clusters with greater blood BHB, blood FA, or liver TG were designated MS and LR within the CTL and KIP treatments, respectively, and clusters with lower lipid metabolites were designated LS and MR within the CTL and KIP treatments, respectively. Parity distribution across clusters was 8 second parity, 2 third parity (MS cluster), 1 fourth parity (LR cluster), and 1 fifth parity (LS cluster).

2.3. *RNA Isolation, Library Preparation, Sequencing, and Mapping*

Liver tissue samples (~50 mg) were homogenized into fine powders in liquid nitrogen using a mortar and pestle. Sample RNA was extracted following the miRNeasy protocol with a QIAcube instrument (Qiagen, Foster City, CA, USA). The RNA integrity number was determined via Bioanalyzer using the RNA 6000 nano kit (Agilent, Santa Clara, CA, USA) for all samples ($n = 48$). To improve overall RNA integrity for the experiment, one cow per cluster was dropped due to excessively low integrity for at least one DRTC. For the remaining samples ($n = 36$), RNA integrity was 6.0 ± 1.2 (SD). Library preparation for RNA sequencing was done using the Illumina TruSeq Ribo-Zero gold kit (Illumina, San Diego, CA, USA) following the manufacturer's instructions. For each sample, 1 µg of total RNA was used as the input. The fragment distribution of prepared libraries was assessed via Bioanalyzer using the DNA 1000 kit (Agilent, Santa Clara, CA, USA). Quantification of prepared libraries was performed using a Kapa quantification kit (Kapa biosystems, Darmstadt, Germany) with a QuantStudio5 quantitative PCR instrument (Thermo Fisher, Waltham, MA, USA). Libraries were further normalized to ensure equal quantities before sequencing. Normalized, pooled samples were sequenced on an Illumina NextSeq 500 instrument (Illumina, San Diego, CA, USA) to obtain paired-end, 2×75 bp reads using a 150 high-output kit. Quality of reads was assessed via FastQC (https://www.bioinformatics.babraham.ac.uk/projects/fastqc/ accessed on 28 December 2019). Before sequence alignment, raw reads were filtered to remove those shorter than 50 bp. The mean number of paired-end reads per library was $34,137,191 \pm 646,323$ (SEM). The ribosomal depletion of sample libraries coupled with the deep read count suggests that our results will be comparable to poly-A-enriched preparation methods, despite the relatively low library RIN values [40]. For alignment, the genome reference and annotation files for *Bos taurus* (release 106, ARS-UCD 1.2) were downloaded from the NCBI (https://www.ncbi.nlm.nih.gov/assembly/GCF_002263795.1 accessed on 14 March 2020) for use. Raw reads from the whole transcriptome RNA-Seq libraries were aligned to the *Bos taurus* reference genome using STAR (2.5.2b). Read count quantification was done using cufflinks [41] with a sorted bam file generated by STAR as the input file. The expression levels of mRNAs in each sample were normalized to fragments per kilobase of transcript per million mapped reads (FPKM) by cufflinks [41].

2.4. Statistical Analysis

Due to the dependence of the retrospective cluster assignments on the original dietary treatment (CTL or KIP), the complete data (n = 12, selected cows) were divided into datasets based on original dietary treatment; therefore, all statistical analyses compared either LS vs. MS or LR vs. MR.

Analysis of biometric (i.e., BCS), productive (i.e., milk energy yield), and metabolite data (i.e., serum BHB) was performed using the SAS (version 9.4; SAS Institute Inc., Cary, NC) procedures UNIVARIATE and GLIMMIX. Several response variables had non-Gaussian distributions based on the Shapiro–Wilk test ($p < 0.05$). For those responses, data transformations were systematically evaluated, and transformations providing Gaussian distributions were selected either empirically, by Shapiro–Wilk test ($p > 0.05$), or subjectively, by histogram visualization (when empirical solutions were not found). The bimodal nature of calculated net energy balance necessitated downstream analysis to be performed on prepartum and postpartum timepoints separately. Linear mixed models were used to evaluate responses for evidence of differences between clusters (LS vs. MS or LR vs. MR, n = 3 cows/cluster), using the systematic model-building procedure [35]. The typical fixed effects included cluster, time, and cluster × time; the random effects included cow, cow nested within week of lactation (models with subsampling), and repeated measures of cow across time (when applicable). All models were assessed for improvement by additional covariates (i.e., previous lactation, 305 d mature equivalent milk yield, parity, and −28 DRTC measurement), controlling for heterogeneous variance, and the use of alternative variance–covariance structures (variance components were default) [35]. Preplanned contrasts were performed to compare clusters (LS vs. MS and MR vs. LR) across postpartum timepoints (+1 to +56 DRTC), and were reported when they provided new information with respect to the LMM. Fixed effects with $p \leq 0.05$ were considered to have significant evidence for differences, while effects with $0.05 < p \leq 0.10$ were considered to have marginal evidence for differences. Whenever a cluster × time effect had some evidence for a difference ($p \leq 0.10$), we made simple-effect comparisons of treatments within timepoints and corrected for multiplicity using the Bonferroni method. Treatment means are expressed as least squares means, and the 95% confidence intervals denoted as (lower limit, upper limit).

Differential gene expression analysis was done using the Cuffdiff function of Cufflinks [41] within each DRTC for LS vs. MS and MR vs. LR. P-values were corrected for multiplicity by false discovery rate [42], and are hereafter referred to as Q-values. The Database for Annotation, Visualization, and Integrated Discovery (DAVID) web-based software was used to evaluate gene ontologies and EMPs [43,44] for DEGs within a DRTC comparison. Genes supplied to the test list (termed gene list by DAVID) had $Q \leq 0.10$ within the respective DRTC timepoint comparison of clusters. Instead of comparing a gene list to all genes in the *Bos taurus* genome, a customized background list was supplied to DAVID for the respective DRTC timepoint comparison of clusters. These customized background lists included all sequenced genes successfully tested within the respective DRTC comparison. Fisher's exact statistics were extracted and corrected for multiplicity by false discovery rate [42]. Our evidence criteria for DEGs, gene ontologies, and EMPs were $Q \leq 0.05$ and $0.05 < Q \leq 0.10$ for significant and marginal evidence, respectively.

3. Results

3.1. Phenotypic Characterization of Clusters

Biometric indicators of obesity—BW and BCS—were not different between LS and MS cows ($p \geq 0.85$; Figure 1a,c). Milk energy output and dry matter intake (DMI) did not differ between susceptibility clusters ($p = 0.59$ and $p = 0.41$, respectively; Supplementary Table S2).

Thus, net energy balance did not differ between susceptibility clusters pre- ($p = 0.89$; Supplementary Table S2) or postpartum ($p = 0.28$; Figure 1e). Plasma glucose (Figure 2a) and FA concentration (Figure 2b) did not differ ($p = 0.62$ and $p = 0.39$, respectively) between LS and MS cows either. Serum BHB concentration showed marginal evidence of being greater ($p = 0.08$) for the MS cows than for the LS cows (Figure 2c); a single MS cow was diagnosed with clinical ketosis (blood BHB $\geq$ 3.0 mmol/L). In addition, liver TG content was greater ($p = 0.02$; Figure 3a) for the MS cows than for the LS cows.

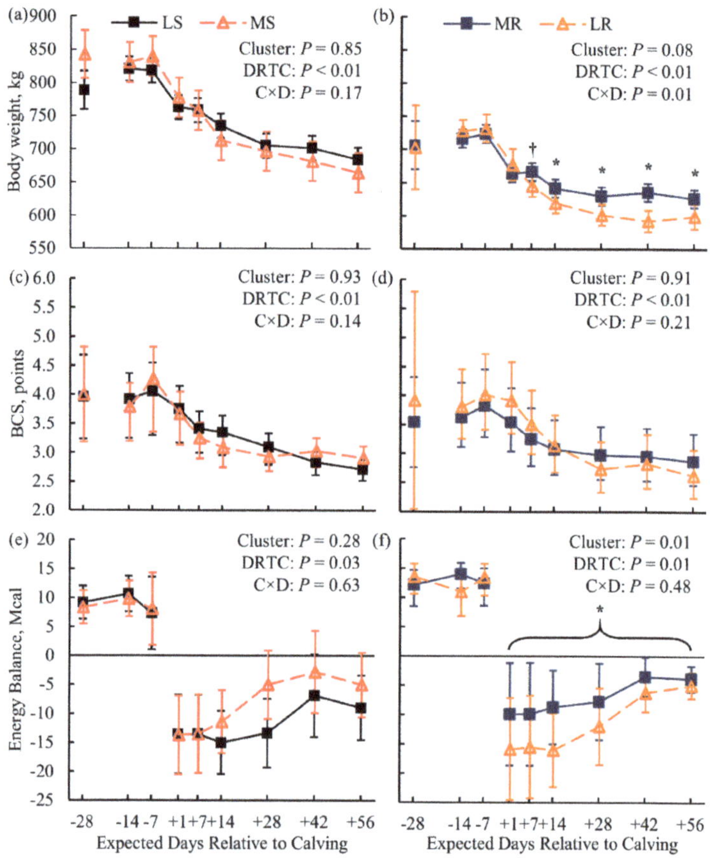

Figure 1. Body weight (panels **a** and **b**), body condition score (BCS; panels **c** and **d**), and calculated energy balance (panels **e** and **f**) for dairy cows clustered based on postpartum lipid metabolites within the original dietary treatment. Left-hand panels compare cows less (LS) or (MS) susceptible to lipid-related metabolic disorders, while right-hand panels compare cows more (MR) or less (LR) resistant to lipid-related metabolic disorders ($n = 3$ cows per cluster). Data are presented as least squares means (points) and their 95% confidence intervals (error bars); the exceptions are the arithmetic mean and 95% confidence interval for the −28 DRTC timepoint covariates (panels **a**–**d**). Statistics for the fixed effects of cluster, day relative to calving (DRTC), and their interaction (C × D) across the experimental period (panels **a**–**d**) or the postpartum period (panels **e** and **f**) are displayed in the top-right corner of each panel. Significant (*, $p \leq 0.05$; Bonferroni adjusted) or marginal evidence (†, $p = 0.07$; Bonferroni adjusted) for the effects of cluster within DRTC or contrast of cluster across postpartum samples (bracketed) are indicated within their respective panels.

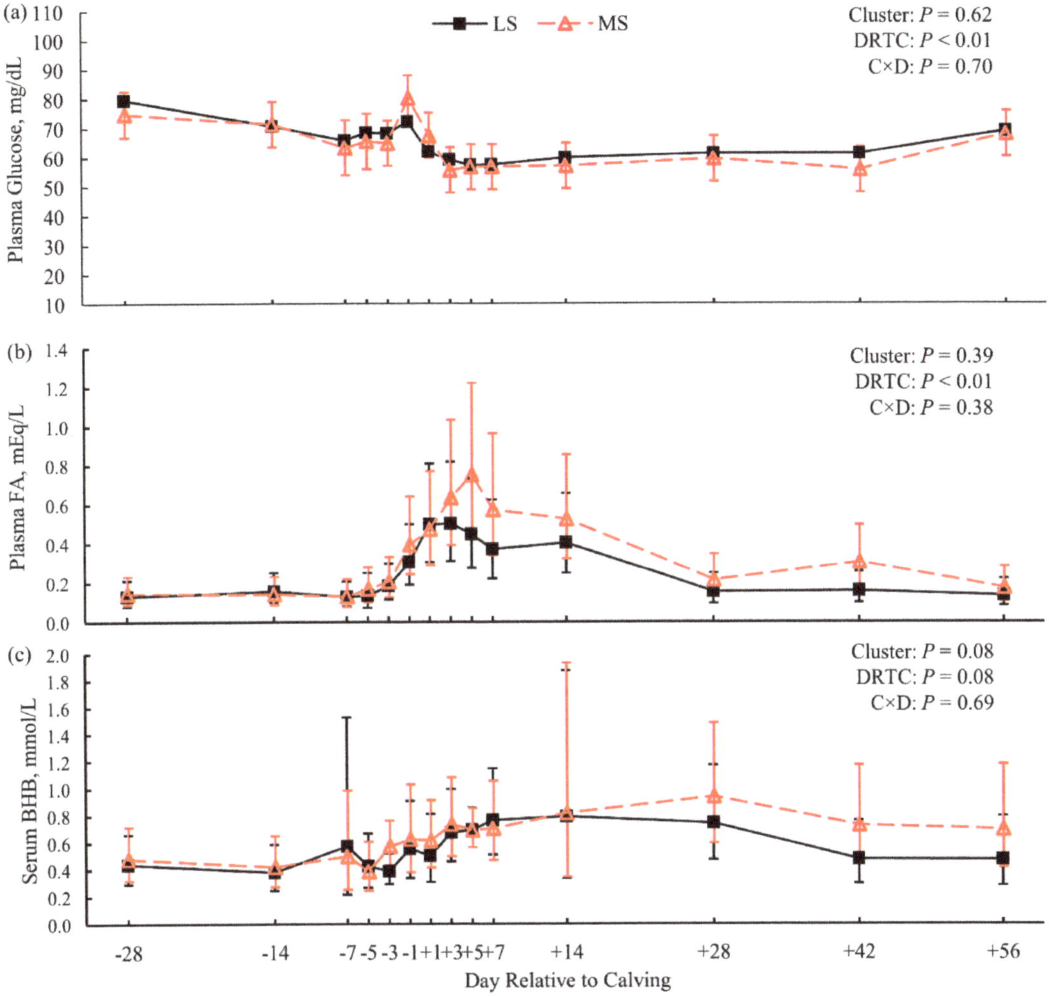

Figure 2. Blood fraction concentrations of glucose (panel **a**), fatty acids (FA; panel **b**), and β-hydroxybutyrate (BHB; panel **c**) for dairy cows less (LS) or more (MS) susceptible to lipid-related metabolic disorders ($n = 3$ cows per cluster). Data are presented as least squares means (points) and their 95% confidence intervals (error bars). Statistics for the fixed effects of cluster, day relative to calving (DRTC), and their interaction (C × D) are displayed in the top-right corner of each panel. Contrasts of cluster across postpartum samples did not reveal additional evidence of differences ($p > 0.10$; panels **a** and **b**).

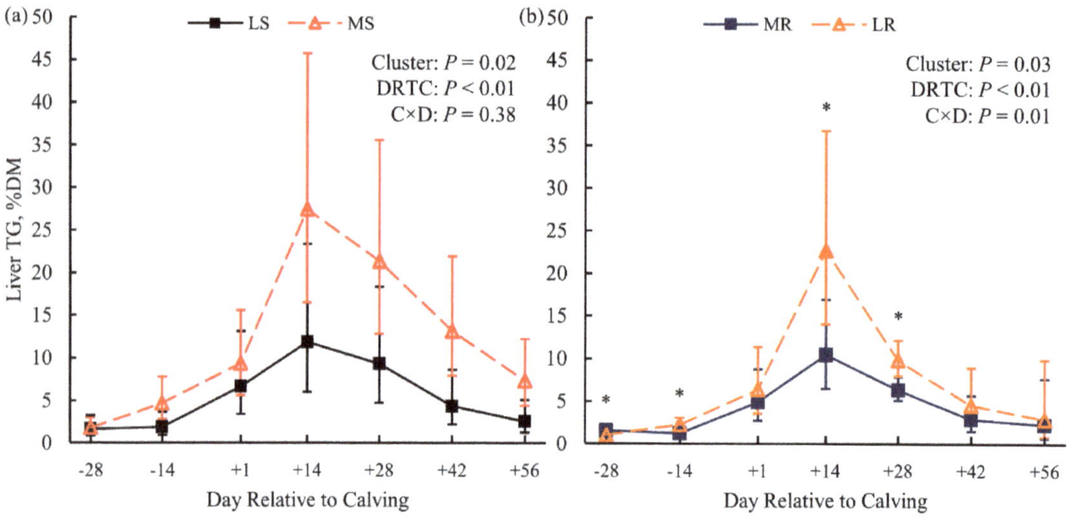

Figure 3. Liver triglyceride (TG) content for dairy cows clustered based on postpartum lipid metabolites within the original dietary treatment. Panel (**a**) depicts cows less (LS) or more (MS) susceptible to lipid-related metabolic disorders, while panel (**b**) compares cows more (MR) or less (LR) resistant to lipid-related metabolic disorders (n = 3 cows per cluster). Data are presented as least squares means (points) and their 95% confidence intervals (error bars). Statistics for the fixed effects of cluster, day relative to calving (DRTC), and their interaction (C × D) are displayed in the top-right corner of each panel. Significant evidence (*, $p \leq 0.02$; Bonferroni adjusted) for simple effects of cluster within DRTC are indicated in their respective panels.

For the resistance clusters within the KIP treatment, cows were individually feed-restricted after the +14 DRTC sampling until blood BHB ≥ 3.0 mmol/L. The number of days feed-restricted until achieving the threshold was 4, 8, and 13 d for the MR cows, and 0, 2, and 8 d for the LR cluster cows. A single LR cow was diagnosed with displaced abomasum and hypocalcemia by the herd veterinarian. Resistance clusters did not differ in BCS (p = 0.91; Figure 1d), but the MR cows had greater BW from +7 to +56 DRTC ($p \leq 0.07$ for Bonferroni-adjusted simple effects; Figure 1b) compared to LR cows. Milk energy output and milk lactose yield were similar ($p > 0.99$ and p = 0.52, respectively; Supplementary Table S3) between resistance clusters. Marginal evidence of greater postpartum DMI (p = 0.09, contrast) for MR cows compared to LR cows was observed, contributing to the significantly (p = 0.01) attenuated negative energy balance observed for MR cows compared to LR cows (Figure 1f). Plasma glucose concentrations showed marginal evidence of being greater postpartum (p = 0.07, contrast) for MR cows than for LR cows (Figure 4a). Compared to their LR contemporaries, MR cows showed marginal evidence of lower concentrations of postpartum plasma FA (p = 0.06, contrast; Figure 4b) and serum BHB (p = 0.10; Figure 4c), as well as significant evidence of lower liver TG (p = 0.03; Figure 3b).

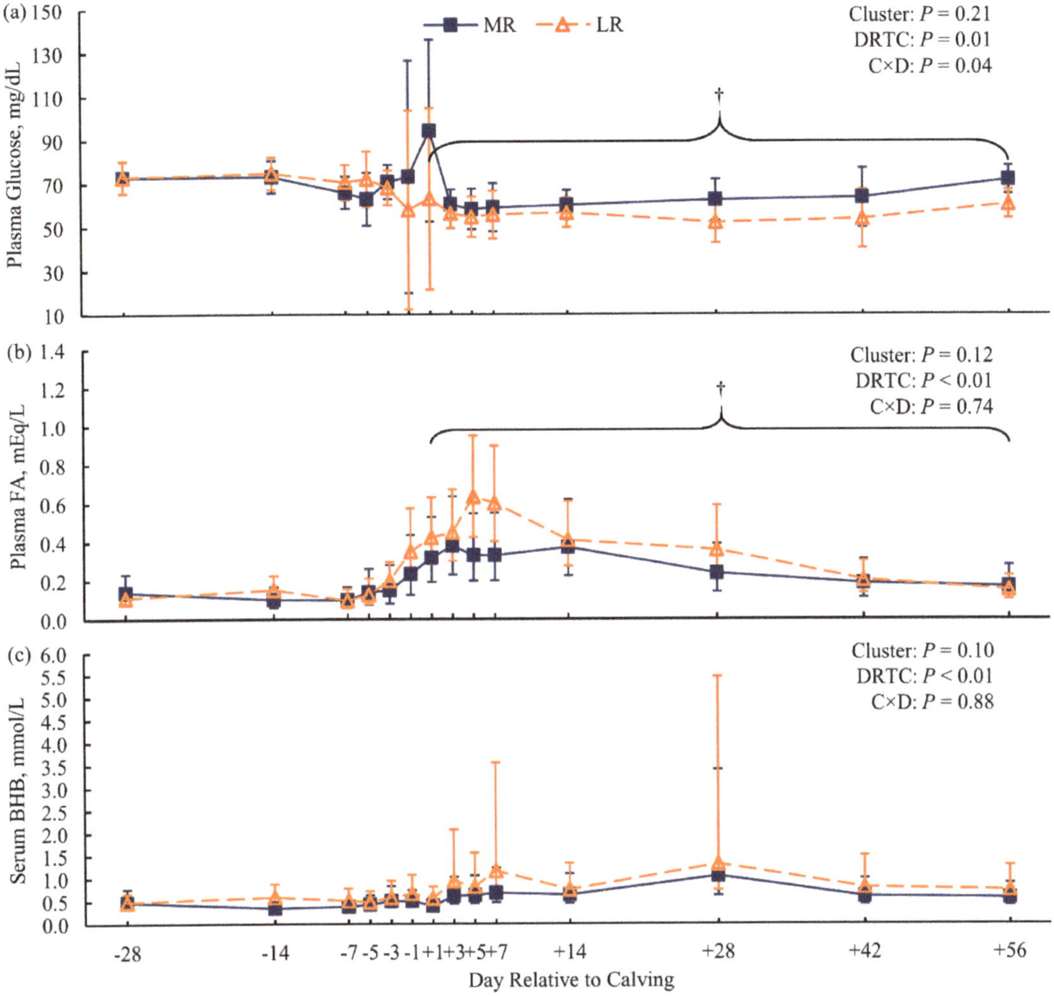

Figure 4. Blood fraction concentrations of glucose (panel **a**), fatty acids (FA; panel **b**), and β-hydroxybutyrate (BHB; panel **c**) for dairy cows more resistant (MR) or less (LR) resistant to lipid-related metabolic disorders (n = 3 cows per cluster). Data are presented as least squares means (points) and their 95% confidence intervals (error bars). Statistics for the fixed effects of cluster, day relative to calving (DRTC), and their interaction (C × D) are displayed in the top-right corner of each panel. Marginal evidence (†; p = 0.07, panel **a**; p = 0.06, panel **b**) for contrasts of cluster across postpartum samples (bracketed) are indicated within their respective panels.

3.2. Differentially Expressed Genes and Enriched Metabolic Pathways

For the comparison of LS and MS clusters, the count of genes with FPKM > 0 was 13,151 genes at −28 DRTC, 13,011 genes at +1 DRTC, and 13,211 genes at +14 DRTC tested for differential expression (Table 1). Genes with significant (marginal) evidence on −28, +1, and +14 DRTC totaled 165 (58), 116 (31), and 199 (39), respectively (Supplementary Microsoft Excel File). A selection of genes with significant or marginal evidence for differential expression across all DRTC within the susceptibility cluster comparisons is listed in Table 2. Metabolic pathways enriched within the DEGs with significant (marginal) evidence numbered 0 (2), 4 (3), and 5 (2) for −28, +1, and +14 DRTC, respectively (Table 3).

Table 1. The numbers of genes tested within days relative to calving (DRTC; diagonal) that were shared within and across comparison of clusters [1].

		Susceptibility			Resistance		
	DRTC	**−28**	**+1**	**+14**	**−28**	**+1**	**+14**
Susceptibility	−28	13,151	12,752	12,837	12,833	12,757	12,519
	+1	21,009	13,011	12,784	12,670	12,744	12,512
	+14	20,891	20,978	13,211	12,768	12,774	12,620
Resistance	−28	20,962	20,939	20,834	13,139	12,736	12,518
	+1	20,874	21,001	20,828	20,865	13,150	12,525
	+14	21,065	21,198	21,103	21,076	21,071	12,718

[1] The number of tested genes shared (pairwise) across DRTC and across cluster comparisons (susceptibility or resistance) are above the diagonal, while the number of untested genes shared are below the diagonal. There were 34,411 genes that were potentially tested across the *Bos taurus* genome (release 106, ARS-UCD 1.2).

Table 2. Fold change [1] of genes with significant or marginal evidence of differential expression at all days relative to calving in liver samples from cows less or more susceptible to lipid-related metabolic disorders [2].

Gene	Symbol	−28	+1	+14
BOLA class I histocompatibility antigen, alpha chain BL3-6	*BOLA*	−1.3	−1.3	−1.4
Major histocompatibility complex, class II, DQ beta	*BOLA-DQB*	−2.2	−3.0	−3.0
Coiled-coil domain-containing 80	*CCDC80*	−5.0	−1.4	−3.4
C-C motif chemokine 14 precursor	*CCL14*	−1.7	−1.9	−1.6
B-cell receptor CD22	*CD22*	1.4	1.1	0.9
Acetylcholine receptor subunit epsilon	*CHRNE*	2.3	1.3 †	1.5
C-type lectin domain family 4 member F	*CLEC4F*	−1.3	−1.3 †	−1.5
GTPase IMAP family member 6	*GIMAP6*	−1.3	−1.7	−1.7
GTPase IMAP family member 8	*GIMAP8*	−2.2	−2.0	−2.4
ISG15 ubiquitin-like modifier	*ISG15*	−1.6	−1.0 †	1.1
Methylenetetrahydrofolate dehydrogenase 1-like	*MTHFD1L*	−1.4	−1.3 †	−1.6
Olfactory receptor 4 × 2	*OR4S1*	1.0	0.9 †	0.8
Prodynorphin	*PDYN*	−2.4	−2.9	−2.2
Secreted frizzled-related protein 2	*SFRP2*	5.6	4.4	2.0
Pulmonary surfactant-associated protein A precursor	*SFTPA1*	−3.7	−2.1	−1.6
Teneurin transmembrane protein 1	*TENM1*	−2.1	−1.3 †	−0.9

[1] Values represent the log_2-transformed fold change within each timepoint. Positive values indicate greater expression for more susceptible than for less susceptible cows, and vice versa for negative values. [2] Genes were considered to have significant evidence of differential expression within days relative to calving (−28, +1, and +14) when $Q \leq 0.05$ (*p*-value corrected for multiplicity by false discovery rate); meanwhile, marginal evidence (†) was declared at $0.05 < Q \leq 0.10$.

Table 3. Kyoto Encyclopedia of Genes and Genomes metabolic pathways enriched within the genes with significant or marginal evidence of differential expression in liver samples from cows less or more susceptible to lipid-related metabolic disorders [1].

DRTC	ID	Metabolic Pathway	FE	*Q*-Value	Gene Symbols [2]
	bta04514	Cell adhesion molecules (CAMs)	5	0.098	*MGC126945, ITGAV, NRCAM, LOC534578, CD22, BOLA-DQB, LOC100848815*
−28	bta04145	Phagosome	4	0.098	*SEC61G, SFTPA1, MRC2, MGC126945, ITGAV, SEC61B, BOLA-DQB*
	bta03320	PPAR signaling pathway	6	0.136	*SCD, CYP8B1, APOA5, SLC27A6, FADS2*
	bta05204	Chemical carcinogenesis	7	0.023	*LOC615303, CYP1A1, GSTA4, GSTT1, GSTA1, CYP2C87*
	bta00480	Glutathione metabolism	8	0.026	*ODC1, GSTA4, GSTT1, ANPEP, GSTA1*
	bta00980	Metabolism of xenobiotics by cytochrome P450	7	0.026	*LOC615303, CYP1A1, GSTA4, GSTT1, GSTA1*
+1	bta05164	Influenza A	4	0.04	*RSAD2, OAS1X, DDX39B, DDX58, LOC784541, BOLA-DQB, LOC100848815*
	bta04141	Protein processing in endoplasmic reticulum	3	0.079	*CKAP4, HSPA5, HYOU1, DNAJB11, MOGS, SEC23B, WFS1*
	bta00982	Drug metabolism-cytochrome P450	6	0.079	*LOC615303, GSTA4, GSTT1, GSTA1*
	bta04514	Cell adhesion molecules (CAMs)	4.0	0.085	*NRCAM, LOC534578, CD22, BOLA-DQB, LOC100848815*
	bta04512	ECM-receptor interaction	9	6.3×10^{-5}	*COL6A3, COL6A1, COL5A1, COL5A3, COL3A1, COL1A1, COL1A2, COL5A2, VWF, TNXB*
	bta04974	Protein digestion and absorption	10	6.8×10^{-5}	*COL6A3, COL6A1, COL5A1, COL5A3, COL3A1, COL1A1, COL1A2, COL5A2, COL15A1*
	bta04611	Platelet activation	4.0	0.046	*COL5A1, COL5A3, ADCY1, COL3A1, COL1A1, COL1A2, COL5A2, VWF*
+14	bta04510	Focal adhesion	3	0.046	*COL6A3, COL6A1, COL5A1, COL5A3, COL3A1, COL1A1, COL1A2, COL5A2, VWF, TNXB*
	bta05146	Amoebiasis	5	0.046	*COL5A1, COL5A3, ADCY1, COL3A1, COL1A1, COL1A2, COL5A2*
	bta04923	Regulation of lipolysis in adipocytes	7	0.058	*PTGER3, ADCY1, FABP4, LIPE, IRS2*
	bta04151	PI3K–Akt signaling pathway	3	0.058	*IFNAR1, COL6A3, COL6A1, COL5A1, COL5A3, LPAR1, COL3A1, COL1A1, COL1A2, COL5A2, VWF, TNXB*

[1] Enrichment analysis was performed within days relative to calving (DRTC) using the Database for Annotation, Visualization, and Integrated Discovery (version 6.8) by comparing a list of genes with significant or marginal evidence ($Q \leq 0.10$; *p*-value adjusted for multiplicity by the false discovery rate method) to a custom background list including all tested genes within a DRTC. Fold enrichment (FE) and Fisher's exact statistics were extracted; tests were corrected for multiplicity via the false discovery rate method (*Q*-value). [2] Annotation of gene transcripts and affiliated gene symbols are based on the *Bos taurus* reference genome (release 106, ARS-UCD 1.2).

Comparing the MR and LR cows, the count of genes with FPKM > 0 was 13,139 genes at −28 DRTC, 13,150 genes at +1 DRTC, and 12,718 genes at +14 DRTC tested for differential expression (Table 1). Significant (marginal) evidence for differential expression was found for 127 (30), 142 (50), and 102 (31) genes at −28, +1, and +14 DRTC, respectively (Supplementary Microsoft Excel File). A selection of genes with significant or marginal evidence for differential expression across all DRTC within the resistance cluster comparisons is listed in Table 4. Metabolic pathways enriched within the DEGs with significant (marginal) evidence numbered 0 (5), 6 (2), and 0 (0) for −28, +1, and +14 DRTC, respectively (Table 5).

Table 4. Fold change [1] of genes with significant or marginal evidence of differential expression at all days relative to calving in liver samples from cows more or less resistant to lipid-related metabolic disorders [2].

Gene	Symbol	−28	+1	+14
Major histocompatibility complex, class II, DQ beta	*BOLA-DQB*	1.9	2.1	2.6
GTPase IMAP family member 4	*GIMAP4*	−0.8 †	−1.2	−1.2
17-Beta-hydroxysteroid dehydrogenase 13	*HSD17B13*	1.2	1.8	1.2
Interferon-induced protein 44	*IFI44*	1.3	1.8	1.3
Interferon alpha-inducible protein 6	*IFI6*	1.8	2.6	1.0
MHC Class I JSP.1	*JSP.1*	0.8 †	1.1	0.7 †
Myomesin 1	*MYOM1*	0.9	1.8	0.8
2′,5′-Oligoadenylate synthetase 1, 40/46kDa	*OAS1X*	2.3	1.8	1.4
Proto-oncogene tyrosine-protein kinase ROS	*ROS1*	−1.1	−1.4	−3.0

[1] Values represent the $\log_2$-transformed fold change within each timepoint. Positive values indicate greater expression for less resistant than for more resistant cows, and vice versa for negative values. [2] Genes were considered to have significant evidence for differential expression within days relative to calving (−28, +1, and +14) when $Q \leq 0.05$ (p-value corrected for multiplicity by false discovery rate); meanwhile, marginal evidence (†) was declared at $0.05 < Q \leq 0.10$.

Table 5. Kyoto Encyclopedia of Genes and Genomes metabolic pathways enriched within the genes with significant or marginal evidence of differential expression in liver samples from cows more or less resistant to lipid-related metabolic disorders [1].

DRTC	Metabolic Pathway	FE	*Q*-Value	Gene Symbols [2]
−28	Cytokine–cytokine receptor interaction	4.6	0.099	*PF4, CXCR2, CXCR1, IL1R2, CCR1, CSF2RB*
	Cell adhesion molecules (CAMs)	4.5	0.099	*VCAN, JSP.1, SELL, LOC534578, BOLA-DQB*
	Viral myocarditis	6.4	0.099	*JSP.1, MYH7, CD55, BOLA-DQB*
	Chemokine signaling pathway	3.5	0.099	*PF4, NCF1, CXCR2, CXCR1, CCR1, CCL24*
	Hematopoietic cell lineage	5.7	0.099	*MS4A1, LOC515418, IL1R2, CD55*
+1	Metabolic pathways	14.1	0.041	*IDI1, HAL, SDS, LOC511161, KYAT1, ALPI, FOLH1B, LOC615045, BCO1, OAT, GCSH, ASAH2, PLB1, FDPS, FUT1, HDC, AKR1B10, MBOAT2, ACSS2, MTHFD1L, ADH4, ENPP3, B4GALT4, PLA2G2A, CYP2J2, LOC100125266, CYP2B6*
	Arachidonic acid metabolism	2.6	0.041	*PLB1, LOC615045, PLA2G2A, CYP2B6, CYP2J2*
	Linoleic acid metabolism	2.1	0.041	*PLB1, LOC615045, PLA2G2A, CYP2J2*
	Autoimmune thyroid disease	2.1	0.041	*JSP.1, LOC524810, LOC100300716, BOLA-DQB*
	Allograft rejection	2.1	0.041	*JSP.1, LOC524810, LOC100300716, BOLA-DQB*
	Viral myocarditis	2.6	0.041	*JSP.1, LOC524810, LOC100300716, CD55, BOLA-DQB*
	Intestinal immune network for IgA production	2.1	0.083	*LOC524810, PIGR, LOC100300716, BOLA-DQB*
	alpha-Linolenic acid metabolism	1.6	0.086	*PLB1, LOC615045, PLA2G2A*

[1] Enrichment analysis was performed within days relative to calving (DRTC) using the Database for Annotation, Visualization, and Integrated Discovery (version 6.8) by comparing a list of genes with significant or marginal evidence ($Q \leq 0.10$; *p*-value adjusted for multiplicity by the false discovery rate method) to a custom background list including all tested genes within a DRTC. Fold enrichment (FE) and Fisher's exact statistics were extracted; tests were corrected for multiplicity by the false discovery rate method (*Q*-value). [2] Annotation of gene transcripts and affiliated gene symbols are based on the *Bos taurus* reference genome (release 106, ARS-UCD 1.2).

4. Discussion

The purpose of this work is to provide novel insight into the genes and metabolic pathways integral to the pathology of LRMD. To that end, we retrospectively identified groups of periparturient dairy cows within a relatively normal dietary scenario and within a metabolically challenging dietary scenario, with apparently different metabolic statuses. These differences in metabolic status were determined via *k*-means clustering of cows based on postpartum liver and blood characterization as indicators of HYK and FLS [17,38,45]. It is important to acknowledge that the sample size for the cluster comparisons was relatively small (n = 3 cows/cluster), which can impede the detection of response differences or limit generalizability. Nevertheless, evidence for differences in lipid biomarkers—particularly liver TG and blood BHB—were found between clusters. Although nutrient partitioning is a normal physiological adaptation that may support feed efficiency, metabolic health, and productivity [46–49], dysregulation or imbalance can reflect metabolic disorders, which are the focus herein. Sequencing of the liver transcriptomes at several peripartum timepoints suggested numerous genes and metabolic pathways associated with LRMD at one or more timepoints. Therefore, this discussion will only concern the phenotypic characterization of the clusters within the original dietary treatment, along with a selection of genes, gene families, and metabolic pathways.

4.1. Phenotypic Characterization of Clusters

Cows within the CTL treatment were subject to typical pre- and postpartum diets for dairy cows, without additional imposed challenges [35]. Differential regulation of genes and metabolic pathways may underlie facets of liver metabolism that may predispose an individual cow to the progression of LRMD. The principal metabolic differences between the LS and MS cows were the greater serum BHB (p = 0.08; Figure 2c) and liver TG (p = 0.02; Figure 4a) concentrations observed for the MS cows. As mentioned, these metabolites serve as the primary biomarkers for HYK and FLS, with greater concentrations suggesting pathology [17,38,45]. Thus, the MS cluster appears to be in a less favorable metabolic condition and prone to LRMD. Greater plasma FA concentration is also a biomarker of LRMD, and would be expected in the MS cluster [39,50]. Although plasma FA were numerically greater for MS cows postpartum (Figure 2b), they was not significantly different from the LS cows. This lack of difference in FA is corroborated by the similar BW and BCS between the susceptibility clusters peripartum (Figure 1), suggesting unappreciable differences in the lipolysis of adipose TG between clusters. Additionally, these clusters did not differ in DMI, milk energy output, energy balance (Figure 1e), or plasma glucose (Figure 2a). Although these differences are in contrast to the present dogma of insufficient nutrient supply and over-mobilization of body energy reserves leading to LRMD [17,18,45], it is important to remember that the lack of differences here may be due to the small sample size with regard to metabolite analysis. Conversely, these data may suggest that some cows are susceptible to onset of LRMD despite the lack of energetic challenges that are classically associated with the dogma, and may be susceptible due to independent risk factors.

The KIP treatment imposed on the cows in the resistance clusters (MR and LR) was intended to predispose cows to LRMD. All cows in the KIP treatment progressed to a clinical HYK blood BHB threshold (BHB $\geq$ 3.0 mmol/L). Thus, the aim of metabolic clustering was to identify cows more resistant to the KIP treatment, with lower metabolite concentrations (plasma FA, serum BHB, and liver TG) or a greater number of feed restriction days until blood BHB $\geq$ 3.0 mmol/L. Consistent with this goal, dairy cows in the MR cluster had lower concentrations of serum BHB (p = 0.10; Figure 4c) and liver TG postpartum (p = 0.03; Figure 4b) compared to the LR cluster. Furthermore, MR cows required more feed restriction days to achieve BHB $\geq$ 3.0 mmol/L. Together, these data suggest that cows in the MR cluster were more resistant to LRMD than LR cows. Compared to LR cows, MR cows had greater postpartum BW (Figure 1b) and less negative energy balance postpartum (Figure 1f), suggesting that the MR cows did not mobilize as much of their endogenous energy reserves to support lactation nutrient requirements as the LR cows. Greater plasma

FA concentration for the LR cows (Figure 4b) corroborates greater lipolysis of adipose tissue TG. The greater plasma glucose concentration for the MR cows postpartum (Figure 4a), with similar milk lactose output, may suggest that MR cows had greater hepatic gluconeogenesis or decreased peripheral glucose utilization compared to LR cows [1,8,51]. Overall, the MR cows appeared to experience a more favorable adaptation to lactation when a dietary challenge was imposed than the LR cows.

4.2. Inferred Differential Regulation of the Liver Transcriptome

Mobilization of adipose tissue TG as FA and their linear uptake by the liver results in a tremendous supply of liver FA for metabolism in dairy cows [52,53]. Oxidation of these FA in the hepatocyte mitochondria and peroxisomes to produce energy can also produce reactive oxygen species (ROS). Accumulation of hepatic ROS in early postpartum dairy cows can induce oxidative stress, characterized by excessive oxidation of proteins and lipids and induction of apoptosis [54,55]. Glutathione is a protein that can serve as an antioxidant, protecting cells from ROS-induced oxidative damage [26,55,56]. In the present experiment, the KEGG pathway glutathione metabolism was enriched in the DEGs identified in the comparison of susceptibility clusters on +1 DRTC (Table 3). The LS cows had greater expression of genes in the glutathione-centered antioxidant defense system than MS cows: *GSTA4* (2.6-fold), *GSTT1*(2.0-fold), and *GSTA1* (2.5-fold); these genes catalyze glutathione conjugations with electrophilic compounds (i.e., drugs, xenobiotics, lipid hydroperoxides), and some have glutathione peroxidase activity [57–59]. Thus, the LS cows appear to have improved their antioxidation capacity compared to their MS contemporaries.

Eicosanoids, a subcategory of oxylipids, are signaling molecules made by the enzymatic or non-enzymatic oxidation of arachidonic acid or other polyunsaturated FA [54,60]. These molecules can have either pro- or anti-inflammatory properties, while influencing oxidative stress [54]. Eicosanoids can directly promote oxidative stress through the production of ROS, or indirectly via the production of ROS during biosynthesis. Some eicosanoids—such as 15-deoxy-delta 12 and 14-Prostaglandin J2—exert antioxidant effects by directly or indirectly targeting and decreasing ROS production by other cellular metabolic processes (i.e., mitochondrial oxidation) [54,60]. At +1 DRTC, the KEGG pathways linoleic acid metabolism, alpha-linoleic acid metabolism, and arachidonic acid metabolism were enriched for the comparison of the LRMD resistance cluster (Table 5). All of the specific DEGs were expressed more in the livers of LR cows: *PLB1* (36.8-fold), *PLA2* (infinity (∞)-fold), *PLA2G2A* (∞-fold), *CYP2B6* (2.6-fold), and *CYP2J2* (4.3-fold). The inferred promotion of eicosanoid synthesis may be a byproduct of the greater ROS production and pathology of LRMD. However, the greater eicosanoid production may be an adaptation by the LR cows to accelerate the termination of the liver's inflammatory state.

The role of the immune system in the progression of metabolic syndromes is an emerging scientific field in dairy science [61,62]. In humans, there appears to be crosstalk between metabolism and the immune response, which furthers the progress of non-alcoholic fatty liver to steatohepatitis and cirrhosis [63]. In the present experiment, many of the most consistent DEGs across the DRTC were the immunity-related genes involved in expressing components of the major histocompatibility complex (MHC; classes I and II) and genes in the interferon-inducible protein (IFI) family for LRMD susceptibility and resistance comparisons.

The MHC molecules are responsible for the presentation of antigens on the cell surface, and recruit other mechanisms of the innate immune response when antigens represent "non-self" proteins [64]. Expression of MHC molecules—particularly MHC class II—has been implicated in the pathology of liver metabolic and inflammatory disease in humans [63,64]. In our data, the expression of MHC elements was found in the cell adhesion molecules pathway at −28 and +1 DRTC for the susceptibility clusters and −28 DRTC for the resistance cluster comparison. Of note, *bovine lymphocyte antigen-DQB (BOLA-DQB)*—an MHC class II protein—was one of the few DEGs observed at every DRTC for the LRMD susceptibility and resistance cluster comparisons. Liver tissue is composed of several cell types, including hepatocytes, hepatic stellate cells, Kupffer cells, and sinusoidal cells [65]. Expression of MHC class II molecules is typically restricted to professional antigen-presenting cell types such as Kupffer cells, and not hepatocytes [66]; however, hepatocytes from humans with clinical hepatitis do express MHC class II molecules [67], which may be capable of attracting CD4+ T lymphocytes [64]. It is theorized that ROS-induced peroxidation of cellular components, such as proteins and lipids, can induce MHC expression and presentation of the oxidized component antigens to the cell surface [63]. The subsequent recruitment of CD4+ T lymphocytes and cytokine expression promote the apoptosis and clearance of ROS-damaged hepatocytes by the immune system [63,64]. Even though hepatocytes are the predominant liver cell type, differential expression of these genes may be found within Kupffer cells. Fatty acid binding to Kupffer Toll-like receptors has been demonstrated to promote macrophage recruitment through the C-JNK and NF-κB pathways, promoting liver lipid synthesis and mitochondrial dysfunction [68,69]. Interestingly, *BOLA-DQB* and other MHC components generally had greater expression for the LS and LR cows than for the MS and MR cows, respectively. Considering that the LS group was the metabolically preferable group within the susceptibility cluster, we did not expect the relative directionality of these DEGs to be similar to LR—the less preferable resistance cluster. It may be the case that cows not experiencing additional dietary challenges benefit from sensitive clearance of ROS-damaged hepatocytes, while cows challenged with additional prepartum energy may suffer from an excessive quantity of ROS-damaged cells or oversensitive presentation of antigens, resulting in disproportional apoptosis and clearance, but this would require further examination.

Interferons are cytokines that are best known for their secretion in response to viral infections, but have been implicated in the pathology of non-alcoholic fatty liver disease in humans [70,71]. The IFI and other interferon-stimulated genes are downstream effectors of interferon [72]. Similar to the MHC genes, IFI genes—specifically *IFI6*, *IFI27*, *IFI44*, and *IFI44L*—had greater relative expression (range of 2.0- to 5.6-fold) for LS and LR cows than for MS and MR cows, respectively, across all DRTC. It is possible that the IFI genes serve as the upstream regulators promoting the expression of the MHC molecules [72], and promote the clearance of ROS-damaged hepatocytes, as previously discussed.

Serum amyloid A (SAA) is an acute phase response protein associated with the inflammatory cascade that has been previously associated with HYK [73–75]. The isoforms SAA1 and SAA2 are generally considered to be the predominant proinflammatory proteins expressed in the liver, while SAA3 has generally been viewed as an adipokine [73,75]. Expression of *SAA3* was significantly greater for LS than MS cows at −28 DRTC (6.1-fold), but LS cows had significantly lower *SAA3* expression (3.2-fold) at +1 DRTC than MS cows. At +14 DRTC, the LS cluster had greater expression of *SAA2* (6.1-fold), *SAA4* (1.7-fold), and *LOC104968478* (21.1-fold) than the MS cluster. The LR cluster had consistently greater expression of all SAA isoforms than the MR cluster, ranging from 4- to 9.8-fold. Comparison of resistance clusters is consistent with previous works associating greater abundance of SAA and states with LRMD [74,76]. However, the susceptibility cluster comparison suggests a more nuanced regulation of SAA in the absence of additional dietary challenges. It has been previously suggested that a priming of the inflammatory response may exert protective effects in peripartum dairy cows [77], which the relative expression pattern of SAA3 for the susceptibility clusters may support.

4.3. Insight into LRMD Pathology

There is considerable variation in the metabolic health of dairy cows peripartum; this is evident in the variable prevalence of LRMD and their comorbidities [12,38,78]. In addition, the variation in the occurrence and severity of LRMD is still evident when cows are subjected to a dietary challenge [32,35]. Our retrospective clustering of dairy cows based on lipid metabolites within the CTL and KIP dietary treatments empirically identified groups of cows with divergent metabolic health, as previously discussed. The apparent presence of these divergent groups within both dietary conditions indicated two potential control points for LRMD pathology: susceptibility to LRMD incidence, and resistance to LRMD severity. Even though the gross pathology of LRMD pertaining to lipid metabolism and gluconeogenesis has been investigated, there has been limited investigation into the divergent metabolic regulation responsible for these LRMD susceptibility or resistance control points. The evaluation of the liver transcriptome through RNA sequencing allowed for a more holistic evaluation of what unique and shared metabolic pathways underpin these control points [79].

The genes and metabolic pathways presented in this work appear to be centered around hepatic adaptation to the lipotoxic effects of ROS through the innate immune system and inflammatory response. As discussed, the intersection of nutrient metabolism, immunity, and inflammation in the peripartum physiological adaptations and LRMD pathology has been a growing cross-disciplinary topic in dairy science [62,76]. These observations support further investigation into immunometabolism—specifically on the contributions of glutathione metabolism, eicosanoid metabolism, and MHC molecules in liver tissue. The observed regulation for these pathways is summarized in Figure 5. Interestingly, the susceptibility and resistance comparisons each had a unique inferred metabolic pathway at +1 DRTC. Glutathione metabolism was unique to the susceptibility comparison, while eicosanoid metabolism was a unique to the resistance comparison (Table 3; Table 5). Of course, these differences could be due to differences in the expressed genes available for testing between these +1 DRTC comparisons (Table 1). The relevance of MHC molecules and the innate immune response is apparently shared across the susceptibility and resistance comparisons. However, the relative expression of the DEGs and EMPs (i.e., BOLA–DQB) was greater for the metabolically preferable susceptibility cluster (LS) and the metabolically less preferable resistance cluster (LR). This discordant direction of expression across LRMD susceptibility and resistance comparisons suggests that experimental conditions—especially induction protocols—require nuanced interpretation. While some differentially regulated genes and pathways may be shared between "natural" LRMD development and induction, the observed response directionality may reflect adaptive mechanisms to mitigate the severity of LRMD rather than their progression or risk. Additionally, there may be unique mechanisms to the natural pathology of LRMD, as well as mechanisms that resist the progression of LRMD.

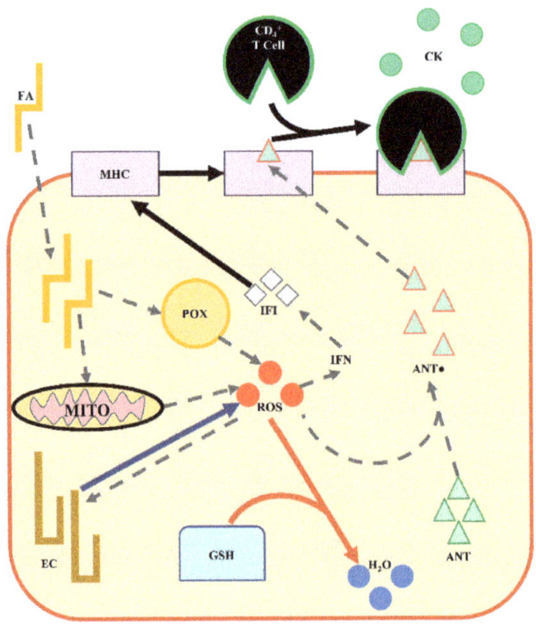

Figure 5. A working model of differentially regulated metabolic pathways in the liver of dairy cows that contribute to the pathology of lipid-related metabolic disorders. Fatty acids (FA) entering the hepatocyte are oxidized in the mitochondria (MITO) and peroxisomes (POX), producing energy and reactive oxygen species (ROS). Interferon (IFN) production is promoted by ROS, stimulating major histocompatibility complex (MHC) expression. Antigens (ANT) are oxidized (ANT•) by ROS. The ANT• are presented by the MHC promoting CD4+ T lymphocyte recruitment and cytokine (CK) production. Eicosanoids (EC) are formed by FA oxidation by ROS, and may promote ROS production. Glutathione (GSH) reduces ROS and other oxidized products. Arrows demonstrate the directionality and specificity of metabolic pathways: pathways not differentially regulated are grey dashed lines (—), pathways upregulated in cows less resistant to LRMD are blue (——), pathways upregulated in cows less susceptible to LRMD are red (——), and pathways generally upregulated in cows less susceptible and less resistant are solid black (——).

5. Conclusions

There is substantial individual variation in the metabolic responses of dairy cows to peripartum conditions, suggesting that the underpinning regulation of key metabolic pathways may confer susceptibility or resistance to LRMD. We empirically grouped multiparous Holstein cows within normal and challenged peripartum dietary conditions based on their blood lipid metabolite profiles and liver TG contents. This approach revealed two metabolic health groups within each dietary condition, suggesting differential susceptibility to LRMD incidence or resistance to LRMD induced by a dietary challenge. These metabolic differences were realized in the liver transcriptomes of these cow groups, with the inferred differential metabolism highlighting the role of the inflammatory response and innate immunity in LRMD pathology. Novel insights included the differential regulation of MHC molecules and IFI, which may aid the response to the ROS-induced cellular damage that occurs in liver tissue peripartum. Furthermore, the contributions of liver glutathione and eicosanoid metabolism to LRMD pathology immediately postpartum appear to be of greater biological importance relative to dietary condition—normal and challenged, respectively. Future work should build on the contributions of these specific mechanisms in liver cell types and delineate their dependence on specific dietary conditions.

Supplementary Materials: The following are available online at https://www.mdpi.com/article/10.3390/ani11092558/s1. Supplementary Table S1. Ingredient and nutrient composition of pre- and postpartum experimental diets., Supplementary Table S2. Least squares means (LSM) and 95% confidence intervals (CI) of phenotypic responses for cow less (LS) or more susceptible (MS) to lipid-related metabolic disorders., Supplementary Table S3. Least squares means (LSM) and 95% confidence intervals (CI) of phenotypic responses for cows more (MR) or less resistant (LR) to lipid-related metabolic disorders.

Author Contributions: Conceptualization, H.M.W. and R.S.P.; methodology, R.S.P. and W.L.; formal analysis, R.S.P. and W.L.; investigation, R.S.P., W.L., B.N.M., H.T.H. and H.M.W.; resources, H.M.W. and W.L.; writing—original draft preparation, R.S.P. and H.M.W.; writing—review and editing, R.S.P., H.M.W. and W.L.; visualization, R.S.P.; supervision, H.M.W.; project administration, H.M.W. and W.L.; funding acquisition, R.S.P., W.L. and H.M.W. All authors have read and agreed to the published version of the manuscript.

Funding: This project was supported by the Agricultural Food Research Initiative of the National Institute of Food and Agriculture, USDA, Grant # 2016-67015-24573, USDA Hatch Grant # WIS01877, and Land O'Lakes John Brandt Memorial Scholarship. H.T. Holdorf was supported by Purina Animal Nutrition LLC's (Gray Summit, MO) graduate fellowship program.

Institutional Review Board Statement: This study was conducted in accordance with the guidelines of the University of Wisconsin–Madison College of Agricultural and Life Sciences' Animal Care and Use Committee (IACUC #A005467).

Informed Consent Statement: Not applicable.

Data Availability Statement: Data are available upon request from the corresponding author.

Acknowledgments: The authors acknowledge and appreciate the assistance of D. Rieman, manager at the UW–Madison Dairy Cattle Center, and the UW–Madison Dairy Cattle Instruction and Research Center staff (University of Wisconsin–Madison, Madison, WI, USA). Furthermore, the authors recognize the valued assistance from H.M. White laboratory members S.J. Erb, C.R. Seely, and M.R. Moede, as well as laboratory manager S.J. Bertics (University of Wisconsin-Madison, Madison, WI, USA).

Conflicts of Interest: The authors declare no conflict of interest.

References

1. Bauman, D.E.; Currie, W.B. Partitioning of Nutrients During Pregnancy and Lactation: A Review of Mechanisms Involving Homeostasis and Homeorhesis. *J. Dairy Sci.* **1980**, *63*, 1514–1529. [CrossRef]
2. Wankhade, P.R.; Manimaran, A.; Kumaresan, A.; Jeyakumar, S.; Ramesha, K.; Sejian, V.; Rajendran, D.; Varghese, M.R. Metabolic and immunological changes in transition dairy cows: A review. *Vet. World* **2017**, *10*, 1367–1377. [CrossRef]
3. Wilkens, M.R.; Nelson, C.D.; Hernandez, L.L.; McArt, J.A. Symposium review: Transition cow calcium homeostasis—Health effects of hypocalcemia and strategies for prevention. *J. Dairy Sci.* **2020**, *103*, 2909–2927. [CrossRef]
4. Holtenius, P.; Holetnius, K. New Aspects of Ketone Bodies in Energy Metabolism of Dairy Cows: A Review. *J. Vet. Med. Ser. A* **1996**, *43*, 579–587. [CrossRef]
5. McNamara, J. Regulation of Adipose Tissue Metabolism in Support of Lactation. *J. Dairy Sci.* **1991**, *74*, 706–719. [CrossRef]
6. Gross, J.; Bruckmaier, R. Repeatability of metabolic responses to a nutrient deficiency in early and mid lactation and implications for robustness of dairy cows. *J. Dairy Sci.* **2015**, *98*, 8634–8643. [CrossRef]
7. Weber, C.; Schäff, C.; Kautzsch, U.; Börner, S.; Erdmann, S.; Bruckmaier, R.; Röntgen, M.; Kuhla, B.; Hammon, H. Variable liver fat concentration as a proxy for body fat mobilization postpartum has minor effects on insulin-induced changes in hepatic gene expression related to energy metabolism in dairy cows. *J. Dairy Sci.* **2017**, *100*, 1507–1520. [CrossRef] [PubMed]
8. Baumgard, L.; Collier, R.; Bauman, D. A 100-Year Review: Regulation of nutrient partitioning to support lactation. *J. Dairy Sci.* **2017**, *100*, 10353–10366. [CrossRef]
9. McArt, J.; Nydam, D.; Oetzel, G. Epidemiology of subclinical ketosis in early lactation dairy cattle. *J. Dairy Sci.* **2012**, *95*, 5056–5066. [CrossRef]
10. Mahrt, A.; Burfeind, O.; Heuwieser, W. Evaluation of hyperketonemia risk period and screening protocols for early-lactation dairy cows. *J. Dairy Sci.* **2015**, *98*, 3110–3119. [CrossRef]
11. Jorritsma, R.; Jorritsma, H.; Schukken, Y.; Wentink, G. Relationships between fatty liver and fertility and some periparturient diseases in commercial Dutch dairy herds. *Theriogenology* **2000**, *54*, 1065–1074. [CrossRef]
12. Jorritsma, R.; Jorritsma, H.; Schukken, Y.; Bartlett, P.; Wensing, T.; Wentink, G. Prevalence and indicators of post partum fatty infiltration of the liver in nine commercial dairy herds in The Netherlands. *Livest. Prod. Sci.* **2001**, *68*, 53–60. [CrossRef]

13. Walsh, R.; Walton, J.; Kelton, D.; LeBlanc, S.; Leslie, K.; Duffield, T. The Effect of Subclinical Ketosis in Early Lactation on Reproductive Performance of Postpartum Dairy Cows. *J. Dairy Sci.* **2007**, *90*, 2788–2796. [CrossRef]
14. Bobe, G.; Young, J.; Beitz, D. Invited Review: Pathology, Etiology, Prevention, and Treatment of Fatty Liver in Dairy Cows. *J. Dairy Sci.* **2004**, *87*, 3105–3124. [CrossRef]
15. Pralle, R.; Amdall, J.; Fourdraine, R.; Oetzel, G.; White, H. Hyperketonemia Predictions Provide an On-Farm Management Tool with Epidemiological Insights. *Animals* **2021**, *11*, 1291. [CrossRef]
16. McArt, J.A.A.; Nydam, D.v.; Overton, M.W. Hyperketonemia in early lactation dairy cattle: A deterministic estimate of component and total cost per case. *J. Dairy Sci.* **2015**, *98*, 2043–2054. [CrossRef] [PubMed]
17. Grummer, R.R. Etiology of Lipid-Related Metabolic Disorders in Periparturient Dairy Cows. *J. Dairy Sci.* **1993**, *76*, 3882–3896. [CrossRef]
18. White, H.M. The Role of TCA Cycle Anaplerosis in Ketosis and Fatty Liver in Periparturient Dairy Cows. *Animals* **2015**, *5*, 793–802. [CrossRef] [PubMed]
19. Contreras, L.L.; Ryan, C.M.; Overton, T.R. Effects of Dry Cow Grouping Strategy and Prepartum Body Condition Score on Performance and Health of Transition Dairy Cows. *J. Dairy Sci.* **2004**, *87*, 517–523. [CrossRef]
20. Vailati-Riboni, M.; Meier, S.; Burke, C.R.; Kay, J.K.; Mitchell, M.D.; Walker, C.G.; Crookenden, M.A.; Heiser, A.; Rodriguez-Zas, S.L.; Roche, J.R.; et al. Prepartum body condition score and plane of nutrition affect the hepatic transcriptome during the transition period in grazing dairy cows. *BMC Genom.* **2016**, *17*, 854. [CrossRef] [PubMed]
21. Zenobi, M.; Gardinal, R.; Zuniga, J.; Dias, A.; Nelson, C.; Driver, J.; Barton, B.; Santos, J.; Staples, C. Effects of supplementation with ruminally protected choline on performance of multiparous Holstein cows did not depend upon prepartum caloric intake. *J. Dairy Sci.* **2018**, *101*, 1088–1110. [CrossRef] [PubMed]
22. Oliveira, R.C.; Sailer, K.; Holdorf, H.; Seely, C.; Pralle, R.; Hall, M.; Bello, N.; White, H. Postpartum supplementation of fermented ammoniated condensed whey improved feed efficiency and plasma metabolite profile. *J. Dairy Sci.* **2019**, *102*, 2283–2297. [CrossRef] [PubMed]
23. Defrain, J.; Hippen, A.; Kalscheur, K.; Jardon, P. Feeding Glycerol to Transition Dairy Cows: Effects on Blood Metabolites and Lactation Performance. *J. Dairy Sci.* **2004**, *87*, 4195–4206. [CrossRef]
24. Soltan, M.A. Effect of dietary chromium supplementation on productive and reproductive performance of early lactating dairy cows under heat stress. *J. Anim. Physiol. Anim. Nutr.* **2010**, *94*, 264–272. [CrossRef]
25. Bollatti, J.; Zenobi, M.; Artusso, N.; Lopez, A.; Nelson, C.; Barton, B.; Staples, C.; Santos, J. Effects of rumen-protected choline on the inflammatory and metabolic status and health of dairy cows during the transition period. *J. Dairy Sci.* **2020**, *103*, 4192–4205. [CrossRef]
26. Chandler, T.; White, H.M. Choline and methionine differentially alter methyl carbon metabolism in bovine neonatal hepatocytes. *PLoS ONE* **2017**, *12*, e0171080. [CrossRef]
27. Piepenbrink, M.; Overton, T. Liver Metabolism and Production of Cows Fed Increasing Amounts of Rumen-Protected Choline During the Periparturient Period. *J. Dairy Sci.* **2003**, *86*, 1722–1733. [CrossRef]
28. Loor, J.J.; Dann, H.M.; Guretzky, N.A.J.; Everts, R.E.; Oliveira, R.; Green, C.A.; Litherland, N.B.; Rodriguez-Zas, S.L.; Lewin, H.A.; Drackley, J.K. Plane of nutrition prepartum alters hepatic gene expression and function in dairy cows as assessed by longitudinal transcript and metabolic profiling. *Physiol. Genom.* **2006**, *27*, 29–41. [CrossRef] [PubMed]
29. McCarthy, S.D.; Waters, S.M.; Kenny, D.; Diskin, M.G.; Fitzpatrick, R.; Patton, J.; Wathes, D.C.; Morris, D.G. Negative energy balance and hepatic gene expression patterns in high-yielding dairy cows during the early postpartum period: A global approach. *Physiol. Genom.* **2010**, *42A*, 188–199. [CrossRef] [PubMed]
30. McCabe, M.; Waters, S.; Morris, D.; Kenny, D.; Lynn, D.; Creevey, C. RNA-seq analysis of differential gene expression in liver from lactating dairy cows divergent in negative energy balance. *BMC Genom.* **2012**, *13*, 193. [CrossRef] [PubMed]
31. Shahzad, K.; Bionaz, M.; Trevisi, E.; Bertoni, G.; Rodriguez-Zas, S.L.; Loor, J.J. Integrative Analyses of Hepatic Differentially Expressed Genes and Blood Biomarkers during the Peripartal Period between Dairy Cows Overfed or Restricted-Fed Energy Prepartum. *PLoS ONE* **2014**, *9*, e99757. [CrossRef] [PubMed]
32. Loor, J.J.; Everts, R.E.; Bionaz, M.; Dann, H.M.; Morin, D.E.; Oliveira, R.; Rodriguez-Zas, S.L.; Drackley, J.K.; Lewin, H.A. Nutrition-induced ketosis alters metabolic and signaling gene networks in liver of periparturient dairy cows. *Physiol. Genom.* **2007**, *32*, 105–116. [CrossRef]
33. Akbar, H.; Bionaz, M.; Carlson, D.; Rodriguez-Zas, S.; Everts, R.; Lewin, H.; Drackley, J.; Loor, J. Feed restriction, but not l-carnitine infusion, alters the liver transcriptome by inhibiting sterol synthesis and mitochondrial oxidative phosphorylation and increasing gluconeogenesis in mid-lactation dairy cows. *J. Dairy Sci.* **2013**, *96*, 2201–2213. [CrossRef] [PubMed]
34. Shahzad, K.; Lopreiato, V.; Liang, Y.; Trevisi, E.; Osorio, J.S.; Xu, C.; Loor, J.J. Hepatic metabolomics and transcriptomics to study susceptibility to ketosis in response to prepartal nutritional management. *J. Anim. Sci. Biotechnol.* **2019**, *10*, 1–17. [CrossRef]
35. Pralle, R.S.; Erb, S.J.; Holdorf, H.T.; White, H.M. Greater liver PNPLA3 protein abundance in vivo and in vitro supports lower triglyceride accumulation in dairy cows. *Sci. Rep.* **2021**, *11*, 2839. [CrossRef]
36. Folch, J.; Lees, M.; Sloane Stanley, G.H. A Simple Method for the Isolation and Purification of Total Lipides from Animal Tissues. *J. Biol. Chem.* **1957**, *226*, 497–509. [CrossRef]

37. Maechler, M.; Rousseeuw, P.; Struyf, A.; Hubert, M.; Hornik, K. cluster: Cluster Analysis Basics and Extensions. R Package Version 2.1.2—For New Features, See the 'Changelog' File (in the Package Source). 2021. Available online: https://CRAN.R-project.org/package=cluster (accessed on 27 February 2020).
38. Overton, T.; McArt, J.; Nydam, D. A 100-Year Review: Metabolic health indicators and management of dairy cattle. *J. Dairy Sci.* **2017**, *100*, 10398–10417. [CrossRef]
39. Ospina, P.; Nydam, D.; Stokol, T.; Overton, T. Evaluation of nonesterified fatty acids and β-hydroxybutyrate in transition dairy cattle in the northeastern United States: Critical thresholds for prediction of clinical diseases. *J. Dairy Sci.* **2010**, *93*, 546–554. [CrossRef]
40. Li, S.; Tighe, S.W.; Nicolet, C.M.; Grove, D.S.; Levy, S.; Farmerie, W.; Viale, A.; Wright, C.V.E.; Schweitzer, P.A.; Gao, Y.; et al. Multi-platform assessment of transcriptome profiling using RNA-seq in the ABRF next-generation sequencing study. *Nat. Biotechnol.* **2014**, *32*, 915–925. [CrossRef]
41. Trapnell, C.; Roberts, A.; Goff, L.; Pertea, G.; Kim, D.; Kelley, D.R.; Pimentel, H.; Salzberg, S.L.; Rinn, J.L.; Pachter, L. Differential gene and transcript expression analysis of RNA-seq experiments with TopHat and Cufflinks. *Nat. Protoc.* **2012**, *7*, 562–578. [CrossRef]
42. Benjamini, Y.; Hochberg, Y. Controlling the False Discovery Rate: A Practical and Powerful Approach to Multiple Testing. *J. R. Stat. Soc. Ser. B Stat. Method.* **1995**, *57*, 289–300. [CrossRef]
43. Huang, D.W.; Lempicki, R.A.; Sherman, B.T. Systematic and integrative analysis of large gene lists using DAVID bioinformatics resources. *Nat. Protoc.* **2009**, *4*, 44–57. [CrossRef]
44. Huang, D.W.; Sherman, B.T.; Lempicki, R.A. Bioinformatics enrichment tools: Paths toward the comprehensive functional analysis of large gene lists. *Nucleic Acids Res.* **2009**, *37*, 1–13. [CrossRef]
45. Drackley, J.K. Biology of Dairy Cows During the Transition Period: The Final Frontier? *J. Dairy Sci.* **1999**, *82*, 2259–2273. [CrossRef]
46. Oliveira, R.C.; Erb, S.; Pralle, R.; Holdorf, H.; Seely, C.; White, H. Postpartum fermented ammoniated condensed whey supplementation altered nutrient partitioning to support hepatic metabolism. *J. Dairy Sci.* **2020**, *103*, 7055–7067. [CrossRef]
47. White, H. ADSA Foundation Scholar Award: Influencing hepatic metabolism: Can nutrient partitioning be modulated to optimize metabolic health in the transition dairy cow? *J. Dairy Sci.* **2020**, *103*, 6741–6750. [CrossRef] [PubMed]
48. Weld, K.A.; Oliveira, R.C.; Bertics, S.J.; Erb, S.J.; White, H.M. Hepatic pyruvate carboxylase expression differed prior to hyperketonemia onset in transition dairy cows. *PLoS ONE* **2020**, *15*, e0241929. [CrossRef] [PubMed]
49. Martin, M.J.; Pralle, R.S.; Weigel, K.A.; Zhou, Z.; White, H.M. Fatty acid metabolism may vary in dairy cows with high and low residual feed intake. *J. Dairy Sci.* **2020**, *103* (Suppl. S1), 203.
50. Chapinal, N.; Carson, M.; LeBlanc, S.; Leslie, K.; Godden, S.; Capel, M.; Santos, J.; Overton, M.; Duffield, T. The association of serum metabolites in the transition period with milk production and early-lactation reproductive performance. *J. Dairy Sci.* **2012**, *95*, 1301–1309. [CrossRef]
51. Kvidera, S.; Horst, E.; Abuajamieh, M.; Mayorga, E.; Fernandez, M.S.; Baumgard, L. Glucose requirements of an activated immune system in lactating Holstein cows. *J. Dairy Sci.* **2017**, *100*, 2360–2374. [CrossRef]
52. Emery, R.S.; Liesman, J.S.; Herdt, T.H. Metabolism of Long Chain Fatty Acids by Ruminant Liver. *J. Nutr.* **1992**, *122*, 832–837. [CrossRef] [PubMed]
53. Thompson, G.; Darling, F. The hepatic uptake of individual free fatty acids in sheep during noradrenaline infusion. *Res. Vet. Sci.* **1975**, *18*, 325–327. [CrossRef]
54. Mavangira, V.; Sordillo, L.M. Role of lipid mediators in the regulation of oxidative stress and inflammatory responses in dairy cattle. *Res. Vet. Sci.* **2018**, *116*, 4–14. [CrossRef] [PubMed]
55. Sordillo, L.M.; Aitken, S.L. Impact of oxidative stress on the health and immune function of dairy cattle. *Vet. Immunol. Immunopathol.* **2009**, *128*, 104–109. [CrossRef]
56. Du, X.; Shi, Z.; Peng, Z.; Zhao, C.; Zhang, Y.; Wang, Z.; Li, X.; Liu, G.; Li, X. Acetoacetate induces hepatocytes apoptosis by the ROS-mediated MAPKs pathway in ketotic cows. *J. Cell. Physiol.* **2017**, *232*, 3296–3308. [CrossRef]
57. Strange, R.C.; Spiteri, M.A.; Ramachandran, S.; Fryer, A. Glutathione-S-transferase family of enzymes. *Mutat. Res. Mol. Mech. Mutagen.* **2001**, *482*, 21–26. [CrossRef]
58. Townsend, D.M.; Tew, K.D. The role of glutathione-S-transferase in anti-cancer drug resistance. *Oncogene* **2003**, *22*, 7369–7375. [CrossRef] [PubMed]
59. Tsuchida, S. Glutathione Transferases. In *Encyclopedia of Cancer*, 2nd ed.; Bertino, J.R., Ed.; Academic Press: New York, NY, USA, 2002; pp. 297–307, ISBN 978-0-12-227555-5.
60. Dennis, E.A.; Norris, P. Eicosanoid storm in infection and inflammation. *Nat. Rev. Immunol.* **2015**, *15*, 511–523. [CrossRef]
61. Hotamisligil, G.S. Inflammation and metabolic disorders. *Nature* **2006**, *444*, 860–867. [CrossRef] [PubMed]
62. Sordillo, L.M.; Mavangira, V. The nexus between nutrient metabolism, oxidative stress and inflammation in transition cows. *Anim. Prod. Sci.* **2014**, *54*, 1204–1214. [CrossRef]
63. Vidali, M.; Stewart, S.F.; Albano, E. Interplay between oxidative stress and immunity in the progression of alcohol-mediated liver injury. *Trends Mol. Med.* **2008**, *14*, 63–71. [CrossRef]
64. Herkel, J.; Jagemann, B.; Wiegard, C.; Lazaro, J.F.G.; Lüth, S.; Kanzler, S.; Blessing, M.; Schmitt, E.; Lohse, A.W. MHC class II-expressing hepatocytes function as antigen-presenting cells and activate specific CD4 T lymphocyutes. *Hepatology* **2003**, *37*, 1079–1085. [CrossRef]

65. Chen, J.; Deng, X.; Liu, Y.; Tan, Q.; Huang, G.; Che, Q.; Guo, J.; Su, Z. Kupffer Cells in Non-alcoholic Fatty Liver Disease: Friend or Foe? *Int. J. Biol. Sci.* **2020**, *16*, 2367–2378. [CrossRef] [PubMed]
66. Ting, J.P.-Y.; Trowsdale, J. Genetic Control of MHC Class II Expression. *Cell* **2002**, *109*, S21–S33. [CrossRef]
67. Dienes, H.P.; Hütteroth, T.; Hess, G.; Meuer, S.C. Immunoelectron microscopic observations on the inflammatory infiltrates and HLA antigens in hepatitis B and non-A, Non-B. *Hepatology* **1987**, *7*, 1317–1325. [CrossRef] [PubMed]
68. Han, R.; Zhang, F.; Wan, C.; Liu, L.; Zhong, Q.; Ding, W. Effect of perfluorooctane sulphonate-induced Kupffer cell activation on hepatocyte proliferation through the NF-κB/TNF-α/IL-6-dependent pathway. *Chemosphere* **2018**, *200*, 283–294. [CrossRef]
69. Malhi, H.; Gores, G. Molecular Mechanisms of Lipotoxicity in Nonalcoholic Fatty Liver Disease. *Semin. Liver Dis.* **2008**, *28*, 360–369. [CrossRef] [PubMed]
70. Honda, M.; Sakai, A.; Yamashita, T.; Nakamoto, Y.; Mizukoshi, E.; Sakai, Y.; Yamashita, T.; Nakamura, M.; Shirasaki, T.; Horimoto, K.; et al. Hepatic ISG Expression Is Associated with Genetic Variation in Interleukin 28B and the Outcome of IFN Therapy for Chronic Hepatitis C. *Gastroenterology* **2010**, *139*, 499–509. [CrossRef] [PubMed]
71. Møhlenberg, M.; Terczynska-Dyla, E.; Thomsen, K.L.; George, J.; Eslam, M.; Grønbæk, H.; Hartmann, R. The role of IFN in the development of NAFLD and NASH. *Cytokine* **2019**, *124*, 154519. [CrossRef] [PubMed]
72. Takaoka, A.; Yanai, H. Interferon signalling network in innate defence. *Cell. Microbiol.* **2006**, *8*, 907–922. [CrossRef]
73. Saremi, B.; Mielenz, M.; Rahman, M.; Hosseini, A.; Kopp, C.; Dänicke, S.; Ceciliani, F.; Sauerwein, H. Hepatic and extrahepatic expression of serum amyloid A3 during lactation in dairy cows. *J. Dairy Sci.* **2013**, *96*, 6944–6954. [CrossRef] [PubMed]
74. Abuajamieh, M.; Kvidera, S.K.; Fernandez, M.V.S.; Nayeri, A.; Upah, N.C.; Nolan, E.A.; Lei, S.M.; DeFrain, J.M.; Green, H.B.; Schoenberg, K.M.; et al. Inflammatory biomarkers are associated with ketosis in periparturient Holstein cows. *Res. Vet. Sci.* **2016**, *109*, 81–85. [CrossRef] [PubMed]
75. Sack, G.H. Serum amyloid A—A review. *Mol. Med.* **2018**, *24*, 1–27. [CrossRef]
76. Bradford, B.J.; Swartz, T. Review: Following the smoke signals: Inflammatory signaling in metabolic homeostasis and homeorhesis in dairy cattle. *Animal* **2020**, *14*, s144–s154. [CrossRef] [PubMed]
77. Loor, J. Genomics of metabolic adaptations in the peripartal cow. *Animal* **2010**, *4*, 1110–1139. [CrossRef]
78. Suthar, V.; Canelas-Raposo, J.; Deniz, A.; Heuwieser, W. Prevalence of subclinical ketosis and relationships with postpartum diseases in European dairy cows. *J. Dairy Sci.* **2013**, *96*, 2925–2938. [CrossRef]
79. Loor, J.; Bionaz, M.; Drackley, J.K. Systems Physiology in Dairy Cattle: Nutritional Genomics and Beyond. *Annu. Rev. Anim. Biosci.* **2013**, *1*, 365–392. [CrossRef] [PubMed]

animals

MDPI

Review

Identification of Crucial Genetic Factors, Such as PPARγ, that Regulate the Pathogenesis of Fatty Liver Disease in Dairy Cows Is Imperative for the Sustainable Development of Dairy Industry

Kerong Shi *, Ranran Li, Zhongjin Xu and Qin Zhang

Key Laboratory of Animal Bioengineering and Disease Prevention, College of Animal Science and Technology, Shandong Agricultural University, Tai'an 271018, Shandong, China; liranranone@163.com (R.L.); 18206381011@139.com (Z.X.); qzhang@cau.edu.cn (Q.Z.)

* Correspondence: krshi@sdau.edu.cn

Received: 27 February 2020; Accepted: 1 April 2020; Published: 7 April 2020

Simple Summary: Fatty liver disease frequently occurs in dairy cows, a typical type of non-alcoholic fatty liver disease (NAFLD), resulting in a high culling rate of dairy cows during the perinatal period because of limitations of lactation and reproduction performance of cows with subsequent complications. This has been developing into a worldwide crucial industrial problem. Studies about NAFLD have shown that PPARγ (peroxisome proliferator-activated receptor γ) participates or regulates the fat deposition in liver by affecting the biological processes of hepatic lipid metabolism, insulin resistance, gluconeogenesis, oxidative stress, and inflammation, which all contribute to fatty liver. This review mainly focuses on the understanding of molecular pathogenesis of fatty liver disease in dairy cows by taking PPARγ as an example, so as to provide important information for discovering critical therapeutic targets, such as PPARγ, for fatty liver disease, and contribute to breeding improvement of fatty liver disease-resistant dairy cattle and eventually sustainable development of dairy industry.

Abstract: Frequently occurring fatty liver disease in dairy cows during the perinatal period, a typical type of non-alcoholic fatty liver disease (NAFLD), results in worldwide high culling rates of dairy cows (averagely about 25%) after calving. This has been developing into a critical industrial problem throughout the world, because the metabolic disease severely affects the welfare and economic value of dairy cows. Findings about the molecular mechanisms how the fatty liver disease develops would help scientists to discover novel therapeutic targets for NAFLD. Studies have shown that PPARγ participates or regulates the fat deposition in liver by affecting the biological processes of hepatic lipid metabolism, insulin resistance, gluconeogenesis, oxidative stress, endoplasmic reticulum stress and inflammation, which all contribute to fatty liver. This review mainly focuses on crucial regulatory mechanisms of PPARγ regulating lipid deposition in the liver via direct and/or indirect pathways, suggesting that PPARγ might be a potential critical therapeutic target for fatty liver disease, however, it would be of our significant interest to reveal the pathology and pathogenesis of NAFLD by using dairy cows with fatty liver as an animal model. This review will provide a molecular mechanism basis for understanding the pathogenesis of NAFLD.

Keywords: dairy cows; PPARγ; fatty liver; non-alcoholic fatty liver disease (NAFLD); genetic factor; dairy industry

1. Introduction

Fatty liver syndrome, a typical type of metabolic disorder, frequently occurs in populations of dairy cows in commercial farms throughout the world, which is caused by negative nutrient balance after calving. High-yielding dairy cows, especially for cows with 30 kg of daily milk yield and more, are usually inclined to develop fatty liver syndrome in the early lactation period. There are more risk factors available to developing fatty liver syndrome, such as cows with high body condition score (BCS) and/or high body fat, feed intake decreases around calving, etc.

Our investigation in the last three years indicated that 48.85% of dairy cows (n = 346) within 2 weeks after parturition were diagnosed with light or severe fatty liver disease by randomly selecting 710 Holstein dairy cows from four commercial farms (Zhang et al., unpublished data) [1,2]. The suspected fatty liver cows and/or normal cows were firstly distinguished by applying the previously reported model [3] using the detected values of serum biochemical traits (glucose, Glu; non-esterified fatty acid, NEFA; aspartate transaminase, AST), then representative cows with different suspected extents were biopsied for liver tissue samples for fat deposition amount assessment by oil red staining. It is estimated that 40%–60% of high-yielding dairy cows (daily milk yield > 35 kg) develop moderate to severe fatty liver disease within 2 weeks after calving [4]. Moreover, it is not uncommon for the two short weeks after parturition to account for 50% of morbidity on a dairy farm [5], which is in accordance with our results.

The perinatal disorders including fatty liver disease remain as prevalent now as they were 20 years ago [5], causing high culling rate of dairy cows in their early lactation period, which is becoming a critical concern in modern dairy industry. The average culling rate of dairy cows within 60 days of lactation (during parturition period) in populations is about 24% in USA [6–8], and about 27% in China [9,10]. In clinical practice, increasing blood calcium levels and using anti-inflammatory drugs after delivery in dairy cows could reduce the incidence of this disease and/or decrease the economic loss [11]. However, these strategies cannot either completely change the situation or alleviate contradictions. The incidence of fatty liver disease at the early lactation not only decreases the milk yield of the coming lactation period, but also attenuates the future milk production and reproduction performance because of subsequent health problems of dairy herds [1,2,7,12], such as ketosis, displaced abomasum, mastitis, etc. It is estimated that economic losses due to the treatment of and reduction in milk production by one dairy cow with ketosis accounts for 151–312 US dollars [11].

2. Ethics Approval and Consent to Participate

All the investigation and biopsy processes were carried out in accordance with guidelines issued by the Shandong Agricultural University Animal Care and Use Committee (approval number, SDAUA-2017-044). We have obtained written informed consent from the animal owner to use these animals in the study.

3. Progress of the Pathogenesis Mechanism of Fatty Liver Disease

The etiology of fatty liver in perinatal dairy cows primarily includes the negative nutrient balance and the accumulation of high level of free fatty acids (FFA) in serum or triglycerides (TAG) deposited in liver [1]. Liver, the central organ of organisms, regulates the metabolic balance of carbohydrate, fat, and protein in mammals [13]. After calving, the food intake of dairy cows further decreases, while lactation slowly increases. Thus, the body lactose consumption easily results in the cow to be susceptible to experiencing an insufficient sugar supply, thus promoting fat mobilization in the liver [14,15]. Additionally, energy and substance metabolism is centered in the liver. The increasing fat mobilization promotes gluconeogenesis, increases the blood sugar concentration, and alleviates the negative nutrient balance. At the same time, the enhanced fat mobilization promotes the dramatic increase of non-esterified fatty acid (NEFA) in the liver [16], which is partly re-esterified to synthesize the triglycerides (TAG), a type of very-low-density lipoprotein (VLDL), that is hardly transported out of the liver [17]. Especially for dairy cattle, TAG is excessively accumulated because of its lack for esterase, resulting in susceptibility to fatty liver disease [14,18].

As for non-alcoholic fatty liver disease (NAFLD) occurring in human beings, metabolic disorder syndromes and obesity are also usually the main causes, with increased plasma insulin and fatty acid concentration, elevated fasting aminotransferase (AST/ALT) and/or triglycerides (TAG) level, and also abnormal lipid accumulation in the liver [19–21]. In addition, another of the most important risk factors is histological evidence of hepatic inflammation [22] caused by acute inflammation and subacute inflammation [5]. Thus, dairy cows with fatty liver disease is a typical NAFLD animal model, good for revealing the pathology and pathogenesis of NAFLD.

In recent years, scientists have proposed a "two-hit" theory to explain the pathogenesis mechanisms of NAFLD [23,24]. (1) The "first hit" was caused by insulin resistance (IR). IR can not only strengthen lipolysis of surrounding tissues, but also causes hyperinsulinemia. The lipolysis of adipose tissue results in increased FFA and enhanced TAG synthesis in the liver. The FFA has hepatocellular toxicity, increasing the permeability of cell membrane and disrupting the mitochondrial function by inhibiting related enzymes. (2) The "second hit" was caused by the imbalance between the coexisting systems of oxidation and anti-oxidation in the liver. The increase in lipid peroxidation results in persistent reactive oxygen species (ROS) production. In addition to the pre-existing factors related to the enhanced oxygen stress, other new or additional factors can increase lipid peroxidation for a second hit to the liver, such as inflammatory cytokines, adipokines, endotoxins, and mitochondrial inactivation. The second hit will eventually lead to NASH (non-alcoholic steatohepatitis) progression beyond hepatic steatosis that promotes oxidative stress, inflammation, cell death, and fibrosis. Especially, the inflammation is positively correlated with liver injury and negatively correlated with lipolysis by inhibiting lipase activity, inhibiting the transport of lipoproteins and/or lipids and causing lipid deposition. Moreover, it also induces lipid peroxidation, IR, and cell apoptosis [25,26], aggravating the pathogenesis of NAFLD. (3) Actually, there is even a "third hit" [27], which is cell death and irreversible cell repair of hepatocytes. (4) Additionally, endoplasmic reticulum (ER) stress is another important "hit" in the pathogenesis of NAFLD. The metabolic disorders, such as obesity and diabetes, can cause ER stress, leading to the accumulation of incorrectly folded proteins (unfolded protein response, UPR) and affecting the normal physiological functions of liver cells. It is worth mentioning that ER stress can activate SREBP (sterol-regulatory element binding protein), promoting the transcription of acetyl-CoA carboxylase (ACC) and fatty acid synthase (FAS), resulting in increased synthesis of TAG and fatty acids in the liver [28,29]. Moreover, the oxidative stress (ROS producing) in liver cells can induce ER stress leading to incorrect protein fold and/or protein modification. Additionally, ER stress also can be the cause of oxidative stress. The biological process of ER stress and oxidative stress interact each other through different pathways, leading to IR and aggravating the NAFLD [30–32].

However, the pathogenesis of NAFLD is still unclear. It was proposed that the abnormality in lipid and lipoprotein metabolism accompanied by chronic inflammation and oxidative stress is the central pathway and/or the major risk factors involved in the pathogenesis of NAFLD [33,34]. It is generally believed that the occurrence of fatty liver is not only closely related to insulin resistance and disorder of fat metabolism, but also related to biological processes, such as glycometabolism disorder, oxidative stress, and intracellular inflammatory response [23,35]. Moreover, these processes are correlated and/or coordinated with each other and accelerate the progress of NAFLD [23,36]. There are numerous factors involved in the pathogenesis of NAFLD. For example, PPARα (peroxisome proliferator-activated receptor α) and/or PPARγ, microsomal triglyceride transport protein (MTP), apolipoprotein (apoB) play important roles in hepatic lipid metabolism, lipid transport, and secretion [35,37]; adipocytokines (Tumor necrosis factor, leptin, and adiponectin), cytokines (Interleukin-6, IL-6; glucagon-like peptide-1, GLP-1; fibroblast growth factor 19, FGF-19; fibroblast growth factor 21, FGF-21; growth-hormone-releasing hormone, GHRH; etc.), and toll-like receptors (TLRs) participate in insulin resistance (IR), oxidative stress, and inflammatory response, and mediate cell apoptosis/necrosis to promote liver fibrosis [35,38–40]; microRNAs (such as mir-107 and miR-103) are reported to regulate insulin resistance. Recently, RG-125 (also named AZD4076), the antagonist against microRNA-103/107, has entered the phase I clinical trial of NASH (non-alcoholic steatohepatitis) treatment [39].

Accordingly, the pathogenesis of NAFLD is usually related to abnormal hepatic lipid metabolism, gluconeogenesis, IR, oxidative stress, and inflammation [23,39–43]. The causal relationship and the underlying molecular mechanisms of these biological processes still remains unclear. However, the discovery of some important regulatory factors/genes that regulate all these biological processes would be helpful to reveal the molecular pathogenesis of NAFLD.

Recent studies have shown that PPARγ (peroxisome proliferator-activated receptor γ) participates or regulates lipid metabolism in the liver by affecting the biological processes of hepatic lipid metabolism, insulin resistance, gluconeogenesis, oxidative stress, and inflammation. In the following content, PPARγ will be taken as an example to elaborate the crucial regulatory role of certain critical genetic factors in the hepatic liver metabolism and therefore lipid deposition, via direct and/or indirect pathways. Rosiglitazone (RGZ) and other thiazolidinedione (TZD) synthetic ligands of PPARγ are insulin sensitizers that have been used for the treatment of type II diabetes [44]. As for the PPARγ related substances, apigenin, a food-derived compound, were reported significantly ameliorated NAFLD and obesity-induced metabolic syndrome by acting as a PPARγ modulator through Nrf2 (nuclear factor E2-related factor 2), inhibiting the lipid metabolism and oxidative stress abnormity [45,46].

4. Molecular Regulatory Effects of PPARγ on the Pathogenesis of Fatty Liver Disease

PPARs are nuclear hormone receptors that belong to the steroid hormone superfamily, playing an important role in regulating various intracellular metabolic processes [47]. PPARs comprise of three subtypes (PPARα, PPARβ, and PPARγ) that are encoded by multiple genes [48,49]. PPARα, PPARβ, and PPARγ contain 468, 441, and 479 amino acid residues, respectively [50,51]. Although they are highly homologous regions regarding their similar amino acid sequence and protein structure, PPARα, PPARβ, and PPARγ exhibit different tissue expression specificity and selectivity for different ligands [52]. PPARγ is mainly expressed in adipose and immune tissues, with low expression in the liver, kidneys, and cartilage [53]. However, PPARγ is an important participant in lipid metabolism and lipogenesis [54,55] and significantly affects hepatic glucose and lipid metabolism, adipocyte differentiation, and inflammatory responses. Studies have shown that PPARγ is activated by peroxisome proliferators as well as endogenous fatty acids and their derivatives. In addition, PPARγ participates in lipid metabolism [48] via increasing the uptake and storage of lipids as TAG in adipose tissues and highly expressing in brown fat and white fat, thereby proving its importance in adipocyte differentiation [56].

The development of fatty liver disease correlates with lipid mobilization, hepatic metabolism of free fatty acids, IR, oxidative stress, and gluconeogenesis. Studies have shown that PPARγ directly participates in lipid metabolism, thereby affecting lipid synthesis, oxidation, and transport in the liver. Moreover, PPARγ regulates the metabolic processes, such as insulin resistance (IR), inflammation, and gluconeogenesis, indirectly affecting lipid metabolism in the liver (Appendix A, Table A1). As shown in Figure 1, PPARγ regulates different pathways through different factors, impacting on hepatic glucose and lipid metabolism. The following contents describe the molecular mechanisms and/or signaling pathways that are regulated by PPARγ, so as to reveal the dysregulation of lipid metabolism in hepatocytes and therefore contribution to fatty liver disease. It will be beneficial for understanding the pathogenesis of fatty liver disease in dairy cows and provide a molecular basis for the treatment and prediction of the disease.

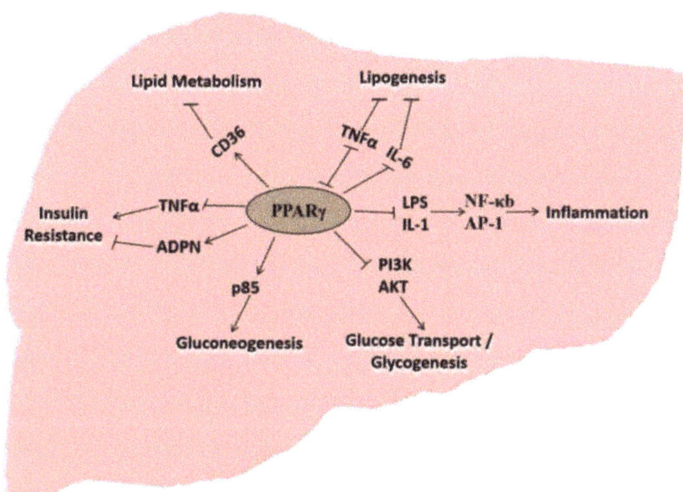

Figure 1. Peroxisome proliferator-activated receptor γ (PPARγ) participates in regulation of liver associated biological processes through different signaling factors and/or cytokines. Arrows indicate positive regulation and blunt arrows indicate negative regulation.

5. PPARγ Directly Regulates Lipid Metabolism in the Liver

PPARγ regulates target gene expression in adipocytes, participating in adipocyte differentiation [52,57] and regulating lipid metabolism [57–59], and primarily regulates signal transduction in pancreatic islet cells, greatly contributing to the occurrence and development of NAFLD. In addition to its significance in adipocyte differentiation, PPARγ is crucial in mediating lipid oxidation and lipogenesis [60,61]. Leptin-deficient obese (ob/ob) mice with inhibited liver specific PPARγ exhibit significant alleviation of fat deposition in liver, but along with aggravated hyperglycemia and IR [62,63]. Deletion of PPARγ in hepatocytes and macrophages prevent mice from developing hepatic steatosis [64,65]. The expression of PPARγ and lipogenesis-related genes, such as the fatty acid transporter *CD36*, are significantly elevated in the liver during hepatic steatosis, along with increased TAG level in the liver [66] (Figure 1). The above findings suggest that PPARγ activation promotes lipid deposition in the liver [67].

The downstream target of PPARγ, CD36, is a scavenger receptor class B that is primarily involved in the membrane transport of medium and long chain fatty acids. In macrophages, PPARγ and RXR (retinoid X receptor) form a heterodimer that binds to the specific PPARγ response element located in the promoter region of *CD36* gene and therefore increases *CD36* mRNA (messanger ribonuclei acid) level [68]. Compared to the mice fed with a normal diet, mice fed with high-fat diet after 4 weeks of administration showed significant upregulation of PPARγ2 and increased TAG content in hepatocytes. In contrast, silencing PPARγ2 using an adenoviral siRNA vector inhibits the expression of *CD36* and subsequently reduces TAG content in liver [69]. In experimental models without adipose tissues, liver PPARγ participates in lipid and glucose metabolism. However, in the presence of adipose tissues, liver PPARγ has a minimal effect on glucose metabolism in the liver [70], suggesting that PPARγ mainly regulates lipid metabolism in the liver.

The observations above indicate that PPARγ regulates the expression of the downstream target (*CD36*), thereby enabling TAG deposition and affecting lipid metabolism in the liver and plays a vital role in the occurrence of fatty liver disease (Wang et al., unpublished data; Figure 1). Our results also indicated that the downstream target genes of PPARγ were regulated by scaffold protein menin and NAD^+ (nicotinamide adenine dinucleotide)-dependent class III histone deacetylase sirtuin 1 (SIRT1) at the transcription level, affecting lipid deposition in hepatocytes (Li et al., unpublished data).

6. PPARγ Indirectly Participates in Lipid Metabolism via Lipid Oxidation

Activated PPARγ regulates the expression of oxidative stress-related factors. After PPARγ ligand activation, PPAR/RXR undergoes conformational changes and associates with coactivators binding to PPARγ response elements locates in target genes, thereby regulating the transcription of these target genes [45,71,72]. Cytokines such as TNFα (tumor necrosis factor α), IFN-γ (interferon-γ), IL-1 (interleukin-1), IL-2 (interleukin-2), and IL-6 (interleukin-6) inhibit lipogenesis in adipose tissues. However, activated PPARγ can decreases cytokine levels, thereby promoting lipogenesis and causing inflammation reaction [73,74] (Figure 1). Activated PPARγ expression in macrophages is confirmed to play anti-inflammatory roles in humans and rodents [74,75]. Moreover, PPARγ ligands stimulate macrophage apoptosis [74,76,77]. Lipopolysaccharides are an important etiological factor for the systemic inflammatory response disease. PPARγ activation inhibits the activity of lipopolysaccharide and IL-1 by reducing DNA (deoxyribonucleic acid) binding and transcription by nuclear factor-κB (NFκB), activator protein-1, and signal transducer and activator of transcription 1, thereby directly inhibiting the expression of pro-inflammatory genes and indirectly promoting liver lipogenesis resulting in excessive lipid accumulation in the liver [78–80]. PPARγ decreases the expression of inflammatory cytokines TNFα and IL-1, and therefore suppressing the inflammation-inducing effects and oxidative stress. In contrast, these inflammation factors may inhibit lipogenesis. Reducing the level of these inflammatory factors indirectly promotes lipogenesis and stimulates lipid deposition, causing the incidence of fatty liver disease.

7. PPARγ Indirectly Participates in Lipid Metabolism via Insulin Resistance

Insulin resistance (IR) refers to impaired normal physiological function of insulin, therefore the target organs have reduced sensitivity to insulin. Studies have shown that IR is the main cause of NAFLD [81,82]. As a result of IR, serum fatty acid levels are increased and produce damaging reactive oxygen species (ROS). Oxidative stress may be exacerbated further by ultrastructural mitochondrial lesions, which impair respiratory chain function. Insulin inhibits β-oxidation in the mitochondria through modifying the protein acetylation [83–85], resulting in the accumulation of lipids in hepatocytes.

A study found that individuals with PPARγ deletions suffer from early-onset type II diabetes accompanied with severe IR [50]. Adipocytokines, such as inflammatory factors (TNF-α and IL-6), released during obesity by adipose tissues affect host tissues and target organs (liver) and cause IR in the liver (Figure 1). Hence, PPARγ is an important candidate for IR suppression. Upon PPARγ activation by its ligand, insulin sensitivity of cells increases, promoting the expression of sterol regulatory element-binding protein-1, fatty acid-binding protein-4, and lipases in adipocytes, and increasing free fatty acid uptake and de novo synthesis of TAG in liver tissues [61,86,87].

Adiponectin (ADPN) is an endogenous bioactive protein that is secreted by adipocytes. ADPN is an insulin-sensitizing hormone that alleviates IR. PPARγ activation increases ADPN mRNA stability and promotes ADPN synthesis, stability, and release [88]. PPARγ also upregulates the expression of insulin-dependent glucose transporter 4 (GLUT4) and phosphoinositide 3-kinase (PI3K) in adipose and muscle tissues, induces the expression of casitas B-lineage lymphoma (CBLB) that participates in insulin signaling, and inhibits the expression of suppressor of cytokine signaling 3 that participates in IR to maintain insulin sensitivity in adipose tissues and skeletal muscles [65,88–90]. PPARγ regulates insulin signal transduction and the expression of ADPN and inflammatory proteins to modulate the degree of hepatic IR (Figure 1), playing an important regulatory role in the accumulation of TAG in the liver.

8. PPARγ Indirectly Participates in Lipid Metabolism via Gluconeogenesis

Gluconeogenesis is the process in which non-carbohydrate precursors are converted to glucose or glycogen. Diabetes and fatty liver disease involve gluconeogenesis and dysregulated glycogenolysis resulting in increased hepatic glucose output [91,92].

PPARγ is an important target for treating disorders involving lipid metabolism, IR, and gluconeogenesis [82,86]. A reduction in PPARγ expression in the liver increases the expression of hexokinase and reduces that of gluconeogenesis enzymes, such as phosphoenolpyruvate carboxykinase (PEPCK) and glucose-6-phosphatase (G6P), thereby inhibiting hepatic gluconeogenesis and deposition of TAG in the liver [93]. Activated PPARγ enhances the expression of the phosphoinositide 3-kinase subunit p85 and promotes hepatic gluconeogenesis (Figure 1) [94]. Activated PPARγ also increases the synthesis of glucagon, influx of amino acids into hepatocytes, and conversion to glucose via gluconeogenesis. Accordingly, PPARγ directly and indirectly regulates the expression of enzymes and proteins involved in hepatic gluconeogenesis. Thus, PPARγ is an important upstream regulator increasing the incidence of diabetes and fatty liver disease [95].

In summary, PPARγ is a transcription factor that regulates adipocyte differentiation, lipid accumulation, fatty acid oxidation and synthesis, oxidative stress, IR, and expression of the gluconeogenesis-related genes, therefore modulating the biological processes involved in hepatic lipid metabolism (Figure 1). Accordingly, PPARγ is an important protein that regulates the pathogenesis of NAFLD. Investigations on the agonists and antagonists of PPARγ might provide novel ideas for the development of drugs against NLFLD.

9. Conclusions and Outlook

NAFLD is a clinico-pathologically defined process associated with metabolic syndrome [23] and fundamentally pin-pointed to the pathogenesis of lipid metabolism, causing, for example, type II diabetes, obesity, either development of cardiovascular diseases or cirrhosis, and hepatocellular carcinoma (HCC), threatening the health of humans and animals. The disease is characterized histologically steatosis and other parenchymal changes, ranging from inflammation to hepatocyte apoptosis/necrosis to fibrosis. NAFLD has become a globally occurring chronic disease that threatens the health of both humans and animals [20,21,96]. However, there is no efficacious drug available that can directly be used to treat NAFLD. Cellular stress and immune reactions, as well as lipid metabolism, had been implicated in the pathogenesis of in animal NAFLD models [23]. There was ample evidence of the positive effects of dietary antioxidant polyphenols, carotenoids, and glucosinolates on the reversion of NAFLD [34,97], although the mechanism of their action was not yet fully elucidated. Discovery and revealing the mechanisms of the important regulatory factors/genes, such as PPARγ, that regulate the pathogenesis of this disease is pertinent and imperative for preventing and treating fatty liver disease in humans and animals. Understanding the molecular mechanisms of the pathogenesis of fatty liver disease will further enhance our understanding of NAFLD, developing safer and effective therapeutics to prevent and/or treat fatty liver disease in humans and animals.

Fatty liver syndrome is a typical type of NAFLD frequently occurring in dairy cows in the perinatal period, caused by negative nutrient balance after calving. The fatty liver disease that occurs in dairy cows is a good animal model to reveal the pathology and pathogenesis of NAFLD. More recent study suggests that impaired hepatic mitochondrial function (such as protein lysine acetylation) is closely associated with fatty liver disease during early lactation in dairy cows [43,84]. Determining the crucial genetic factors, such as PPARγ, will also provide essential clues in breeding improvement of fatty liver disease-resistant dairy cattle, eventually contributing to sustainable development of dairy industry.

Author Contributions: K.S., R.L., and Z.X. participated in the manuscript writing; R.L. and Z.X. contributed to the investigation of metabolic disorder in farms and analyzed the data; Q.Z. and K.S. led the core concept of the review; Q.Z. participated in content examination and discussion. All authors have read and agreed to the published version of the manuscript.

Funding: The work was supported by the Key Project of Agricultural Fine Breeding of Shandong Province (2019LZGC011, 2016LZGC030), National Natural Science Foundation of China (31402054) and Natural Science Foundation of Shandong Province (ZR2013CM013).

Acknowledgments: We thank Jinming Huang (Dairy Cattle Research Center, Shandong Academy of Agricultural Sciences), Yundong Gao and Jianbin Li (Shandong OX Liverstock Breeding Co., Ltd), Dongxiao Sun, Yachun

Wang, and Ying Yu (China Agricultural University) for their assistance in investigations of fatty liver disease in commercial farms. We thank Elseviewer for its linguistic assistance during the preparation of this manuscript.

Conflicts of Interest: The authors declare that they have no conflicts of interest concerning this review.

Appendix A

Table A1. Biological processes associated with NAFLD are regulated by PPARγ through different signaling factors/proteins.

Biological Processes	Proteins or Signaling Factors That Interact with PPARγ		
	Protein Name	Protein Description or Abbreviation	References
Lipid Metabolism	NRF2	Nuclear factor E2-related factor 2	[45]
	CD36	Fatty acid translocase	[66]
	RXR	Retinoid X receptor	[68]
Oxidative Stress	NRF2	Nuclear factor E2-related factor 2	[45,46]
	IFN-γ	Interferon-γ	[74]
	IL-1	Interleukin-1	[74]
	IL-2	Interleukin-2	[74]
	IL-6	Interleukin-6	[74]
	LPS	Lipopolysaccharide	[78]
	NF-κB	Nuclear factor kappa B	[79]
	AP-1	Activator protein-1	[78]
	STAT-1	Signal transducers and activators of transcription 1	[78]
	TNF-α	Tumor necrosis factor α	[86]
Insulin Resistance	SREBP-1	Sterol-regulatory element binding protein-1	[86]
	TNF-α	Tumor necrosis factor α	[86]
	ADPN	Adiponectin	[88]
	CBLB	Casitas B-lineage lymphoma	[88]
	SOCS3	Suppressor of cytokine signaling 3	[88]
	AKT	Protein kinase B	[89]
	GLUT4	Glucose transport protein 4	[89]
	PI3K	Phosphoinositide 3-kinase	[89]
Endoplasmic Reticulum Stress	ATF6	Activating Transcription Factor 6	[29]
	GRP78	Glucose regulated protein 78	[30]
	IRE1α	Inositol-requiring enzyme-1α	[31]
	NRF2	Nuclear factor E2-related factor 2	[45]
	TNF-α	Tumor necrosis factor α	[86]
Gluconeogenesis	HK	Histinine kinase	[93]
	PEPCK	Phosphoenolpyruvate carboxykinase	[93]
	G6P	Glucose-6-phosphate	[93]
	PI3K	Phosphoinositide 3-kinase	[94]

NAFLD, non-alcoholic fatty liver disease; PPARγ, peroxisome proliferator-activated receptor γ.

References

1. Shi, K.R.; Niu, F.; Zhang, Q.; Ning, C.; Yue, S.J.; Hu, C.Z.; Xu, Z.J.; Wang, S.X.; Li, R.R.; Hou, Q.L.; et al. Identification of whole-genome significant single nucleotide polymorphisms in candidate genes associated with serum biochemical traits in Chinese Holstein cattle. *Fornt. Genet.* **2020**, *11*, 163. [CrossRef]
2. Yan, Z.G.; Wang, Z.H.; Zhang, Q.; Yue, S.J.; Yin, B.; Jiang, Y.L.; Shi, K.R. Identification of whole-genome significant single nucleotide polymorphisms in candidate genes associated with body conformation traits in Chinese Holstein cattle. *Anim Genet.* **2020**, *51*, 141–152. [CrossRef]
3. Reid, I.M.; Rowlands, G.J.; Dew, A.M.; Collins, R.A.; Roberts, C.J.; Manston, R. The relationship between1 post-parturient fatty liver and blood composition in dairy cows. *J. Agric. Sci.-Camb.* **1983**, *101*, 473–480. [CrossRef]
4. Starke, A.; Schmidt, S.; Haudum, A.; Scholbach, T.; Wohlsein, P.; Beyerbach, M.; Rehage, J. Evaluation of portal blood flow using transcutaneous and intraoperative Doppler ultrasonography in dairy cows with fatty liver. *J. Dairy Sci.* **2011**, *94*, 2964–2971. [CrossRef] [PubMed]
5. Bradford, B.J.; Yuan, K.; Farney, J.K.; Mamedova, L.K.; Carpenter, A.J. Invited review: Inflammation during the transition to lactation: New adventures with an old flame. *J. Dairy Sci.* **2015**, *98*, 6631–6650. [CrossRef]
6. Godden, S.; Rapnicki, P.; Stewart, S.; Fetrow, J.; Johnson, A.; Bey, R.; Farnsworth, R. Effectiveness of an internal teat seal in the prevention of new intramammary infections during the dry and early-lactation periods in dairy cows when used with a dry cow intramammary antibiotic. *J. Dairy Sci.* **2003**, *86*, 3899–3911. [CrossRef]
7. Carvalho, M.R.; Peñagaricano, F.; Santos, J.E.P.; DeVries, T.J.; McBride, B.W.; Ribeiro, E.S. Long-term effects of postpartum clinical disease on milk production, reproduction, and culling of dairy cows. *J. Dairy Sci.* **2019**, *102*, 11701–11717. [CrossRef]
8. Quenon, J.; Ingrand, S.; Magne, M. Managing the transition from purebred to rotational crossbred dairy cattle herds: Three technical pathways from a retrospective case-study analysis. *Animal* **2020**, *1*, 1–11. [CrossRef] [PubMed]
9. Li, S.L.; Huang, W.M.; Tian, Y.J.; Cao, Z.J. Energy metabolism and its regulation of perinatal dairy cows. *Prog. J. Nutr.* **2012**, 170–176.
10. Vries, A.B.; Huang, H.W.; Zou, Y.; Cao, Z.J. Economic benefit analysis of prolonging the service life of dairy cattle. *China Dairy Cattle* **2014**, *12*, 68–75.
11. Carpenter, A.J.; Ylioja, C.M.; Vargas, C.F.; Mamedova, L.K.; Mendonca, L.G.; Coetzee, J.F.; Hollis, L.C.; Gehring, R.; Bradford, B.J. Hot topic: Early postpartum treatment of commercial dairy cows with nonsteroidal antiinflammatory drugs increases whole-lactation milk yield. *J. Dairy Sci.* **2016**, *99*, 672–679. [CrossRef] [PubMed]
12. Yue, S.J.; Zhao, Y.Q.; Gu, X.R.; Yin, B.; Jiang, Y.L.; Wang, Z.H.; Shi, K.R. A genome-wide association study suggests new candidate genes for milk production traits in Chinese Holstein cattle. *Anim. Genet.* **2017**, *48*, 677–681. [CrossRef] [PubMed]
13. Koo, S.H. Nonalcoholic fatty liver disease: Molecular mechanisms for the hepatic steatosis. *Clin. Mol. Hepatol.* **2013**, *19*, 210–215. [CrossRef]
14. Katoh, N. Relevance of apolipoproteins in the development of fatty liver and fatty liver-related peripartum diseases in dairy cows. *J. Vet. Med. Sci.* **2002**, *64*, 293–307. [CrossRef]
15. Jorritsma, R.; Jorritsma, H.; Schukken, Y.H. Relationships between fatty liver and fertility and some periparturient diseases in commercial Dutch dairy herds. *Theriogenology* **2000**, *54*, 1065–1074. [CrossRef]
16. Hu, Z.Y.; Yin, Z.Y.; Lin, X.Y.; Yan, Z.G.; Wang, Z.H. Effects of feeding fatty acid calcium and the interaction of forage quality on production performance and biochemical indexes in early lactation cow. *J. Anim. Physiol. Anim. Nutr.* **2015**, *99*, 899–904. [CrossRef]
17. Farid, A.S.; Honkawa, F.; Fath, E.M. Serum paraoxonase-1 as biomarker for improved diagnosis of fatty liver in dairy cows. *BMC Vet. Res.* **2013**, *9*, 73. [CrossRef]
18. Pullen, D.L.; Liesman, J.S.; Emery, R.S. A species comparison of liver slice synthesis and secretion of triacylglycerol from nonesterified fatty acids in media. *J. Anim. Sci.* **1990**, *68*, 1395–1399. [CrossRef]
19. Nasr, P.; Ignatova, S.; Kechagias, S.; Ekstedt, M. Natural history of nonalcoholic fatty liver diasease: A prospective follow-up study with serial biopsies. *Hepatol. Res.* **2018**, *2*, 199–210.

20. Demir, M.; Lang, S.; Steffen, H.M. Nonalcoholic fatty liver disease—Current status and future directions. *J. Digest Dis.* **2015**, *16*, 541–557. [CrossRef]
21. Du, H.T.; Wang, C.Y.; Wang, X.P.; Ma, M.W.; Li, F.C. The effects of dietary α-linolenic acid on growth performance, meat quality, fatty acid composition, and liver relative enzyme mRNA expression of growing meat rabbits. *J. Anim. Feed Sci.* **2013**, *22*, 122–129. [CrossRef]
22. Bellentani, S. The epidemiology of non-alcoholic fatty liver disease. *Liver Int.* **2017**, *37*, 81–84. [CrossRef] [PubMed]
23. Tiniakos, D.G.; Vos, M.B.; Brunt, E.M. Nonalcoholic fatty liver disease: Pathology and pathogenesis. *Annu. Rev. Pathol.-Mech.* **2010**, *5*, 145–171. [CrossRef]
24. Day, C.P. Non-alcoholic fatty liver disease: A massive problem. *J. Clin. Med.* **2011**, *11*, 176–178. [CrossRef]
25. Clavien, P.; Camargo, C.; Gorczynski, R.; Washington, M.; Levy, G.; Langer, B.; Greig, P. Reasctant cytokines and neutrophil adhesion after warm ischemia in cirrhotic and noncirrhotic human livers. *Hepatology* **1996**, *23*, 1456. [CrossRef] [PubMed]
26. Duvnjak, M.; Lerotić, I.; Barsić, N.; Tomasić, V.; Virović, J.L.; Velagić, V. Pathogenesis and management issues for non-alcoholic fatty liver disease. *World J. Gastroenterol.* **2007**, *13*, 4539–4550. [CrossRef]
27. Jou, J.; Choi, S.S.; Diehl, A.M. Mechanisms of disease progression in nonalcoholic fatty liver disease. *Semin. Liver Dis.* **2008**, *28*, 370–379. [CrossRef]
28. Dela Peña, A.; Leclercq, I.; Field, J.; George, J.; Jones, B.; Farrell, G. NF-kappaB activation, rather than TNF, mediates hepatic inflammation in a murine dietary model of steatohepatitis. *Gastroenterology* **2005**, *129*, 1663–1674. [CrossRef] [PubMed]
29. Winnay, J.N.; Boucher, J.; Mori, M.A.; Ueki, K.; Kahn, C.R. A regulatory subunit of phosphoinositide 3-kinaseincreases the nuclear accumulation of X-box binding protein-1 to modulate the unfolded protein response. *Nat. Med.* **2010**, *16*, 438–445. [CrossRef]
30. Bjornsson, E.; Angulo, P. Non-alcoholic fatty liver disease. *Gastroenterology* **2007**, *l42*, 1023–1130. [CrossRef]
31. Wang, S.; Chen, Z.; Lam, V.; Han, J.; Hassler, J.; Finck, B.N.; Davidson, N.O.; Kaufman, R.J. IRE1α-XBP1S induces PDI expression to increase MTP activity for hepatic VLDL assembly and lipid homeostasis. *Cell Metabolism.* **2012**, *16*, 473–486. [CrossRef]
32. Bartlett, P.J.; Gaspers, L.D.; Pierobon, N.; Thomas, A.P. Calcium-dependent regulation of glucose homeostasis in the liver. *Cell Calcium.* **2014**, *55*, 306–316. [CrossRef]
33. Tacer, K.F.; Rozman, D. Nonalcoholic fatty liver disease: Focus on lipoprotein and lipid deregulation. *J. Lipid Res.* **2011**, *2011*, 783976.
34. Ferramosca, A.D.; Giacomo, M.; Zara, V. Antioxidant dietary approach in treatment of fatty liver: New insight and updates. *World J. Gastroenterol.* **2017**, *23*, 4146–4157. [CrossRef]
35. Adams, L.A.; Angulo, P.; Lindor, K.D. Nonalcoholic fatty liver disease. *Can. Med. Assoc. J.* **2005**, *172*, 899–905. [CrossRef] [PubMed]
36. Dowman, J.K.; Tomlinson, J.W.; Newsome, P.N. Pathogenesis of non-alcoholic fatty liver disease. *QJM* **2010**, *103*, 71–83. [CrossRef]
37. Liu, L.; Fu, C.; Li, F. Acetate affects the process of lipid metabolism in rabbit liver, skeletal muscle and adipose tissue. *Animals* **2019**, *9*, 799. [CrossRef]
38. Lu, W.J.; Liu, X.F.; Zhao, X.C.; Guo, X.Q.; Kang, M.J.; Xu, B.H. Identification and antioxidant characterisation of thioredoxin-like1 from Apis cerana cerana. *Apidologie* **2012**, *43*, 737–752. [CrossRef]
39. Rotman, Y.; Sanyal, A.J. Current and upcoming pharmacotherapy for non-alcoholic fatty liver disease. *Gut* **2017**, *66*, 180–190. [CrossRef] [PubMed]
40. Younossi, Z.M.; Loomba, R.; Rinella, A.E.; Bugianesi, E.; Marchesini, G.; Neuschwander-Tetri, B.A.; Serfaty, L.; Negro, F.; Caldwell, S.H.; Ratziu, V.; et al. Current and future therapeutic regimens for nonalcoholic fatty liver disease and nonalcoholic steatohepatitis. *Hepatology* **2018**, *68*, 361–371. [CrossRef] [PubMed]
41. Machado, M.; Cortez-Pinto, H. Non-alcoholic fatty liver disease and insulin resistance. *Eur. J. Gastroenterol. Hepat.* **2005**, *17*, 823–826. [CrossRef]
42. Fraenkel, E.; Lazurova, I.; Feher, J. Role of lipid peroxidation in non-alcoholic steatohepatitis. *Orvosi Hetilap.* **2004**, *145*, 611–618. [PubMed]
43. He, X.W.; Gao, J.; Hou, H.; Qi, Z.; Chen, H.; Zhang, X.X. Inhibition of mitochondrial fatty acid oxidation contributes to development of nonalcoholic fatty acid liver disease induced by environmental cadmium exposure. *Environ. Sci. Technol.* **2019**, *53*, 13992–14000. [CrossRef]

44. Wang, S.; Dougherty, E.J.; Danner, R.L. PPARγ signaling and emerging opportunities for improved therapeutics. *Pharmacol. Res.* **2016**, *111*, 76–85. [CrossRef] [PubMed]
45. Feng, X.; Yu, W.; Li, X.; Zhou, F.; Zhang, W.; Shen, Q.; Li, J.; Zhang, C.; Shen, P. Apigenin, a modulator of PPARgamma, attenuates HFD-induced NAFLD by regulating hepatocyte lipid metabolism and oxidative stress via Nrf2 activation. *Biochem. Pharmacol.* **2017**, *136*, 136–149. [CrossRef]
46. Cheng, Q.; Jiang, S.; Huang, L.; Wang, Y.; Yang, W.; Yang, Z.; Ge, J. Effects of zearalenone-induced oxidative stress and Keap1-Nrf2 signaling pathway-related gene expression in the ileum and mesenteric lymph nodes of post-weaning gilts. *Toxicology* **2020**, *429*, 152337. [CrossRef]
47. Zhu, Y.; Kan, L.; Qi, C.; Kanwar, Y.S.; Yeldandi, A.V.; Rao, M.S.; Reddy, J.K. Isolation and characterization of peroxisome proliferator-activated receptor (PPAR) interacting protein (PRIP) as a coactivator for PPAR. *J. Biol. Chem.* **2020**, *275*, 13510–13516. [CrossRef]
48. Latruffe, N.; Vamecq, J. Peroxisome proliferators and peroxisome proliferator activated receptors (PPARs) as regulators of lipid metabolism. *Biochimie* **1997**, *79*, 81–94. [CrossRef]
49. Mano, H.; Kimura, C.; Fujisawa, Y.; Kameda, T.; Watanabe-Mano, M.; Kaneko, H.; Kumegawa, M. Cloning and function of rabbit peroxisome proliferator-activated receptor delta/beta in mature osteoclasts. *J. Biol. Chem.* **2020**, *275*, 8126–8132. [CrossRef] [PubMed]
50. Vamecq, J.; Latruffe, N. Medical of significance peroxisome proliferator-activated receptors. *Lancet* **1999**, *354*, 141–148. [CrossRef]
51. Fu, C.Y.; Liu, L.; Li, F.C. Acetate alters the process of lipid metabolism in rabbits. *Animal* **2018**, *12*, 1895–1902. [CrossRef] [PubMed]
52. Guan, Y.; Zhang, Y.; Breyer, M.D. The Role of PPARs in the Transcriptional Control of Cellular Processes. *Drug News Perspect.* **2002**, *15*, 147–154. [CrossRef]
53. Bordji, K.; Grillasca, J.P.; Gouze, J.N.; Magdalou, J.; Schohn, H.; Keller, J.M.; Terlain, B. Evidence for the presence of peroxisome proliferator-activated receptor (PPAR) alpha and gamma and retinoid Z receptor in cartilage. *J. Biol. Chem.* **2000**, *275*, 12243–12250. [CrossRef] [PubMed]
54. Grygiel-Gorniak, B. Peroxisome proliferator-activated receptors and their ligands: Nutritional and clinical implications a review. *J. Nutr.* **2014**, *13*, 13–17. [CrossRef] [PubMed]
55. Gupta, M.; Mahajan, V.K.; Mehta, K.S.; Chauhan, P.S.; Rawat, R. Peroxisome proliferator-activated receptors (PPARs) and PPAR agonists: The "future" in dermatology therapeutics? *Arch. Dermatol. Res.* **2015**, *307*, 767–780. [CrossRef]
56. Cui, J.X.; Zeng, Y.Q.; Wang, H.; Chen, W.; Du, J.F.; Chen, Q.M.; Hu, Y.X.; Yang, L. The effects of DGAT1 and DGAT2 mRNA expression on fat deposition in fatty and lean breeds of pig. *Livest Sci.* **2011**, *140*, 292–296. [CrossRef]
57. Wang, H.; Wang, J.; Yang, D.D.; Liu, Z.L.; Zeng, Y.Q.; Chen, W. Expression of lipid metabolism genes provides new insights into intramuscular fat deposition in Laiwu pigs. *Asian Aystral. J. Anim.* **2020**, *33*, 390–397. [CrossRef]
58. Filip-Ciubotaru, F.; Foia, L.; Manciuc, C.; Grigore, C. PPARs: Structure, mechanisms of action and control. *Note I. Rev. Med. Chir. Soc. Med. Nat. Iasi.* **2011**, *115*, 477–484.
59. Yu, X.L.; Guo, X.Q.; Kang, M.J.; Liu, L.; Xu, B.H. Identification and expression analysis of a putative fatty acid-binding protein gene in the Asian honeybee, Apis cerana cerana. *J. Insect Sci.* **2013**, *13*, 101. [CrossRef]
60. Kurosaki, E.; Nakano, R.; Shimaya, A.; Yoshida, S.; Ida, M.; Suzuki, T.; Shibasaki, M.; Shikama, H. Differential effects of YM440 a hypoglycemic agent on binding to a peroxisome proliferator-activated receptor gamma and its transactivation. *Biochem. Pharmacol.* **2003**, *65*, 795–805. [CrossRef]
61. Wang, Y.; He, J.Z.; Yang, W.X.; Muhantay, G.; Chen, Y.; Xing, J.M.; Liu, J.Z. Correlation between heart-type fatty acid-binding protein gene polymorphism and mrna expression with intramuscular fat in baicheng-oil chicken. *Asian Austral. J. Anim.* **2015**, *28*, 1380. [CrossRef]
62. Liu, L.; Zhu, L. Effect of 24 h fasting on gene expression of AMPK, appetite regulation peptides and lipometabolism related factors in the hypothalamus of broiler chicks. *Asian Austral. J. Anim.* **2012**, *25*, 1300–1308. [CrossRef] [PubMed]
63. Zhang, X.; Ji, R.; Sun, H.; Peng, J.; Ma, X.; Wang, C.; Fu, Y.; Bao, L.; Jin, Y. Scutellarin ameliorates nonalcoholic fatty liver disease through the PPARγ/PGC-1α-Nrf2 pathway. *Free Radical. Res.* **2018**, *52*, 198–211. [CrossRef]
64. Moran-Salvador, E. Role for PPARgamma in obesity-induced hepatic steatosis as determined by hepatocyte- and macrophage-specific conditional knockouts. *Faseb. J.* **2011**, *25*, 2538–2550. [CrossRef]

65. Liu, L.; Xu, S.H.; Wang, X.J.; Jiao, H.C.; Lin, H. Peripheral insulin doesn't alter appetite of broiler chicks. *Asian Austral. J. Anim.* **2016**, *29*, 1294–1299. [CrossRef]
66. Rangwala, S.M.; Lazar, M.A. Peroxisome proliferator-activated receptor gamma in diabetes and metabolism. *Trends Pharmacol. Sci.* **2004**, *25*, 331–336. [CrossRef]
67. Larter, C.Z.; Yeh, M.M.; Williams, J.; Bell-Anderson, K.S.; Farrell, G.C. MCD-induced steatohepatitis is associated with hepatic adiponectin resistance and adipogenic transformation of hepatocytes. *J. Hepatol.* **2008**, *49*, 407–416. [CrossRef] [PubMed]
68. Nagy, L.; Tontonoz, P.; Alvarez, J.G.; Chen, H.; Evans, R.M. Oxidized LDL regulates macrophage gene expression through ligand activation of PPARgamma. *Cell* **1998**, *93*, 229–240. [CrossRef]
69. Yamazaki, T.; Shiraishi, S.; Kishimoto, K.; Miura, S.; Ezaki, O. An increase in liver PPARgamma2 is an initial event to induce fatty liver in response to a diet high in butter: PPARgamma2 knockdown improves fatty liver induced by high-saturated fat. *J. Nutr. Biochem.* **2011**, *22*, 543–553. [CrossRef]
70. Gavrilova, O.; Haluzik, M.; Matsusue, K.; Cutson, J.J.; Johnson, L.; Dietz, K.R.; Nicol, C.J.; Vinson, C.; Gonzalez, F.J.; Reitman, M.L. Liver peroxisome proliferator-activated receptor gamma contributes to hepatic steatosis, triglyceride clearance, and regulation of body fat mass. *J. Biol. Chem.* **2003**, *278*, 34268–34276. [CrossRef]
71. Yan, H.R.; Meng, F.; Jia, H.H.; Guo, X.Q.; Xu, B.H. The identification and oxidative stress response of a zeta class glutathione s-transferase (GSTZ1) gene from apis cerana cerana. *J. Insect Physiol.* **2012**, *58*, 782–791. [CrossRef] [PubMed]
72. Dubaisi, S.; Fang, H.; Kocarek, T.A. Transcriptional regulation of human cytosolic sulfotransferase 1C3 by peroxisome proliferator-activated receptor gamma in LS180 human colorectal adenocarcinoma cells. *Mol. Pharmacol.* **2016**, *90*, 562–569. [CrossRef] [PubMed]
73. Liu, L.; Li, H.P.; Li, X.L.; Jiao, H.C.; Lin, H.; Ardashir, S.; Wang, Y.F.; Song, Z.G. Effects of acute heat stress on gene expression of brain-gut neuropeptides in broiler chickens (Gallus gallus domesticus). *Anim. Sci. J.* **2013**, *91*, 5194–5201.
74. Khazaei, M. Effect of peroxisome proliferator-activated receptor gamma on inflammatory markers. *ARYA Atheroscler* **2015**, *11*, 261–262.
75. Liu, L.; Fu, C.; Li, F.C. Dietary niacin supplementation suppressed hepatic lipid accumulation in rabbits. *Asian Austral. J. Anim.* **2016**, *29*, 1748–1755. [CrossRef]
76. Dong, X.S.; Zhai, R.N.; Liu, Z.L.; Lin, X.Y.; Wang, Z.H.; Hu, Z.Y. The effect of intravenous infusions of glutamine on duodenal cell autophagy and apoptosis in early-weaned calves. *Animals* **2019**, *9*, 404. [CrossRef]
77. Zhai, R.N.; Dong, X.S.; Feng, L.; Li, S.L.; Hu, Z.Y. The effect of heat stress on autophagy and apotosis of rumen, abomasum, duodenum, liver and kidney cells in calves. *Animals* **2019**, *9*, 85. [CrossRef] [PubMed]
78. Ricote, M.; Li, A.C.; Willson, T.M.; Kelly, C.J.; Glass, C.K. The peroxisome proliferator-activated receptor-gamma is a negative regulator of macrophage activation. *Nature* **1998**, *391*, 79–82. [CrossRef]
79. Genini, D.; Carbone, G.M.; Catapano, C.V. Multiple interactions between peroxisome proliferators-activated receptors and the ubiquitin-proteasome system and implications for cancer pathogenesis. *PPAR Res.* **2008**, *2008*, 195065. [CrossRef]
80. Li, R.G.; Wang, X.P.; Wang, C.Y.; Ma, M.W.; Li, F.C. Growth performance, meat quality and fatty acid metabolism response of growing meat rabbits to dietary linoleic acid. *Asian Austral. J. Anim.* **2012**, *25*, 1169–1177. [CrossRef]
81. Liu, L.; Wang, X.J.; Jiao, H.C.; Zhao, J.P.; Lin, H. Glucocorticoids inhibited hypothalamic target of rapamycin in high fat diet-fed chicks. *Poultry Sci.* **2015**, *94*, 2221–2227. [CrossRef]
82. Yang, Z.; Zhang, H.; Niu, Y.; Zhang, W.; Zhu, L.; Li, X.; Lu, S.; Fan, J.; Li, X.; Ning, G.; et al. Circulating periostin in relation to insulin resistance and nonalcoholic fatty liver disease among overweight and obese subjects. *Sci. Rep. UK* **2016**, *6*, 37886. [CrossRef] [PubMed]
83. Iyer, A.; Fairlie, D.P.; Brown, L. Lysine acetylation in obesity, diabetes and metabolic disease. *Immunol. Cell Biol.* **2012**, *90*, 39–46. [CrossRef]
84. Garcia-Roche, M.; Casal, A.; Mattiauda, D.A.; Ceriani, M.; Jasinsky, A.; Mastrogiovanni, M.; Trostchansky, A.; Carriquiry, M.; Cassina, A.; Quijiano, C. Impaired hepatic mitochondrial function during early lactation in dairy cows: Association with protein lysine acetylation. *PLoS ONE* **2019**, *14*, e0213780. [CrossRef] [PubMed]

85. Zhang, Y.Q.; Zhou, F.Y.; Bai, M.Y.; Liu, Y.; Zhang, L.; Zhu, Q.; Bi, Y.; Ning, G.; Zhou, L.; Wang, X. The pivotal role of protein acetylation in linking glucose and fatty acid metabolism to β-cell function. *Cell Death Dis.* **2019**, *10*, 66. [CrossRef] [PubMed]
86. Lalloyer, F.; Staels, B. Fibrates, glitazones, and peroxisome proliferator-activated receptors. *Arterioscl. Throm. Vas.* **2010**, *30*, 894–899. [CrossRef] [PubMed]
87. Yuan, X.J. Expression of androgen receptor and estrogen receptor-alpha in the developing pituitary gland of male sheep lamb. *Anim. Reprod Sci.* **2011**, *127*, 164–168.
88. Shi, H.; Cave, B.; Inouye, K.; Bjorbaek, C.; Flier, J.S. Overexpression of suppressor of cytokine signaling 3 in adipose tissue causes local but not systemic insulin resistance. *Diabetes* **2006**, *55*, 699–707. [CrossRef]
89. Armoni, M.; Kritz, N.; Harel, C.; Bar-Yoseph, F.; Chen, H.; Quon, M.J.; Karnieli, E. Peroxisome proliferator-activated receptor-gamma represses GLUT4 promoter activity in primary adipocytes, and rosiglitazone alleviates this effect. *J. Biol. Chem.* **2003**, *278*, 30614–30623. [CrossRef]
90. Yang, L.J.; Zhou, M.; Huang, L.B.; Yang, W.R.; Yang, Z.B.; Jiang, S.Z.; Ge, J.S. Zearalenone promotes follicle growth through modulation of Wnt-1/β-catenin signaling pathway and expression of estrogen receptor genes in ovaries of post-weaning piglets. *J. Agr. Food Chem.* **2018**, *66*, 7899–7906. [CrossRef]
91. Moller, D.E. New drug targets for type 2 diabetes and the metabolic syndrome. *Nature* **2011**, *414*, 821–827. [CrossRef] [PubMed]
92. Jiang, F.G.; Lin, X.Y.; Yan, Z.G.; Hu, Z.Y.; Wang, Y.; Wang, Z.H. Effects of forage source and particle size on feed sorting, milk production and nutrient digestibility in lactating dairy cows. *J. Anim. Physiol. Anim. Nutr.* **2018**, *102*, 1472–1481. [CrossRef]
93. Goto, M.; Yoshioka, T.; Battelino, T.; Ravindranath, T.; Zeller, W.P. TNFalpha decreases gluconeogenesis in hepatocytes isolated from 10-day-old rats. *Pediatr. Res.* **2001**, *49*, 552–557. [CrossRef] [PubMed]
94. Bo, Q.F.; Sun, X.M.; Liu, J.; Sui, X.M.; Li, G.X. Antitumor action of the peroxisome proliferator-activated receptor-γ agonist rosiglitazone in hepatocellular carcinoma. *Oncol. Lett.* **2015**, *10*, 3554. [CrossRef]
95. Tong, E.K.; England, L.; Glantz, S.A. Changing conclusions on secondhand smoke in a sudden infant death syndrome review funded by the tobacco industry. *Pediatrics* **2005**, *115*, 356–366. [CrossRef]
96. Lou, X.M.; Li, J.; Zhang, X.X.; Wang, J.M.; Wang, C.F. Variations in fatty acid composition of Laoshan goat milk from partum to 135 days postpartum. *Anim. Sci. J.* **2018**, *89*, 1628–1638. [CrossRef] [PubMed]
97. Anania, C.; Perla, F.M.; Olivero, F.; Pacifico, L.; Chiesa, C. Mediterranean diet and nonalcoholic fatty liver disease. *World J. Gastroenterol.* **2018**, *24*, 2083–2094. [CrossRef]

animals

MDPI

Article

Low Expression of Sirtuin 1 in the Dairy Cows with Mild Fatty Liver Alters Hepatic Lipid Metabolism

Yu Li [1,2], Suping Zou [1], Hongyan Ding [1], Ning Hao [1], Yingying Huang [1], Jishun Tang [3], Jianbo Cheng [1], Shibin Feng [1], Jinchun Li [1], Xichun Wang [1,*] and Jinjie Wu [1,*]

1 College of Animal Science and Technology, Anhui Agricultural University, Hefei 230036, China; lydhy2014@ahau.edu.cn (Y.L.); 18855952196@163.com (S.Z.); dinghy1988@163.com (H.D.); hao621065@163.com (N.H.); huangyingying@ahau.edu.cn (Y.H.); chengjianbofcy@163.com (J.C.); luyifsb@163.com (S.F.); jinchunli64@163.com (J.L.)

2 Anhui Province Key Laboratory of Veterinary Pathobiology and Disease Control, Hefei 230036, China

3 Institute of Animal Husbandary and Veterinary Medicine, Anhui Academy of Agriculture Sciences, Hefei 230031, China; tjs157@163.com

* Correspondence: wangxichun@ahau.edu.cn (X.W.); wjj@ahau.edu.cn (J.W.)

Received: 22 February 2020; Accepted: 24 March 2020; Published: 27 March 2020

Simple Summary: Sirtuin 1 (SIRT1), a NAD-dependent histone deacetylase, is involved in oxidative stress and lipid metabolism regulation. Limited studies exist regarding the role of SIRT1 in lipid metabolism disorder in periparturient dairy cows. This study explores the effect of hepatic steatosis on the expression of the SIRT1 gene and protein and the proteins encoded by the genes downstream to it, all of which are involved in lipid metabolism in the liver. Control cows (n = 6, parity 3.0 ± 2.0, milk production 28 ± 47 kg/d) and mild fatty liver cows (n = 6, parity 2.3 ± 1.5, milk production 20 ± 6 kg/d) were retrospectively selected based on liver triglycerides (TG) content (% wet liver). The present study indicates that low SIRT1 expression caused by hepatic steatosis promotes hepatic fatty acid synthesis and inhibits fatty acid β-oxidation. We believe that our study makes a significant contribution to the literature because it demonstrates that hepatic steatosis is associated with increased hepatic fatty acid synthesis, inhibited fatty acid β-oxidation and reduced lipid transport.

Abstract: Dairy cows usually experience negative energy balance coupled with an increased incidence of fatty liver during the periparturient period. The purpose of this study was to investigate the effect of hepatic steatosis on the expression of the sirtuin 1 (SIRT1), along with the target mRNA and protein expressions and activities related to lipid metabolism in liver tissue. Control cows (n = 6, parity 3.0 ± 2.0, milk production 28 ± 7 kg/d) and mild fatty liver cows (n = 6, parity 2.3 ± 1.5, milk production 20 ± 6 kg/d) were retrospectively selected based on liver triglycerides (TG) content (% wet liver). Compared with the control group, fatty liver cows had greater concentrations of cholesterol and TG along with the typically vacuolated appearance and greater lipid droplets in the liver. Furthermore, fatty liver cows had greater mRNA and protein abundance related to hepatic lipid synthesis proteins sterol regulatory element binding proteins (SREBP-1c), long-chain acyl-CoA synthetase (ACSL), acyl-CoA carbrolase (ACC) and fatty acid synthase (FAS) and lipid transport proteins Liver fatty acid binding protein (L-FABP), apolipoprotein E (ApoE), low density lipoprotein receptor (LDLR) and microsomal TG transfer protein (MTTP) ($p < 0.05$). However, they had lower mRNA and protein abundance associated with fatty acid β-oxidation proteins SIRT1, peroxisome proliferator-activated receptor co-activator-1 (PGC-1α), peroxisome proliferator–activated receptor-α (PPARα), retinoid X receptor (RXRα), acyl-CoA 1 (ACO), carnitine palmitoyltransferase 1 (CPT1), carnitine palmitoyltransferase 2 (CPT2) and long- and medium-chain 3-hydroxyacyl-CoA dehydrogenases (LCAD) ($p < 0.05$). Additionally, mRNA abundance and enzyme activity of enzymes copper/zinc superoxide dismutase (Cu/Zn SOD), catalase (CAT), glutathione peroxidase (GSH-Px) and manganese superoxide dismutase (Mn SOD) decreased and mRNA and protein abundance of p45 nuclear factor-erythroid 2 (p45 NF-E2)-related factor 1 (Nrf1), mitochondrial transcription factor A

(TFAM) decreased ($p < 0.05$). Lower enzyme activities of SIRT1, PGC-1α, Cu/Zn SOD, CAT, GSH-Px, SREBP-1c and Mn SOD ($p < 0.05$) and concentration of reactive oxygen species (ROS) were observed in dairy cows with fatty liver. These results demonstrate that decreased SIRT1 associated with hepatic steatosis promotes hepatic fatty acid synthesis and inhibits fatty acid β-oxidation. Hence, SIRT1 may represent a novel therapeutic target for the treatment of the fatty liver disease in dairy cows.

Keywords: dairy cow; fatty liver; lipid metabolism; oxidative stress; SIRT1

1. Introduction

Dairy cows undergo negative energy balance (NEB) when they transition from late gestation to early lactation, during which ketosis, fatty liver and metritis are likely to occur [1,2]. The occurrence of fatty liver in dairy cows increases treatment costs and culling and decreased milk production [3]. In addition, fatty liver disease develops when increased infectious and metabolic diseases are likely to occur [4,5]. For these reasons, fatty liver has become a major international health burden in dairy cows.

The protein sirtuin 1 (SIRT1), an NAD^+-dependent deacetylase, and deacetylates histones including transcription factors lead to the regulation of metabolism, oxidative stress and cellular survival [6]. SIRT1 is highly sensitive to intracellular redox status and provides cells with the ability to tolerate oxidative stress. SIRT1 protects cells from oxidative stress by increasing the activity of catalase [7–9]. Oxidative stress was increased by SIRT1 inhibitor (Ex 527) treatment and decreases in SIRT1 expression were observed in neurons [10]. Ex 527 attenuated the activity of histone deacetylase (HDAC) and increased the degree of myocardial injury during oxidative stress [11]. Accumulating evidence suggests that SIRT1 plays a protective role in the process of oxidative stress and that oxidative stress was induced by excessive fat mobilization in perinatal cows. We observed in our studies that oxidative stress was induced by high non-esterified fatty acid (NEFA) concentrations in dairy cows [12,13]. To date, the effect of SIRT1 on the oxidative stress in dairy cows remains unclear.

It was reported that SIRT1 plays a critical role in lipid metabolism by modulating the activity of transcription factors [14]. In mice and obese patients fed with a high-fat diet, the activity and expression of SIRT1 were significantly decreased, which attenuated the mobilization of fatty acids and promoted the occurrence of metabolic disorders [15–17]. In addition, mounting evidence indicates that a long-term, high-calorie-diet induced liver steatosis is promoted by increasing the expression of SIRT1 in animals [18,19]. Here, we report that the knocked down expression or overexpression of SIRT1 results in changes in the lipid/cholesterol (Chol) levels in the serum and liver, and causes accumulation of lipids in the liver, a process leading to hepatic steatosis [20,21]. Evidence is emerging that SIRT1 plays a vital role in lipid metabolism in the liver. However, the underlying molecular mechanisms, the effect of hepatic steatosis on the expression of SIRT1 genes and proteins and the expressions and activities of downstream lipid-metabolism-related proteins and key enzymes, have not yet been fully clarified.

Therefore, we aimed to investigate the change of expression levels of SIRT1 and downstream lipid-metabolism-related proteins in the pathogenesis of fatty liver disease in dairy cows, thereby providing a theoretical and experimental basis for revealing the pathogenesis of fatty liver in dairy cows and searching for therapeutic targets.

2. Materials and Methods

2.1. Ethics Statement

This study was conducted following the recommendations in the Guide for Care and Use of Laboratory Animals of the National Institutes of Health. All experimental procedures were approved by the Institutional Animal Care and Use Committee of Anhui Agricultural University (permit number:

20170624). All surgeries were performed under anesthesia and all efforts were made to minimize suffering. Briefly, cows were anesthetized with thiamylal sodium along the midline skin of the abdomen. Then the abdominal cavity was opened, the liver was obtained and placed on a sterile bench treated with sterile 0.9% sodium chloride solution for removing bloodstains on the surface.

2.2. *Animals*

Lactating Holstein multiparous cows with the same breed, age and having similar milk production characteristics and body condition scores were selected from a commercial dairy farm located in Hefei city, Anhui province, China. The rectal temperature, respiratory rate and pulse rate for each cow was performed by a skilled veterinarian to ensure the cows had no other co-morbidities. Clinical observations and other disease conditions were recorded (Table 1). The cows were fed ad libitum with a total mixed ration (TMR) forage that met the animals' nutrient requirements (Table 2) and had free access to get water. A total of 34 dairy cows were screened by liver biopsy using a liver puncture needle based on the hepatic TG content (% wet liver, <1% was considered to be healthy cows, 1–10% was considered to be mild fatty liver cows), which is the standard for fatty liver diagnosis [22,23]. Equal numbers of control (n = 6) and fatty liver cows and were retrospectively selected based on liver triglycerides (TG) content. Finally, 12 dairy cows were slaughtered for collecting liver tissue samples and processed as part of the normal work of a commercial abattoir (Hefei, China).

Table 1. Clinical observations of control and mild fatty liver cows.

Groups	Milk Ketones	Feed Intake	Milk Production	Health Status	Reproductive Performance	Other Disease Conditions
Control	0	0	0	0	0	0
Mild fatty liver	+	−	−	−	−	0

The symbols + and − mean positive and negative association, respectively. The 0 means no association.

Table 2. The basic diet formulation, %.

Item	Content	
	Prenatal	**Postpartum**
Silage	31.4	40.0
Guinea grass	23.4	
Corn	19.6	35.0
Wheat bran	10.0	8.0
Soybean meal	2.0	5.0
Sunflower	11.5	8.0
NaCl	0.8	1.0
Premix1	1.3	1.8
NaHCO3		1.2
Total	100.00	100.00
	Energy (%) [1]	
NEL(MJ/Kg) [2]	5.7	6.7
Crude protein	11.3	15.2
Neutral detergent fiber	50.2	33.45
Acid detergent fiber	28.5	17.2
Ca	0.3	0.7
P	0.3	0.5
NFC Non fiber carbohydrate	28.0	40.4
RDP Neutral detergent fiber	7.0	7.4
NFC/RDP Nonfiber carbohydrate/ Neutral detergent fiber	4.0	5.5

[1] The premix provided the following per kg of diets: VA 200,000 IU, VD 70,000 IU, VE 1000 IU, Fe 2000 mg, Cu 600 rng, Zn 2400 mg, Mn 1300 mg, I 6 mg, Se 17 mg, CO 7 mg; [2] NEL was a calculated value and others were measured values.

2.3. Average Milk Production Collection and Milk Components Analysis

As described in the previous study, the daily dry matter intake (DMI) for individual cows was calculated by subtracting the orts from the feed offered [24]. Cows were milked at 06:00, 14:00 and 20:00 and milk production was recorded at each milking. At d 70 in milk, the liver was collected. The milk aliquots were stored at 4 °C until analysis. Milk components include fat, protein, lactose, total solids and milk urea nitrogen (MUN) concentrations were determined by mid-infrared spectrophotometry method on a MilkoTMScan (MilkoScan Type FT120, Foss Electric, Hillerød, Denmark). Somatic cell counts (SCC) were conducted on a Fossomatics 5000 (Foss Analytical A/S; Foss Electric, Hillerød, Denmark). Both 4% fat corrected milk (FCM) and energy corrected milk (ECM) were calculated and the equations were as follows: 4% FCM = 0.4 × milk (kg) + 15 × fat (kg) and ECM = 0.327 × milk (kg) + 12.95 × fat (kg) + 7.20 × protein (kg). Feed efficiency was calculated as the daily milk yield/kg of feed DMI on an individual cow basis.

2.4. Liver chemistry measurement

The liver Chol and TG measurements were performed as previously described [25]. In brief, frozen liver tissue (50 mg) was homogenized with 1 mL lysis buffer. The Chol contents were measured with Chol assay kit, while TG contents were determined using TG kits (IDEXX Vet Tests, Westbrook, ME, USA). Total protein concentration was measured using bicinchoninic acid (BCA) method.

2.5. ELISA Assay for the Contents of SIRT1 and Downstream Lipid- metabolism-related Proteins and Key Enzymes

Liver samples were homogenized in ice-cold lysis buffer (0.05 M-phosphosaline, pH 7.4, containing 0.025 M EDTA, 0.08% sodium azide and 0.05% Triton X-100) supplemented with protease inhibitor cocktail (P1860, Sigma-Aldrich Co., St. Louis, MO, USA). Homogenates were incubated for 10 min at 4 °C and centrifuged at 14,000× g at 4 °C for 15 min. The supernatants were separated into two aliquots (500 µL each). One was immediately used for analysis, while the other was stored at −80 °C until further analysis. SIRT1, SREBP-1c, PGC-1α and the concentrations of CAT, Cu/Zn-SOD, Mn-SOD, GSH-Px, GSH, GSSG and ROS (oxidation and antioxidation key indexes) levels in liver supernatants were determined by ELISA provided by Shanghai Bluegene Biotech Co., Ltd. according to the manufacture's protocol. Total protein concentration was measured using the bicinchoninic acid (BCA) method.

2.6. Liver Histology

Tissue samples obtained from biopsy or necropsy were fixed in 10% formalin and embedded in paraffin. Sections were prepared with a thickness of 2–3 µm, stained with hematoxylin/eosin (H&E), and examined. Accumulation of triglyceride (TG) content in the liver was visualized by Oil Red O (Sigma-Aldrich) staining. Slides were viewed and images were taken using an Olympus BX41TF System Microscope (Olympus Corporation, Tokyo, Japan).

2.7. Total RNA Isolation and qRT-PCR

The liver was homogenized in TRIzol reagent (Invitrogen Life Technologies, Grand Island, NY) and total RNA was isolated according to the manufacturer's instructions. The RNA concentration and quality were measured by K5500 Micro-Spectrophotometer (Kaiao, Beijing, China). RNA was reverse-transcribed into cDNA according to the Reverse Transcription Systems instructions (TaKaRa, Dalian, China). According to the GenBank sequence, the primer sequences of the target genes were designed using the software Primer Premier 5.0 and β-actin was used as a reference gene (Table 3). The amplification products were analyzed by 1.5% agarose gel electrophoresis and a gel imaging and analysis system (UVItec, Cambridge, UK). The mRNA expression levels were determined by quantitative reverse-transcription polymerase chain reaction (qPCR) via an ABI prism 7500 Real-Time PCR system (Applied Biosystems). The mRNA relative abundance was calculated according to the method of Pfaffl and was normalized to the mean expression of β-actin and results (fold changes) were expressed as $2^{-\Delta\Delta Ct}$ [26].

Table 3. Primers sequences qRT-PCR used in this experiment.

Genes [1]	Primer Sequences (5′-3′)	Gene Bank Accession no.	Amplicon (bp)	Annealing Temperature (°C)
SIRT1	Forward: TATGGAGTGACATAGAGTGTGCT Reverse: GTCGCTACACCACTTCAATCC	XM_015461011.1	143	57
SREBP-1c	Forward:CGACACCACCAGCATCAACCACG Reverse: GCAGCCCATTCATCAGCCAGACC	NM_001113302.1	119	62
ACCα	Forward: TGCTGAATATCCTCACGGAGCT Reverse: CGACGTTTCGGACAAGATGAGT	NM_174224.2	212	60
FAS	Forward: ACAGCCTCTTCCTGTTTGACG Reverse: CTCTGCACGATCAGCTCGAC	NM_174662.2	226	59
ACSL	Forward: TCGGAACTGAAGCCATCACC Reverse: GCCTCGTTCCAGCAGATCAC	NM_001076085.1	173	63
CPT-1	Forward: ACGCCGTGAAGTATAACCCT Reverse: CCAAAAATCGCTTGTCCCTT	NM_001304989.1	119	60
CPT-2	Forward: TGAACATCCTCTCCATCTGG Reverse: GGTCAACAGCAACTACTACG	NM_001045889.2	188	58
ACO	Forward: TACGGAGGGATGAGGAGTGT Reverse: TCTCAGGAAGCGAGTTTGG	NM_001205495.1	143	64
LCAD	Forward: GGTCCACAGCACAGACTTGGT Reverse: GGAATTGGCTAGGCTTGTGATC	NM_001076936.1	151	56
L-FABP	Forward: AAGTACCAAGTCCAGACCCAG Reverse: CACGATTTCCGACACCC	NM_175817.3	111	61
LDLR	Forward: GCTGTTCTGCCTTTCTCCTT Reverse: ACTTTCTCCCCTGACCCTTG	NM_001166530.1	228	65
Apo-100	Reverse: GATACTCAGAACGGAGCAAT Forward: GCACCAATCAGATAACAGGA	XM_019969506.1	222	58
ApoE	Reverse: TCCTGAATGACCTGGGTGTTG Forward: TCTGTGGGTTGCCGTGGTG	XM_005219148.3	219	62
MTTP	Reverse: CAGTTTGCAGCCTTGGTTCTG Forward: TTCAAAAGCACCGAGAGCGTT	NM_001101834.1	201	56
PGC-1α	Reverse: GACCACAAATGATGACCCTC Forward: GGTTTGGCTTGTAGATGTT	NM_177945.3	123	60
Nrf1	Reverse: TTTTAGTAACCCTGATGGC Forward: GCTTGCGTTGTCTGGATG	NM_001098002.2	198	57
TFAM	Reverse: TGGCACATCACAGGTAAA Forward: GTTCCTCCCAAGATTTCA	NM_001034016.2	137	63
MnSOD	Reverse: TTCAATAAGGAGCAGGGAC Forward: CAGTGTAAGGCTGACGGTTT	NM_201527.2	234	64
Cu/Zn SOD	Reverse: GAAGAGAGGCATGTTGGAGA Forward: CCAATTACACCACGAGCCAA	NM_174615.2	221	60
CAT	Reverse: AGATACTCCAAGGCGAAGGTG Forward: AAAGCCACGAGGGTCACGAAC	NM_001035386.2	120	61
GSH-Px	Reverse: GCGGGAGCAGGACTTCTACGA Forward: CCCGATAGTGCTGGTCTGTGAA	NM_001101113.2	137	65
PPARα	Reverse: GGGTTTTCTTAGGCTTTT Forward: AGTCCATCCCTGGGTTTG	NM_001034036	176	60
RXRα	Reverse: GGCAGATGTTGGTGACGGG Forward: GGCGAGAGCGAGGTGGAGT	NM_001304343	163	62
β-actin	Reverse: GCCCTGAGGCTCTCTTCCA Forward:GCGGATGTCGACGTCACA	NM_173979.3	101	59

[1] SIRT1 = sirtuin 1; SREBP-1c = sterol regulatory element binding transcription factor 1; ACCα = acetyl-CoA carboxylase alpha; FAS = fatty acid synthetase; ACSL = acyl-CoA synthetase long chain family member; CPT1 = carnitine palmitoyltransferase 1; CPT2 = carnitine palmitoyltransferase 2; ACO = acyl coenzyme A oxidase; LCAD = long chain acyl-CoA dehydrogenase; L-FABP = liver fatty acid binding proteins; LDLR = low density lipoprotein receptor; ApoB100 = apolipoprotein B 100; ApoE = apolipoprotein E; MTTP = microsomal triglyceride transfer protein; PGC-1α = peroxisome proliferator-activated receptor γ coactivator-1 alpha; Nrf1 = nuclear respiratory factor 1; TFAM = transcription factor A, mitochondrial; MnSOD = superoxide dismutase 2, mitochondrial; Cu/Zn SOD = superoxide dismutase 1; CAT = catalase; GSH-Px = glutathione peroxidase 7; PPARα = peroxisome proliferator-activated receptor alpha; RXRα = retinoid X receptor alpha.

2.8. Western Blot Analysis

Liver tissue homogenates were solubilized in SDS sample buffer. As previously described, liver total protein was extracted by a commercial kit (Sangon Biotech, China) [12]. Protein concentrations were measured by BCA method. A total of 50 μg protein for each sample was separated by SDS-PAGE. The required gels were cut and transferred into a PVDF membrane (Shanghai Jinsheng Biological Engineering Co, Shanghai, China) with electrophoresis buffer. By using appropriate antibodies (Abs), immunoreactive bands were visualized by a gel imaging system (Bio-Rad, Hercules, CA, USA). Primary antibodies against ACO (YN0401), FAS (YM1224), ACCα (YT0074), p-ACCα (YP0620), ACSL1 (YN0827), LDLR (YN2236), RXRα (YN0018), Nrf1 (YT3188) and TFAM (YT2916) were purchased from ImmunoWay Biotechnology Company (Newark, DE, USA). Primary antibodies against CPT1 (15184-1-AP) and CPT2 (26555-1-AP) were purchased from Proteintech Group (Chicago, IL, USA). Primary antibodies against PPARα (SC-1985), SREBP-1c (SC-365513) and ApoE (SC-31822) were purchased from Santa Cruz Biotechnology (Santa Cruz, CA, USA). Primary antibody against SIRT1 (D1D7) was purchased from Cell Signaling Technology (Danvers, MA, USA). Antibodies against PGC-1α (ab54481) was purchased from Abcam (Cambridge, MA, USA). Horseradish peroxidase-conjugated secondary antibodies were purchased from Wuhan Boster Biological Engineering Co. (Wuhan, China). The results were analyzed using Quantity One software (Bio-Rad, Hercules, CA, USA). The protein levels were normalized by blots against β-actin.

2.9. Statistical Analysis

Data are presented as mean ± SEM. Differences between the mean values of the normally distributed and homogeneity of variance data were analyzed using a two-tailed Student's t-test. The results are exploratory and the differences were considered significant at $p < 0.05$ or $p < 0.01$ in all the studies.

3. Results

3.1. Milk Production, Milk Component and SCC

As shown in Table 4, the DMI and milk yield in the fatty liver group were significantly lower than those in the control group ($p < 0.01$). The milk protein and MUN were significantly reduced ($p < 0.01$) and SCCs were significantly increased ($p < 0.01$) in the fatty liver group.

Table 4. Milk production and milk component of control and fatty liver dairy cows.

Parameter	Control	Fatty Liver	*p*-Value
DMI, kg/d	23.17 ± 3.24	18.45 ± 2.84	<0.01
Milk production, kg/d	28 ± 7	20 ± 6	<0.01
Milk component			
Fat, %	3.48 ± 0.35	3.33 ± 0.32	0.39
Protein, %	3.25 ± 0.21	2.91 ± 0.34	<0.01
Lactose, %	4.96 ± 0.11	4.88 ± 0.13	0.91
MUN [1] (mg/100 mL)	13.76 ± 1.91	11.52 ± 1.43	0.01
SCC (×10^3/mL)	112.35 ± 20.23	365.61 ± 43.25	<0.01

[1] MUN, milk urea nitrogen; SCC, somatic cell count.

3.2. Chol and TG Concentrations in the Control and Fatty Liver Dairy Cows

Compared with the control group, the concentrations of Chol (Figure 1A) and TG (Figure 1B) were increased in the fatty liver group.

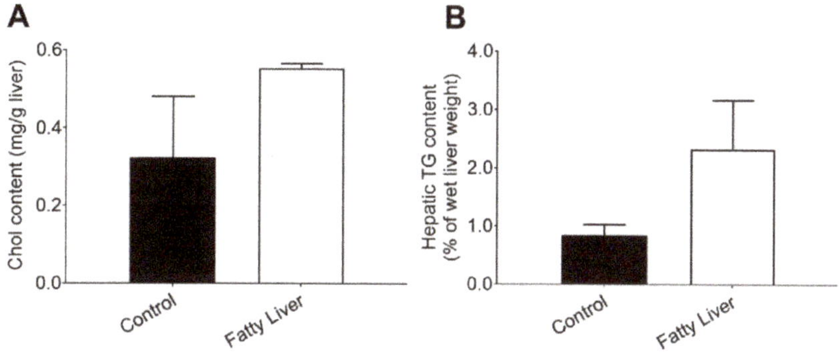

Figure 1. Cows with fatty livers exhibit over the induction of hepatic lipid synthesis (**A**,**B**). Dairy cows were classified according to the hepatic triglyceride (TG) content into a control group (n = 6) and a fatty liver group (n = 6) dairy cows (B); The data presented are the mean ± SEM. * $p < 0.05$; ** $p < 0.01$.

3.3. Histological Analysis of Liver

Isolated cow livers were subjected to H&E staining to evaluate their hepatic steatosis status. It was characterized by a typically vacuolated appearance of the lipid-laden hepatocytes in the H&E-stained sections of the liver (Figure 2A,B) and confirmed by the demonstration of more intrahepatic lipid on oil red O staining of the frozen sections (Figure 2C,D). Steatosis was primarily centrilobular and either microvesicular or mixed microvesicular/ macrovesicular. In fatty liver cows, the oil red O stained lipid droplets occupied an area larger and more than that which was in the control group cows. These results suggest that cows with fatty liver have a hepatic injury.

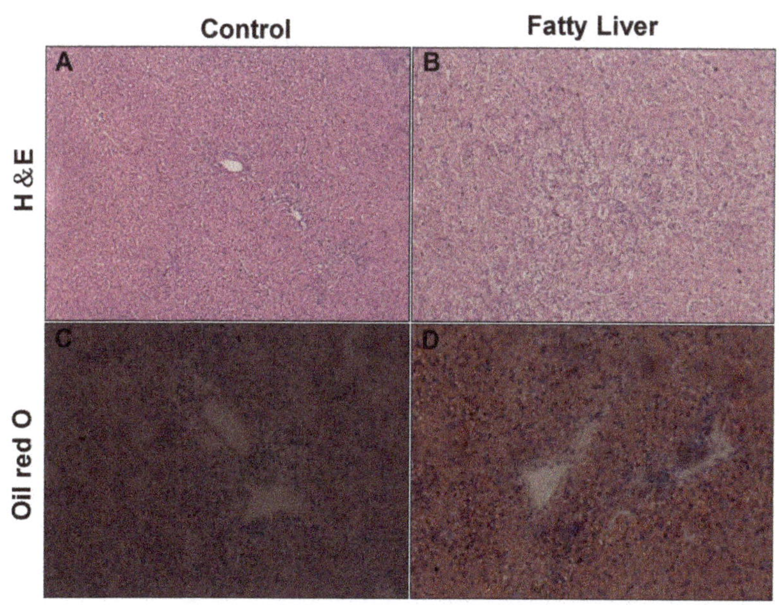

Figure 2. Assessment of hepatic histology in the liver of the control group and fatty liver group. (**A**,**B**) Hematoxylin and Eosin staining of liver sections from the control group and fatty liver group. Original magnification: 20× (**C**,**D**) Liver sections were stained with Oil Red O and Hematoxylin stain for nuclei. Original magnification: 20×.

3.4. *Effects of Hepatic Steatosis on Enzyme Activities of SIRT1, SREBP-1c, PGC-1α and Redox Index*

The effects of hepatic steatosis on hepatic SIRT1, sterol regulatory element binding transcription factor 1 (SREBP-1c), peroxisome proliferator-activated receptor γ coactivator-1 alpha (PGC-1α), oxidation and antioxidation activities in the liver are shown in Figure 3. The SIRT1, catalase (CAT) and glutathione peroxidase (GSH-Px) activities of the fatty liver group were significantly reduced ($p < 0.05$). The enzyme activities of Cu/Zn superoxide dismutase (SOD), Mn SOD and the content of glutathione (GSH) in the fatty liver group were significantly lower than those in the control group ($p < 0.01$). The GSSG (Figure 3I) contents in the fatty liver were higher than those in the control group liver ($p < 0.01$). The cow livers in the fatty liver group generated more reactive oxygen species (ROS) compared to the cow livers in the control group ($p < 0.01$, Figure 3J). Besides, mRNA abundance of CAT, GSH-Px and Mn SOD have a greater decrease ($p < 0.01$, Table 5) compared with the fatty liver group.

Table 5. Genes mRNA abundance for antioxidation activity of control and fatty liver cows (means ± SEM).

Genes	Control (n = 6)	Fatty Liver (n = 6)
Cu/Zn SOD	1.00 ± 0.12	0.31 ± 0.09
CAT	1.00 ± 0.06	0.21 ± 0.15 **
GSH-Px	1.00 ± 0.20	0.09 ± 0.06 **
Mn SOD	1.00 ± 0.13	0.06 ± 0.03 **

Note: ** represent $p < 0.01$.

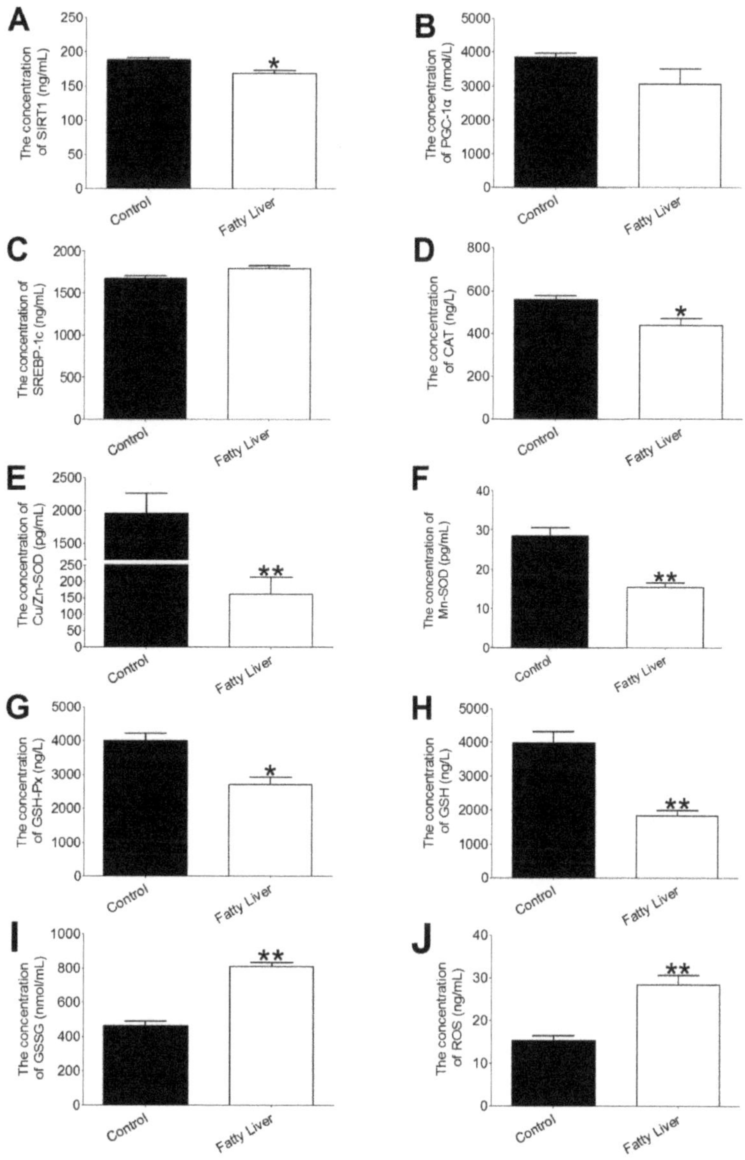

Figure 3. Effect of hepatic steatosis on the activities of hepatic sirtuin 1 (SIRT1), sterol regulatory element binding proteins (SREBP-1c), peroxisome proliferator-activated receptor γ coactivator-1 alpha (PGC-1α), catalase (CAT), copper-and zinc-containing superoxide dismutase (Cu/Zn SOD), manganese superoxide dismutase (Mn-SOD), glutathione peroxidase (GSH-Px) and the concentrations of glutathione (GSH), reduce glutathione disulfide (GSSG) and reactive oxygen species (ROS). Effect of lipid deposition on SIRT1, SREBP-1c and PGC-1α concentrations in control (n = 6) and fatty liver (n = 6) group dairy cows (**A–C**); Effect of lipid deposition on antioxidant index concentrations in control (n = 6) and fatty liver (n = 6) group dairy cows (**D–H**); Effect of lipid deposition on oxidative indexes concentrations in dairy cows (**H–J**). The data presented are the mean ± SEM. * $p < 0.05$; ** $p < 0.01$.

3.5. *Effect of Hepatic Steatosis on the Protein and mRNA Abundance of Fatty Acid Oxidation Proteins*

To explore the effect of hepatic steatosis on hepatic lipid metabolism, we measured the protein and mRNA abundance of different enzymes involved in lipid acid oxidation process. As shown in Figure 4A–C, compared with the control group, a greater downregulation of SIRT1 took place in liver tissues of the fatty liver group ($p < 0.01$). Additionally, hepatic steatosis (fatty liver group) led to a lower protein abundance of SIRT1 related transcription factors PGC-1α and peroxisome proliferator-activated receptor alpha (PPARα) ($p < 0.05$ or $p < 0.01$). Subsequently, the transcription factors target genes retinoid X receptor alpha (RXRα), acyl coenzyme A oxidase (ACO), carnitine palmitoyltransferase 1 (CPT1), carnitine palmitoyltransferase 2 (CPT2), nuclear respiratory factor 1 (Nrf1) and transcription factor A (TFAM) decreased in the fatty liver group ($p < 0.05$ or $p < 0.01$).

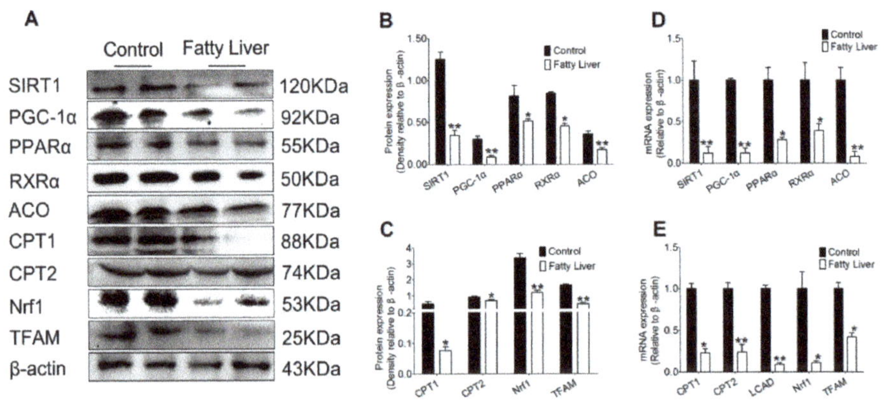

Figure 4. Lipid deposition hinders the fatty acid oxidation pathway in cow liver. (**A–C**) Levels of the protein levels of SIRT1, PGC-1α, peroxisome proliferator-activated receptor alpha (PPARα), retinoid X receptor alpha (RXRα), acyl coenzyme A oxidase (ACO), carnitine palmitoyltransferase 1 (CPT1), carnitine palmitoyltransferase 2 (CPT2), nuclear respiratory factor 1 (Nrf1) and transcription factor A (TFAM) in control (n = 6) and fatty liver (n = 6) group dairy cows were determined by western blotting. (**D,E**) Total RNA was extracted from liver samples and the expressions of genes involved in the fatty acid oxidation pathway in control (n = 6) and fatty liver (n = 6) group dairy cows were determined by real-time RT-PCR. The data presented are the mean ± SEM. * $p < 0.05$; ** $p < 0.01$.

Meanwhile, mRNA abundance of fatty acid oxidation proteins, including SIRT1, PGC-1α, PPARα, RXRα, ACO, CPT1, CPT2, long chain acyl-CoA dehydrogenase (LCAD), Nrf1 and TFAM were observed. As shown in Figure 4D, mRNA abundance of SIRT1 and its downstream transcription factors PGC-1αand PPARα decreased in the liver tissue of the fatty liver group compared with the control group ($p < 0.05$ or $p < 0.01$). Furthermore, compared with the control group, hepatic steatosis (fatty liver group) led to decreased mRNA abundance of RXRα, ACO, CPT1, CPT2, LCAD, Nrf1 and TFAM ($p < 0.05$ or $p < 0.01$).

3.6. *Hepatic Steatosis Enhances Protein and mRNA Abundance of Hepatic Lipogenesis*

As shown in Figure 5A,B, compared with the control group, hepatic steatosis (fatty liver group) increased protein abundance of SIRT1 related transcription factor SREBP-1c ($p < 0.01$). Moreover, the transcription factors target genes including acyl-CoA synthetase long chain family member 1 (ACSL1) and fatty acid synthetase (FAS) increased in liver tissue of fatty liver group ($p < 0.01$). It is noteworthy that, protein abundance of acetyl-CoA carboxylase alpha (ACCα) being greater increase compared with the control group, hepatic steatosis (fatty liver group) led to lower p-ACCα and p-ACCα/ACCα.

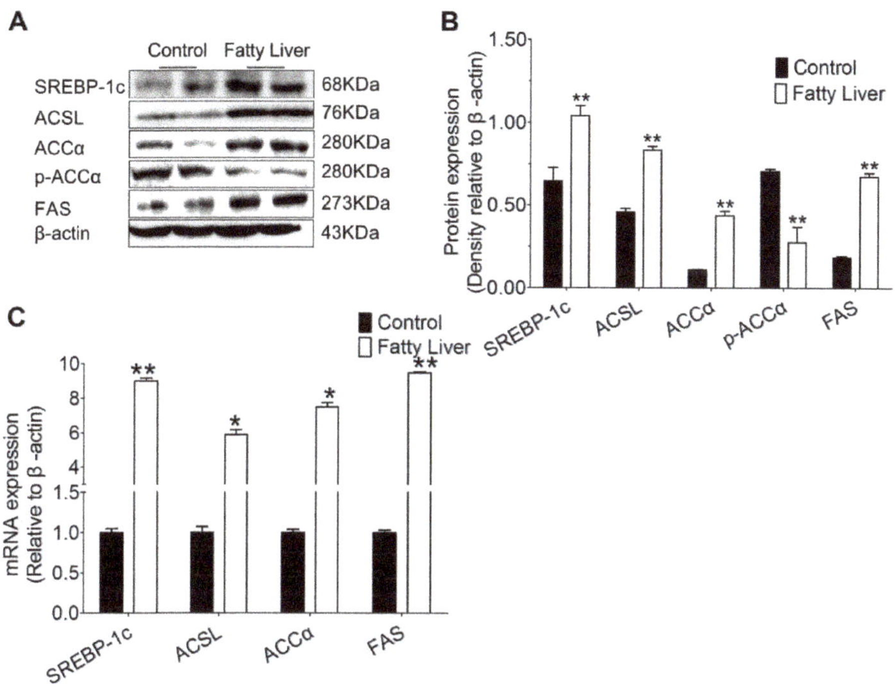

Figure 5. Lipid deposition enhances the hepatic lipogenesis pathway in the cow liver. (**A**,**B**) Protein levels of SREBP-1c, acyl-CoA synthetase long chain family member 1 (ACSL1), acetyl-CoA carboxylase alpha (ACCα), p-ACCα and fatty acid synthetase (FAS) in control (n = 6) and fatty liver (n = 6) group dairy cows were determined by western blotting. (**C**) Total RNA was extracted from liver samples and the expressions of the genes involved in free fatty acid (FFA) and triglycerides (TG) biosynthesis in control (n = 6) and fatty liver (n = 6) group dairy cows were determined by real-time RT-PCR. The data presented are the mean ± SEM. * $p < 0.05$; ** $p < 0.01$.

Compared with the control group, mRNA abundance of SREBP-1c and its target genes ACSL and FAS had a greater increase in liver tissue of the fatty liver group ($p < 0.01$, Figure 5C). Hepatic steatosis (fatty liver group) also led to increased mRNA abundance of ACCα compared with control.

3.7. Effect of Hepatic Steatosis on Hepatic Lipid Transport Related Proteins

We also found that the protein abundance of hepatic lipid transport including apolipoprotein E (ApoE) and low density lipoprotein receptor (LDLR) in the fatty liver group were increased, compared with the control group ($p < 0.05$ or $p < 0.01$, Figure 6A,B). Liver tissue in the fatty liver group had greater mRNA abundance of liver fatty acid binding proteins (L-FABP), apolipoprotein E (ApoE), low density lipoprotein receptor (LDLR) and microsomal triglyceride transfer protein (MTTP) compared with the control group ($p < 0.01$, Figure 6C). However, mRNA abundance of apolipoprotein B 100 (ApoB100) decreased in the liver tissue of the fatty liver group.

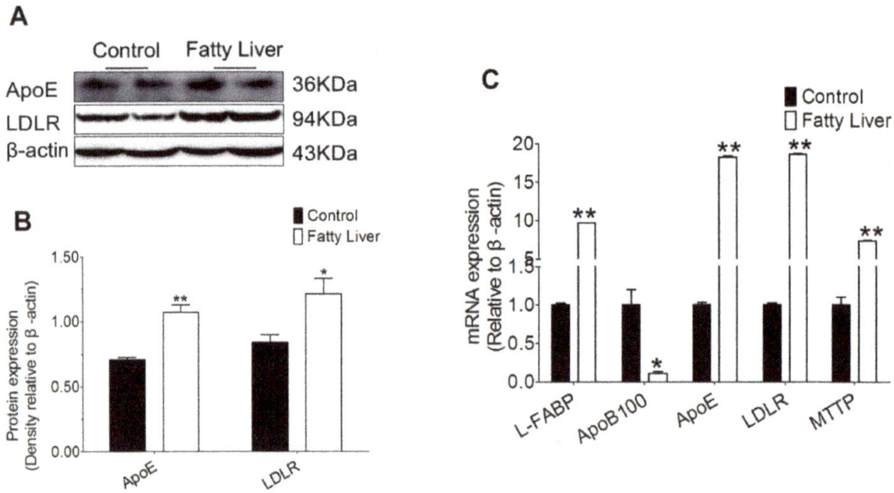

Figure 6. Lipid deposition promotes the lipid transport pathway in the cow liver. (**A**,**B**) Levels of the proteins apolipoprotein E (ApoE) and low density lipoprotein receptor (LDLR) in control (n = 6) and fatty liver (n = 6) group dairy cows were determined by western blotting. (**C**) Total RNA was extracted from liver samples and the expressions of the genes involved in the lipid transport pathway in control (n = 6) and fatty liver (n = 6) group dairy cows were determined by real-time RT-PCR. The data presented are the mean ± SEM. * $p < 0.05$; ** $p < 0.01$.

4. Discussion

The present study provides the basis for further animal and clinical studies to validate the mechanism, namely SIRT1, as a nicotinamide adenine dinucleotide (NAD^+, NADH)-dependent class III protein deacetylase, which is a key regulator of energy homeostasis in response to nutrient availability. Negative energy balance is a normal incidence in dairy cows during the transition from late gestation to early lactation [27]. It occurs because the energy demands for early milk production cannot be completely met by feed intake [3], during which most infectious and metabolic diseases are likely to occur, such as hepatic steatosis [1,4]. Our study showed that fatty liver significantly decreased DMI, milk yield, the milk protein and MUN, and significantly increased SCC, which indicates that fatty liver decreased the milk production and negatively affect milk quality. Furthermore, the protein, as well as the levels of Chol and TG were higher in the livers of fatty liver dairy cows. The liver is of immediate importance in maintaining health and plays a key role in the development of impaired metabolic regulation [28]. Previous studies have highlighted the importance of SIRT1 in cell metabolism, but mainly focused on the upstream regulation of SIRT1, and did not explore the mechanisms regarding SIRT1, lipid metabolism and oxidative stress [29–31]. In the present study, our results showed that hepatic steatosis decreased SIRT1 activity that facilitated hepatic fatty acid synthesis and inhibited fatty acid oxidation and lipid transport.

While SIRT1 is mainly localized in the nucleus, it is also present in the cytosol [30]. It deacetylates a variety of protein targets [32]. Hepatic SIRT1 deficiency in mice impairs lipid metabolism and results in hepatic steatosis [31,33]. Treatment with resveratrol, a SIRT1 activator, ameliorates fatty liver with a reduction in the expression of lipogenic enzymes in mice exhibiting obesity and insulin resistance [34–36]. Numerous studies have demonstrated that SIRT1 can act as a signaling molecule that promotes lipid metabolism by affecting the expressions of the genes controlling fatty acid synthesis, fatty acid oxidation and lipid transport [37]. It was reported that fatty livers facilitate the development of hepatic steatosis by upregulating primarily the lipogenic pathway genes via the SREBP-1c signaling pathway and down-regulating the lipid oxidizing genes predominantly via the SIRT1 and PGC-1α signaling pathways, which is in agreement with our results. Concomitant with the occurrence of hepatic steatosis, oxidative stress is an established risk factor for the development of lipid metabolism disorders at least partly through the SIRT1-SREBP-1c/ PGC-1α signaling pathway.

Previous studies suggest that excess lipid accumulation is associated with fatty liver through oxidative stress [38,39]. Also, it is found that oxidative stress participates in liver disease both in humans and animal models [40,41]. Oxidative stress is caused by an imbalance between the generation of ROS and the defense capabilities of antioxidants [42]. Previous studies on dairy cows have also demonstrated that elevated NEFA during the periparturient period largely activates oxidative stress [43]. Further results from in vivo study indicate that the primary approaches to prevent the occurrence of fatty liver disease in cows are counteracted oxidative damage [44]. However, most studies focus on the blood oxidative and antioxidative levels to assess the oxidative stress status of dairy cows. Fortunately, the oxidative status of ketotic dairy cows had been evaluated by liver biopsies [12]. As shown in the present study, the activities of hepatic CAT and GSH-Px and the GSH content had a greater reduction in dairy cows with fatty livers. However, the hepatic GSSG and ROS contents were significantly increased in dairy cows with fatty livers compared with the control group. Herein, our results indicated that dairy cows with fatty livers showed redox imbalance, which was in agreement with a previous report.

Fatty liver disease mainly occurs when fat accumulation is caused by impairing fatty acid oxidation along with increased lipid synthesis [44]. PPARαregulates the lipid oxidation genes expression involved in lipid oxidation, including ACO, CTP1 and CPT2 [45,46]. ACO has been considered as the rate-limiting enzyme for fatty acid oxidation [47]. CTP1 and CPT2 are thought to be the key enzymes in the process of transferring fatty acids into the mitochondria for β-oxidation [45]. In bovine hepatocytes treated 0.15, 0.30 and 0.45 mM non-esterified fatty acids (NEFA) induced the increased PPARα, ACO, CPT1 and CPT2 relative mRNA expression [48]. However, in another vivo study, high-fat diet induced the decreased expression of ACO and CPT2, but had a slight decrease in the CPT2 expression in mice [49]. In the present study, we demonstrated that fatty liver led to a greater reduction of PPARα, ACO, CPT1 and CPT2 protein and mRNA abundance. Our data indicate that hepatic steatosis impaired fatty acid oxidation. PGC-1α plays a crucial role in energy metabolism regulation and stimulates mitochondrial biogenesis via regulating NRF1 and TFAM [50,51]. RXRα was recognized as the receptor of PGC-1α [52]. LCAD, a mitochondrial enzyme, is involved in the branched chain and unsaturated fatty acids oxidation. In this study, the protein and mRNA abundance of hepatic PGC-1α, NRF1, TFAM and RXRα was decreased in fatty liver dairy cows compared with the control group. However, it was reported that the protein levels of PGC-1α decreased in mice with fatty liver followed by reduction of NRF1 and TFAM, the mRNA abundance of PGC-1α increased [53]. The discrepancies between the previous study and our results attribute to different animal species. The mRNA abundance of LCAD had a greater reduction in the dairy cows with fatty livers. These results show that it is highly likely that PGC-1α, NRF1, TFAM and LCAD are associated with fatty liver disease in dairy cows.

SREBP-1c is considered as an important activator of lipid synthesis by regulating lipogenic genes ACSL, ACCα and FAS [54]. ACSL is a key enzyme and involved in the first step of fatty acid metabolism [55]. ACCα is thought to synthesis long chain fatty acids [56]. FAS, a key lipogenic enzyme, catalyzes the long chain fatty acid synthesis [57]. Moreover, the protein and mRNA abundance of SREBP-1c and its downstream gene ACSL, ACCα and FAS were significantly increased, but the phosphorylation of ACCα decreased. In the present study, the mRNA abundance of L-FABP, a lipid transport protein, markedly increased in the dairy cows with fatty liver. ApoE, ApoB100 and MTT are used to the lipid synthesis and VLDL assembly [58]. Otherwise, ApoE and ApoB100 play a key role in cholesterol transport by binding to LDLR [59]. In this study, the protein and mRNA abundance of ApoE and LDLR and the mRNA abundance of MTP and ApoB100 significantly increased in the dairy cows with fatty liver. Our results demonstrated lipid transport was affected after hepatic steatosis happened to dairy cows. It reports that dysregulation of lipid metabolic pathways results in the development of hepatic steatosis and contributes to the development of chronic hepatic inflammation, insulin resistance and liver damage [60]. The aforementioned studies showing that hepatic steatosis decreased SIRT1 activity suggest that the SIRT1-SREBP-1c/PGC-1α signaling pathway could influence oxidative stress. Nevertheless, whether lipid deposition affects the oxidative stress through the SIRT1-SREBP-1c/PGC-1α signaling pathway remains to be investigated in the future study.

Based on these results it can be concluded that SIRT1/PGC-1α/SREBP-1c and redox are involved in fatty liver dairy cows by promoting hepatic fatty acid synthesis, impairing fatty acid β-oxidation and reducing lipid transport. Limitations of the present study is the relatively small sample size. However, this study provides a foundation for future investigations of lipid metabolism disorder disease during the transition period.

5. Conclusions

These results demonstrate that decreased SIRT1 associated with hepatic steatosis promotes hepatic fatty acid synthesis, inhibits fatty acid β-oxidation and reduces lipid transport. Hence, SIRT1 may represent a novel therapeutic target for the treatment of fatty liver disease in dairy cows.

Author Contributions: Performed experiments and drafted the manuscript by Y.L., S.Z. and H.D.; performed experiments and analyzed the data Y.L., S.Z., H.D., N.H., Y.H., and S.F.; contributed to the experimental design and manuscript J.T., J.C., J.L., X.W., and J.W.; conceived the idea, designed the experiment, and finalized the manuscript, X.W., and J.W. All authors have read and agreed to the published version of the manuscript.

Funding: This work was supported by the National Natural Science Foundation of China (Nos. 31873029 and 31502136), the Anhui province Natural Science Foundation of China (1608085QC56), the collection, analysis of this study was supported by the Project of Modern Agricultural Industry and Technology System of Anhui Province (AHCYJSTX-07), Graduate Innovation Fund Project of Anhui Agricultural University (2017yjs-16) and Agricultural Industry Chief Expert Studio of Hefei City (dairy cows farming). The authors would like to thank the staff at the dairy cow farm in Anhui Province for their kind help in conducting the animal part of the present study. We thank the Biotechnology center of Anhui Agricultural University (Hefei, China) and the laboratory team of the Clinical Veterinary Medicine (Anhui Agricultural University, China). Finally, the authors would like to thank Jihong Dong for her assistance in analyzing the real-time PCR and western blot data presented in this paper.

Conflicts of Interest: The authors have no conflicts of interest to declare.

References

1. Havlin, J.M.; Robinson, P.H.; Garrett, J.E. Niacin feeding to fresh dairy cows: Immediate effects on health and milk production. *Anim. Prod. Sci.* **2017**, *57*, 1069–1078. [CrossRef]
2. Li, Y.; Ding, H.; Wang, X.; Liu, L.; Huang, D.; Zhang, R.; Guo, L.; Wang, Z.; Li, X.; Liu, G. High levels of acetoacetate and glucose increase expression of cytokines in bovine hepatocytes, through activation of the NF-κB signalling pathway. *J. Dairy Res.* **2016**, *83*, 51–57. [CrossRef] [PubMed]
3. McArt, J.A.A.; Nydam, D.V.; Oetzel, G.R.; Overton, T.R.; Ospina, P.A. Elevated non-esterified fatty acids and β-hydroxybutyrate and their association with transition dairy cow performance. *Vet. J.* **2013**, *198*, 560–570. [CrossRef] [PubMed]
4. Birkenfeld, A.L.; Shulman, G.I. Nonalcoholic fatty liver disease, hepatic insulin resistance, and type 2 diabetes. *Hepatology* **2014**, *59*, 713–723. [CrossRef] [PubMed]
5. Sun, X.; Yuan, X.; Chen, L.; Wang, T.; Wang, Z.; Sun, G.; Li, X.; Li, X.; Liu, G. Histamine induces bovine rumen epithelial cell inflammatory response via NF-κB pathway. *Cell. Physiol. Biochem.* **2017**, *42*, 1109–1119. [CrossRef]
6. Chung, S.W.; Yao, H.W.; Caito, S.; Hwang, J.-W.; Arunachalam, G.; Rahman, I. Regulation of SIRT1 in cellular functions: Role of polyphenols. *Arc. Biochem. Biophys.* **2010**, *501*, 79–90. [CrossRef]
7. Cheng, Y.; Takeuchi, H.; Sonobe, Y.; Jin, S.; Wang, Y.; Horiuchi, H.; Parajuli, B.; Kawanokuchi, J.; Mizuno, T.; Suzumura, A. Sirtuin 1 attenuates oxidative stress via upregulation of superoxide dismutase 2 and catalase in astrocytes. *J. Neuroimmunol.* **2014**, *269*, 38–43. [CrossRef]
8. Hasegawa, K.; Wakino, S.; Yoshioka, K.; Tatematsu, S.; Hara, Y.; Minakuchi, H.; Washida, N.; Tokuyama, H.; Hayashi, K.; Itoh, H. Sirt1 protects against oxidative stress-induced renal tubular cell apoptosis by the bidirectional regulation of catalase expression. *Biochem. Bioph. Res. Co.* **2008**, *372*, 51–56. [CrossRef]

9. Chan, S.-H.; Hung, C.-H.; Shih, J.-Y.; Chu, P.-M.; Cheng, Y.-H.; Lin, H.-C.; Tsai, K.-L. SIRT1 inhibition causes oxidative stress and inflammation in patients with coronary artery disease. *Redox Bio.* **2017**, *13*, 301–309. [CrossRef]
10. De, D.; Halder, D.; Shin, I.; Kim, K.K. Small molecule-induced cellular conversion. *Chem. Soc. Rev.* **2017**, *46*, 6241–6254. [CrossRef]
11. Wu, D.; Hu, Q.; Liu, X.; Pan, L.; Xiong, Q.; Zhu, Y.Z. Hydrogen sulfide protects against apoptosis under oxidative stress through SIRT1 pathway in H9c2 cardiomyocytes. *Nitric Oxide* **2015**, *46*, 204–212. [CrossRef] [PubMed]
12. Du, X.; Shi, Z.; Peng, Z.; Zhao, C.; Zhang, Y.; Wang, Z.; Li, X.; Liu, G.; Li, X. Acetoacetate induces hepatocytes apoptosis by the ROS-mediated MAPKs pathway in ketotic cows. *J. Cell. Physiol.* **2017**, *232*, 3296–3308. [CrossRef] [PubMed]
13. Li, Y.; Ding, H.Y.; Wang, X.C.; Feng, S.B.; Li, X.B.; Wang, Z.; Liu, G.W.; Li, X.W. An association between the level of oxidative stress and the concentrations of NEFA and BHBA in the plasma of ketotic dairy cows. *J. Anim. Physiol. An. N.* **2017**, *100*, 844–851. [CrossRef] [PubMed]
14. Thakran, S.; Sharma, P.; Attia, R.R.; Hori, R.T.; Deng, X.; Elam, M.B.; Park, E.A. Role of sirtuin 1 in the regulation of hepatic gene expression by thyroid hormone. *J. Biol. Chem.* **2013**, *288*, 807–818. [CrossRef]
15. Li, H.; Xu, M.; Lee, J.; He, C.; Xie, Z. Leucine supplementation increases SIRT1 expression and prevents mitochondrial dysfunction and metabolic disorders in high-fat diet-induced obese mice. *Am. J. Physiol. Endoc. M.* **2012**, *303*, E1234–E1244. [CrossRef]
16. Jeon, B.T.; Kim, K.E.; Shin, H.J.; Yi, C.-o.; Hah, Y.-S.; Kim, W.-H.; Lee, S.-I.; Roh, G.S. Myeloid-specific deletion of SIRT1 increases hepatic steatosis and hypothalamic inflammation in mice fed a high-fat diet. *Metab. Brain Dis.* **2014**, *29*, 635–643. [CrossRef]
17. Rao, Y.; Yu, H.; Gao, L.; Lu, Y.; Xu, Z.; Liu, H.; Gu, L.; Ye, J.; Huang, Z. Natural alkaloid bouchardatine ameliorates metabolic disorders in high-fat diet-fed mice by stimulating the sirtuin 1/liver kinase B-1/AMPK axis. *Brit. J. Pharmacol.* **2017**, *174*, 2457–2470. [CrossRef]
18. Deng, X.; Chen, L.; Li, N. The expression of SIRT1 in nonalcoholic fatty liver disease induced by high-fat diet in rats. *Liver Int.* **2007**, *27*, 708–715. [CrossRef]
19. Xu, F.; Gao, Z.; Zhang, J.; Rivera, C.A.; Yin, J.; Weng, J.; Ye, J. Lack of SIRT1 (Mammalian Sirtuin 1) activity leads to liver steatosis in the SIRT1+/- mice: A role of lipid mobilization and inflammation. *Endocrinology* **2010**, *151*, 2504–2514. [CrossRef]
20. Li, X.; Zhang, S.; Blander, G.; Jeanette, G.T.; Krieger, M.; Guarente, L. SIRT1 deacetylates and positively regulates the nuclear receptor LXR. *Mol. Cell* **2007**, *28*, 91–106. [CrossRef]
21. Pfluger, P.T.; Herranz, D.; Velasco-Miguel, S.; Serrano, M.; Tschöp, M.H. Sirt1 protects against high-fat diet-induced metabolic damage. *Proc. Natl Acad. Sci. USA* **2008**, *105*, 9793–9798. [CrossRef] [PubMed]
22. Zhu, Y.W.; Guan, Y.; Loor, J.J.; Sha, X.Y.; Coleman, D.N.; Zhang, C.; Du, X.L.; Shi, Z.; Li, X.B.; Wang, Z.; et al. Fatty acid-induced endoplasmic reticulum stress promoted lipid accumulation in calf hepatocytes, and endoplasmic reticulum stress existed in the liver of severe fatty liver cows. *J. Dairy Sci.* **2019**, *102*, 7359–7370. [CrossRef] [PubMed]
23. Dong, J.H.; Loor, J.J.; Zuo, R.K.; Chen, X.Y.; Liang, Y.S.; Wang, Y.Z.; Shu, X.; Sun, X.D.; Jia, H.D.; Liu, G.W.; et al. Low abundance of mitofusin 2 in dairy cows with moderate fatty liver is associated with alterations in hepatic lipid metabolism. *J. Dairy Sci.* **2019**, *102*, 7536–7547. [CrossRef] [PubMed]
24. Cheng, J.B.; Bu, D.P.; Wang, J.Q.; Sun, X.Z.; Pan, L.; Zhou, L.Y.; Liu, W. Effects of rumen-protected γ-aminobutyric acid on performance and nutrient digestibility in heat-stressed dairy cows. *J. Dairy Sci.* **2014**, *97*, 5599–5607. [CrossRef] [PubMed]
25. Cheng, J.; Liu, C.; Hu, K.; Greenberg, A.; Wu, D.; Ausman, L.M.; McBurney, M.W.; Wang, X.-D. Ablation of systemic SIRT1 activity promotes nonalcoholic fatty liver disease by affecting liver-mesenteric adipose tissue fatty acid mobilization. *BBA-Mol. Basis Dis.* **2017**, *1863*, 2783–2790. [CrossRef] [PubMed]
26. Pfaffl, M.W. A new mathematical model for relative quantification in real-time RT-PCR. *Nucleic Acids Res.* **2001**, *29*, e45. [CrossRef]
27. Ospina, P.A.; McArt, J.A.; Overton, T.R.; Stokol, T.; Nydam, D.V. Using nonesterified fatty acids and β-hydroxybutyrate concentrations during the transition period for herd-level monitoring of increased risk of disease and decreased reproductive and milking performance. *Vet. Clin. Food A.* **2013**, *29*, 387–412. [CrossRef]

28. Ntambi, J.M. *Hepatic de Novo Lipogenesis and Regulation of Metabolism*; Springer: New York, NY, USA, 2015.
29. Cantó, C.; Gerhart-Hines, Z.; Feige, J.N.; Lagouge, M.; Noriega, L.; Milne, J.C.; Elliott, P.J.; Puigserver, P.; Auwerx, J. AMPK regulates energy expenditure by modulating NAD+ metabolism and SIRT1 activity. *Nature* **2009**, *458*, 1056–1060. [CrossRef]
30. Han, C.; Wan, H.; Ma, S.; Liu, D.; He, F.; Wang, J.; Pan, Z.; Liu, H.; Li, L.; He, H. RETRACTED: Role of mammalian sirtuin 1 (SIRT1) in lipids metabolism and cell proliferation of goose primary hepatocytes. *Mol. Cell. Endocrin.* **2014**, *382*, 282–291. [CrossRef]
31. Purushotham, A.; Schug, T.T.; Xu, Q.; Surapureddi, S.; Guo, X.; Li, X. Hepatocyte-specific deletion of SIRT1 alters fatty acid metabolism and results in hepatic steatosis and inflammation. *Cell Metab.* **2009**, *9*, 327–338. [CrossRef]
32. Yu, J.; Auwerx, J. Protein deacetylation by SIRT1: An emerging key post-translational modification in metabolic regulation. *Pharmacol. Res.* **2010**, *62*, 35–41. [CrossRef] [PubMed]
33. Wang, R.-H.; Kim, H.-S.; Xiao, C.; Xu, X.; Gavrilova, O.; Deng, C.-X. Hepatic Sirt1 deficiency in mice impairs mTorc2/Akt signaling and results in hyperglycemia, oxidative damage, and insulin resistance. *J. Clin. Invest.* **2011**, *121*, 4477–4490. [CrossRef] [PubMed]
34. Ajmo, J.M.; Liang, X.; Rogers, C.Q.; Pennock, B.; You, M. Resveratrol alleviates alcoholic fatty liver in mice. *Am. J. Physiol. Gastr. L.* **2008**, *295*, G833–G842. [CrossRef] [PubMed]
35. Baur, J.A.; Pearson, K.J.; Price, N.L.; Jamieson, H.A.; Lerin, C.; Kalra, A.; Prabhu, V.V.; Allard, J.S.; Lopez-Lluch, G.; Lewis, K. Resveratrol improves health and survival of mice on a high-calorie diet. *Nature* **2006**, *444*, 337–342. [CrossRef] [PubMed]
36. Shang, J.; Chen, L.-l.; Xiao, F.-x.; Sun, H.; Ding, H.-c.; Xiao, H. Resveratrol improves non-alcoholic fatty liver disease by activating AMP-activated protein kinase. *Acta Pharmacol. Sin.* **2008**, *29*, 698–706. [CrossRef]
37. Rodgers, J.T.; Puigserver, P. Fasting-dependent glucose and lipid metabolic response through hepatic sirtuin 1. *Proc. Natl. Acad. Sci. USA* **2007**, *104*, 12861–12866. [CrossRef]
38. Satapati, S.; Kucejova, B.; Duarte, J.A.G.; Fletcher, J.A.; Reynolds, L.; Sunny, N.E.; He, T.; Nair, L.A.; Livingston, K.; Fu, X. Mitochondrial metabolism mediates oxidative stress and inflammation in fatty liver. *J. Clin. Invest.* **2015**, *125*, 4447–4462. [CrossRef]
39. Du, X.L.; Chen, L.; Huang, D.; Peng, Z.C.; Zhao, C.X.; Zhang, Y.M.; Zhu, Y.W.; Wang, Z.; Li, X.W.; Liu, G.W. Elevated apoptosis in the liver of dairy cows with ketosis. *Cell. Physiol. Biochem.* **2017**, *43*, 568–578. [CrossRef]
40. Shaw, J.E.; Sicree, R.A.; Zimmet, P.Z. Global estimates of the prevalence of diabetes for 2010 and 2030. *Diabetes Res. Clin. Pr.* **2010**, *87*, 4–14. [CrossRef]
41. Oliveira, C.P.M.S.; da Costa Gayotto, L.C.; Tatai, C.; Della Bina, B.I.; Janiszewski, M.; Lima, E.S.; Abdalla, D.S.P.; Lopasso, F.P.; Laurindo, F.R.M.; Laudanna, A.A. Oxidative stress in the pathogenesis of nonalcoholic fatty liver disease, in rats fed with a choline-deficient diet. *J. Cell. Mol. Med.* **2002**, *6*, 399–406. [CrossRef]
42. Lee, J.W.; Davis, J.M. Future applications of antioxidants in premature infants. *Curr. Opin. Pediatr.* **2011**, *23*, 161. [CrossRef] [PubMed]
43. Bernabucci, U.; Ronchi, B.; Lacetera, N.; Nardone, A. Influence of body condition score on relationships between metabolic status and oxidative stress in periparturient dairy cows. *J. Dairy Sci.* **2005**, *88*, 2017–2026. [CrossRef]
44. Bobe, G.; Young, J.W.; Beitz, D.C. Invited review: Pathology, etiology, prevention, and treatment of fatty liver in dairy cows. *J. Dairy Sci.* **2004**, *87*, 3105–3124. [CrossRef]
45. Qiu, L.X.; Wu, X.C.; Chau, J.F.L.; Szeto, I.Y.Y.; Tam, W.Y.; Guo, Z.S.; Chung, S.K.; Oates, P.J.; Chung, S.S.M.; Yang, J.Y. Aldose reductase regulates hepatic peroxisome proliferator-activated receptor α phosphorylation and activity to impact lipid homeostasis. *J. Biol. Chem.* **2008**, *283*, 17175–17183. [CrossRef] [PubMed]
46. Sugden, M.C.; Bulmer, K.; Gibbons, G.F.; Knight, B.L.; Holness, M.J. Peroxisome-proliferator-activated receptor-α (PPARα) deficiency leads to dysregulation of hepatic lipid and carbohydrate metabolism by fatty acids and insulin. *Biochem. J.* **2002**, *364*, 361–368. [CrossRef]
47. Eaton, S. Control of mitochondrial β-oxidation flux. *Prog. Lipid Res.* **2002**, *41*, 197–239. [CrossRef]
48. Li, X.W.; Li, X.B.; Chen, H.; Lei, L.; Liu, J.X.; Guan, Y.; Liu, Z.X.; Zhang, L.; Yang, W.T.; Zhao, C.X.; et al. Non-esterified fatty acids activate the AMP-activated protein kinase signaling pathway to regulate lipid metabolism in bovine hepatocytes. *Cell Biochem. Biophy.* **2013**, *67*, 1157–1169. [CrossRef]

49. Noh, H.; Lee, H.; Kim, E.; Mu, L.; Rhee, Y.K.; Cho, C.-W.; Chung, J. Inhibitory effect of a Cirsium setidens extract on hepatic fat accumulation in mice fed a high-fat diet via the induction of fatty acid β-oxidation. *Biosci. Biotech. Bioch.* **2013**, *77*, 1424–1429. [CrossRef]
50. Laubenthal, L.; Hoelker, M.; Frahm, J.; Dänicke, S.; Gerlach, K.; Südekum, K.H.; Sauerwein, H.; Häussler, S. Mitochondrial DNA copy number and biogenesis in different tissues of early-and late-lactating dairy cows. *J. Dairy Sci.* **2016**, *99*, 1571–1583. [CrossRef]
51. Laubenthal, L.; Ruda, L.; Sultana, N.; Winkler, J.; Rehage, J.; Meyer, U.; Dänicke, S.; Sauerwein, H.; Häussler, S. Effect of increasing body condition on oxidative stress and mitochondrial biogenesis in subcutaneous adipose tissue depot of nonlactating dairy cows. *J. Dairy Sci.* **2017**, *100*, 4976–4986. [CrossRef]
52. Schreiber, S.N.; Knutti, D.; Brogli, K.; Uhlmann, T.; Kralli, A. The transcriptional coactivator PGC-1 regulates the expression and activity of the orphan nuclear receptor estrogen-related receptor α (ERRα). *J. Biol. Chem.* **2003**, *278*, 9013–9018. [CrossRef] [PubMed]
53. Aharoni-Simon, M.; Hann-Obercyger, M.; Pen, S.; Madar, Z.; Tirosh, O. Fatty liver is associated with impaired activity of PPAR γ-coactivator 1 α (PGC1 α) and mitochondrial biogenesis in mice. *Lab. Invest.* **2011**, *91*, 1018–1028. [CrossRef] [PubMed]
54. Porstmann, T.; Griffiths, B.; Chung, Y.-L.; Delpuech, O.; Griffiths, J.R.; Downward, J.; Schulze, A. PKB/Akt induces transcription of enzymes involved in cholesterol and fatty acid biosynthesis via activation of SREBP. *Oncogene* **2005**, *24*, 6465–6481. [CrossRef] [PubMed]
55. Coleman, R.A.; Lewin, T.M.; Van Horn, C.G.; Gonzalez-Baró, M.R. Do long-chain acyl-CoA synthetases regulate fatty acid entry into synthetic versus degradative pathways? *J. Nutr.* **2002**, *132*, 2123–2126. [CrossRef] [PubMed]
56. Wakil, S.J.; Stoops, J.K.; Joshi, V.C. Fatty acid synthesis and its regulation. *Annu. Rev. Biochem.* **1983**, *52*, 537–579. [CrossRef] [PubMed]
57. Postic, C.; Girard, J. Contribution of de novo fatty acid synthesis to hepatic steatosis and insulin resistance: Lessons from genetically engineered mice. *J. Clin. Invest.* **2008**, *118*, 829–838. [CrossRef]
58. Greenow, K.; Pearce, N.J.; Ramji, D.P. The key role of apolipoprotein E in atherosclerosis. *J. Mol. Med.* **2005**, *83*, 329–342. [CrossRef]
59. Michaely, P.; Zhao, Z.Z.; Li, W.P.; Garuti, R.; Huang, L.J.; Hobbs, H.H.; Cohen, J.C. Identification of a VLDL-induced, FDNPVY-independent internalization mechanism for the LDLR. *EMBO J.* **2007**, *26*, 3273–3282. [CrossRef]
60. Fabbrini, E.; Magkos, F. Hepatic steatosis as a marker of metabolic dysfunction. *Nutrients* **2015**, *7*, 4995–5019. [CrossRef]

animals

MDPI

Article

Fatty Acid Profile and Enterolactone Content of Early and Commercial Milk of Dairy Cows Supplemented with Flaked Flaxseed during the Dry Period

Lucia Bailoni [1], Laura Da Dalt [1], Gianfranco Gabai [1], Elisa Giaretta [1,*], Nadia Guzzo [1] and Roberto Mantovani [2]

[1] Department of Comparative Biomedicine and Food Science (BCA), University of Padova, Viale dell'Università 16, 35020 Legnaro, Italy; lucia.bailoni@unipd.it (L.B.); laura.dadalt@unipd.it (L.D.D.); gianfranco.gabai@unipd.it (G.G.); nadia.guzzo@unipd.it (N.G.)

[2] Department of Agronomy, Food, Natural resources, Animals and Environment (DAFNAE), University of Padova, Viale dell'Università 16, 35020 Legnaro, Italy; roberto.mantovani@unipd.it

* Correspondence: elisa.giaretta@unipd.it; Tel.: +39-0498272585

Received: 20 November 2020; Accepted: 10 December 2020; Published: 15 December 2020

Simple Summary: Several studies have been published on n-3 fatty acids enrichment of food of animal origin by the use of specific supplements in animal feeding. Flaxseed (*Linum usitatissimum*) is known as an excellent source of unsaturated fatty acids, particularly of alpha-linolenic acid. Moreover, most of the plant lignans contained in flaxseeds are converted by the animal in mammalian lignans, such as enterodiol and enterolactone. Thus, milk with elevated enterolactone may be an efficient strategy to optimize the effects of lignans on human health. This study aimed to investigate the effect of flacked flaxseed supplementation on dairy cow during the pre-partum period. The increase of n-3 on milk obtained from cow fed the flaxseed diet was detected only in the first days of lactation. In addition, an increase of enterolactone on milk from flaxseed fed cows was observed only at 15th day after calving; a higher amount of enetrolactone was detected in plasma and milk from the 15th day after calving, independently from the treatment group. This study suggests that the duration of n-3 carry-over into the milk is dependent on the n-3 feeding source concentration. Finally, more studies are needed to investigate enterolactone metabolism in ruminants.

Abstract: Various supplementations in animal feeding have been investigate in order to enrich food of animal origin with n-3 fatty acids. Although the effects of flaxseeds inclusion in diets for lactating dairy have already been assessed, few studies have focused on this n-3 source supplementation during the transition period. The aim of this work was to evaluate the effects of flacked flaxseed (200 g/head/day; 2.13% DM) dietary treatment during the dry period on milk yield and quality in the 30 days after calving. In addition, the enterolactone content in plasma (before and after calving) and in milk of cows fed diets supplemented or not with flaxseed was considered. The study demonstrated that the carry-over effect on the milk profile of C18:2, C18:3 n-3, and C20:5 n-3 was significantly higher in flaxseed diet than in the control one at 4th day of lactation. A significant increase of enterolactone on milk from flaxseed fed cows was observed only at 15 sampling day. The quick modification in fatty acid (FA) profile of the milk in the first few days of lactation suggests that the carry over effect from pre-calving flaxseed feeding at this concentration was very short lasting.

Keywords: flaxseed; dairy cows; dry period; enterolactone; milk fatty acids

1. Introduction

In human nutrition, the beneficial effects of n-3 fatty acids (FA; i.e., a decreased risk of cardiovascular diseases, hypertension, and arthritis) are well known from years and dietary inclusion of these nutrients is recommended [1,2]. As a consequence, a large number of studies have been published on n-3 FA enrichment of food of animal origin (milk, meat, eggs, and fish) by the use of specific supplements in animal feeding [3–6]. Flaxseed (*Linum usitatissimum*) is known as an excellent source of unsaturated fatty acids (FA), particularly of alpha-linolenic acid (ALA, C18:3n-3), a member of n-3 FA [2]. A large number of studies have reported the results of flaxseeds inclusion in diets for lactating dairy cows finalized to enrich milk [7–9] and derivate [10] in n-3 FA. In addition, several studies have shown that the incorporation of dietary n-3 into ovarian structures and bull sperm influence the reproductive features in terms of follicular and embryo development, and bull semen quality [11]. It was also demonstrated that the supplementation of n-3 FA during the early postpartum period modifies the endocannabinoid system in the bovine endometrium [12]. Feeding source of n-3 FA can modulate the chemical composition and the functional proprieties of the immune cells, which could affect the immune response in several ways [11].

To our knowledge, most of the studies have analyzed how flaxseed supplementation during only the dry period could impact on subsequent reproductive performance, bloody metabolites, and milk yield and quality during the subsequent lactation [13–15]. However, few studies have focused on the effects of flaxseed supplementation during the transition period on milk composition and FA metabolism [16–18].

Several studies demonstrated the beneficial effects of lignans on human health [7]. Plant lignans contained in flaxseed (about 370 mg/100 dry matter; DM) have a lower bioavailability in the human intestine compared with the mammalian lignans synthetized in the rumen [19]. The secoisolariciresinol diglucoside (SDG), which represents over 95% of the total lignans in flaxseed, could be converted by the rumen microorganisms [20] in enterodiol and enterolactone (EL). Recent studies demonstrated that flaxseed dietary treatment could increment the amount of milk EL [7], which in this way, could be considered as a valid source of mammalian lignans.

Following this rationale, the objectives of the present experiment were i) to evaluate the effects on milk yield and quality in early lactation and commercial milk due to the supplementation of dairy cows with flacked flaxseed during the dry period and ii) to evaluate the EL content in plasma and milk of cows fed diets supplemented or not with flaxseed during the dry period.

2. Materials and Methods

2.1. Ethic Statement

All experimental procedures were carried out according to Italian law on animal care (Legislative Decree No. 26 of 14 March 2014) and approved by the ethical committee at the University of Padova (approval number 70/2019).

2.2. Animals and Feeding

The study was conducted in a commercial dairy farm located in Crespano del Grappa in the province of Treviso, northeast of Italy. Italian Friesian cows with a close date of drying off were divided in two homogeneous groups for parity (2.25 ± 1.72), days open (117 ± 66), and milk production level in the current lactation (9.858 ± 1.598 kg of mature equivalent milk). Couples of cows at the beginning of the dry off period were randomly assigned to receive a control (CTR) or an experimental flaxseed (FLAX) supplement diet by assigning them to different pens to allow the proper feeding supplementation. At the end of the trial, cows used in the experiment were 38 in the CTR and 35 in the FLAX group. During the dry period, the cows of the FLAX group were fed a diet supplemented with 200 g/head/day (i.e., 2.13% of DM) of flacked flaxseed (Cortal Extrasoy, Cittadella, Padova, Italy) and the CTR group received a mix of soybean and corn meals in order to obtain isoenergetic and isoproteic rations. The steaming-up treatment was managed in both groups following the customary farm procedures. After calving, lactating cows of both groups were fed ad libitum the same total mixed ration (TMR). Diets of dry and lactating cows were formulated in order to satisfy the nutritional requirements reported by the National Research Council—NRC [21]. During the trial, cows had free access to water and were fed once daily.

2.3. Data and Sample Collection

Samples of the single ingredients belonging to each diet (except the flaked flaxseed) were collected every 2 months, immediately transferred to the laboratory and, if fresh, subjected to pre-drying, before the analyses (overall 4 samplings were carried out during the experiment). The chemical composition of CTR and FLAX diets during the dry period and lactation diet after calving was calculated on the basis of the results of the chemical analyses of feed and the changes of the diet formulation over time. Diets were prepared and distributed as total mixed rations (TMR) using a mixed wagon (Table 1).

Table 1. Chemical composition (mean ad standard deviation, SD) of control (CTR) and flaxseed supplemented (FLAX) diets used during the trial expressed as % on DM basis.

Chemical Composition	Dry Period				Lactation Period	
	CTR [1]		FLAX [2]			
	Mean	SD	Mean	SD	Mean	SD
DM, %	49.0	8.47	49.4	8.49	54.3	4.70
Crude protein	12.2	0.76	12.4	0.73	16.7	0.30
Ether extract	2.64	0.64	3.10	0.68	4.80	0.60
Ash	6.82	1.08	6.75	1.06	5.86	0.71
NDF	49.1	6.56	48.5	6.47	34.0	4.10
ADF	28.1	4.30	27.7	4.22	18.9	2.91
Starch	12.2	2.85	12.0	2.79	24.2	2.30
MFU [3] no./kg DM	0.78	0.04	0.78	0.04	097	0.02

[1] CTR: control diet, [2] FLAX: experimental flaxseed supplemented diet, [3] MFU: milk forage unit; calculated using published value of feed ingredients (INRA. 1984).

Flacked flaxseed was collected at each change of the batch (for a total of 7 samples) to be analyzed for chemical composition and FA profile (Table 2).

Table 2. Descriptive statistics (mean ad standard deviation, SD) for chemical composition (% on DM basis) and fatty acid profile (% of total FA) of the flacked flaxseed used in the study (n = 7 samples).

Chemical Composition	Mean	SD
DM, %	93.2	1.69
Crude protein	22.3	1.37
Ether extract	44.1	2.05
Ash	3.25	0.47
NDF	13.4	0.75
ADF	7.13	0.23
MFU [1] no./kg DM	0.98	-
Fatty acid profile		
SFA	10.9	0.75
MUFA	16.8	2.80
PUFA	67.9	3.46
n-3	53.9	3.65
C18:3 n3	53.8	3.63
n-6	15.9	0.46
n-6/n-3	0.31	0.03

[1] MFU: milk forage unit; calculated using published value of feed ingredients (INRA. 1984).

Daily milk yield was recorded in the first 30 days after calving using an automatic milking system for a herringbone parlor coupled with an automatic recording software (ALPROTM by DeLaval©, Tumba, Sweden). Individual milk samples (500 mL/each) from the morning milking were collected from each cow at about 4, 15, and 30 days after calving to be analyzed for composition, FA profile, and the enterolactone content.

Individual milk samples (500 mL/each) from the morning milking were collected from each cow at 4.1 ± 1.5, 15.5 ± 1.6, and 30.9 ± 1.6 days after calving. Blood samples were collected from each animal at about 54.9 ± 6.7, 32.2 ± 10.6, 9.2 ± 2.5, and 3.1 ± 1.5 days before the expected calving and at 4.1 ± 1.5, 15.5 ± 1.6, and 30.9 ± 1.6 days after calving. The samples were obtained from the jugular vein in vacuum tubes containing K3-EDTA anticoagulant and without anticoagulant. Plasma and serum were obtained by centrifugation (1.500× *g* for 15 min at 4 °C) and an aliquot of each sample was frozen and stored at −20 °C for enterolactone content analysis.

2.4. Sample Analysis

Samples of feeds were analyzed for dry matter (DM: # 934.01; [22]), N (# 976.05; [22]), EE (# 920.29; [22]), and ash (# 942.05; [22]). Neutral detergent fiber (NDF), inclusive of residual ash, was determined with α-amylase using the Ankom220 Fiber Analyzer (Ankom Technology, Macedon, NY, USA). Acid detergent fiber (ADF), inclusive of residual ash, was determined sequentially after NDF determination [23]. Starch content was determined after hydrolysis to glucose [22] by liquid chromatography [24]. The flacked flaxseed fat was extracted by accelerated solvent extraction (ASE 200, Dionex Corp., Sunnyvale, CA, USA) using petroleum ether. The fatty acid profile of the flaxseed samples was analyzed by GC with flame-ionization detector (7890A GC system, Agilent Technologies, Milan, Italy) using two columns in series and equipped with a modulator (Agilent G3486 A CFT), an automatic sampler (Agilent 7693), and a specific machine software (Agilent Chem Station). This instrument was chosen because the double column allows separating and identifying each FA on a 2-dimensional basis [25]. The first column was a 75 m × 180 μm (internal diameter) × 0.14 μm film thickness column (23348U, Supelco, Bellefonte, PA, USA), the second was a 3.8 m × 250 μm (internal diameter) × 0.25 μm film thickness column (J&W 19091-L431, Agilent Technologies, Santa Clara, USA). Both columns used H2 as carrier gas at a flow rate of 0.22 mL/min.

The early and commercial milk samples were analyzed for fat, protein, casein, and lactose contents using the FIL-IDF procedure [26] with MilkoScan™ FT1 apparatus (Foss Electric, DK-3400, Hillerød, Denmark). Milk urea nitrogen was measured automatically by the conduct metric-enzymatic method (CL 10 micro analyzer, Eurochem, Roma, Italy). Somatic cell count was carried out through a FossomaticTM 5000 (Foss Electric, DK-3400, Hillerød, Denmark) according to the standard FIL-IDF148a [27], and transformed in logarithmic terms using the following equation: SCS = 3 + $\log_2$ (SCC/100.000).

Milk fat was extracted by accelerated solvent extraction (ASE 200, Dionex Corp., Sunnyvale, CA, USA) using petroleum ether:isopropanol (3:2 vol/vol), and the extracted fat was analyzed for FA composition using a two-dimensional gas-chromatography instrument (Agilent 7890A, Agilent Technologies, Milan, Italy) as described above.

Enterolactone concentration in plasma and milk samples was measured following hydrolysis and solvent extraction. Briefly, 200 μL of plasma and 500 μL milk were hydrolyzed respectively in 200 or 500 μL of acetate buffer (0.1 M. pH 5.0) containing 0.2 U/mL of β-glucuronidase and 2 U/mL of arylsulfatase (Sigma-Aldrich, S. Louis, MO, USA) as described by Stumpf et al., 2000 [28]. Samples were incubated overnight for plasma and 1.5 h for milk at 37 °C. Only the milk samples were washed with 3 mL of hexane to remove lipids as described by [29]. Then, enterolactone was extracted twice with 2 mL of diethyl ether in both plasma and milk. The dry extracts were dissolved in 200 and 500 μL of assay buffer respectively (TR-FIA Enterolactone kit, Labmaster, Finland) and assayed in duplicate by a commercial DELFIA method (TR-FIA Enterolactone kit, Labmaster, Finland). For the final assay, extracts corresponding to 20 μL of plasma/milk were used. Fluorescence was read in a multilabel reader (Victor X4 2030, Perkin-Elmer Instruments, Norwalk, CT, USA).

2.5. Statistical Analysis

Daily milk production, milk quality data, and fatty acid profile of milk were analyzed using a hierarchical linear model for repeated measures implemented through the PROC MIXED of SAS [30], according to the following linear model:

$$y_{ijklm} = \mu + D_i + P_j + DP_{ij} + C_{k:ij} + S_l + DS_{il} + PS_{jl} + e_{ijklm}, \quad (1)$$

where μ is the overall mean; D_i is the fixed effect of diet (i = 2 levels: CTR and FLAX); P_j is the fixed effect of parity (j = 3 levels: 2nd lactation, 3rd lactation, and more than 3rd lactations); DP_{ij} is the fixed effect of the interaction between diet i and parity j; $C_{k:ij}$ is the random effect of the cow within DP $\sim N(0, \sigma^2_c)$ used as error term for D, P, and DP; S_l is the fixed effect of sampling days (l = 1–30 levels for milk yield; l = 3 levels i.e., 4, 15, and 30 days after calving for milk quality and fatty acids); DS_{il} is the fixed effect of interaction between diet i and sampling days l; PS_{jl} is the fixed effect of interaction between parity j and sampling days l; e_{ijklm} is the random residual term $\sim N(0, \sigma^2_e)$.

Data on enterolactone concentration was analyzed using the same hierarchical linear model as above, but considering also individual SCS as linear covariate.

In all statistical analyses, $p < 0.05$ indicates significance and $p < 0.10$ and ≥ 0.05 indicate a tendency toward significance

When multiple comparison tests were carried out for selected variables, the Bonferroni correction method was applied [30].

3. Results

The results are reported considering the milk yield during the first month of lactation, the quality and the fatty acids profile of milk, and the content of enterolactone in plasma and milk.

3.1. Milk Yield

The production of early and commercial milk during the first month of lactation was similar for the cows fed with or without flaxseed during the dry period (Figure 1). The overall milk yield of cows of the control group (CTR) and the cows receiving flaxseed (FLAX) was respectively 989 ± 41 kg and 998 ± 43 kg ($p \geq 0.05$).

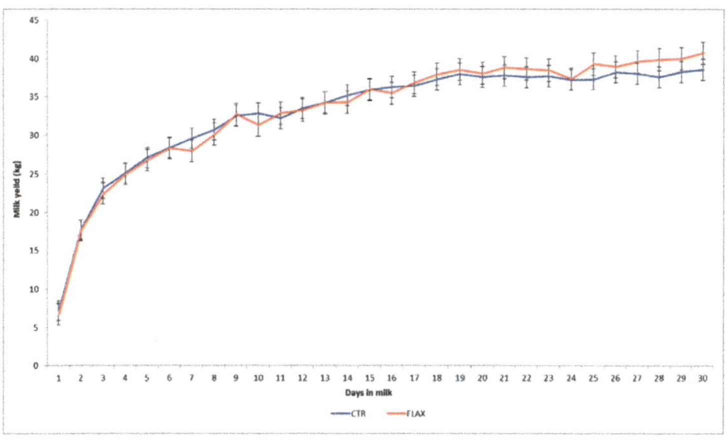

Figure 1. Least square means for control (CTR; solid line) and flaxseed supplemented (FLAX; dotted line) diets on the early and commercial milk production of Holstein cows during the first 30 d of the subsequent lactation.

There were no differences in overall milk yield produced during the first 30 days after calving between the cows of 2nd, 3rd, or more than 3rd lactation ($p \geq 0.05$). The effect of interaction between diet and parity was not significant.

3.2. *Quality of Early and Commercial Milk*

The effect of the diet, sampling day, and parity on the quality of early and commercial milk is shown in Table 3.

There were no significant differences in all parameters of milk quality between the CTR and FLAX diet (Table 3).

On the contrary, the sampling day affected significantly ($p < 0.001$) all quality parameters of early and commercial milk (Table 3). The fat, protein, and casein content decreased passing from 4 (early milk) to 15 and 30 days in milk ($p < 0.001$). The lactose content increased from 4 to 15 and 30 days of lactation ($p < 0.05$). The somatic cell score (SCS) of early and commercial milk produced after 4 DIM was higher than those of milk sampled subsequently ($p < 0.001$). Lastly, the milk urea nitrogen (MUN) increased passing from 4 to 15 and 30 day of sampling ($p < 0.001$).

The differences in early and commercial milk quality parameters were not significant ($p \geq 0.05$) among early and commercial milk of cows different in parity (Table 3).

All effects of the first order interactions were not significant.

3.3. *Fatty Acid Profile of Early and Commercial Milk*

The Effect of the Diet, Sampling Day and Parity on the Fatty Acid Profile of Early and Commercial Milk Is Shown in Table 4

No differences ($p \geq 0.05$) were observed in saturated fatty acids (SFA) and in monounsaturated fatty acids (MUFA) contents between CTR and FLAX experimental groups, except for C17:1 n-10. Considering the polyunsaturated fatty acids (PUFA), the levels of C18:2 (linoleic acid), C18:3 n3 (alpha linolenic acid, ALA), and C20:5 n3 (eicosapentaenoic acid, EPA) were higher ($p < 0.05$) in early and commercial milk obtained by cows receiving flaxseed during dry period (FLAX) than that produced by cows of the control (CTR) group. Otherwise, C20:3 n6 was significantly lower in FLAX than in CTR. Consequently, the total amount of n3 fatty acid was higher in FLAX than in CTR group ($p = 0.02$) and the ratio between n6 and n3 fatty acids resulted more favorable ($p < 0.01$) in the early and commercial milk obtained from cows receiving flax during the dry period.

The sampling day affected significantly ($p < 0.001$) most of the fatty acids reported in Table 4. The content of the total saturated FA was lower at 30 days in milk than at the beginning of lactation (day 4), while the pattern of monounsaturated FA was opposite ($p < 0.001$). The content of polyunsaturated FA was lower in early milk (4 day sampling) compared with the subsequent samplings ($p < 0.001$). ALA and total n3 fatty acid content resulted higher in the intermediate sampling (15 d) than in the first and last samplings ($p < 0.001$). The n6:n3 ratio resulted more favorable in the intermediate sampling (15 d) than at 4 or 30 d of sampling ($p < 0.001$).

The effect of parity was not significant ($p \geq 0.05$) for the most of fatty acids showed in Table 4, excluding C18:0 (stearic acid), C12:1, C18:1 trans n-7, and C18:1 cis n-2 ($p < 0.05$). The conjugated linoleic acid isomer (CLA) named C18:2 cis-9, trans-11 resulted higher ($p < 0.05$) in early milk (first sampling) than in the two other samplings (15 and 30 d).

No interactions between dietary treatment and sampling period were observed in the fatty acids analyzed, except for ALA, the sum of n3 fatty acids, and the n6:n3 ratio, which were higher ($p < 0.001$) in the early milk (4 day of sampling) of the FLAX group than of the CTR one (Figure 2A–C).

The fatty acids profile was unaffected by the interaction between diet and parity. The effect of the interaction between sampling day and parity was significant for some polyunsaturated fatty acids (data not shown).

Table 3. Effect of flaxseed supplementation during dry period, sampling day, and parity on the early and commercial milk quality of Holstein cows during the first 30 d of the subsequent lactation.

Item	Diet			Sampling Day				Parity				Residual
	CTR [3]	FLAX [4]	*p*	4	15	30	*p*	2nd	3rd	>3rd	*p*	
Milk quality												
Fat, %	4.65	4.79	0.614	6.25	4.01	3.88	<0.001	4.79	4.77	4.59	0.796	1.518
Protein, %	3.55	3.54	0.881	4.40	3.20	3.03	<0.001	3.62	3.49	3.53	0.504	0.396
Casein, %	2.75	2.75	0.968	3.35	2.50	2.40	<0.001	2.83	2.70	2.74	0.302	0.228
Lactose, %	4.72	4.69	0.431	4.47	4.79	4.85	<0.001	4.75	4.67	4.68	0.175	0.021
SCS [1], score	3.04	2.89	0.648	3.91	2.53	2.46	<0.001	2.56	2.97	3.37	0.078	2.352
MUN [2], mg/dL	18.77	18.70	0.941	15.52	19.71	20.99	<0.001	18.71	18.16	19.35	0.619	24.206

[1] SCS = somatic cell score; [2] MUN = milk urea nitrogen; [3] CTR = control group; [4] FLAX = flaxseed group.

Table 4. Effect of flaxseed supplementation during dry period, sampling day, and parity on the fatty acid profile in early and commercial milk of Holstein cows during the first 30 d of the subsequent lactation.

Item	Diet			Sampling Day				Parity				Residual
	CTR [4]	FLAX [5]	*p*	4	15	30	*p*	2nd	3rd	>3rd	*p*	
SFA [1], % of total FA [2]												
C4:0	3.05	3.09	0.492	3.29	3.08	2.84	<0.001	3.05	3.15	3.01	0.216	0.123
C6:0	1.88	1.89	0.735	1.69	2.03	1.94	<0.001	1.93	1.85	1.87	0.256	0.056
C8:0	1.05	1.04	0.963	0.84	1.15	1.14	<0.001	1.09	1	1.05	0.122	0.028
C10:0	2.17	2.14	0.778	1.61	2.3	2.56	<0.001	2.27	2.02	2.18	0.117	0.196
C11:0	0.08	0.08	0.535	0.03	0.11	0.1	<0.001	0.09	0.07	0.08	0.137	0.001
C12:0	2.43	2.39	0.747	1.83	2.5	2.9	<0.001	2.54	2.24	2.45	0.116	0.271
C13:0	0.12	0.12	0.508	0.06	0.15	0.15	<0.001	0.13	0.11	0.12	0.133	0.002
C14:0	8.92	8.89	0.894	8.16	8.87	9.69	<0.001	9.22	8.71	8.79	0.175	2.096
C15:0	0.89	0.86	0.509	0.62	0.94	1.06	<0.001	0.91	0.85	0.85	0.251	0.037

Table 4. *Cont.*

Item	Diet			Sampling Day				Parity				Residual
	CTR [4]	FLAX [5]	p	4	15	30	p	2nd	3rd	>3rd	p	
C16:0	29.4	29.1	0.627	30.5	28.3	28.9	<0.001	29.1	29.5	29.1	0.785	6.037
C17:0	0.7	0.7	0.904	0.73	0.72	0.65	<0.001	0.7	0.7	0.7	0.889	0.008
C18:0	11.1	11.1	0.902	11.5	10.9	10.9	0.074	10.9	11.8	10.5	0.012	2.501
C19:0	0.06	0.06	0.72	0.04	0.06	0.08	<0.001	0.06	0.06	0.05	0.222	0.001
C20:0	0.15	0.15	0.466	0.11	0.17	0.18	<0.001	0.15	0.16	0.15	0.089	0.001
C22:0	0.04	0.04	0.352	0.02	0.04	0.06	<0.001	0.04	0.04	0.04	0.318	0
Others SFA [1]	2.2	2.17	0.432	1.81	2.43	2.31	<0.001	2.23	2.14	2.2	0.227	0.05
MUFA [3] % of total FA [2]												
C10:1	0.18	0.18	0.459	0.1	0.2	0.24	<0.001	0.18	0.16	0.19	0.041	0.002
C12:1	0.06	0.06	0.427	0.03	0.06	0.08	<0.001	0.06	0.05	0.06	0.026	0
C14:1	0.36	0.33	0.101	0.46	0.57	0.01	<0.001	0.35	0.32	0.36	0.228	0.016
C16:1	0.11	0.1	0.815	0.18	0.07	0.06	<0.001	0.11	0.1	0.1	0.586	0.004
C17:1n-10	0.02	0.01	0.041	0.02	0.02	0.01	<0.001	0.02	0.01	0.02	0.21	0
C17:1n-9	0.02	0.01	0.15	0.01	0.02	0.02	<0.001	0.01	0.01	0.01	0.973	0
C17:1 cis n-7	0.4	0.41	0.567	0.42	0.44	0.35	<0.001	0.4	0.38	0.42	0.3	0.005
C18:1 trans n-7	1.4	1.48	0.472	0.95	1.72	1.66	<0.001	1.62	1.42	1.28	0.028	0.544
C18:1 cis n-9	23.7	23.9	0.828	26.1	23.5	21.7	<0.001	23.1	23.7	24.57	0.231	10.01
C18:1n-6	0.32	0.32	0.995	0.22	0.36	0.39	<0.001	0.33	0.33	0.3	0.283	0.005
C18:1 cis n-2	0.06	0.06	0.964	0.04	0.07	0.08	<0.001	0.06	0.06	0.05	0.038	0
C20:1n-12	0.1	0.1	0.338	0.06	0.12	0.12	<0.001	0.1	0.1	0.1	0.695	0.001
Others MUFA [3]	4.68	4.75	0.554	4.41	4.5	5.23	<0.001	4.67	4.64	4.83	0.471	0.214
PUFA [6], % of total FA												
C18:2 trans-9, cis-12 [5]	0.031	0.035	0.424	0.03	0.04	0.03	0.346	0.04	0.03	0.03	0.034	0.001
C18:2n-6 cis	2.243	2.239	0.927	2.29	2.24	2.19	0.124	2.23	2.15	2.34	0.064	0.074
C18:2	0.022	0.017	0.04	0.03	0.02	0.01	<0.001	0.02	0.02	0.02	0.348	0

Table 4. *Cont.*

Item	Diet			Sampling Day				Parity				Residual
	CTR [4]	FLAX [5]	*p*	4	15	30	*p*	2nd	3rd	>3rd	*p*	
C18:2 cis-9, trans-11 CLA [7]	0.329	0.329	0.994	0.26	0.37	0.35	<0.001	0.34	0.31	0.32	0.03	0.005
C18:2 trans-10, cis-12 CLA [7]	0.014	0.014	0.833	0.01	0.02	0.01	<0.001	0.01	0.01	0.01	0.862	0
C18:3n-3	0.334	0.353	0.059	0.33	0.37	0.33	<0.001	0.34	0.33	0.35	0.317	0.002
C18:3n-6	0.05	0.046	0.219	0.04	0.05	0.05	0.001	0.05	0.05	0.05	0.213	0
C20:2	0.044	0.04	0.395	0.03	0.03	0.07	<0.001	0.04	0.04	0.04	0.831	0.001
C20:3n-6	0.12	0.108	0.035	0.1	0.12	0.12	<0.001	0.11	0.12	0.11	0.424	0.001
C20:4n-6	0.21	0.211	0.95	0.23	0.22	0.18	<0.001	0.21	0.21	0.22	0.284	0.002
C20:5n-3	0.048	0.055	0.018	0.05	0.06	0.05	0.072	0.05	0.05	0.05	0.992	0
Others PUFA [3]	0.99	1.01	0.444	0.82	1.09	1.1	<0.001	1.01	0.97	1.02	0.426	0.022
SFA [1]	64.2	63.9	0.695	62.8	63.8	65.5	<0.001	64.5	64.4	63.1	0.327	13.32
MUFA [3]	31.4	31.7	0.731	33	31.7	30	<0.001	31.1	31.3	32.3	0.393	12.14
PUFA [6]	4.42	4.47	0.609	4.22	4.61	4.51	<0.001	4.45	4.3	4.57	0.121	0.184
SFA [1]/(MUFA [3] + PUFA [6])	1.85	1.81	0.598	1.73	1.8	1.95	<0.001	1.86	1.86	1.75	0.358	0.086
n-6	2.66	2.65	0.886	2.71	2.66	2.59	0.048	2.63	2.57	2.77	0.048	0.082
n-3	0.5	0.53	0.022	0.5	0.55	0.5	<0.001	0.52	0.51	0.52	0.757	0.004
n-6/n-3	5.35	5.04	0.006	5.53	4.86	5.19	<0.001	5.11	5.08	5.4	0.043	0.3

[1] SFA = saturated fatty acids; [2] FA = FA; [3] MUFA = monounsaturated FA; [4] CTR = control group; [5] FLAX = flaxseed group; [6] PUFA = polyunsaturated FA; [7] CLA = conjugated linoleic acid isomers.

Figure 2 reports the interaction between diets and sampling day effects, highlighting the significant differences between treatments within sampling days.

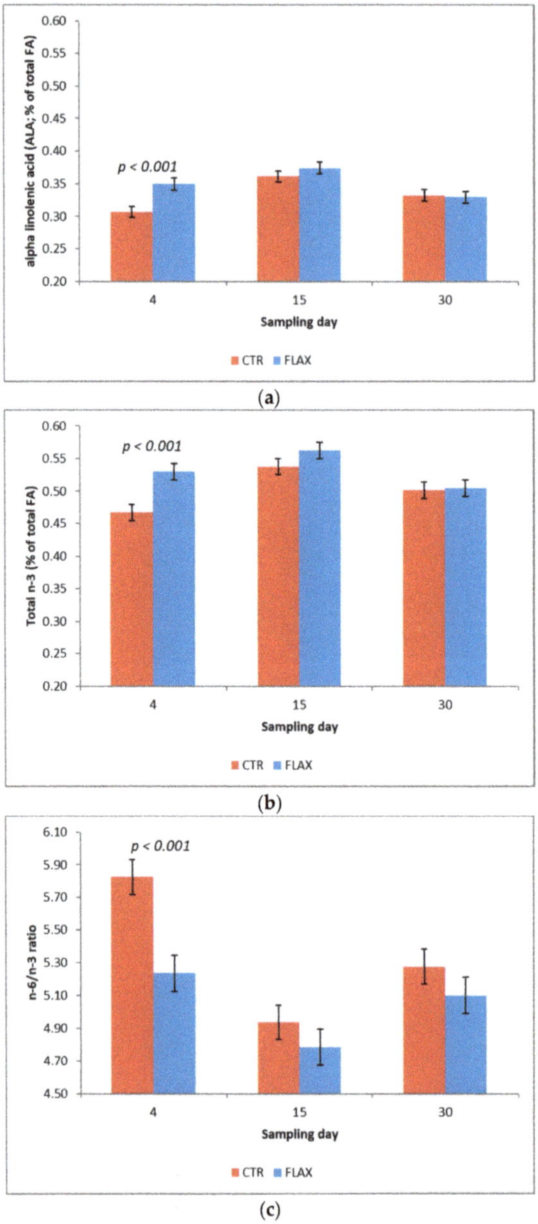

Figure 2. Least square means for control (CTR; dark red bars) and flaxseed supplemented diets (FLAX; blue bars): (**a**) alpha linolenic acid (ALA); (**b**) total n-3 fatty acids contents (expressed as % on the total fatty acids); and (**c**) n-6: n-3 ratio in the early and commercial milk. Differences between treatments within sampling day have been reported when significant ($p < 0.05$ or less).

3.4. Enterolactone in Plasma and Early and Commercial Milk

The enterolactone concentration (EL, nmol/L) was detected in plasma and in early and commercial milk (Figure 3). No significant differences were observed between diets for the concentration of enterolactone in plasma, although the effect of sampling day was significant ($p < 0.001$; data not shown). The EL concentration in plasma samples collected before calving (4 sampling days) was lower than those of samples collected during the first month of lactation (3 sampling days; $p < 0.001$; data not presented). The parity of cows had no effect of EL. All interactions of first order were not significant. Therefore, the supplementation of FLAX did not affect the EL concentration during the dry period and the subsequent lactation. The only exception was for the interaction between diet and sampling day. Indeed, EL concentration in plasma was greater in FLAX fed than in CTR cows at d 15 after calving ($p < 0.001$; Figure 3a).

There were no significant effects ($p \geq 0.05$) on EL in milk for the two experimental groups (CTR and FLAX). On the contrary, the EL concentration increased significantly ($p < 0.001$) in early and commercial milk from 4 to 15 days of lactation, remaining high at 30 day after calving. No differences were observed among the cows of different parity. All interactions of first order were not significant.

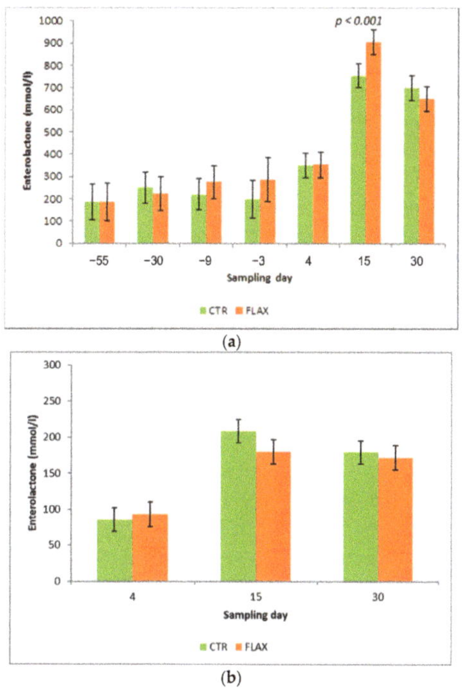

Figure 3. Concentration (nmol/L) of enterolactone in Holstein cows fed either with flacked flaxseed or a control diet during the dry period: (**a**) concentration in plasma (before and 30 d after calving); (**b**) concentration in milk during 30 d after calving. Differences between treatments within sampling day have been reported when significant ($p < 0.05$).

4. Discussion

Supplementation with n-3 FA of cattle feeding is an effective approach for improving the nutritional value of animal products such as meat and milk [3,15,31,32]. Many studies [9,10,17,32] demonstrated that flaxseed supplementation in cows during lactation or transition periods significantly increase the amount of n-3 FA in milk, which is otherwise relatively poor in these FA [33]. Nevertheless, very few studies have been conducted on the effects of pre-partum n-3 supplementation on milk composition [13,15].

During early lactation, dairy cows experience a negative energy balance associated with mobilization of body reserves. In this period, dairy cows are in negative energy balance, causing extensive body-fat mobilization and their incorporation into milk fat. Thus, milk FA synthesized de novo by the mammary gland is lower, compared to mid or late lactation [15,34] and depends on FA composition of adipose tissue [35]. n-3 FA cannot be synthesized de novo by mammals and is referred to as essential fatty acids, which must be acquired with the diet [36]. Our study demonstrated that the carry-over effect on the milk profile of C18:2, C18:3 n-3, and C20:5 n-3 (EPA) was significantly higher in FLAX diet than in the CTR one on the first sampling at 4th day of lactation. Present results are in agreement with early studies reporting that extruded or crushed linseeds (2.5% or 3.3% DM) supplemented on transition period increased 18:3 n-3, and total n-3 concentrations in early milk, but did not exert any effect after 7 weeks of lactation [32,37]. In contrast with our results, a study [15] demonstrated that the carry-over effects persisted for 7 weeks of lactation (10 weeks after the withdrawal of extruded flaxseed supplementation, 2.9% DM). However, the greater amount of n-3 was observed during the first week with respect to the other lactation phases. In contrast, another study has shown that restricted diet with canola, linola, or flax rolled seeds supplementation (8% DM) in the 4 last weeks of gestation had no effects on fatty acids milk compositions and metabolic peri-partum responses [13].

These different results could be related to the different flaxseed treatments and concentration used in these studies [38]. Lipids in extruded flaxseed are considered to be at least partially rumen-protected and thus can overtake the microbial bio-hydrogenation [39,40], while unprocessed oilseeds have a less influence on supplying n-3 on meat or milk [35,41]. Although various studies have been performed on the effect of different processing treatment (grounding vs. term-flaking) on cows performance and feed utilization [42,43], the lipid and FA metabolism of enriched n-3 sources subjected to different physical treatments has not yet been deepened. Regarding the amount of oilseed supplementation, in our FLAX diet, flaked flaxseeds was added at 2.13% of DM, thus in a lower concentration than the previous studies. This difference and the length of supplementation, which only covered the dry period, may have determined the differences due to flaxseed supplementation compared to the other studies.

The quick variation in FA profile of the milk in the first few days of lactation observed in our study suggests that the carry-over effect from pre-calving flaxseed feeding at this concentration was very short lasting. In addition, an interesting result of our study is the significant increase of EPA, resulting from ALA elongation, on early milk of cows fed the FLAX diet [44]. Previous findings have shown that conversion of linoleic acid (LA) and alpha-linolenic acid (ALA) to their higher chain homologues (DHA and EPA) in humans depends on the ratio of ingested n-6 and n-3 fatty acids [44]. Twenty-carbon PUFAs are precursors of eicosanoids that regulate the inflammatory and immune responses through pro- and anti-inflammatory activities [36]. In this sense, n-3 supplementation on pre-partum diet could be considered as a valid approach to improve the nutritional value of early milk and consequently calf health. Indeed, as recently demonstrated by [45], pre-partum fat feeding modified the FA profile of early milk to a greater degree than plasma of newborns, suggesting that the metabolism and transfer of essential FA from the mammary gland to the early milk has a less tight regulation than those from the placenta to the fetus. Thus, providing early milk with selected FA through pre-calving dietary supplementation could be an efficient strategy to improve calf health status immediately after birth.

Our results demonstrated that, independently from treatment group, plasma and milk EL significantly increase 15 days after calving and it remains high for the consecutive 15 days postpartum. According to [19] study, plant and mammalian lignans have different bioavailability in humans: while the first must be converted to EL by the colon bacteria, the others can be passively absorbed along the human intestine. Therefore, milk with an elevated amount of EL may be an efficient strategy to optimize the effects of lignans on human health [46]. According to our results, a significant increase of EL on milk from FLAX-fed cows was observed only at 15 sampling day. Four dose-response studies using flax hulls (FH), flax meal (FM), and whole flaxseed (WF) have been conducted to date [7,46–48].

In these studies, higher concentration of milk EL was observed with FH or FM diet supplementation compared with WF feeding. In addition, a significant increase of milk EL concentration when cows were fed the greatest amount of flaxseed supplementation (15% FM or 20% FH on DM basis) [8]. However, these higher amounts of flaxseed products compared with that used in our study may have multiple adverse effects associated with the excess intake of crude protein or crude fat. Indeed, the excess of N due to environmental and milk production concerns associated with excess intake of crude protein or crude fat, depending on the flax source used [8]. A recent study demonstrated that calves fed EL-enriched milk (481 nmol/L) showed a faster and higher EL increase on plasma, compared with those fed milk enriched with a lower amount of EL (123 nmol/L). Further research is needed to better understand the physiological EL concentration on milk and plasma and its metabolism before and after calving, in order to find the balanced intake of oilseed supplements on dairy cow dietary treatment.

5. Conclusions

Fatty acid profiles of early milk samples were influenced by flaked flaxseed supplementation provided to dry cows. However, the effect obtained from the concentration used in this study (2.13% DM) was very short lasting. These results suggest that different concentrations and the various treatments, to which flaxseed are subjected, significantly affect the n-3 carry over into the milk. Differences of EL plasma concentration before and after calving should be better investigated, in order to understand EL metabolism in ruminants and how it could be efficiently manipulated by dietary treatments.

Author Contributions: Conceptualization. G.G., L.B. and R.M.; methodology. R.M., L.B.; formal analysis. L.D.D.; data curation. N.G.; writing—original draft preparation E.G., G.G., L.B., and R.M.; writing—review and editing. E.G., G.G., L.B. and R.M.; funding acquisition. G.G. All authors have read and agree to the published version of the manuscript.

Funding: This research was funded by a National Grant PRIN 2010–2011 "Sustainability between conventional dairy farm systems and longevity: models of dry period management and biomarkers of stress and aging in the mammary gland".

Conflicts of Interest: The authors declare no conflict of interest.

References

1. Sinclair, A.J.; Attar-Bashi, N.M.; Li, D. What is the role of αt-linolenic acid for mammals? *Lipids* **2003**, *37*, 1113–1123. [CrossRef] [PubMed]
2. Parikh, M.; Maddaford, T.G.; Austria, J.A.; Aliani, M.; Netticadan, T.; Pierce, G.N. Dietary flaxseed as a strategy for improving human health. *Nutrients* **2019**, *11*, 1171. [CrossRef]
3. Mordenti, A.L.; Brogna, N.; Canestrari, G.; Bonfante, E.; Eusebi, S.; Mammi, L.M.E.; Giaretta, E.; Formigoni, A. Effects of breed and different lipid dietary supplements on beef quality. *Anim. Sci. J.* **2019**, *90*, 619–627. [CrossRef]
4. Giaretta, E.; Mordenti, A.; Palmonari, A.; Brogna, N.; Canestrari, G.; Belloni, P.; Cavallini, D.; Mammi, L.; Cabbri, R.; Formigoni, A. NIRs calibration models for chemical composition and fatty acid families of raw and freeze-dried beef: A comparison. *J. Food Compos. Anal.* **2019**, *83*, 103257. [CrossRef]
5. Ehr, I.J.; Persia, M.E.; Bobeck, E.A. Comparative omega-3 fatty acid enrichment of egg yolks from first-cycle laying hens fed flaxseed oil or ground flaxseed. *Poult. Sci.* **2017**, *96*, 1791–1799. [CrossRef] [PubMed]
6. Wall, R.; Ross, R.P.; Fitzgerald, G.F.; Stanton, C. Fatty acids from fish: The anti-inflammatory potential of long-chain omega-3 fatty acids. *Nutr. Rev.* **2010**, *68*, 280–289. [CrossRef] [PubMed]
7. Petit, H. V Review: Feed intake, milk production and milk composition of dairy cows fed flaxseed. *Can. J. Anim. Sci.* **2010**, *90*, 115–127. [CrossRef]
8. Brito, A.F.; Zang, Y. A review of lignan metabolism, milk enterolactone concentration, and antioxidant status of dairy cows fed flaxseed. *Molecules* **2019**, *24*, 41. [CrossRef]
9. Isenberg, B.J.; Soder, K.J.; Pereira, A.B.D.; Standish, R.; Brito, A.F. Production, milk fatty acid profile, and nutrient utilization in grazing dairy cows supplemented with ground flaxseed. *J. Dairy Sci.* **2019**, *102*, 1294–1311. [CrossRef]

10. Cattani, M.; Mantovani, R.; Schiavon, S.; Bittante, G.; Bailoni, L. Recovery of n-3 polyunsaturated fatty acids and conjugated linoleic acids in ripened cheese obtained from milk of cows fed different levels of extruded flaxseed. *J. Dairy Sci.* **2014**, *97*, 123–135. [CrossRef]
11. Moallem, U. Invited review: Roles of dietary n-3 fatty acids in performance, milk fat composition, and reproductive and immune systems in dairy cattle. *J. Dairy Sci.* **2018**, *101*, 8641–8661. [CrossRef] [PubMed]
12. Dirandeh, E.; Ghaffari, J. Effects of feeding a source of omega-3 fatty acid during the early postpartum period on the endocannabinoid system in the bovine endometrium. *Theriogenology* **2018**, *121*, 141–146. [CrossRef]
13. Hayirli, A.; Keisler, D.H.; Doepel, L. Peripartum responses of dairy cows to prepartal feeding level and dietary fatty acid source. *J. Dairy Sci.* **2011**, *94*, 917–930. [CrossRef] [PubMed]
14. Badiei, A.; Aliverdilou, A.; Amanlou, H.; Beheshti, M.; Dirandeh, E.; Masoumi, R.; Moosakhani, F.; Petit, H.V. Postpartum responses of dairy cows supplemented with n-3 fatty acids for different durations during the peripartal period. *J. Dairy Sci.* **2014**, *97*, 6391–6399. [CrossRef] [PubMed]
15. Lerch, S.; Pires, J.A.A.; Delavaud, C.; Shingfield, K.J.; Pomiès, D.; Martin, B.; Chilliard, Y.; Ferlay, A. Rapeseed or linseed in dairy cow diets over 2 consecutive lactations: Effects on adipose fatty acid profile and carry-over effects on milk fat composition in subsequent early lactation. *J. Dairy Sci.* **2015**, *98*, 1005–1018. [CrossRef] [PubMed]
16. Santschi, D.E.; Lacroix, R.; Durocher, J.; Duplessis, M.; Moore, R.K.; Lefebvre, D.M. Prevalence of elevated milk β-hydroxybutyrate concentrations in Holstein cows measured by Fourier-transform infrared analysis in Dairy Herd Improvement milk samples and association with milk yield and components. *J. Dairy Sci.* **2016**, *99*, 9263–9270. [CrossRef] [PubMed]
17. Do Prado, R.M.; Palin, M.F.; do Prado, I.N.; dos Santos, G.T.; Benchaar, C.; Petit, H.V. Milk yield, milk composition, and hepatic lipid metabolism in transition dairy cows fed flaxseed or linola. *J. Dairy Sci.* **2016**, *99*, 8831–8846. [CrossRef] [PubMed]
18. Gandra, J.R.; Mingoti, R.D.; Barletta, R.V.; Takiya, C.S.; Verdurico, L.C.; Freitas, J.E.; Paiva, P.G.; Jesus, E.F.; Calomeni, G.D.; Rennó, F.P. Effects of flaxseed, raw soybeans and calcium salts of fatty acids on apparent total tract digestibility, energy balance and milk fatty acid profile of transition cows. *Animal* **2016**, *10*, 1303–1310. [CrossRef]
19. Saarinen, N.M.; Smeds, A.; Mäkelä, S.I.; Ämmälä, J.; Hakala, K.; Pihlava, J.M.; Ryhänen, E.L.; Sjöholm, R.; Santti, R. Structural determinants of plant lignans for the formation of enterolactone in vivo. *J. Chromatogr. B Anal. Technol. Biomed. Life Sci.* **2002**, *777*, 311–319. [CrossRef]
20. Gagnon, N.; Côrtes, C.; Da Silva, D.; Kazama, R.; Benchaar, C.; Dos Santos, G.; Zeoula, L.; Petit, H.V. Ruminal metabolism of flaxseed (*Linum usitatissimum*) lignans to the mammalian lignan enterolactone and its concentration in ruminal fluid, plasma, urine and milk of dairy cows. *Br. J. Nutr.* **2009**, *102*, 1015–1023. [CrossRef]
21. Council, N.R. *Nutrient Requirements of Dairy Cattle*; National Academies Press: Washington, DC, USA, 2001.
22. AOAC. *Official Methods of Analysis of AOAC International*; Association of Official Analysis Chemists International: Rockville, MD, USA, 1990; Volume II, pp. 1058–1059. ISBN 0935584544.
23. Robertson, J.B.; Van Soest, P.J. The detergent system of analysis. In *The Analysis of Dietary Fiber in Food*; James, W.P.T., Theander, O., Eds.; Marcel Dekker: New York, NY, USA, 1981.
24. Bouchard, J.; Chornet, E.; Overend, R.P. High-performance liquid chromatographic monitoring of carbohydrate fractions in partially hydrolyzed corn starch. *J. Agric. Food Chem.* **1988**, *36*, 1188–1192. [CrossRef]
25. Pellattiero, E.; Cecchinato, A.; Tagliapietra, F.; Schiavon, S.; Bittante, G. The use of 2-dimensional gas chromatography to investigate the effect of rumen-protected conjugated linoleic acid, breed, and lactation stage on the fatty acid profile of sheep milk. *J. Dairy Sci.* **2015**, *98*, 2088–2102. [CrossRef] [PubMed]
26. International Dairy Federation. *International IDF Standard 141C. Determination of Milk Fat, Protein and Lactose Content. Guidance on the Operation of Mid-Infrared Instruments*; International Dairy Federation: Brussels, Belgium, 2000.
27. International Dairy Federation. *International IDF Standard 148A. Milk Enumeration of Somatic Cells*; International Dairy Federation: Brussels, Belgium, 1995.
28. Stumpf, K.; Uehara, M.; Nurmi, T.; Adlercreutz, H. Changes in the time-resolved fluoroimmunoassay of plasma enterolactone. *Anal. Biochem.* **2000**, *284*, 153–157. [CrossRef] [PubMed]
29. Gagnon, N.; Côrtes, C.; Petit, H.V. Weekly excretion of the mammalian lignan enterolactone in milk of dairy cows fed flaxseed meal. *J. Dairy Res.* **2009**, *76*, 455–458. [CrossRef] [PubMed]
30. SAS Institute. *SAS User's Guide: Basics*; SAS Institute Inc.: Cary, NC, USA, 2009.

31. Hao, M.; Beta, T. Qualitative and quantitative analysis of the major phenolic compounds as antioxidants in barley and flaxseed hulls using HPLC/MS/MS. *J. Sci. Food Agric.* **2012**, *92*, 2062–2068. [CrossRef] [PubMed]
32. Santschi, D.E.; Wettstein, H.-R.; Leiber, F.; Witschi, A.-K.M.; Kreuzer, M. Colostrum and milk fatty acids of dairy cows as influenced by extruded linseed supplementation during the transition period. *Can. J. Anim. Sci.* **2009**, *89*, 383–392. [CrossRef]
33. Bailoni, L.; Prevedello, G.; Schiavon, S.; Mantovani, R.; Bittante, G. CLA content and n-3/n-6 ratio in dairy milk as affected by farm size and management. In *Indicators of Milk and Beef Quality*; Hocquette, J.F., Gigli, S., Eds.; EAAP: Rome, Italy, 2005; pp. 333–338.
34. Palmquist, D.L.; Denise Beaulieu, A.; Barbano, D.M. Feed and animal factors influencing milk fat composition. *J. Dairy Sci.* **1993**, *76*, 1753–1771. [CrossRef]
35. Chilliard, Y.; Ferlay, A.; Mansbridge, R.M.; Doreau, M. Ruminant milk fat plasticity: Nutritional control of saturated, polyunsaturated, trans and conjugated fatty acids. *Anim. Res.* **2000**, *49*, 181–205. [CrossRef]
36. Lee, J.M.; Lee, H.; Kang, S.B.; Park, W.J. Fatty acid desaturases, polyunsaturated fatty acid regulation, and biotechnological advances. *Nutrients* **2016**, *8*, 23. [CrossRef]
37. Leiber, F.; Hochstrasser, R.; Wettstein, H.R.; Kreuzer, M. Feeding transition cows with oilseeds: Effects on fatty acid composition of adipose tissue, colostrum and milk. *Livest. Sci.* **2011**, *138*, 1–12. [CrossRef]
38. Rabiee, A.R.; Breinhild, K.; Scott, W.; Golder, H.M.; Block, E.; Lean, I.J. Effect of fat additions to diets of dairy cattle on milk production and components: A meta-analysis and meta-regression. *J. Dairy Sci.* **2012**, *95*, 3225–3247. [CrossRef] [PubMed]
39. Petit, H.V.; Rioux, R.; Ouellet, D.R. Milk production and intake of lactating cows fed raw or extruded peas. *J. Dairy Sci.* **1997**, *80*, 3377–3385. [CrossRef]
40. Doreau, M.; Laverroux, S.; Normand, J.; Chesneau, G.; Glasser, F. Effect of linseed fed as rolled seeds, extruded seeds or oil on fatty acid rumen metabolism and intestinal digestibility in cows. *Lipids* **2009**, *44*, 53–62. [CrossRef] [PubMed]
41. Turner, T.D.; Mitchell, A.; Duynisveld, J.; Pickova, J.; Doran, O.; McNiven, M.A. Influence of oilseed supplement ranging in n-6/n-3 ratio on fatty acid composition and δ5-, δ6-desaturase protein expression in steer muscles. *Animal* **2012**, *6*, 1973–1984. [CrossRef]
42. Safaei, K.; Ghorbani, G.R.; Alikhani, M.; Sadeghi-Sefidmazgi, A.; Yang, W.Z. Response of lactating dairy cows to degree of steam-flaked barley grain in low-forage diets. *J. Anim. Physiol. Anim. Nutr.* **2017**, *101*, e87–e97. [CrossRef]
43. Santos, J.E.P.; Huber, J.T.; Theurer, C.B.; Nussio, L.G.; Tarazon, M.; Santos, F.A.P. Nutrition, feeding, and calves: Response of lactating dairy cows to steam-flaked sorghum, steam-flaked corn, or steam-rolled corn and protein sources of differing degradability. *J. Dairy Sci.* **1999**, *82*, 728–737. [CrossRef]
44. Harnack, K.; Andersen, G.; Somoza, V. Quantitation of alpha-linolenic acid elongation to eicosapentaenoic and docosahexaenoic acid as affected by the ratio of n6/n3 fatty acids. *Nutr. Metab.* **2009**, *6*, 1–11. [CrossRef]
45. Garcia, M.; Greco, L.F.; Favoreto, M.G.; Marsola, R.S.; Martins, L.T.; Bisinotto, R.S.; Shin, J.H.; Lock, A.L.; Block, E.; Thatcher, W.W.; et al. Effect of supplementing fat to pregnant nonlactating cows on colostral fatty acid profile and passive immunity of the newborn calf. *J. Dairy Sci.* **2014**, *97*, 392–405. [CrossRef]
46. Petit, H.V.; Gagnon, N. Concentration of the mammalian lignans enterolactone and enterodiol in milk of cows fed diets containing different concentrations of whole flaxseed. *Animal* **2009**, *3*, 1428–1435. [CrossRef]
47. Petit, H.V.; Gagnon, N. Production performance and milk composition of dairy cows fed different concentrations of flax hulls. *Anim. Feed Sci. Technol.* **2011**, *169*, 46–52. [CrossRef]
48. Schogor, A.L.B.; Palin, M.F.; Santos, G.T.; Benchaar, C.; Petit, H.V. β-glucuronidase activity and enterolactone concentration in ruminal fluid, plasma, urine, and milk of Holstein cows fed increased levels of flax (Linum usitatissimum) meal. *Anim. Feed Sci. Technol.* **2017**, *223*, 23–29. [CrossRef]

animals

MDPI

Article

The Capacity of Holstein-Friesian and Simmental Cows to Correct a Negative Energy Balance in Relation to Their Performance Parameters, Course of Lactation, and Selected Milk Components

Ilona Strączek, Krzysztof Młynek * and Agata Danielewicz

Faculty of Agrobioengineering and Animal Sciences, University of Natural Sciences and Humanities in Siedlce, 08-110 Siedlce, Poland; ilonastraczek1984@gmail.com (I.S.); agata.danielewicz@uph.edu.pl (A.D.)
* Correspondence: krzysztof.mlynek@uph.edu.pl

Simple Summary: The aim of the study was to analyse the ability of Simmental (SIM) and Holstein-Friesian (HF) cows to correct a negative energy balance (NEB). NEB dynamics were assessed based on the content of NEFA in the blood; a reduction in body condition score; and levels of C16:0, C18:0 and C18:1 in the milk. The efficiency of liver metabolism was evaluated based on the content of BHBA in the blood and urea in the milk. The rate of changes was analysed during lactation, with assessments of daily yield, production at peak lactation and its duration, and changes in selected milk components. The results indicated that the most significant changes took place up to the peak of lactation. During this time, the values for parameters characterizing NEB were similar in both breeds. After the peak of lactation, the body condition score of SIM cows was restored more quickly. HF cows, on the other hand, achieved greater milk production and reached peak lactation earlier, but they were less capable of correcting the NEB, as indicated by the higher content of non-esterified fatty acid and β-hydroxybutyrate as well as C16:0, C18:0, and C18:1 in most cases. Their milk also contained more urea. The dynamics of NEB were found to be linked to the level of leptin, which has an anorectic effect. The results of the study indicate the great potential of Simmental cows and may facilitate the search for solutions for the more efficient exploitation of their potential.

Citation: Strączek, I.; Młynek, K.; Danielewicz, A. The Capacity of Holstein-Friesian and Simmental Cows to Correct a Negative Energy Balance in Relation to Their Performance Parameters, Course of Lactation, and Selected Milk Components. *Animals* **2021**, *11*, 1674. https://doi.org/10.3390/ani11061674

Academic Editors: João Simões and Marina Von Keyserlingk

Received: 14 April 2021
Accepted: 1 June 2021
Published: 4 June 2021

Abstract: A significant factor in improving the performance of dairy cows is their physiological ability to correct a negative energy balance (NEB). This study, using Simmental (SIM) and Holstein-Friesian (HF) cows, aimed to assess changes in NEB (non-esterified fatty acid; body condition score; and C16:0, C18:0, and C18:1) and its effect on the metabolic efficiency of the liver (β-hydroxybutyrate and urea). The effects of NEB on daily yield, production at peak lactation and its duration, and changes in selected milk components were assessed during complete lactation. Up to peak lactation, the loss of the body condition score was similar in both breeds. Subsequently, SIM cows more efficiently restored their BCS. HF cows reached peak lactation faster and with a higher milk yield, but they were less able to correct NEB. During lactation, their non-esterified fatty acid, β-hydroxybutyrate, C16:0, C18:0, C18:1, and urea levels were persistently higher, which may indicate less efficient liver function during NEB. The dynamics of NEB were linked to levels of leptin, which has anorectic effects. Its content was usually higher in HF cows and during intensive lactogenesis. An effective response to NEB may be exploited to improve the production and nutritional properties of milk. In the long term, it may extend dairy cows' productive life and increase lifetime yield.

Keywords: cattle; body condition score; fatty acids; urea; β-hydroxybutyrate; metabolism; urea in milk

1. Introduction

During the post-partum period, dramatically increasing milk yield leads to the destabilization of the energy balance [1]. This process involves the liver, in which a sharp increase in metabolism, including of lipids and glycogen, takes place [2,3], as does the detoxification of ammonia. One of the consequences of the disturbance in homeostasis, on average up to the peak of lactation, is a temporary decrease in feed intake. This initiates changes associated with a negative energy balance (NEB) [4–6]. Lipolysis mainly leads to the release of palmitic (C16:0), stearic (C18:0), and oleic (C18:1) acid [7,8]. These increase the pool of non-esterified fatty acid supplied to the liver [9], in which they are oxidized and distributed in the form of low-density lipoproteins [2]. By inducing spontaneous lipolysis, NEB can adversely affect the metabolic efficiency of the liver, causing the excessive generation of β-hydroxybutyrate (BHBA) and the retention of triglycerides in the hepatocytes [10,11]. This affects the functions of the liver [12], in which the pathways of lipid metabolism, the detoxification of ammonia, and glucogenesis cross. The interactions of these pathways can impair the metabolism of these compounds [13], which may increase losses of nitrogen with the milk. This is not without significance for the ammonia detoxification pathway in the liver. An increase in ammonia levels in the blood may induce excessive lipid retention in the liver [14]. During NEB, this may result in the excessive catabolism of amino acids in the liver and increased losses of nitrogen into the environment. Studies by Knob et al. (2021) [6] and Lopreiato et al. (2019) [15] showed that the metabolic changes associated with NEB can also be influenced by the breed of cows. Therefore, the production potential of popular dairy cattle breeds remains an important area of research—especially regarding the effect of their physiological responses on production parameters and the duration of their productive life. The aim of the study was to analyse the ability of Simmental and Holstein-Friesian cows to correct a negative energy balance. We hypothesized that a less dynamic NEB at the start of lactation makes it possible to obtain more favourable production effects over a complete lactation.

2. Materials and Methods

The study was carried out on two breed groups (BGs) of dairy cows: 25 Holstein-Friesian (HF) cows at farm A, and 25 Simmental (SIM) cows at farm B. During the lactation period, the data were collected from the same cows. All cows in the experiment were subjected to the same number of periodic observations. The average yield (LSM ± SE) of the cows was 8390 kg ± 48 for HF and 7956 kg ± 38 for SIM, and the average lactation number (LSM ± SE) was 2.5 ± 0.23 for HF and 2.9 ± 0.43 for SIM. The cows were from two farms and constituted separate experimental groups. They were evaluated on average from the 6th day postpartum. The length of lactation (number of days) was defined as the period from calving to the day milking was discontinued, when average production had fallen to 10 kg of milk. The experiment was lasted until the completion of lactation.

2.1. Feeding and Housing of Cows

In both breed groups, the test cows were kept tethered. The cows had direct access to feed and water (open drinkers), and their living conditions met the requirements of good production practice.

Nutrient requirements were established based on information about feed quality—chemical analyses [16], the approximate body weight of the breed groups, and forecast milk production for the analysed stages of lactation: days 6–100 (SL I), 101–200 (SL II), and >200 (SL III). The nutritional value of the feed was determined by chemical analyses performed a few days prior to each stage of lactation. The percentages of particle sizes in the feed were determined at the same time. The body condition score (BCS) of each cow was also assessed before each stage of lactation using a 5-point scale (LSM ± SE): HF = 2.72 ± 0.42 and SIM = 2.78 ± 0.54. A loss of body condition was expressed as a percentage (LBCS%) in relation to the body condition score determined 5 days before calving. The average from three independent scores was calculated. The BCS was used as a subjective indicator of NEB. Nutrient requirements and the balancing of the diet were determined according to feeding standards for ruminants [17] and the INRAtion software, version 2.xx. HF cows were fed in a total mixed ration (TMR) system, with the ingredients mixed in a feed wagon. The SIM groups were fed in a partial mixed ration (PMR) system, in which the feed components were placed directly on the feed platform and mixed. The cows were fed three times a day (on average every 8 h) using feed pushing. The basal portion of the feed ration was calculated for cows with a body weight of 650 kg and expected milk production of 25 L. The production mix was introduced in cows exceeding this production value. The composition of the diet was similar in both herds. Table 1 shows the average content of nutrients in the daily ration of the experimental dairy cows in each stage of lactation. The ingredients of the daily cow diet are presented in Table 2.

Table 1. Chemical components and balancing of the diet in each stage of lactation and breed group (day/cows).

Nutrient Components	Stage of Lactation (SL)/Breed Group (BG)					
	I		II		III	
	HF	SIM	HF	SIM	HF	SIM
Crude protein (%)	16.37	15.86	16.93	16.05	14.94	12.52
Dry matter (%)	42.39	41.85	44.09	42.33	38.73	35.85
Crude fibre (%)	19.17	18.52	18.46	18.41	20.60	20.42
Crude fat (%)	2.51	2.31	2.46	2.19	2.69	2.31
Crude ash (%)	8.01	7.84	8.01	7.89	8.08	7.95
Starch (%)	22.71	21.50	22.64	21.99	21.39	21.52
Acid detergent fibre—ADF (%)	22.77	21.24	22.20	21.56	24.10	22.98
Neutral detergent fibre—NDF (%)	39.49	38.74	38.57	37.44	41.54	39.42
Physically effective NDF—peNDF(%)	30.58	29.77	28.34	28.11	36.09	35.72
UFL	21.53	20.52	23.51	22.41	17.68	15.85
PDIN (g)	2459	2341	2766	2527	1854	1698
PDIE (g)	2201	2148	2459	2374	1684	1577
Energy (MJ NEL):						
Requirement	151.8	148.5	166.8	159.8	124.9	115.9
Intake	150.3	147.2	167.4	160.9	125.9	116.9
Balance	−1.5	−1.3	+0.6	+1.1	+0.2	+0.9
Dry matter intake—DMI (kg/day)	21.49	23.75	23.89	25.07	19.54	20.09

Stage of lactation: days 6–100 (SL I), 101–200 (SL II), and >200 (SL III); HF—Holstein-Friesian; SIM—Simmental; PDIN—protein digested in the small intestine, calculated from feed nitrogen (N) available in the rumen, PDIE—protein digested in the small intestine, calculated from feed energy (E) available in the rumen.

Table 2. Ingredients of the dairy cow diet in each stage of lactation and breed group (herd).

Nutrient Components	Stage of Lactation (SL)/Breed Group (BG)					
	I		II		III	
	HF	SIM	HF	SIM	HF	SIM
Maize silage (kg)	22.5	22.5	22.5	22.5	22.5	22.5
Haylage (kg)	12.4	12.4	12.4	12.4	12.4	12.4
Ground rapeseeds (kg)	1.45	1.5	1.03	0.9	-	-
Straw (kg)	-	-	0.2	0.2	0.5.	0.5
Hay (kg)	0.5	0.5	0.8	0.8	1.6	1.6
Production mix * (kg)	6.0	5.4	6.0	5.4	3.5	3.5

Stage of lactation: days 6–100 (SL I), 101–200 (SL II), and >200 (SL III); HF—Holstein-Friesian; SIM—Simmental; * Composition of the production mix (%): crushed maize kernels—15.5; barley—10.0; triticale—10.0; oats—16.0; Krowimix 18–2.5; ground rapeseeds—18.0; NaCl—0.3; $CaCO_3$—2.3; mineral compound supplement—0.7.

The average amount of uneaten feed in each group was calculated based on weighing twice a month on average in each study period. On this basis, the average dry matter intake (DMI) was calculated. All diet components were included in the DMI calculation. The average percentages of particle sizes in the diet (PSPS sieves) were as follows: >19 mm (7%), 8–19 mm (52%), 4–8 mm (19%), and ≤4 mm (22%). About two weeks before calving, the cows received a preparatory diet. To rule out the effect of the farm, a cluster analysis that took the actual energy intake and DMI (considered to be the main factors inducing

NEB) into account was performed. The analysis showed high similarity between the farms for these parameters during the stages of lactation: 0.782–0.921 in the case of housing and 0.789–0.829 for the diets used on the farms.

2.2. Sample Collection and Analyses

Milk samples of about 250 mL were collected using a calibrated milk meter (DeLaval) that simultaneously measured the amount of milk. Milk from morning and evening milking was combined into one sample. The samples were stored in refrigerated conditions (4 °C ± 0.5). Milk was collected, on average, from the 6th day postpartum until the end of lactation: in SL I, at approximately 30, 60, and 90 days of lactation (3 times); in SL II, at approximately 130 and 170 days; and in SL III, at about 230 and 270 days, for a total of 350 milk samples. The contents of protein, fat, lactose, dry matter, and urea in the milk were determined using the Bentley Combi 150 (Bentley Instruments, Inc., Chaska, MN, USA). The analysis was performed in a laboratory accredited by the Polish Accreditation Centre.

The fatty acid (FA) profile of the milk fat was determined by gas chromatography (Agilent 6890 N). Fat was extracted by the Röse-Gottlieb method [18]. The transmethylation of FA to methyl esters (FAME) was carried out at 70 °C ± 0.5 (Thermo heat block). The GLC modules were an autosampler, a split/splitless injector (split 1:5), and a flame ionization detector (FID). Separation was carried out on a 100 m, 0.250 mm column (HP-88; SN:UST458414H, Agilent Technologies Inc., Santa Clara, CA, USA). The temperature programme; injector and FID had the following run: 250 °C; furnace—95 °C (5 min); 120 °C (15 °C/min—15 min); 210 °C (25 °C/min—30 min); and 250 °C (20 °C/min—5 min). Carrier gas flow (He) was set at 5.0 mL/min. The identification of FA and determination of their percentages were based on retention times (reference Supelco 37.No:47885-U; Sigma Aldrich, St. Louis, MO, USA) in the Agilent Tech GC Chemstation A09.03 software. The following FA groups were distinguished: SCFAs—short-chain fatty acids; LCFAs—long-chain fatty acids; SFAs—saturated fatty acids; MUFAs—monounsaturated fatty acids; PUFAs—polyunsaturated fatty acids; and UFAs—unsaturated fatty acids. In addition, the contents of the (C18:0), oleic (C18:1), and palmitic (C16:0) acids in the structure of milk fat were examined. Among others, these FAs were aNEB markers.

Blood for analysis was drawn before morning feeding from the jugular vein. Due to the potential effect of stress on milk yield, blood was taken 24 h after milk was collected for analysis. Blood was collected at 30, 60, and 90 days of lactation (SL I), at 130 and 170 days (SL II), and at 230 and 270 days (SL III) for a total of 350 samples: 3 (SL) × 2 (BG) × 2 (samples) × 25 (animals). Test tubes with sodium fluoride and sodium heparin (Medlab-Products Ltd., Raszyn, Poland) were refrigerated. Blood for glucose determination was placed in ice. Glucose content in the blood was measured using original Randox kits (Randox Laboratories Ltd., Crumlin, UK) and a UV–Vis spectrophotometer (Varian Inc., Palo Alto, CA, USA). Samples for the determination of BHBA (β-hydroxybutyrate) were centrifuged at 1500× *g* at 4 °C for 20 min. The supernatant was collected and stored at −75 °C ± 1.0 until BHBA analysis using original Randox kits (Randox Laboratories Ltd., Crumlin, UK) and a UV–Vis spectrophotometer (Varian Inc., Palo Alto, CA, USA). The plasma samples were analysed for leptin levels using a bovine-specific ELISA kit (EIAab, Wuhan, China). The share of NEFA was determined as the sum of C16:0, C18:0, and cis-9 C18:1. All samples were measured in triplicate.

2.3. Statistical Analysis

The statistical analysis of the results was performed with the Statistica 12.0 software. Analysis of variance was carried out in a general linear model (GLM) with repeated measures. For each BG, HF and SIM, the model included the effect of the three periods of lactation (SL): days 6–100 (SL I), 101–200 (SL II), and >200 (SL III); the interaction between the BG and SL; and the random effect of the cow. The effect of feeding technique (TMR or PMR) was verified by cluster analysis using the k-means method. The model took the actual energy intake, DMI, and effect of farm into account. The results are presented as means (LSM) and standard error (SEM). The significance of differences between means was estimated by Duncan's test at $p \leq 0.05$. Correlations (r) between selected parameters were estimated using the Pearson correlation model ($p \leq 0.05$).

3. Results

HF cows usually had a higher daily milk production (DMP) than SIM cows (Table 3). In the stage with the highest milk production (SL I), the loss of body condition was similar in cows from both breed groups. During this period, the average LBCS was −11.80% ($p \leq 0.05$). SIM cows, however, were able to restore their BCS faster, as from SL II, their LBCS was on average −2.1 percentage point (p.p.) smaller ($p \leq 0.05$). Up to SL II, the DMP level remained similar. After this period, a significant decrease in DMP was noted (SL III). The average difference was 7.9 kg ($p \leq 0.05$). As lactation progressed, a slight increase in the BCS was observed ($p \leq 0.05$). In the first two stages of lactation, however, the restoration of body condition was slower. This was indicated by the differences in the LBCS between analysed stages of lactation: 1.2 between SL I and SL II and 7.7 p.p. between SL II and SL III ($p \leq 0.05$).

Table 3. Production parameters and body condition of cows depending on the breed group and stage of lactation.

Parameters	Breed Group (BG)/Stage of Lactation (SL)						SEM	p
	HF			SIM				
	I	II	III	I	II	III		
Number of cows (n)	25	25	25	25	25	25		SL
Day of lactation (day)	40 ^{c}	165 ^{b}	285 ^{a}	45 ^{c}	162 ^{b}	282 ^{a}	6.4	*
DMP (kg)	30.9 ^{a}	29.8 ^{a}	21.5 ^{c}	27.6 ^{b}	27.3 ^{b}	20.5 ^{c}	0.4	*
BCS (points)	2.40 ^{d}	2.41 ^{d}	2.63 ^{b}	2.45 ^{d}	2.52 ^{c}	2.71 ^{a}	0.3	*
LBCS (%)	−11.8 ^{a}	−11.4 ^{a}	−3.3 ^{c}	−11.9 ^{a}	−9.3 ^{b}	−2.5 ^{c}	0.2	*

Significance of the differences in the results within the breed group (BG): ^{abcd} $p \leq 0.05$ and stages of lactation (SL): * $p \leq 0.05$; stage of lactation: days 6–100 (SL I), 101–200 (SL II), and >200 (SL III); HF—Holstein-Friesian; SIM—Simmental; DMP—daily milk production; BCS—body condition score; LBCS—loss of BCS relative to BCS at 5 days before calving.

The data in Table 4 show that over the course of lactation, there was a downward trend in the contents of NEFA and BHBA ($p \leq 0.05$), whose levels were usually higher in HF cows. Differences between breed groups were usually greater in the first two stages of lactation than in SL III—on average, 19.65 μmol $\cdot$L^{-1} in the case of NEFA and 0.121 mmol$\cdot$L^{-1} for BHBA ($p \leq 0.05$). The changes in glucose content were also greater during these stages. Glucose content decreased up to SL II and then increased ($p \leq 0.05$). In the SIM group, however, it decreased less (0.052 mmol$\cdot$L^{-1}; $p \leq 0.05$) while the content of leptin, which exerts an anorectic effect, decreased more (0.19 ng$\cdot$ml^{-1}; $p \leq 0.05$). Leptin content was also influenced by the SL. The differences between the stages of lactation indicated the greatest decrease in leptin content up to SL II, on average 0.14 ng$\cdot$mL^{-1} ($p \leq 0.05$), after which it remained at a similar level until the end of lactation. A higher urea nitrogen in milk (MUN) content, on average by 7.9 mmol$\cdot$L^{-1} ($p \leq 0.05$), was noted in the HF cows in SL I (Table 4). Generally, in subsequent stages of lactation, its level decreased ($p \leq 0.05$), and in SL III it, was similar in both groups. Compared to that of HF cows, the milk of SIM cows had a higher DM content. However, in both BGs, there were no significant differences that indicated the influence of SL.

The milk of the SIM cows contained more SCFA (Table 5). The differences relative to HF ranged from 0.23 to 0.74 p.p. ($p \leq 0.05$) and were smaller up to SL II. Changes in the content of SCFA during lactation showed a downward trend ($p \leq 0.05$). The reverse trend was noted for LCFA, whose content was higher in the milk of HF cows, on average from 0.86 p.p (SL I) to 3.39 p.p (SL III). Higher contents of C16:0, C18:0, and C18:1 (treated as markers of NEB) were noted in the milk of HF cows. The greatest changes were observed in SL I and II. The differences between breed groups ranged from 0.12 p.p. in the case of C18:0 in SL III to 1.51 p.p for C16:0 in SL I ($p \leq 0.05$). Differences between stages of lactation indicated that as lactation progressed, the share of C16:0 and C18:0 gradually declined ($p \leq 0.05$). The levels of MUFA and PUFA, including C18:2, in the milk of the breed groups were only slightly higher in SIM cows and after the peak of lactation. In the case of these fractions, however, no clear patterns were confirmed (Table 5). The same was true for UFA. In the case of SFA, C18:1, and C18:2, gradual decreases in their contents in milk were observed over the course of lactation. The differences between stages were confirmed at $p \leq 0.05$ for SFA and $p \leq 0.05$ for C18:1 and C18:2. A different trend was noted in the case of the effect of lactation on PUFA and UFA. Their contents increased over the course of lactation, which was confirmed at $p \leq 0.05$ in both cases.

Table 4. Parameters of energy metabolism and contents of milk components depending on the breed group and stage of lactation.

Parameters	Breed Group (BG)/Stage of Lactation (SL)						SEM	p
	HF			SIM				
	I	II	III	I	II	III		
Blood samples (n)	75	50	50	75	50	50		SL
NEFA (μmol L^{-1})	248.4 [a]	211.5 [c]	178.9 [e]	229.5 [b]	191.1 [d]	182.8 [e]	2.8	*
BHBA (mmol L^{-1})	1.021 [a]	0.978 [b]	0.743 [c]	0.955 [b]	0.776 [c]	0.647 [d]	0.02	*
Glucose (mmol L^{-1})	2.429 [c]	2.360 [d]	2.638 [a]	2.559 [b]	2.507 [c]	2.632 [a]	0.01	*
Leptin (ng ml^{-1})	2.74 [a]	2.65 [b]	2.50 [c]	2.59 [b]	2.40 [c]	2.43 [c]	0.02	*
Milk samples (n)	75	75	75	75	75	75		
MUN (mmol L^{-1})	181.3 [a]	167.5 [c]	164.3 [c]	173.4 [b]	164.7 [c]	162.7 [c]	0.71	*
Fat/Protein—F/P (%)	1.23	1.21	1.21	1.22	1.23	1.21	0.54	ns
Lactose (%)	5.01	4.78	5.19	4.97	4.85	5.12	0.32	*
Dry matter—DM (%)	12.35 [b]	12.23 [b]	12.52 [b]	12.70 [a]	12.97 [a]	12.86 [a]	0.08	ns

Significance of the differences in the results within the breed group (BG): [abcde] $p \leq 0.05$ and stages of lactation (SL): * $p \leq 0.05$, ns—not significant; stage of lactation (SL): days 6–100 (SL I), 101–200 (SL II), and >200 (SL III); HF—Holstein-Friesian; SIM—Simmental; BCS—body condition score; BHBA—β-hydroxybutyrate; NEFA—non-esterified fatty acid (FA); MUN—urea nitrogen in milk.

Table 5. Contents of the main fatty acid fractions in milk depending on the breed group and stage of lactation.

Parameters	Breed Group (BG)/Stage of Lactation (SL)						SEM	p
	HF			SIM				
	I	II	III	I	II	III		
Number of samples (n)	75	75	75	75	75	75		SL
SCFA (%)	9.52 [b]	8.67 [c]	7.75 [e]	9.73 [a]	8.86 [c]	8.49 [d]	0.05	*
LCFA (%)	59.79 [b]	60.53 [a]	60.77 [a]	58.93 [c]	58.99 [c]	57.38 [d]	0.16	ns
Palmitic acid C16:0 (%)	25.81 [a]	24.90 [b]	22.09 [c]	24.30 [b]	24.23 [b]	21.86 [c]	0.15	*
Stearic acid C18:0 (%)	11.98 [a]	11.59 [b]	11.38 [c]	11.67 [b]	11.26 [d]	11.23 [d]	0.44	*
SFA (%)	68.93 [b]	69.23 [a]	68.24 [b]	68.29 [b]	67.57 [c]	65.64 [d]	0.17	*
MUFA (%)	27.93	26.74	28.19	28.59	28.97	28.95	0.51	ns
Oleic acid C18:1 (%)	2.83 [a]	2.37 [b]	2.20 [bc]	2.67 [a]	2.16 [c]	2.02 [c]	0.04	*
PUFA (%)	1.57 [c]	1.68 [c]	1.94 [a]	1.63 [c]	1.79 [b]	1.96 [a]	0.01	*
Linoleic acid C18:2 (%)	1.68 [a]	1.71 [a]	1.42 [c]	1.74 [a]	1.57 [b]	1.53 [b]	0.02	*
UFA (%)	29.50 [a]	28.42 [b]	30.13 [a]	30.23 [a]	30.76 [a]	30.93 [a]	0.12	*

Significance of the differences in the results within the breed group (BG): [abcde] $p \leq 0.05$ and stages of lactation (SL): * $p \leq 0.05$, ns—not significant; stage of lactation (SL): days 6–100 (SL I), 101–200 (SL II), and >200 (SL III); HF—Holstein-Friesian; SIM—Simmental; LCFA—long-chain fatty acid; SCFA—short-chain FA; SFA—saturated FA; MUFA—monounsaturated FA; PUFA—polyunsaturated FA; UFA—unsaturated FA.

The results presented in Table 6 indicate that the day of peak lactation and milk production on that day (DMP-PY) were negatively associated with the content of leptin in the blood and BSC. This was indicated by the correlation coefficients, which ranged from

−0.318 to −0.452 ($p \leq 0.05$). These parameters, however, were positively correlated with the contents of NEFA, C16:0, C18:0, and C18:1, whose release from the adipocytes was increased during NEB. These values ranged from 0.286 to 0.543 ($p \leq 0.05$). These markers were positively correlated with BHBA (0.797–0.295; $p \leq 0.05$) and MUN (0.711 and 0.449; $p \leq 0.05$). The content of leptin, which exerts an anorectic effect, was positively correlated with the LBCS (0.399) and markers of NEB (0.547–0.629; $p \leq 0.05$). A negative correlation was noted between leptin and MUN (−0.455; $p \leq 0.05$).

Table 6. Selected parameters of production and energy metabolism, as well as correlations between them.

Parameter	LSM ± SD	Value of Correlation Coefficient, Significant at $p \leq 0.05$						
		DMP	BCS	Leptin	NEFA	C16:0	C18:1	C18:0
Lactation (days)	307 ± 9.2	−0.464	0.462	−0.260	−0.418	−0.462	−0.333	−0.426
Peak yield (days)	65.3 ± 8.1	0.426	−0.452	−0.358	0.543	0.385	0.429	0.337
DMP-PY (kg)	33.4 ± 4.2	0.482	−0.437	−0.318	0.342	0.428	0.382	0.286
LBCS (%)	−8.4 ± 3.4	0.758	-	0.399	0.637	0.745	0.567	0.385
NEFA (μmol L^{-1})	207.1 ± 18.3	0.805	−0.471	0.586	-	0.789	0.718	0.399
BHBA (mmol L^{-1})	0.853 ± 0.217	0.810	−0.513	0.524	0.797	0.794	0.751	0.295
Glucose (mmol L^{-1})	2.519 ± 0.251	−0.831	0.238	−0.566	−0.698	−0.793	−0.696	−0.250
MUN (mmol L^{-1})	168.9 ± 18.5	0.676	−0.530	−0.455	0.711	0.685	0.643	0.449
C16:0 (%)	23.86 ± 3.34	0.975	−0.577	0.629	0.789	-	0.816	0.353
C18:1 (%)	2.37 ± 0.16	0.831	−0.438	0.547	0.717	0.816	-	0.274

DMP—daily milk production; DMP-PY—daily milk production at peak yield; MUN—urea nitrogen in milk LBCS—loss of BCS relative to BCS at 5 days before calving; BCS—body condition score; BHBA—β-hydroxybutyrate; NEFA—non-esterified fatty acid (FA); MUN—urea nitrogen in milk.

4. Discussion

The correction of disturbances of energy homeostasis induced by intensive lactogenesis is a major challenge for the physiology of dairy cows. A lower feed intake at this time intensifies NEB and increases lipolysis [4,19]. Though a considerable portion of NEFA in cows is derived from the diet, Gross et al. (2013) [4] indicated that the intensity of daily production also has a major impact. In their opinion, the rate of change in the NEB and triglycerides, including NEFA, released from the adipocytes increases up to the peak of lactation, thus increasing the NEFA level. The consequence of this is of the inability of liver to completely use acetyl CoA from the beta oxidation of NEFA for gluconeogenesis because other intermediates of the Krebs cycle are limited for the excessive load of acetyl CoA. Finally, this situation may be conducive to the generation of excess BHBA in the liver and affects the further course of lactation [13,20]. In a study by Knob et al. (2021) [6], the BCS of Simmental cows was, on average, 1 point higher than that of Holstein cows during lactation. As in the present study, Knob et al. (2021) [6] observed the greatest loss of the BCS up to the peak of production. However, in Simmental and Simmental crossbreds, body condition relative to the BCS before calving was reduced by 12–18%, while in Holsteins, it fell by 20–24%. The stronger body condition of Simmental cows was explained by research of Yan et al. (2006) [21] and Ledinek et al. (2018) [22], in which nutrients were more evenly distributed between the lactogenesis and restoration of the BCS in cows with lower production potential. In the present study, we observed a similar relationship between NEB markers and daily yield in both breeds. The observed trends were confirmed by their significant correlations with DMP and the length of lactation. Knob et al. (2021) [6] found that the genetic group influenced yield at the peak of lactation and the time when the peak was attained. In comparison with Holstein cows, the production potential of the Simmental breed was much lower, as these authors noted a reduction in daily yield by about the fourth week of lactation. In our study, the reduction in DMP occurred much later. This difference could be explained by the much lower milk yield of the breeds analysed in our research and the study by Knob et al. (2021) [6]. Our study also showed a greater production potential in HF cows. SIM cows, on the other hand, had a greater capacity to correct the NEB, and after the peak of lactation, their BCS was more quickly restored. In this respect, the results of our study were consistent with those reported by Knob et al. (2021) [6]. Furthermore, Knob et al. (2021) [6] showed that the interaction of breed and the stage of lactation only affects NEB markers (NEFA and BHBA) to a certain extent. In contrast with our results, Knob et al. [6] found that these differences began to vanish from the fourth week of lactation. A study by Ingvartsen and Andersen (2000) [23] showed that due to genetic determinants, feed consumption in early lactation has a minor influence on milk production. Yan et al. (2006) [21], however, found that diet quality remains a significant factor by maintaining body condition during lactation and allowing cows to meet their genetic potential. According to Friggens et al. (2007) [24], changes in milk production during lactation are mainly determined by hormone metabolism, which determines the distribution of energy resources released from fat stores. In a study by Ledinek et al. (2018) [22], differences in the yield between breeds began to vanish in late lactation. Like Friggens et al. (2007) [24], they observed a decrease in hormones involved in milk secretion. Ledinek et al. (2018) [22] and Friggens et al. (2007) [24] observed a decrease in NEFA and an increase in glucose content as lactation progressed. In light of research by Ledinek et al. [22], Yan et al. (2006) [21], and Friggens et al. (2007) [24], the effects of breed and lactation on the content of the anorectic hormone leptin noted in our study are not without significance. Ledinek et al. (2018) [22] noted a higher BCS and a faster restoration of body condition to the state immediately after calving in cows of the Fleckvieh breed in comparison to HF. The differences in leptin content noted between breeds in our study and its positive correlation with the LBCS indicated that its anorectic function may delay the restoration of the BCS. According to Ledinek et al. (2018) [22] and Yan et al. (2006) [21], genetic potential is not without influence on changes in the BCS. In a study by Stengarde et al. (2008) [25], the trends in glucose content were similar to those noted in our study, and its level depended, in part, on the intensity of lactogenesis. Our study also showed that lower glucose content is accompanied by a greater loss of the BCS and a higher NEFA content. These results were in contrast to those reported by Knob et al. [6]. Though they found the highest glucose content in Simmental cows, there was no significant difference in comparison with the Holstein breed. The function of the NEB and dry matter intake were also affected by the level of glucose present in the body. Glucose content is generally lower during periods of limited feed intake and at high concentrations of NEFA [20]. Šamanc et al. (2015) [26] showed that glucose concentration was usually lower in cows with a greater loss of the BCS. In their research, the glucose content in these cows was 0.16–0.72 mmol L^{-1} lower and NEFA content was 0.12–0.20 mmol L^{-1} higher. This could be explained by the reduced glucogenic efficiency of liver cells, especially in cows with a greater reduction in the BCS and with a higher rate of BHBA production [27]. This is in agreement with research by Loor et al. (2007) [28], in which induced ketosis led to a marked decrease in glucose content and a sudden increase in NEFA. Drackley et al. (2001) [3], however, did not find glucose concentration to be dependent on NEB. In that study, the glucose levels in groups of cows differing in NEFA (as an indicator of NEB) content were similar, and no relationship was shown between the contents of glucose and insulin. The correlation coefficients obtained in our study indicated that the glucose level in the blood was negatively correlated with leptin. It was also strongly positively correlated with the BCS and strongly negatively correlated with DMP and markers of NEB. Cows with a greater NEB—especially in terms of BCS loss and increased levels of C16:0, C18:0, and C18:1—had lower glucose levels. The higher levels of these FAs may be explained by research by Stoop et al. (2009) [29], in which cows with NEB had increased contents of C16:0 and C18:0. According to the authors, this may indicate an increased mobilization of body fat reserves to provide substrates for the de novo synthesis of FA. Our study also showed that NEB can cause changes in the content of certain FA fractions, and the magnitude of these changes may be associated with both the lactation stage and breed. Research by Weber et al. (2013) [20] and Bastin et al. (2011) [30] also showed an increase in the proportion of LCFA, mainly up to the peak of lactation. After this time, they observed a stronger increase in the content of SFAs, including C18:0. In our study, we noted a slight decrease in this fraction as lactation progressed. However, we found that the stage of lactation affected the share of UFAs and that it was higher in the milk of Simmental cows than in the milk of HF. Roche et al. (2009) [31] and Ducháček et al. (2020) [19] demonstrated that the milk of cows with a mild NEB and a less severe loss of the BCS had a lower content of SFA. This may explain the results of our study, in which SFA content was lower following the period of intensive milk production and the reduction in the NEB. The differences we noted between breeds may, to some extent, be explained by research by Bastin et al. (2011) [30] that showed that changes in the structure of milk fat may be influenced by individual traits and genetic potential. The effect of the rate of production, resulting from the genetic potential of the breed, on the FA profile was also confirmed by Samková et al. (2014) [32] and Młynek et al. (2021) [33]. On the other hand, studies by Kay et al. (2005) [27] found that selection for milk yield was not the main factor determining the FA profile of milk. Petit and Côrtes (2010) [34] reported a positive correlation between SFA and an increase in the level of BHBA produced in the liver, but they noted these changes in cows fed ground flax seeds. This diet caused an increase in the share of MUFAs and PUFAs in the milk. In our study, in which both breeds received a similar diet, Simmental cows produced milk with a more favourable FA profile and also had lower BHBA and NEB markers. Furthermore, the milk of Simmental cows usually contained less urea (MUN), another product of metabolic changes in the liver [35]. However, this situation may have been influenced by the greater intake of biodegradable nitrogen, as the HF cows, due to their greater production, may have ingested more concentrate feed to compensate for their greater milk yield. MUN content is determined by numerous factors [36,37]. Studies by Kohn et al. (2002) [38] and Spek (2013) [39] indicated that a large portion of it arises from protein supplied with the diet. Following the detoxification of ammonia in the liver, it can then be eliminated, in part with the milk [40]. Of course, maintaining a balance between the intake of protein and energy in the diet plays a major role [41]. Our study, however, indicated that NEB also has a significant effect, as we observed higher MUN values in cows with faster BCS losses and higher contents of NEB markers, which were metabolized, in part, in the liver. The NEB's function is also associated with the level of glucose present in the body. Glucose content is generally lower during periods of limited feed intake and at high concentrations of NEFA [20]. Šamanc et al. (2015) [26] showed that the glucose concentration was usually lower in cows with a greater loss of the BCS. In these cows, glucose content was 0.16–0.72 mmol L lower and the NEFA content was 0.12–0.20 mmol L higher. This can be explained by the reduced glucogenic efficiency of liver cells, especially in cows with a greater reduction in the BCS and a higher rate of BHBA production [40]. This is in agreement with research by Loor et al. (2007) [28], in which induced ketosis led to a marked decrease in glucose content and a sudden increase in NEFA. Drackley et al. (2001) [3], however, did not find glucose concentration to be dependent on NEB. In that study, the glucose level in groups of cows differing in NEFA content was similar, and no relationship was shown between the contents of glucose and insulin. In contrast with our study, Blum et al. (1983) [42] showed no differences in blood glucose levels between genetic groups, including HF and Simental. In their opinion, variation in glucose content is mainly influenced by the stage of lactation. However, a study by Djokovic et al. (2011) [43] indicated that it also depends on daily milk production. This may explain the differences obtained in our study between the HF and Simmental breeds, which had different intensities of lactogenesis in the first stage of lactation. In our study, higher glucose levels were usually noted in cows with smaller losses of the BCS and lower levels of NEFA and BHBA. At the same time, leptin content was lower in these cows, which may indicate a link with stronger appetite [5].

5. Conclusions

Breed influenced daily milk production up to about 100 days of lactation. After this time and until the end of lactation, the differences between NEB markers were much smaller. Cows of the SIM and HF breeds had similar dynamics of body condition loss up to the peak of lactation. In the SIM breed, however, the BCS was more quickly restored to the level noted immediately before calving. The lower urea content in the milk of SIM cows may be explained by their slightly lower production, as well as by their NEB markers. This may indicate differences arising from the potential of these breeds in terms of the physiological capacity to cope with NEB. The results indicated that the anorectic effect of leptin may play an important role in determining parameters characterizing NEB because higher leptin levels were noted during periods of intensive lactogenesis and in HF cows. The results provide a source of information that can be useful in finding solutions that could lead to the more efficient exploitation of the production potential of popular breeds of dairy cows.

Author Contributions: Conceptualization, I.S. and K.M.; Data curation, I.S. and A.D.; Formal analysis, I.S., K.M. and A.D.; Investigation, I.S.; Methodology, K.M.; Writing—original draft, I.S.; Writing—review & editing, K.M. All authors have read and agreed to the published version of the manuscript.

Funding: Statutory research 53/20/B and Implementation project "Casein program" SM SPOMLEK, POLAND.

Institutional Review Board Statement: The Third Local Ethics Committee in Warsaw, Poland (approval no. 70/2013).

Informed Consent Statement: Not applicable.

Data Availability Statement: The data presented in this study are available on request from the corresponding author. The data are not publicly available due to: The research was carried out in private farms. Due to the privacy of farmers, the data will be made available in the form of a database.

Conflicts of Interest: The authors declare no conflict of interest.

References

1. Bobe, G.; Young, J.; Beitz, D. Invited Review: Pathology, Etiology, Prevention, and Treatment of Fatty Liver in Dairy Cows. *J. Dairy Sci.* **2004**, *87*, 3105–3124. [CrossRef]
2. Drackley, J.; Donkin, S.; Reynolds, C. Major Advances in Fundamental Dairy Cattle Nutrition. *J. Dairy Sci.* **2006**, *89*, 1324–1336. [CrossRef]
3. Drackley, J.K.; Overton, T.R.; Douglas, G.N. Adaptations of Glucose and Long-Chain Fatty Acid Metabolism in Liver of Dairy Cows during the Periparturient Period. *J. Dairy Sci.* **2001**, *84*, E100–E112. [CrossRef]
4. Gross, J.; Schwarz, F.; Eder, K.; Van Dorland, H.; Bruckmaier, R. Liver fat content and lipid metabolism in dairy cows during early lactation and during a mid-lactation feed restriction. *J. Dairy Sci.* **2013**, *96*, 5008–5017. [CrossRef] [PubMed]
5. Młynek, K.; Głowińska, B.; Salamończyk, E. The relationship of body condition indicators with the level of plasma cocaine and amphetamine regulated transcript and energy metabolism indicators in dairy cows. *Acta Vet. Brno* **2019**, *88*, 11–18. [CrossRef]
6. Knob, D.; Neto, A.T.; Schweizer, H.; Weigand, A.; Kappes, R.; Scholz, A. Energy Balance Indicators during the Transition Period and Early Lactation of Purebred Holstein and Simmental Cows and Their Crosses. *Animals* **2021**, *11*, 309. [CrossRef]
7. Rukkwamsuk, T.; Geelen, M.; Kruip, T.; Wensing, T. Interrelation of Fatty Acid Composition in Adipose Tissue, Serum, and Liver of Dairy Cows During the Development of Fatty Liver Postpartum. *J. Dairy Sci.* **2000**, *83*, 52–59. [CrossRef]
8. Loften, J.R.; Linn, J.G.; Drackley, J.K.; Jenkins, T.C.; Soderholm, C.G.; Kertz, A.F. Invited review: Palmitic and stearic acid metabolism in lactating dairy cows. *J. Dairy Sci.* **2014**, *97*, 1–14. [CrossRef]
9. Puppel, K.; Gołębiewski, M.; Solarczyk, P.; Grodkowski, G.; Ślósarz, J.; Kunowska-Ślósarz, M.; Balcerak, M.; Przysucha, T.; Kalińska, A.; Kuczyńska, B. The relationship between plasma β-hydroxybutyric acid and conjugated linoleic acid in milk as a biomarker for early diagnosis of ketosis in postpartum Polish Holstein-Friesian cows. *BMC Vet. Res.* **2019**, *15*, 367. [CrossRef]
10. McNamara, J.P. Integrating Genotype and Nutrition on Utilization of Body Reserves During Lactation of Dairy Cattle. In *Ruminant Physiology*; Cronje, P.B., Ed.; CAB Int.: London, UK, 2000; pp. 353–370. [CrossRef]
11. Brickner, A.E.; Pires, J.A.A.; Gressley, T.F.; Grummer, R.R. Effects of abomasal lipid infusion on liver triglyceride accumulationand adipose lipolysis during fatty liver induction in dairy cows. *J. Dairy Sci.* **2009**, *92*, 4954–4961. [CrossRef]
12. Kuhla, B.; Metges, C.C.; Hammon, H.M. Endogenous and dietary lipids influencing feed intake and energy metabolism of periparturient dairy cows. *Domest. Anim. Endocrinol.* **2016**, *56*, 2–10. [CrossRef]

13. White, H.M. The Role of TCA Cycle Anaplerosis in Ketosis and Fatty Liver in Periparturient Dairy Cows. *Animals* **2015**, *5*, 793–802. [CrossRef] [PubMed]
14. Zhu, L.; Armentano, L.; Bremmer, D.; Grummer, R.; Bertics, S. Plasma Concentration of Urea, Ammonia, Glutamine Around Calving, and the Relation of Hepatic Triglyceride, to Plasma Ammonia Removal and Blood Acid-Base Balance. *J. Dairy Sci.* **2000**, *83*, 734–740. [CrossRef]
15. Lopreiato, V.; Minuti, A.; Trimboli, F.; Britti, D.; Morittu, V.; Cappelli, F.P.; Loor, J.; Trevisi, E. Immunometabolic status and productive performance differences between periparturient Simmental and Holstein dairy cows in response to pegbovigrastim. *J. Dairy Sci.* **2019**, *102*, 9312–9327. [CrossRef]
16. AOAC International. *Official Methods of Analysis of AOAC International*, 18th ed.; AOAC International: Gaithersburg, MD, USA, 2007.
17. Strzetelski, J.E.D. National Research Institute of Animal Production INRA Norm of Ruminant Nutrition. In *The Nutritional Value of French and National Feed fur Ruminants*; Instytut Zootechniki PIB: Kraków, Poland, 2009. (In Polish)
18. Official Methods of Analysis. International Method IDF–ISO–AOAC. Nr 905.02. Gravimetric Method (Röse–Gottlieb). Available online: https://www.aoac.org/official-methods-of-analysis-21st-edition-2019 (accessed on 28 December 2020).
19. Ducháček, J.; Stádník, L.; Ptáček, M.; Beran, J.; Okrouhlá, M.; Gašparík, M. Negative Energy Balance Influences Nutritional Quality of Milk from Czech Fleckvieh Cows due Changes in Proportion of Fatty Acids. *Animals* **2020**, *10*, 563. [CrossRef]
20. Weber, C.; Hametner, C.; Tuchscherer, A.; Losand, B.; Kanitz, E.; Otten, W.; Singh, S.P.; Bruckmaier, R.M.; Becker, F.; Kanitz, W.; et al. Variation in fat mobilization during early lactation differently affects feed intake, body condition, and lipid and glucose metabolism in high-yielding dairy cows. *J. Dairy Sci.* **2013**, *96*, 165–180. [CrossRef] [PubMed]
21. Yan, T.; Mayne, C.S.; Keady, T.W.J.; Agnew, R.E. Effects of dairy cow genotype with two planes of nutrition on energy partitioning between milk and body tissue. *J. Dairy Sci.* **2006**, *89*, 1031–1042. [CrossRef]
22. Ledinek, M.; Gruber, L.; Steininger, F.; Zottl, K.; Royer, M.; Krimberger, K.; Mayerhofer, M.; Egger-Danner, C.; Fuerst-Waltl, B. Analysis of lactating cows in commercial Austrian dairy farms: Diet composition, and influence of genotype, parity and stage of lactation on nutrient intake, body weight and body condition score. *Ital. J. Anim. Sci.* **2018**, *18*, 202–214. [CrossRef]
23. Ingvartsen, K.L.; Andersen, J.B. Integration of Metabolism and Intake Regulation: A Review Focusing on Periparturient Animals. *J. Dairy Sci.* **2000**, *83*, 1573–1597. [CrossRef]
24. Friggens, N.; Berg, P.; Theilgaard, P.; Korsgaard, I.; Ingvartsen, K.; Lovendahl, P.; Jensen, J. Breed and Parity Effects on Energy Balance Profiles Through Lactation: Evidence of Genetically Driven Body Energy Change. *J. Dairy Sci.* **2007**, *90*, 5291–5305. [CrossRef]
25. Stengarde, L.; Traven, M.; Emanuelson, U.; Holtenius, K.; Hultragen, J.; Niskanen, R. Metabolic profiles in five high-producing Swedish dairy herds with a history of abomasal displacement and ketosis. *Acta Vet. Scand.* **2008**, *50*, 31–41. [CrossRef]
26. Šamanc, H.; Gvozdić, D.; Fratrić, N.; Kirovski, D.; Djoković, R.; Sladojević, Ž.; Cincović, M. Body condition score loss, hepatic lipidosis and selected blood metabolites in Holstein cows during transition period. *Anim. Sci. Pap. Rep.* **2015**, *33*, 35–47.
27. Kay, J.K.; Weber, W.J.; Moore, C.E.; Bauman, D.E.; Hansen, L.B.; Chester-Jones, H.; Crooker, B.A.; Baumgard, L.H. Effects of week of lactation and genetic selection for milk yield on milk fatty acid composition in Holstein cows. *J. Dairy Sci.* **2005**, *88*, 3886–3893. [CrossRef]
28. Loor, J.J.; Everts, R.E.; Bionaz, M.; Dann, H.M.; Morin, D.E.; Oliveira, R.; Rodriguez-Zas, S.L.; Drackley, J.K.; Lewin, H.A. Nutrition-induced ketosis alters metabolic and signaling gene networks in liver of periparturient dairy cows. *Physiol. Genomics* **2007**, *32*, 105–116. [CrossRef]
29. Stoop, W.M.; Bovenhuis, H.; Heck, J.M.L.; van Arendonk, J.A.M. Effect of lactation stage and energy status on milk fat composition of Holstein-Friesian cows. *J. Dairy Sci.* **2009**, *92*, 1469–1478. [CrossRef] [PubMed]
30. Bastin, C.; Gengler, N.; Soyeurt, H. Phenotypic and genetic variability of production traits and milk fatty acid content across days in milk for Walloon Holstein first-parity cows. *J. Dairy Sci.* **2011**, *94*, 4152–4163. [CrossRef] [PubMed]
31. Roche, J.R.; Friggens, N.C.; Kay, J.K.; Fisher, M.W.; Stafford, K.J.; Berry, D.P. Invited review: Body condition score and its association with dairy cow productivity, health, and welfare. *J. Dairy Sci.* **2009**, *92*, 5769–5801. [CrossRef]
32. Samková, E.; Čertíková, J.; Špička, J.; Hanuš, O.; Pelikánová, T.; Kvač, M. Eighteen-carbon fatty acids in milk fat of Czech Fleckvieh and Holstein cows following feeding with fresh lucerne (*Medicago sativa* L.). *Anim. Sci. Pap. Rep.* **2014**, *32*, 209–218.
33. Młynek, K.; Danielewicz, A.; Strączek, I. The Effect of Energy Metabolism Up to the Peak of Lactation on the Main Fractions of Fatty Acids in the Milk of Selected Dairy Cow Breeds. *Animals* **2021**, *11*, 112. [CrossRef] [PubMed]
34. Petit, H.V.; Côrtes, C. Milk production and composition, milk fatty acid profile, and blood composition of dairy cows fed whole or ground flaxseed in the first half of lactation. *Anim. Feed Sci. Tech.* **2010**, *158*, 36–43. [CrossRef]
35. Kauffman, A.; St-Pierre, N. The Relationship of Milk Urea Nitrogen to Urine Nitrogen Excretion in Holstein and Jersey Cows. *J. Dairy Sci.* **2001**, *84*, 2284–2294. [CrossRef]
36. Roy, B.; Mehla, R.K.; Sirohi, S.K. Influence of Milk Yield, Parity, Stage of Lactation and Body Weight on Urea and Protein Concentration in Milk of Murrah Buffaloes. *Asian Australas. J. Anim. Sci.* **2003**, *16*, 1285–1290. [CrossRef]
37. Sawa, A.; Bogucki, M.; Krężel-Czopek, S. Effect of some factors on relationships between milk urea levels and cow fertility. *Arch. Anim. Breed.* **2011**, *54*, 468–476. [CrossRef]
38. Kohn, R.A.; Kalscheur, K.F.; Russek-Cohen, E. Evaluation of Models to Estimate Urinary Nitrogen and Expected Milk Urea Nitrogen. *J. Dairy Sci.* **2002**, *85*, 227–233. [CrossRef]

39. Spek, J.W.; Dijkstra, J.; Van Duinkerken, G.; Bannink, A. A review of factors influencing milk urea concentration and its relationship with urinary urea excretion in lactating dairy cattle. *J. Agric. Sci.* **2013**, *151*, 407–423. [CrossRef]
40. Nousiainen, J.; Shingfield, K.; Huhtanen, P. Evaluation of Milk Urea Nitrogen as a Diagnostic of Protein Feeding. *J. Dairy Sci.* **2004**, *87*, 386–398. [CrossRef]
41. Geerts, N.; De Brabander, D.; Vanacker, J.; De Boever, J.; Botterman, S. Milk urea concentration as affected by complete diet feeding and protein balance in the rumen of dairy cattle. *Livest. Prod. Sci.* **2004**, *85*, 263–273. [CrossRef]
42. Blum, J.W.; Kunz, P.; Leuenberger, H.; Gautschi, K.; Keller, M. Thyroid hormones, blood plasma metabolites and haematological parameters in relationship to milk yield in dairy cows. *Anim. Sci.* **1983**, *36*, 93–104. [CrossRef]
43. Djokovic, R.; Ilic, Z.; Kurcubic, V.; Petrovic, M.; Doskovic, V. Functional and morphological state of the liver in Simmental dairy cows during transitional period. *Revue Méd. Vét.* **2011**, *162*, 574–579.

animals

MDPI

Article

Milk Beta-Hydroxybutyrate and Fat to Protein Ratio Patterns during the First Five Months of Lactation in Holstein Dairy Cows Presenting Treated Left Displaced Abomasum and Other Post-Partum Diseases

Mariana Alves Caipira Lei [1] and João Simões [2,*]

[1] Department of Zootechnics, School of Agricultural and Veterinary Sciences, University of Trás-os-Montes and Alto Douro (UTAD), 5000-801 Vila Real, Portugal; marianalei@utad.pt

[2] Department of Veterinary Sciences, School of Agricultural and Veterinary Sciences, University of Trás-os-Montes and Alto Douro (UTAD), 5000-801 Vila Real, Portugal

* Correspondence: jsimoes@utad.pt; Tel.: +351-259-350-666

Simple Summary: This study aimed to evaluate the 5-month pattern (averaged days in milk; DIM1 to 5) of milk beta-hydroxybutyrate (BHB) concentration and fat to protein content (F:P) ratio patterns from Holstein cows presenting postpartum diseases which have been treated. Cows presenting left displaced abomasum (LDA) and concomitant diseases within the first three months had higher concentrations of BHB than the control group (cows without diseases) in the first, but not in the second month postpartum. The F:P ratio had a similar evolution pattern also for DIM2. Animals with LDA were four to six times more likely to have a F:P ratio ≥1.29 than the control group during DIM1 and DIM2, respectively. Moderate and high correlations were also observed between the F:P ratio and BHB in DIM1 and DIM2, respectively. We concluded that animals suffering from LDA within the first three months postpartum have a significantly higher concentration of BHB and F:P ratio in milk than cows without postpartum diseases during the first two months. The treated cows with LDA quickly recovered normal levels, up to DIM3. The F:P ratio is a viable and economic indicator, mainly between the first two months postpartum, to estimate BHB concentration and energy balance in cows presenting LDA and in recovery.

Citation: Lei, M.A.C.; Simões, J. Milk Beta-Hydroxybutyrate and Fat to Protein Ratio Patterns during the First Five Months of Lactation in Holstein Dairy Cows Presenting Treated Left Displaced Abomasum and Other Post-Partum Diseases. *Animals* **2021**, *11*, 816. https://doi.org/10.3390/ani11030816

Academic Editor: Thomas W. Swerczek

Received: 25 January 2021
Accepted: 12 March 2021
Published: 14 March 2021

Abstract: The main objective of the present study was to evaluate the beta-hydroxybutyrate (BHB) and fat to protein content (F:P) ratio patterns in the milk of Holstein cows with postpartum diseases throughout the first five months of lactation. This prospective study was performed at Vestjyske Dyrlaeger ApS (Nørre Nebel, Denmark). The milk fat, protein, and BHB were evaluated in the Danish Eurofins laboratory according to the monthly averaged days in milk (DIM1 to 5). According to clinical records, five groups were formed: A (control group; cows without diseases; $n = 32$), B (cows with left displaced abomasum -LDA- and concomitant diseases; $n = 25$); C (cows with other diseases up to DIM3; $n = 13$); D (cows with foot disorders up to DIM3; $n = 26$); and E (cows with disease manifestations in DIM4 and DIM5; $n = 26$). All the sick cows were treated after diagnosis, and laparoscopy was performed on cows with LDA. In group B, a higher concentration of BHB (0.18 ± 0.02 mmol/L; $p < 0.001$) was observed than in the control group (0.07 ± 0.02 mmol/L; $p < 0.001$) in DIM1, presenting an odds ratio (OR) = 8.9. In all groups, BHB decreased to 0.03–0.05 mmol/L ($p < 0.05$) since DIM3. The F:P ratio was higher in group B (1.77 ± 0.07) than in group A (1.32 ± 0.06; $p < 0.05$) in DIM1. A similar profile is observed in DIM2. It was observed that animals in group B were four to six times more likely to have a F:P ratio ≥1.29 during DIM1 (OR = 4.0; 95% CI:1.3–14.4; $p = 0.01$) and DIM2 (OR = 5.9; 95% CI %:1.9–21.9; $p < 0.01$), than cows in group A. There were also moderate and high correlations between the F:P ratio and the BHB for DIM1 ($r = 0.57$; $r^2 = 0.33$; RSD = 0.09; $p < 0.001$) and DIM2 ($r = 0.78$; $r^2 = 0.60$; RSD = 0.07; $p < 0.001$), respectively. We concluded that animals affected by LDA in the postpartum period have a higher concentration of BHB in milk in DIM1 and all treated animals quickly recover BHB levels up to DIM3. The F:P ratio is a viable and economic indicator, mainly in DIM1 and DIM2, to estimate BHB concentration and energy balance in cows with LDA and other postpartum diseases.

Keywords: milk beta-hydroxybutyrate; fat to protein content ratio; left displaced abomasum; negative energy balance; postpartum diseases

1. Introduction

The presence of high concentrations of ketone bodies in the blood—namely beta-hydroxybutyrate (BHB), acetoacetate, and acetone—is called hyperketonemia or ketosis and is one of the most harmful and damaging metabolic diseases in dairy cows, resulting from negative energy balance (NEB), especially during early lactation [1]. In Europe, a prevalence of subclinical ketosis has been reported (serum BHB $\geq$ 1.2 mmol/L) between 11.2 and 36.6% from 2nd to 15th days in milk (DIM) [2], similar to that observed in other continents, and where the prevalence of cows developing serum BHB $\geq$ 3.0 mmol/L, and related to clinical ketosis, was 3.4% on average [3].

Clinical ketosis, displacement of the abomasum (DA), metritis, and lameness are more likely to occur in dairy cows with hyperketonemia levels [2]. It is reasonable to expect that metabolic (oxidative) stress NEB at the beginning of lactation associated with hyperketonemia, will have a depressive effect on the immune system of the affected animals, increasing their susceptibility to the occurrence of pathologies such as metritis and laminitis, in this initial phase [2,4]. Moreover, Holstein cows, as other breeds such as Jerseys, Brown Swiss, Guernseys, Ayrshires, and Simmental-Red Holsteins commonly present DA [5]. Left side DA (LDA), with heritability estimated at 0.30 on Holstein cows [5], is more frequent than right side DA with implications for animal health and welfare, representing significant financial losses for dairy farmers.

LDA is common in high-producing dairy cows, mainly during the first month of lactation [6]. A surgical approach is usually required to treat this condition. A one-step laparoscopy-guided abomasopexy (Christiansen modified technique) treatment of LDA is a minimally invasive technique that allows the confirmation of the LDA and the evaluation of eventual adhesions between the abomasum and the left abdominal wall or rumen. More importantly, it can be performed completely in a standing cow, without the need to put it in recumbency, and is easier and faster than two-step laparoscopy [7–9]. According to Wapenaar and Roberts (2017), the survival rate ranges from 73 to 88% in six months [10].

The determination, by spectrophotometry, of the BHB concentration has been considered the gold standard method [1]. However, in recent years, due to the high correlation between the concentration of BHB in blood and milk (about 10 times less concentration), this latter fluid has served as a viable sample to determine this metabolite [11,12], being of common use in dairy farms. Another indirect method of assessing the energy balance is the fat to protein content (F:P) ratio [13]. An increase in the F:P ratio coincides with periods of NEB associated with increased mobilization of lipid reserves, which stimulates hyperketonemia and the fat content of milk [14]. According to Heuer et al. F:P ratio threshold between 1.35 and 1.50 can predict cows in energy deficit [13], but these values are not consensual [15]. Moreover, the F:P ratio of milk reflects the health status of the cow if the cow has already suffered or is suffering from the effects of a disease, so it can potentially serve as an indicator of energy deficit, including estimating hyperketonemia [16] and for monitoring energy balance during the first few months after delivery. Furthermore, van Knegsel et al. [17] and Denis-Robichaud et al. [18] suggested a F:P ratio threshold of $\geq$1.5 and $\geq$1.3 [18], respectively, to predict cows with hyperketonemia. The serum BHB concentration was used in these last studies.

Although numerous studies on the association between hyperketonemia and postpartum diseases, such as some of the above, have been carried out, we have not found relevant studies investigating the metabolic pattern of BHB measured by milk, or that of the F:P ratio preceding and following its treatment. The present study's main objective is to determine the profiles of BHB in milk and the F:P ratio, during the first five months of lactation, in laparoscopic-treated LDA Holstein dairy cows. Other general postpartum diseases were also assessed.

2. Materials and Methods

2.1. Animals and Study Design

One hundred and twenty-two Holstein cows, at first (n = 43), second (n = 33), and third (n = 46) lactation, belonging to 20 dairy farms assisted by the veterinary clinic Vestjyske Dyrlaeger ApS (Denmark) were considered in the present prospective study. All animals calved between June 2019 and March 2020. All farms were managed for balanced diets.

Initially, we selected cows that presented LDA up to the first three months of lactation and that were treated by one-step laparoscopy-guided abomasopexy technique [7–9].

To create the control group, for each of the previously selected animals, between 1 and 4 animals from the same farm were selected to ensure equal environmental, nutritional, and management effects. In each holding, each pair or group pre-filled, cumulatively, the following requirements:

1. The same number of lactations;
2. Approximately the same date of delivery (±10 days) or the same DIM (±10 days). Cows were moved to the maternity unit 7 to 10 days before giving birth; so cows from each pair, who at some point during this period shared this park, were under the same effect of environmental and management factors;
3. Be under the same nutritional and management plan;
4. Present similar body condition.

In some cases of 2nd or 3rd lactation, the animals—for which at least one pair was not found with the same number of lactations—were paired with animals of 3rd or 2nd lactation, respectively, which met the remaining selection and matching criteria cumulatively.

In farms, a composite milk sample of each animal was collected from the respective milking machine or robot in a 30 mL container, without preservatives, which was previously prepared and identified by barcodes. The samples were immediately refrigerated and sent to the laboratory by CKR transporter (Aarhus, Denmark; see https://www.ryk-fonden.dk/). All lactating cows were sampled monthly with an interval as close to 30–31 days as possible, for the first five months throughout the study period, and according to the farms' milk control management. The averaged DIM was calculated for each month according to the sequential milk samples collection and classified as DIM1 to DIM5. The milk BHB, protein, and fat were measured using Fourier-transform infrared spectroscopy (FTIR) methods at Eurofins Steins Laboratorium (Vejen, Denmark; see https://www.eurofins.dk/). Finally, data of milk production, percentages of fat and protein in milk, and BHB concentration in milk, as well somatic cells count, were assessed from test-day records using the DMS Dyreregistrering platform. All complete records (n = 610) were considered.

With the animals gathered for this study, and recording the clinical evolution during the five months postpartum, five different groups were formed:

- Group A: control group, i.e., cows without any apparent pathology during the 5 months of study (n = 32).
- Group B: cows with LDA (n = 25). The proportion of cows with LDA observed in DIM1 (40%) was similar to DIM2 (56%; $p > 0.05$) and greater than in DIM3 (4%; $p < 0.001$). In these animals, other pathologies (i.e., comorbidities) were observed, namely lameness (n = 22), clinical ketosis (n = 11; 6 and 5 in DIM1 and DIM2, respectively; $p > 0.05$), mastitis (n = 6), inflammation of unknown origin (n = 3), metritis (n = 6), digestive or respiratory disease (n = 3), hypocalcemia (n = 2), and uterine torsion (n = 1). It was observed that 33.3% (18/54), 38.9% (21/54), and 27.8% (15/54) of these diseases occurred in DIM1, DIM2, and DIM3, respectively ($p > 0.05$). Despite LDA, all cows of this group suffering lameness also presented one of the above-reported disease. All diseases were treated.
- Group C: cows with other diseases (that not LDA) up to third milk control (n = 13). The treated diseases were metritis (n = 6), lameness (n = 6), mastitis (n = 5), hypocalcemia

($n = 2$), and clinical ketosis ($n = 1$). It was observed that 60% (12/20; $p < 0.001$), 15% (3/20), and 25% (5/20) of the diseases occurred at DIM1, DIM2, and DIM3, respectively.
- Group D: cows with only foot disorders up to third milk control ($n = 26$). The distribution of foot problems was 15.4%, 42.3%, and 42.3% at DIM1, DIM2, and DIM3, respectively ($p = 0.06$)
- Group E: cows with disease manifestations only in fourth and fifth milk control ($n = 26$). The treated diseases were lameness ($n = 20$), mastitis ($n = 4$), pneumonia ($n = 3$), and inflammation of unknown origin ($n = 1$). Similar proportions (50%) occurred in DIM4 and DIM5.

2.2. Statistical Analysis

Differences in disease proportions between groups were evaluated using the chi-square test.

One-way ANOVA was used to evaluate averaged DIM differences between groups or between monthly milk sample collections.

ANOVA with repeated measures was used, based on the repetition of the milk samples through the software StatView® 5.0 for Windows (SAS Institute, Cary, NC, USA), according to the following linear model:

$$y_{ij} = \mu + \alpha_i + \beta_j + (\alpha\beta)_{ij} + \epsilon_{ijk} \quad (1)$$

where, y_{ij} is the value obtained from milk yield, milk BHB, milk protein, milk fat, or F/P ratio, μ is the overall mean, αi is the fixed effect of the ith level of group (i = A, B, C, D, E), β_j is the repeated effect jth level of DIM (j = 1, 2, 3, 4, 5), $(\alpha\beta)_{ij}$ is the ij interaction effect and ϵ_{ijk} is the random error.

The Bonferroni/Dunn post hoc test was used to evaluate differences between pairs from one way ANOVA analysis and all models.

A receiver operating characteristic (ROC) curve was made to estimate F:P ratio threshold to detect cows with $\geq$0.14 mmol/L BHB, i.e., cows presenting ketosis according to Renaud et al. [12]. The odds ratio (OR) involving these thresholds were evaluated using the Wald test. Pearson's correlations between BHB concentration and F:P ratio were also assessed. The software JMP® 14 for Windows (SAS Institute, Cary, NC, USA) was used for these last purposes.

The results were presented as mean $\pm$ SD (standard deviation). The level of statistical significance for all results was $p < 0.05$.

3. Results

No differences of the averaged DIM were observed between groups ($p > 0.05$) in each sampling period when milk was successively sampled each month (DIM1: 20.1 $\pm$ 12.5; DIM2: 56.5 $\pm$ 19.7; DIM3: 89.1 $\pm$ 26.1; DIM4: 119.5 $\pm$ 25.9; DIM5: 155.2 $\pm$ 26.0 days; $p < 0.001$). The descriptive analysis of milk yield and milk contents is reported in Table 1.

Table 1. Descriptive analysis of milk yield and milk contents from all 610 milk samples.

Milk Parameter	Mean ± SD	95% Confidence Interval
Milk yield (kg)	37.0 ± 9.5	36.2–37.7
Fat (%)	4.09 ± 0.93	4.01–4.16
Protein (%)	3.41 ± 0.29	3.39–3.43
BHB (mmol/L)	0.06 ± 0.04	0.05–0.06
SCC (10^3/mL) [1]	236.5 ± 552.6	192.6–280.5

[1] Arithmetic mean. SD: standard deviation; BHB: beta-hydroxybutyrate; SCC: somatic cells count.

3.1. Milk Yield

Contrarily to DIM ($p < 0.05$), no effect of groups was observed on the milk yield ($p > 0.05$). A higher milk yield was observed in DIM2 (37.9 $\pm$ 9.9 kg) than in DIM5

(35.9 ± 8.1 kg). Nevertheless, group × DIM interaction ($p < 0.01$) was also observed: group B had lower milk yield (33.7 ± 13.4 kg) than group E (41.5 ± 11.4 kg; $p < 0.05$), but only on DIM2.

3.2. BHB Concentration

Significant differences ($p < 0.001$) of BHBA were observed between groups, DIM and groups × DIM interaction. According to Table 2, a consistent decrease of BHBA during the first (group C) or second (group B) months were observed.

Table 2. BHB concentration (mmol/L) pattern by group over the first five months of lactation.

Group	DIM1	DIM2	DIM3	DIM4	DIM5
A	0.07 ± 0.02 ^{A,a}	0.06 ± 0.02 ^{a}	0.03 ± 0.01 ^{b}	0.03 ± 0.01 ^{b}	0.04 ± 0.01 ^{a,b}
B	0.18 ± 0.02 ^{B,a}	0.12 ± 0.02 ^{a,b}	0.05 ± 0.01 ^{b}	0.04 ± 0.01 ^{b}	0.04 ± 0.01 ^{b}
C	0.14 ± 0.03 ^{A,B,a}	0.07 ± 0.03 ^{b}	0.04 ± 0.01 ^{b}	0.04 ± 0.01 ^{b}	0.04 ± 0.01 ^{b}
D	0.06 ± 0.02 ^{A}	0.06 ± 0.02	0.05 ± 0.01	0.04 ± 0.01	0.03 ± 0.01
E	0.06 ± 0.02 ^{A}	0.04 ± 0.02	0.04 ± 0.01	0.04 ± 0.01	0.05 ± 0.01

DIM: Days in milk. ^{A,B} different letters within the same column: $p < 0.001$. ^{a,b} different letters within the same line: $p < 0.05$.

No significant difference in milk BHB concentration was observed between DIM1 (0.20 ± 0.06 mmol/L) and DIM2 (0.22 ± 0.06 mmol/L; $n = 14$; $p = 0.60$) in cows presenting LDA in DIM2. Nevertheless, half of the cows ($n = 5$) suffering LDA in DIM1 also presented high milk BHB concentration in DIM2 (0.11 ± 0.03 mmol/L; $n = 5$).

3.3. Fat Content in Milk

Significant differences were found between groups ($p < 0.01$) and DIM ($p < 0.001$), and group × DIM interaction ($p < 0.001$) on milk fat as reported in Table 3.

Table 3. Averaged milk fat (%) pattern by group over the first five months of lactation.

Group	DIM1	DIM2	DIM3	DIM4	DIM5
A	4.59 ± 0.22 ^{A,a}	3.89 ± 0.14 ^{A,b}	3.76 ± 0.10 ^{b}	1.2 ± 0.11 ^{b}	3.76 ± 0.09 ^{b}
B	5.98 ± 0.25 ^{B,a}	4.54 ± 0.16 ^{B,b}	3.86 ± 0.12 ^{b}	1.2 ± 0.12 ^{b}	3.92 ± 0.10 ^{b}
C	5.20 ± 0.34 ^{A,B,a}	4.06 ± 0.22 ^{A,B,b}	3.80 ± 0.16 ^{b}	1.2 ± 0.17 ^{b}	3.59 ± 0.15 ^{b}
D	4.51 ± 0.24 ^{A,a}	3.99 ± 0.16 ^{A,a,b}	3.81 ± 0.12 ^{b}	1.13 ± 0.12 ^{b}	3.87 ± 0.10 ^{a,b}
E	4.66 ± 0.24 ^{A,a}	3.83 ± 0.16 ^{A,b}	3.80 ± 0.12 ^{b}	1.07 ± 0.12 ^{b}	3.66 ± 0.10 ^{b}

DIM: days in milk. ^{A,B} different letters within the same column: $p < 0.05$. ^{a,b} different letters within the same line: $p < 0.05$.

3.4. Protein Content in Milk

There were no significant differences ($p = 0.92$) in the protein between groups. However, there was an effect of the contrast days ($p < 0.001$). The interaction between the two variables was not significant ($p = 0.30$; Table 4).

Table 4. Milk protein content in milk (%) pattern by group over the first five months of lactation.

Group	DIM1	DIM2	DIM3	DIM4	DIM5
A	3.49 ± 0.40 ^{a}	3.27 ± 0.20 ^{b}	3.39 ± 0.21 ^{a,b}	3.44 ± 0.21 ^{a,b}	3.46 ± 0.21 ^{a}
B	3.39 ± 0.43 ^{a,b}	3.26 ± 0.29 ^{a}	3.40 ± 0.26 ^{a,b}	3.471 ± 0.23 ^{a,b}	3.50 ± 0.22 ^{b}
C	3.50 ± 0.42	3.23 ± 0.28	3.30 ± 0.23	3.40 ± 0.23	3.48 ± 0.24
D	3.41 ± 0.32 ^{a,b}	3.26 ± 0.27 ^{a}	3.33 ± 0.25 ^{a,b}	3.43 ± 0.25 ^{a,b}	3.47 ± 0.24 ^{b}
E	3.60 ± 0.40 ^{a}	3.26 ± 0.20 ^{b}	3.31 ± 0.26 ^{b}	3.45 ± 0.25 ^{a,b,c}	3.52 ± 0.26 ^{a,c}

DIM: days in milk. ^{a,b,c} different letters within the same line: $p < 0.05$.

3.5. Fat to Protein Content Ratio

An effect of groups ($p < 0.01$), DIM and groups ($p < 0.001$) × DIM interaction ($p < 0.001$) on the F:P ratio was also observed (Table 5).

Table 5. F:P ratio pattern by group over the first five months of lactation.

Group	DIM1	DIM2	DIM3	DIM4	DIM5
A	1.32 ± 0.06 [A,a]	1.19 ± 0.05 [A,b]	1.19 ± 0.03 [b]	1.09 ± 0.03 [b]	1.09 ± 0.02 [b]
B	1.77 ± 0.07 [B,a]	1.41 ± 0.06 [B,b]	1.14 ± 0.04 [b,c]	1.13 ± 0.04 [c]	1.12 ± 0.03 [c]
C	1.47 ± 0.09 [A,B,a]	1.26 ± 0.08 [A,B,a,b]	1.15 ± 0.05 [b]	1.06 ± 0.05 [b]	1.05 ± 0.04 [b]
D	1.33 ± 0.06 [A,a]	1.23 ± 0.06 [A,B,a,b]	1.15 ± 0.03 [b]	1.13 ± 0.03 [b]	1.14 ± 0.03 [b]
E	1.29 ± 0.06 [A,a]	1.18 ± 0.05 [A,a,b]	1.15 ± 0.03 [b]	1.07 ± 0.03 [b]	1.11 ± 0.03 [b]

DIM: days in milk. [A,B] different letters within the same column: $p < 0.05$. [a,b,c] different letters within the same line: $p < 0.05$.

3.6. Relationships between Beta-Hydroxybutyrate and Fat to Protein Content Ratio or Milk Yield/Contents

It was observed a F:P ratio threshold ≥1.29 to estimate cows with ≥0.14 mmol/L BHB, according to the ROC curve (Figure 1) in DIM1. The sensitivity and specificity were 92.0% and 54%, respectively. Similar values were obtained when overall DIM (1 to 5) were considered.

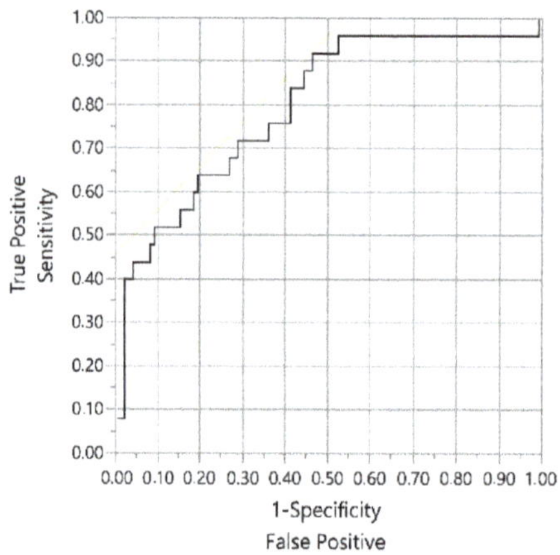

Figure 1. Receiver operating characteristic (ROC) curve to calculate the threshold ratio F:P to predict hyperketonemia (subclinical ketosis). Positive level: beta-hydroxybutyrate ≥ 0.14 mmol/L; $p < 0.001$; AUC (area under the curve) = 0.80.

Group B animals were four to six times more likely to have a F:P ratio ≥ 1.29 than group A cows, in DIM1 and DIM2, respectively (Table 6). Moreover, this group showed about nine times greater chance than group A (control) of presenting hyperketonemia (BHB ≥ 0.14 mmol/L) in DIM1, but not in the remaining periods. Group C animals did not show significant differences in any of the periods.

Table 6. The odds ratio for group B to detect cows with fat to protein content ratio (F:P ratio) ≥ 1.29 or beta-hydroxybutyrate levels ([BHB]) ≥ 0.14 mmol/L (group A as reference).

Parameter	DIM	Odds Ratio	95% CI
F:P ratio ≥ 1.29	1	4.0 **	1.3–14.4
	2	5.9 **	1.9–21.9
BHB ≥ 0.14 mmol/L	1	8.9 ***	2.6–37.2

DIM: days in milk; 95% CI: 95% confidence interval; **: $p < 0.01$; ***: $p < 0.001$.

Moderate and high correlations were observed between the F:P ratio or milk fat and BHB for DIM1 and DIM2 (Table 7).

Table 7. Correlations and regression equations between milk yield, milk contents, or fat to protein content ratio (F:P ratio) and beta-hydroxybutyrate (BHB).

Correlation [1] for (X):	DIM	r	r^2	Regression Equation (Y = BHB/mmol)	RSD
Milk yield (kg)	2	0.24 **	0.06	Y = −0.038 + 0.003 × X (kg)	0.11
Milk fat (%)	1	0.46 ***	0.21	Y = −0.085 + 0.037 × X (%)	0.09
	2	0.71 ***	0.50	Y = −0.331 + 0.099 × X (%)	0.08
Milk protein (%)	2	−0.25 **	0.06	Y = 0.469 − 0.122 × X (%)	0.11
	3	−0.35 ***	0.13	Y = 0.286 − 0.072 × X (%)	0.05
	4	−0.38 ***	0.14	Y = 0.219 − 0.053 × X (%)	0.03
	5	−0.28 **	0.08	Y = 0.173 − 0.038 × X (%)	0.07
F:P ratio	1	0.57 ***	0.33	Y = −0.136 + 0.163 × X	0.09
	2	0.78 ***	0.60	Y = −0.322 + 0.314 × X	0.07

[1] The correlations of the omitted DIM for each parameter were not significant ($p > 0.05$). DIM: days in milk; r: correlation coefficient; regression coefficient; RSD: residual standard deviation; **: $p < 0.01$; ***: $p < 0.001$.

4. Discussion

Cows with LDA (group B) were found to have higher BHB concentrations in milk (0.18 ± 0.02 mmol/L) than the control group (0.07 ± 0.02 mmol/L; $p < 0.001$) in DIM1, decreasing rapidly in the following month, even though the proportion of LDA and clinical ketosis was similar in DIM1 and DIM2. The results also show that the animals recover quickly after treatment, with no significant differences in BHB between groups from DIM3.

The control group kept BHB levels (0.7 mmol/ L) in DIM1 and DIM2 stable, which halved in DIM3, when the peak dry matter intake was reached, indicating that the energy management of the farms in question was adequate to the different groups. The higher concentrations of BHB at DIM1 and DIM2 of group A (control) follow the same trend found by Belay et al. [19]; these authors observed that milk BHB concentrations (obtained by FTIR technique) were higher between 11–61 DIM [19]. Previously, Koeck et al. had already verified the same trend [20]. This is probably due to the increase in dry matter intake and the gradual return to the positive energy balance as the days in milk progress [21]. It is known that the manifestation of clinical ketosis is not necessarily associated with higher serum concentrations of BHB [2] and that the lower capacity of dry matter intake, and as a consequence, lower supply of energy, as well as the process of adaptation of the papillae and rumen microbiome, may have influenced the greater degree of energy deficit in this phase (DIM1).

Although group C did not show any significant difference with the control group in DIM1, a similar profile of decrease in BHB was observed, with significant differences in DIM3. Unlike group B, whose distribution of LDA as well as their concomitant pathologies, were homogeneous ($p > 0.05$) in DIM1 and DIM2, in group C, 60% of the pathologies occurred in DIM1 ($p < 0.001$), with no significant differences to group A or group B. This suggests that although there may be an additive effect of disease association, this effect is largely dependent on the time that has elapsed since the beginning of lactation. The relation of pathologies associated with the beginning of lactation, and the position they occupy in a cause-effect relationship, are complex issues and not yet fully clarified [1]. Most authors believe that cows with ketosis have an increased risk of developing other pathologies at the beginning of lactation [1]. Effectively, Raboisson et al. summarized from multiple studies that the OR of hyperketonemic animals develop different pathologies: 5.4 (3.3–8.8) for clinical ketosis; 3.3 (2.6–4.3) for abomasum displacement; 1.8 (1.5–2.0) for metritis; 1.5 (1.2–1.9) for placental retention; 1.6 (1.2–2.1) for mastitis; 1.4 (1.3–1.6) for somatic cells count duplication; 2.0 (1.6–2.4) for laminitis, and 1.9 (1.6–2.3) for early culling [22]. This same risk, according to McArt et al. [23] and Suthar et al. [2], increases with increasing blood BHB concentration. McArt et al. [23] observed that cows diagnosed with ketosis from the third to the fifth day of lactation were 6.1 (95% CI = 2.3 to 16.0) times more likely to develop displacement of the abomasum than cows diagnosed after the first week. Ketosis probably induces hypoglycemia in multiparous cows [24] and decreased time and rumination activity [25]. Duffield et al. [26] observed that some cows developed pathologies, such as DA, before being diagnosed with ketosis, reinforcing the hypothesis that the pathologies associated with ketosis may be more than the effects of this, the cause. The interaction between ketosis and the DA has been stated as bidirectional, proposing that they can both be a risk factor or a consequence of each other. Inclusively they are both multifactorial diseases and there are already described some risk factors that are common to both pathologies, mostly related to housing, management, and feeding adopted systems [27]. Suthar et al. [2] recorded both diseases at the same time and verified that they were both related to high levels of BHB, proving that they are both directly correlated with NEB. Duffield et al. [26] hypothesized the explanation: the presence of common factors of hyperketonemia and DA with a similar causal etiologic is also related to a poor adaptive response at the onset of lactation conducting to NEB. This same theory has already been addressed regarding other postpartum diseases like mastitis and metritis [28].

In our study, only 11 of the 25 cows with LDA had clinical ketosis in the same month that the pathology was diagnosed, which shows the complexity of demonstrating cause and effect. An interesting finding is the observed decrease of the BHB concentration from DIM1 to DIM2, DIM2 as the period with the higher number of LDA (14 LDA). Duffield et al. [26], McArt et al. [23] and Suthar et al. [2] suggested that ketosis is a risk factor for LDA; It would be expected that the month with the higher number of diagnosed LDA (DIM2 in our study) presents high concentration of BHB, and the OR increases with increasing BHB concentration. In the present study, similar BHB concentrations of the animals that suffered from LDA on DIM2 were observed between DIM1 and DIM2 suggesting that, in these cases, ketosis contributed to LDA etio-pathophysiology. Nevertheless, BHB concentrations remained high in DIM2 for cows suffering LDA in DIM1 suggesting that hyperketonemia could be a consequence of LDA. These findings are in agreement with the bidirectional association between ketosis and LDA referred to in the Stengärde et al. study [27].

We observed that the fat content in milk in the different groups was maximum at the first milk control, a control that mirrors the repercussions of the NEB that can be installed up to three weeks postpartum. Furthermore, and in agreement with Zhang et al. [14], in our study, the maximum percentages of fat content in milk also correspond to the maximum concentrations of BHB in milk for most groups, as we can see through the higher fat content in group B than in group A. Additionally, moderate and higher correlations between fat content in milk and BHB in group B were observed in DIM1 and DM2, respectively. From the second or third milk control, after the NEB, we saw a less abrupt decrease in the fat content in milk in all groups.

During phases of energy deficit, such as the beginning of lactation, there are increased levels of fat content in milk [15] and decreased levels of protein [29]. The high concentrations of fat content in milk in periods of NEB correspond in more than 95% to triglycerides (TG) [14] and are mainly due to the increased mobilization of lipid reserves [15] that lead to the production of a large amount of BHB [30] that induces increased synthesis of TG by mammary epithelial cells [14].

The essential fatty acids to the synthesis of these TG come from the diet or the activity of the ruminal microbiota [31]. In the mammary gland, there is also the "de novo" synthesis in the mammary epithelial cells. It is in the "de novo" synthesis that BHB, mainly from the butyric acid metabolization, gains special importance; in this way, in ruminants, BHB is one of the main precursors (originating about 50%) of the "de novo" synthesized fat in mammary epithelial cells [14].

Regarding the protein content in milk, we observed that there was a decrease between DIM1 and DIM2 for groups A and E, contrary to what was observed in groups B, C, and D in which diseases occur in the first three months of lactation. The decrease in protein content in milk is a normal situation, since NEB worsens by that time, reducing milk protein synthesis. These data are in agreement with the conclusions of Gross et al. [15] that protein evolves inversely to milk productivity and hence it is expected that, at the beginning of lactation when productivity is low, the maximum levels of protein in milk can be verified. In fact, a healthy animal reflects the expected start of lactation and therefore relatively low productivity that increases until the peak of lactation.

The non-decreasing in milk protein in groups B, C, and D, between DIM1 and DIM2, may be related to the lower protein supply in DIM1 due to the potential lower DMI that normally occurs during the manifestation of diseases (anorexia). A significant part of these groups developed some form of the disease during the first month. In some of these cases, there may still be less protein production in milk due to the worsening of NEB, since the increase in circulating non-esterified fatty acids may inhibit the production of somatotropin (growth hormone; Akers) [32]. It should be noted that the amount of milk was not significantly reduced in these groups (except between groups B and E at DIM2) at the time of sample collection, which due to the normal lag between the occurrence of the disease and that the moment of milk sample collection indicates that this production is quickly resumed after treatment.

The amount of fat and protein reflects a cow's energy status/condition [15]. Toni et al. [33] observed that cows with high F:P ratios at the beginning of lactation showed a higher incidence of various pathologies and were, as a result, more often early culled. Similar to the results found by Toni et al. [33], we observed that animals affected by LDA (group B) and various pathologies other than LDA (group C), presented higher F:P ratios in DIM1 than cows in the control group. Furthermore, we found that animals in group B with a F:P ratio $\geq$ 1.29 are about four to six times more likely to have BHB concentrations $\geq$ 0.14 mmol/L, that is, of having subclinical ketosis. These observations are accompanied by moderate and high correlations between the F:P ratio and the BHB for DIM1, in this same group. Zhang et al. [14] observed, in vitro, that animals with high circulating concentrations of BHB had higher TG content, indicating that high concentrations of BHB increase TG synthesis by mammary epithelial cells, justifying why animals with ketosis generally have milk with higher fat content. The increase in the F:P ratio, therefore, coincides evidently with periods of NEB associated with increased mobilization of lipid reserves and consequently with the increase in circulating levels of BHB.

Our results indicate that the determination of the F:P ratio, starting at 1.29, is a useful tool in the management of dairy cattle health, allowing the estimation of subclinical ketosis associated with LDA and other postpartum pathologies with high sensitivity and reasonable specificity. This F:P ratio threshold agrees with the $\geq$1.3 threshold obtained by Denis-Robichaud et al. [18] to predict hyperketonemia (serum BHB $\geq$ 1.4 mmol/L). Equally important, the present study emphasizes that in farms that offer, to their animals, diets with correctly balanced energy, cows with health problems (mainly LDA) are the ones most associated with the development of ketosis and most exposed to greater negative energy balance during the first two months of lactation, especially during the first month. Therefore, cows with high BHB and F:P ratio in milk in DIM1 should be preventively monitored for the diagnosis of the pathologies in question (groups B and C) in the first two months. This strategy allows a timely (earlier) diagnosis and respective treatment. Cows with diseases on the third, fourth, and fifth months of lactation do not have such a significant impact on the evaluated milk parameters (subclinical ketosis and NEB).

5. Conclusions

In conclusion, animals with LDA and/or other pathologies of the beginning of lactation are the ones that should be monitored and require more attention, concerning the BHB concentration, mainly in the first two milk controls. These animals, after treatment, recovered from LDA episodes and showed levels of BHB and other metabolites identical to healthy animals. Our work also suggests that lameness does not significantly influence concentrations of BHB or F:P ratio.

The determination of the F:P ratio profile during the postpartum period seems to be an important and economic indicator that can complement the evaluation of animals affected with LDA and even other postpartum pathologies. Special attention should be given to high-yielding cows presenting a F:P ratio $\geq$ 1.29 which can indicate the presence of a more intense NEB ($\geq$0.14 mmol/L BHB) related to LDA.

Author Contributions: Conceptualization, methodology, formal analysis, data curation, writing—original draft preparation and writing—review and editing were equally performed by M.A.C.L. and J.S. All authors have read and agreed to the published version of the manuscript.

Funding: This research received no external funding.

Institutional Review Board Statement: The study was conducted according to the guidelines of the Declaration of Helsinki, and approved by the Institutional Review Board of University of Trás-os-Montes and Alto Douro (UTAD) with protocol code request n. 2021011553000671 in 15 January 2021 (last update).

Informed Consent Statement: Informed consent was obtained from the board of Vestjyske Dyrlaeger ApS (Denmark) to use their data collected from the DMS Dyreregistrering platform.

Data Availability Statement: Data used in the present study can be requested to Vestjyske Dyrlaeger ApS (Denmark).

Acknowledgments: This paper is obtained from the M.Sc. thesis in Veterinary Medicine of the first author at University of Trás-os-Montes and Alto Douro (UTAD), Vila Real, Portugal. A special acknowledgment to Vestjyske Dyrlaeger ApS (Denmark) that provided the authors with all the help needed to collect the presented data.

Conflicts of Interest: The authors declare no conflict of interest.

References

1. Benedet, A.; Manuelian, C.L.; Zidi, A.; Penasa, M.; De Marchi, M. Invited review: β-hydroxybutyrate concentration in blood and milk and its associations with cow performance. *Animal* **2019**, *13*, 1676–1689. [CrossRef]
2. Suthar, V.S.; Canelas-Raposo, J.; Deniz, A.; Heuwieser, W. Prevalence of subclinical ketosis and relationships with postpartum diseases in European dairy cows. *J. Dairy Sci.* **2013**, *96*, 2925–2938. [CrossRef]
3. Brunner, N.; Groeger, S.; Raposo, J.C.; Bruckmaier, R.M.; Gross, J.J. Prevalence of subclinical ketosis and production diseases in dairy cows in Central and South America, Africa, Asia, Australia, New Zealand, and Eastern Europe. *Transl. Anim. Sci.* **2019**, *3*, 84–92. [CrossRef]
4. Zarrin, M.; Wellnitz, O.; Van Dorland, H.A.; Bruckmaier, R.M. Induced hyperketonemia affects the mammary immune response during lipopolysaccharide challenge in dairy cows. *J. Dairy Sci.* **2014**, *97*, 330–339. [CrossRef]
5. Zerbin, I.; Lehner, S.; Distl, O. Genetics of bovine abomasal displacement. *Vet. J.* **2015**, *204*, 17–22. [CrossRef]
6. Guo, Y.S.; Tao, J.Z.; Xu, L.H.; Wei, F.H.; He, S.H. Identification of disordered metabolic networks in postpartum dairy cows with left displacement of the abomasum through integrated metabolomics and pathway analyses. *J. Vet. Med. Sci.* **2020**, *82*, 115–124. [CrossRef] [PubMed]
7. Barisani, C. Abomasopessilaparoscopica one-step. *Summa Animali Reddito* **2017**, *12*, 9–13.
8. Newman, K.D.; Harvey, D.; Roy, J.P. Minimally invasive field abomasopexy techniques for correction and fixation of left displacement of the abomasum in dairy cows. *Vet. Clin. N. Am. Food Anim. Pract.* **2008**, *24*, 359–382. [CrossRef]
9. Newman, K.D.; Anderson, D.E. Comparison of two techniques for laparoscopic abomasopexy for correction of left displaced abomasum in dairy cows. In Proceedings of the Thirty-Sixth Annual Conference, Columbus, OH, USA, 18–20 September 2003; Smith, R.A., Ed.; Frontier Printers, Inc.: Fort Collins, CO, USA, 2003.
10. Wapenaar, W.; Roberts, J. LDAs in cattle—What method should we use? A report from a BCVA congress workshop. *Livestock* **2017**, *22*, 198–201. [CrossRef]
11. Grelet, C.; Bastin, C.; Gelé, M.; Davière, J.B.; Johan, M.; Werner, A.; Reding, R.; Fernandez Pierna, J.A.; Colinet, F.G.; Dardenne, P.; et al. Development of Fourier transform mid-infrared calibrations to predict acetone, β-hydroxybutyrate, and citrate contents in bovine milk through a European dairy network. *J. Dairy Sci.* **2016**, *99*, 4816–4825. [CrossRef] [PubMed]
12. Renaud, D.L.; Kelton, D.F.; Duffield, T.F. Short communication: Validation of a test-day milk test for β-hydroxybutyrate for identifying cows with hyperketonemia. *J. Dairy Sci.* **2019**, *102*, 1589–1593. [CrossRef]
13. Heuer, C.; Schukken, Y.H.; Dobbelaar, P. Postpartum body condition score and results from the first test day milk as predictors of disease, fertility, yield, and culling in commercial dairy herds. *J. Dairy Sci.* **1999**, *82*, 295–304. [CrossRef]
14. Zhang, M.; Zhang, S.; Hui, Q.; Lei, L.; Du, X.; Gao, W.; Zhang, R.; Liu, G.; Li, X.; Li, X. B-Hydroxybutyrate facilitates fatty acids synthesis mediated by sterol regulatory element-binding protein1 in bovine mammary epithelial cells. *Cell. Physiol. Biochem.* **2015**, *37*, 2115–2124. [CrossRef]
15. Gross, J.; van Dorland, H.A.; Bruckmaier, R.M.; Schwarz, F.J. Performance and metabolic profile of dairy cows during a lactational and deliberately induced negative energy balance with subsequent realimentation. *J. Dairy Sci.* **2011**, *94*, 1820–1830. [CrossRef]

16. King, M.T.M.; Duffield, T.F.; DeVries, T.J. Short communication: Assessing the accuracy of inline milk fat-to-protein ratio data as an indicator of hyperketonemia in dairy cows in herds with automated milking systems. *J. Dairy Sci.* **2019**, *102*, 8417–8422. [CrossRef]
17. van Knegsel, A.T.M.; van der Drift, S.G.A.; Horneman, M.; de Roos, A.P.W.; Kemp, B.; Graat, E.A.M. Short communication: Ketone body concentration in milk determined by Fourier transform infrared spectroscopy: Value for the detection of hyperketonemia in dairy cows. *J. Dairy Sci.* **2010**, *93*, 3065–3069. [CrossRef]
18. Denis-Robichaud, J.; Dubuc, J.; Lefebvre, D.; DesCôteaux, L. Accuracy of milk ketone bodies from flow-injection analysis for the diagnosis of hyperketonemia in dairy cows. *J. Dairy Sci.* **2014**, *97*, 3364–3370. [CrossRef]
19. Belay, T.K.; Svendsen, M.; Kowalski, Z.M.; Ådnøy, T. Genetic parameters of blood β-hydroxybutyrate predicted from milk infrared spectra and clinical ketosis, and their associations with milk production traits in Norwegian Red cows. *J. Dairy Sci.* **2017**, *100*, 6298–6311. [CrossRef]
20. Koeck, A.; Jamrozik, J.; Schenkel, F.S.; Moore, R.K.; Lefebvre, D.M.; Kelton, D.F.; Miglior, F. Genetic analysis of milk β-hydroxybutyrate and its association with fat-to-protein ratio, body condition score, clinical ketosis, and displaced abomasum in early first lactation of Canadian Holsteins. *J. Dairy Sci.* **2014**, *97*, 7286–7292. [CrossRef]
21. Macraea, A.I.; Burrougha, E.; Forresta, J.; Corbishleya, A.; Russella, G.; Shawb, D.J. Prevalence of excessive negative energy balance in commercial United Kingdom dairy herds. *Vet. J.* **2019**, *248*, 51–57. [CrossRef]
22. Raboisson, D.; Mounié, M.; Maigné, E. Diseases, reproductive performance, and changes in milk production associated with subclinical ketosis in dairy cows: A meta-analysis and review. *J. Dairy Sci.* **2014**, *97*, 7547–7563. [CrossRef] [PubMed]
23. McArt, J.A.A.; Nydam, D.V.; Oetzel, G.R. Epidemiology of subclinical ketosis in early lactation dairy cattle. *J. Dairy Sci.* **2012**, *95*, 5056–5066. [CrossRef] [PubMed]
24. Ruoff, J.; Borchardt, S.; Heuwieser, W. Short communication: Associations between blood glucose concentration, onset of hyperketonemia, and milk production in early lactation dairy cows. *J. Dairy Sci.* **2017**, *100*, 5462–5467. [CrossRef] [PubMed]
25. Kaufman, E.I.; Leblanc, S.J.; Mcbride, B.W.; Duffield, T.F.; Devries, T.J. Association of rumination time with subclinical ketosis in transition dairy cows. *J. Dairy Sci.* **2016**, *99*, 5604–5618. [CrossRef] [PubMed]
26. Duffield, T.F.; Lissemore, K.D.; Mcbride, B.W.; Leslie, K.E. Impact of hyperketonemia in early lactation dairy cows on health and production. *J. Dairy Sci.* **2009**, *92*, 571–580. [CrossRef]
27. Stengärde, L.; Hultgren, J.; Tråvén, M.; Holtenius, K.; Emanuelson, U. Risk factors for displaced abomasum or ketosis in Swedish dairy herds. *Prev. Vet. Med.* **2012**, *103*, 280–286. [CrossRef]
28. Duffield, T. Subclinical ketosis in lactating dairy cattle. *Vet. Clin. N. Am. Food Anim. Pract.* **2000**, *16*, 231–253. [CrossRef]
29. Gross, J.J.; Bruckmaier, R.M. Review: Metabolic challenges in lactating dairy cows and their assessment via established and novel indicators in milk. *Animal* **2019**, *13*, s75–s81. [CrossRef]
30. Zhang, G.; Ametaj, B.N. Ketosis an old story under a new approach. *Dairy* **2020**, *1*, 5. [CrossRef]
31. Månsson, H.L. Fatty acids in bovine milk fat fatty acids in bovine milk fat. *J. Food Nutr. Res.* **2008**, *52*, 1–3. [CrossRef]
32. Akers, R.M. Major advances associated with hormone and growth factor regulation of mammary growth and lactation in dairy cows. *J. Dairy Sci.* **2006**, *89*, 1222–1234. [CrossRef]
33. Toni, F.; Vincenti, L.; Grigoletto, L.; Ricci, A.; Schukken, Y.H. Early lactation ratio of fat and protein percentage in milk is associated with health, milk production, and survival. *J. Dairy Sci.* **2011**, *94*, 1772–1783. [CrossRef] [PubMed]

animals

MDPI

Article

Associations between the Bovine Myostatin Gene and Milk Fatty Acid Composition in New Zealand Holstein-Friesian × Jersey-Cross Cows

Ishaku L. Haruna, Yunhai Li, Ugonna J. Ekegbu, Hamed Amirpour-Najafabadi, Huitong Zhou and Jon G. H. Hickford *

Gene-Marker Laboratory, Department of Agricultural Sciences, Faculty of Agriculture and Life Sciences, Lincoln University, Lincoln 7647, New Zealand; Ishaku.Haruna@lincolnuni.ac.nz (I.L.H.); Yunhai.Li@Lincolnuni.ac.nz (Y.L.); Jane.Ekegbu@lincolnuni.ac.nz (U.J.E.); Hamed.Amirpour@lincolnuni.ac.nz (H.A.-N.); zhouh@lincoln.ac.nz (H.Z.)

* Correspondence: jon.hickford@lincoln.ac.nz; Tel.: +64-3-423-0665

Received: 26 May 2020; Accepted: 15 August 2020; Published: 19 August 2020

Simple Summary: The gene that encodes myostatin influences more than one trait, and its expression has been observed in skeletal muscle, as well as the mammary gland. In this study, association analysis revealed that variation in the bovine myostatin gene affects milk fatty acid composition, raising the possibility that this genetic variation may be utilized to increase the amount of unsaturated fatty acid and decrease the amount of saturated fatty acid in milk.

Abstract: The myostatin gene (*MSTN*), which encodes the protein myostatin, is pleiotropic, and its expression has been associated with both increased and decreased adipogenesis and increased skeletal muscle mass in animals. In this study, the polymerase chain reaction, coupled with single strand conformation polymorphism analysis, was utilized to reveal nucleotide sequence variation in bovine *MSTN* in 410 New Zealand (NZ) Holstein-Friesian × Jersey (HF × J)-cross cows. These cows ranged from 3 to 9 years of age and over the time studied, produced an average 22.53 ± 2.18 L of milk per day, with an average milk fat content of 4.94 ± 0.17% and average milk protein content of 4.03 ± 0.10%. Analysis of a 406-bp amplicon from the intron 1 region, revealed five nucleotide sequence variants (*A*–*E*) that contained seven nucleotide substitutions. Using general linear mixed-effect model analyses the *AD* genotype was associated with reduced C10:0, C12:0, and C12:1 levels when compared to levels in cows with the *AA* genotype. These associations in NZ HF × J cross cows are novel, and they suggest that this variation in bovine *MSTN* could be explored for increasing the amount of milk unsaturated fatty acid and decreasing the amount of saturated fatty acid.

Keywords: myostatin gene; variation; milk; fatty acid; cattle

1. Introduction

Improving the efficiency of cattle production systems can be achieved by selecting for fast growing animals with increased muscling that have desirable maternal reproductive traits, good milk production, and good mothering ability. To achieve this, it is important to have an understanding of the genes that underpin muscularity and adiposity.

The gene for the protein myostatin (MSTN; gene *MSTN*), which is also called the growth and differentiation factor 8 gene (*GDF8*) has pleiotropic effects. Its expression has been associated with decreased adipogenesis and increased skeletal muscle mass as a result of decreased secretion of leptin [1–3].

Sequence variation in *MSTN* has been associated with increases in growth and muscling traits in several species. For example, in cattle it has been connected with having increased numbers of muscle fibers (otherwise known as "double-muscling") in a number of breeds [4]. Similarly, sequence variation in the first intron has been found to influence growth and carcass traits such as the yield of leg, loin, and total lean meat in NZ Romney sheep [5].

While there is well-documented evidence describing how variation in *MSTN* is associated with growth and muscle traits in beef cattle breeds, there is little information about the effects of *MSTN* nucleotide sequence variation on milk yield and milk fatty acid (FA) composition. There are suggestions that *MSTN* could affect lactation by affecting the production of fatty acid in milk, and primarily through MSTN deficiency being associated with decreased adipogenesis [2,3,6].

The fat of milk from dairy cattle is approximately 70% saturated fatty acid (SFA), 25% monounsaturated fatty acid (MUFA), and 5% polyunsaturated fatty acid (PUFA) [7]. From a human health perspective, an increase in unsaturated fatty acid (UFA) content and a decrease in SFA content could be considered favorable; and equally from a physical perspective (e.g., increased spread-ability of butter), increased relative levels of UFA would be desirable. Previously, it has been identified that the concentrations of different fatty acids in milk are affected most by four parameters: the diet of the cow [8,9], genetic variation between cows within a breed [10,11], breed differences [12,13], and the number of days in milk [14]. In their report, Strucken et al. [14] established that cows in early lactation stage are characterized by negative energy balance as the dietary energy intake is unable to meet the demands of high milk production in approximately the first 60 days of lactation. To offset this balance, an alternative energy source is needed. This leads to the mobilization of body energy stores to balance the deficit between feed intake, and energy expenditure on maintenance and milk production [15]. However, as a consequence of the cow's body fat being used up for this purpose, other biological pathways are affected, resulting in a change in milk composition. The Holstein-Friesian and the Jersey breeds have one of the most notable differences in terms of the composition of milk fatty acids. Milk from Jersey cows tend to have higher concentrations of some short- and medium-chain length SFA, but lower concentrations of some UFA [16]. These differences could be capitalized upon in order to obtain the preferred fatty acid profile through cross-breeding of these breeds.

In the above context, the objective of this research was to explore the effects of *MSTN* variation on key milk production traits and composition of milk fatty acids in Holstein-Friesian × Jersey (HF × J)-cross (alternatively known as Kiwicross™) dairy cows in Hamilton, New Zealand (NZ). This cross is now the preferred cow for dairy production in NZ and it constitutes more than half of all the dairy cows milked annually. Our working hypothesis is that variation in the gene will affect some milk traits.

2. Materials and Methods

2.1. The Dairy Cattle Investigated and the Collection of Milk Samples from Them

This research was approved by the Lincoln University Animal Ethics Committee (AEC Approval Number 521). This is a mandated and registered committee that was established, and is regularly audited, under the provisions of the New Zealand Government's Animal Welfare Act 1999 (http://www.legislation.govt.nz/act/public/1999/0142/latest/DLM49664.html). A total of 410 HF × J-cross dairy cows (alternatively known as Kiwicross™ cows, in Hamilton, New Zealand), of individually variable and unidentified breed ratio, and of 3 to 9 years of age were investigated. The cows were managed in two herds on the Lincoln University Dairy Farm (Canterbury, New Zealand), and all of them were grazed outdoors at all times on pasture (a blend of white clover and perennial ryegrass). All of the cows calved over the months of August-September (spring in the Southern Hemisphere) and they were then milked twice a day for up to 10 months.

Milk samples for gross milk trait analysis were collected once a month from September to February and the daily milk yield in liters was recorded using Tru-test milk meters (Tru-test Ltd., Auckland, New Zealand) and Fourier-transform infra-red spectroscopy (MilkoScan FT 120 Foss, Hillerød, Denmark) was used to analyze milk samples for fat percentage (%) and protein percentage (%). Average daily milk yield, and average protein and fat percentages were calculated over the 6 months of milk collection. The milk samples for fatty acid (FA) analysis were collected at 148 ± 19 days in milk from each cow in a single afternoon milking in mid-January (the middle of summer in the Southern Hemisphere). These were frozen at −20 °C, and then freeze-dried, prior to being individually ground to a fine powder for component analysis.

2.2. Gas Chromatography of the Fatty Acids in the Milk Samples

The FAs were methylated and extracted in n-heptane, before being analyzed by gas chromatography (GC) as FA methyl esters (FAMEs). The methylation reactions were performed in 10-mL Kimax tubes. Individual freeze-dried and powdered milk samples (0.17 g) were dissolved in 900 μL of n-heptane (100%, AR grade), before 100 μL of internal standard (5 mg/mL of C21:0 methyl ester in n-heptane) and 4.0 mL of 0.5 M NaOH (in 100% anhydrous methanol) were added.

The tubes were vortexed then incubated in a block heater (Ratek Instruments, Australia) at 50 °C for 15 min. After cooling to room temperature, another 2.0 mL of n-heptane and 2.0 mL of deionized water were added to each tube. After vortexing, the tubes were centrifuged (Megafuge 1.0R, Heraeus, Germany) for 5 min at 1500× *g*. The top layer of n-heptane was transferred into a second Kimax tube and 2.0 mL of n-heptane was added to each of the original tubes. The extraction was repeated and the n-heptane aspirates were then pooled. Anhydrous sodium sulphate (10 mg) was added to the n-heptane extracts, to remove any residual water.

The GC analysis was carried out using a Shimadzu GC-2010 Gas Chromatograph (Shimadzu Corporation, Kyoto, Japan) equipped with a flame ionization detector and an AOC-20i auto sampler. The output was analyzed with GC Solution Software (Shimadzu). For analysis, 1 μL of the n-heptane sample extract was injected into a 100 m GC column (250 μm × 0.25 μm capillary column, CP-Select, Varian) with a 1:60 split ratio. The separation was undertaken with a helium carrier gas and was run for 92 min. The temperature of both the injector and detector were set at 250 °C and the thermal profile of the column consisted of 45 °C for 4 min, followed by 27 min at 175 °C (ramped at 13 °C/min), 35 min at 215 °C (ramped at 4 °C/min), and a final "bake-off" at 250 °C for 5 min (ramped at 25 °C/min). The individual FAMEs were identified by the peak retention time compared to commercially obtained external standards (ME61, ME93, BR3, BR2, ME100, GLC411, and GLC463; Laroden AB, Sweden). Quantification of the individual FAMEs was based on peak area assessment and comparison with the internal and external standards. The threshold for peak area determination on the chromatogram was a 500-unit count, with peaks that were under 500-unit count, being ignored. The calculated minimum component of an individual FAME was therefore 0.01 g per 100 g of total FA.

After their individual measurement, the FAs were arranged into various groups and indices. These groups were, short-chain length FAs (SCFA) = C4:0 + C6:0 + C8:0; medium-chain length FAs (MCFA) = C10:0 + C12:0 + C14:0; long-chain length FAs (LCFA) = C15:0 + C16:0 + C17:0 + C18:0 + C19:0 + C20:0 + C22:0 + C24:0; omega 3 FAs = C18:3 cis-9, 12, 15 + C20:5 cis-5,8, 11, 14, 17 + C22:5 cis-7, 10, 13, 16, 19; omega 6 FAs = C18:2 cis-9, 12 + C18:3 cis-6, 9, 12 + C20:3 cis-8, 11, 14 + C20:4 cis-5, 8, 11, 14; monounsaturated FAs (MUFA) = C10:1 + C12:1 + C14:1 cis-9 + C15:1 + C16:1 cis-9 + C17:1 + C18:1 trans-11 + C18:1 cis-9 + C18:1 cis-(10 to 15) + C20:1 cis-5 + C20:1 cis-9 + C20:1 cis-11 + C22:1 trans-13; polyunsaturated FAs (PUFA) = C18:2 trans-9, 12 + C18:2 cis-9,trans-13 + C18:2 cis-9,trans-12 + C18:2 trans-9,cis-12 + C18:2 cis-9, 12 + C18:3 cis-6, 9, 12 + C18:3 cis-9, 12, 15 + CLA + C20:3 cis-8, 11, 14 + C20:4 cis-5, 8, 11, 14 + C20:5 cis-5, 8, 11, 14, 17 + C22:5 cis-7, 10, 13, 16, 19; and total branched FA = C13:0 *iso* + C13:0 *anteiso* + C15:0 *iso* + C15:0 *anteiso* + C17:0 *iso*.

Unsaturated FA indices were also calculated as follows: C12:1 index (C12:1 divided by the sum of C12:0 and C12:1); C16:1 index (C16:1 cis-9 divided by the sum of C16:0 and C16:1 cis-9); and C18:1 index (C18:1 cis-9 divided by the sum of C18:0 and C18:1 cis-9). The method is as described by Li et al. [17], with the un-adjusted mean levels in the 430 cows being calculated and used subsequently in the statistical analyses.

2.3. Blood Sample Collection

Samples of blood were collected from each of the cows studied onto FTA™ cards (Whatman™, Middlesex, UK) by piercing the ear of the animal. This is allowed under a Code of Welfare issued by the NZ Minister of Agriculture, under Section 75 and 76 of the NZ Animal Welfare Act 1999. The samples were allowed to dry in the air and the purification of the DNA from 1.2-mm punches taken from the FTA™ cards was carried out using a two-step procedure described by Zhou et al. [18].

2.4. Polymerase Chain Reaction (PCR) Amplification of a Region of the Cattle Myostain Gene

The intron 1 region of *MSTN* was amplified using forward and reverse primers (5′-catggtactattgttgagag-3′ and 5′-aaggcaaatctattccagg-3′ respectively) adapted from the work of Haruna et al. [19]. The 15-μL reactions contained the purified DNA on a 1.2-mm diameter disc of the FTA™ card, and a content of 0.25 μM for each primer, 150 μM for each dNTP (Eppendorf, Hamburg, Germany), 3.0 mM Mg^{2+}, 0.5 U of *Taq* DNA polymerase (Qiagen, Hilden, Germany), and 1× the buffer supplied with the DNA polymerase enzyme.

The PCR amplifications were carried out in Bio-Rad S1000 thermal cyclers (Bio-Rad, Hercules, CA, USA) and the thermal cycling parameters included an initial denaturation at 94 °C for 2 min, and then 35 repeated cycles of denaturation at 94 °C for 30 s, primer annealing at 58 °C for 30 s, and primer extension at 72 °C for 30 s. Following this process, a final extension step at 72 °C for 5 min was used.

2.5. Single Strand Conformation Polymorphism (SSCP) Analysis

An SSCP technique was employed to detect nucleotide sequence variation in the amplicons obtained from the PCR reactions. A 0.7-μL aliquot of the completed reactions was added to 7 μL of a solution containing 10 mM ethylenediaminetetraacetic acid (EDTA), 0.025% bromophenol blue, 0.025% xylene-cyanol, and 98% formamide. The samples were then placed on a hot plate already set at 95 °C for 5 min for denaturation, followed by immediate cooling on wet ice. They were then loaded onto 14% acrylamide: bisacrylamide (37.5:1) (Bio-Rad) gels that were 16 cm × 18 cm in size. Electrophoresis was undertaken using Protean II xi cells (Bio-Rad) for 19 h at 390 volts and 7 °C room temperature in 0.5 × Tris/Borate/EDTA running buffer.

To detect the SSCP banding patterns, the SSCP gels were stained using the silver-staining method of Byun et al. [20].

2.6. Nucleotide Sequencing and Sequence Analysis

Based on the PCR-SSCP patterns observed nucleotide sequence variation could be identified. Cows that appeared to be homozygous with unique banding patterns were subjected to direct sequencing. For cows that appeared to have heterozygous variant patterns, unique bands were excised from the wet gel and incubated in water at 69 °C for 1 h. A 1-μL aliquot of the water product was pipetted into 14-μL of PCR pre-mixture (as used for the original PCR reactions), re-amplified using the same thermal cycle profile, and subsequently sequenced. This approach has been described in more detail by Gong et al. [21]. The sequences were aligned, and other analyses were undertaken using version 5.2.10 of DNAMAN (Lynnon BioSoft, Vaudreuil, QC, Canada).

2.7. Statistical Analyses

All statistical analyses were performed using IBM SPSS version 22 (IBM, Armonk, NY, USA), and an alpha level of $p < 0.05$ was set as a threshold.

For genotypes with a frequency greater than 5% (thus having adequate sample size per group), the effect of variation in a cow's *MSTN* genotype on gross milk production traits, and the component levels of individual and grouped FAs was tested using general linear mixed-effects models (GLMMs) and multiple pair-wise comparisons (least significant difference tests) with Bonferroni corrections. The age of the cow expressed in an integer value of years (i.e., as a categorical variable in a range from

3 to 9 years of age), the number of days in milk for each cow (DIM; expressed as an integer value, but entered into the model as a continuous trait) and herd (to correct for herd-specific effects) were fitted to the models as fixed explanatory factors.

The model was $Y_{ijkl} = \mu + G_i + A_j + D_k + H_l + e_{ijkl}$ for genotype: where Y_{ijkl} = the observed trait value in the ijklth cow; μ = the mean trait value for a given trait; G_i = the fixed effect of ith *MSTN* genotype; A_j = effect of age (j = 3–9 years); D_k = effect of the number of days the cow has produced milk (DIM: k = 94–186 days); H_l = the fixed effect of lth farm (l = 1 or 2); and e_{ijkl} = random error.

The effect of sire of cow could not be included in the GLMMs, because some semen straws (sire genetics) used in NZ dairy cattle artificial insemination-based breeding approaches contain mixed-sire semen purchased from commercial semen producers. In these cases, it is impossible to ascertain individual sire identity. However, since the straws were mixed-semen straws and because different sires are used for different inseminations, in different years, it is unlikely that sire was a strongly confounding effect. Cow age and herd might also be confounded with sire, but this cannot be confirmed.

3. Results

3.1. Milk Production

Over the time the cows were studied, they produced an average of 22.53 ± 2.18 L of milk per day, with an average milk fat content of 4.94 ± 0.17% and average milk protein content of 4.03 ± 0.10%.

3.2. Identification of Nucleotide Sequence Variation in Bovine MSTN

A 406 bp fragment of the intron 1 region of bovine *MSTN* was amplified and analyzed using the polymerase chain reaction coupled with single strand conformation polymorphism (PCR-SSCP) analyses.

The PCR-SSCP analyses coupled with DNA sequencing revealed five banding patterns (*A-E*) in the region of intron 1 investigated (Figure 1). A total of seven single-nucleotide substitutions (c.373+751G/T, c.373+803T/G, c.373+877A/G, c.373+895G/C, c.374−909C/T, c.374−842G/C, c.374−812A/G) were identified, all of which have been previously reported [19].

3.3. MSTN Genotype Models

The genotypes *AA* (*n* = 149), *AB* (*n* = 87), *AC* (*n* = 48) and *AD* (*n* = 63) occurred at a frequency over 5% and were analyzed in this model. The other genotypes *AE* (*n* = 16), *BB* (*n* = 6), *BC* (*n* = 5), *BD* (*n* = 15), *CC* (*n* = 5), *CD* (*n* = 7), CE (*n* = 1) and *DD* (*n* = 3) had frequencies of less than 5% each, and were not included in the model.

In the genotype models (Table 1), genotype *AD* was associated with a lower C12:0 and C12:1 levels than genotype *AA*.

(a)

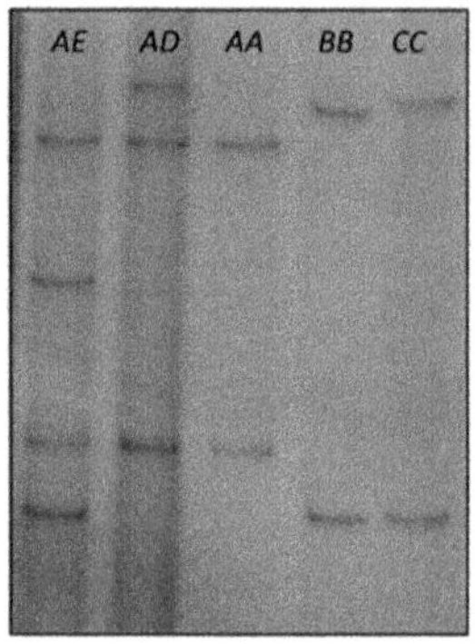

(b)

Nucleotide position	Nucleotide substitution	Variant				
		A	*B*	*C*	*D*	*E*
c.373+751	G/T	G	G	T	G	G
c.373+803	T/G	T	T	G	G	G
c.373+877	A/G	A	A	G	G	A
c.373+895	G/C	G	G	C	G	G
c.374-909	C/T	C	T	T	C	T
c.374-842	G/C	G	G	G	G	C
c.374-812	A/G	A	A	A	G	A

Figure 1. Variation in intron 1 of bovine *MSTN* revealed using PCR-SSCP analysis (**a**) and verified with nucleotide sequencing (**b**). Five PCR-SSCP patterns, representing the five distinct sequence variants (*A–E*, in both homozygous and heterozygous genotypes) are shown.

Table 1. Associations between milk fatty acid methyl ester levels and myostatin genotypes in NZ HF × J cross cows (n = 347).

Individual/Grouped Fatty Acids [1]	Mean ± Standard Error [2] (g/100 g Milk Fatty Acids)				p^3
	AA (*n* = 149)	*AB* (*n* = 87)	*AC* (*n* = 48)	*AD* (*n* = 63)	
C4:0	1.264 ± 0.012	1.252 ± 0.015	1.252 ± 0.020	1.262 ± 0.017	0.871
C6:0	1.566 ± 0.010	1.548 ± 0.013	1.549 ± 0.017	1.547 ± 0.015	0.458
C8:0	1.198 ± 0.009	1.179 ± 0.011	1.188 ± 0.014	1.169 ± 0.012	*0.150*
C10:0	3.299 ± 0.035	3.244 ± 0.043	3.293 ± 0.056	3.164 ± 0.050	*0.097*
C10:1	0.294 ± 0.004	0.282 ± 0.005	0.283 ± 0.007	0.284 ± 0.006	*0.108*
C11:0	0.062 ± 0.002	0.058 ± 0.002	0.056 ± 0.003	0.057 ± 0.002	*0.124*
C12:0	4.029 ± 0.046 [a]	3.939 ± 0.057 [ab]	4.015 ± 0.075 [ab]	3.824 ± 0.067 [b]	**0.046**
C13:0 *iso*	0.084 ± 0.002	0.079 ± 0.002	0.080 ± 0.002	0.078 ± 0.002	*0.076*
C12:1	0.096 ± 0.002 [a]	0.091 ± 0.002 [ab]	0.093 ± 0.003 [ab]	0.088 ± 0.003 [b]	**0.019**
C13:0 *anteiso*	0.038 ± 0.000	0.038 ± 0.001	0.038 ± 0.001	0.039 ± 0.001	0.450
C13:0	0.123 ± 0.002	0.119 ± 0.003	0.117 ± 0.004	0.118 ± 0.004	0.364
C14:0	12.513 ± 0.079	12.478 ± 0.098	12.690 ± 0.128	12.365 ± 0.114	0.266
C14:1 *cis*-9	1.003 ± 0.020	0.959 ± 0.025	0.967 ± 0.033	0.966 ± 0.029	0.390
C15:0 *iso*	0.296 ± 0.002	0.290 ± 0.003	0.299 ± 0.004	0.290 ± 0.004	*0.148*

Table 1. *Cont.*

Individual/Grouped Fatty Acids [1]	Mean ± Standard Error [2] (g/100 g Milk Fatty Acids)				p^3
	AA (*n* = 149)	*AB* (*n* = 87)	*AC* (*n* = 48)	*AD* (*n* = 63)	
C15:0 *anteiso*	0.639 ± 0.009	0.633 ± 0.011	0.647 ± 0.014	0.646 ± 0.012	0.814
C15:0	1.485 ± 0.016	1.473 ± 0.019	1.474 ± 0.026	1.471 ± 0.023	0.922
C15:1	0.283 ± 0.003	0.281 ± 0.004	0.287 ± 0.005	0.286 ± 0.004	0.714
C16:0	37.396 ± 0.288	37.624 ± 0.354	36.999 ± 0.465	37.558 ± 0.412	0.711
C16:1 *cis*-9	1.266 ± 0.023	1.245 ± 0.028	1.273 ± 0.037	1.301 ± 0.033	0.576
C17:0 *iso*	0.555 ± 0.006	0.553 ± 0.008	0.556 ± 0.010	0.553 ± 0.009	0.987
C17:0	0.690 ± 0.006	0.698 ± 0.008	0.693 ± 0.010	0.690 ± 0.009	0.816
C17:1	0.195 ± 0.002	0.194 ± 0.003	0.198 ± 0.004	0.199 ± 0.003	0.639
C18:0	8.547 ± 0.119	8.652 ± 0.146	8.652 ± 0.192	8.607 ± 0.170	0.918
C18:1 *trans*-11	2.756 ± 0.068	2.782 ± 0.083	2.801 ± 0.110	2.794 ± 0.097	0.974
C18:1 *cis*-9	12.833 ± 0.142	12.891 ± 0.175	12.891 ± 0.230	13.146 ± 0.204	0.595
C18:2 *trans*-9,12	0.377 ± 0.006	0.370 ± 0.007	0.391 ± 0.009	0.381 ± 0.008	0.230
C18:2 *cis*-9, *trans*-13	0.288 ± 0.003	0.281 ± 0.004	0.291 ± 0.005	0.292 ± 0.005	0.225
C18:2 *cis*-9, *trans*-12	0.067 ± 0.002	0.064 ± 0.002	0.065 ± 0.003	0.067 ± 0.002	0.670
C18:2 *trans*-9, *cis*-12	0.475 ± 0.011	0.483 ± 0.014	0.492 ± 0.018	0.473 ± 0.016	0.771
C18:2 *cis*-9,12	0.693 ± 0.007	0.677 ± 0.009	0.697 ± 0.012	0.699 ± 0.011	0.279
C19:0	0.140 ± 0.003	0.140 ± 0.003	0.143 ± 0.004	0.142 ± 0.004	0.890
C18:3 *cis*-6,9,12	0.075 ± 0.001	0.073 ± 0.001	0.074 ± 0.002	0.073 ± 0.001	0.511
C18:3 *cis*-9,12,15	0.805 ± 0.010	0.776 ± 0.013	0.818 ± 0.017	0.800 ± 0.015	*0.144*
CLA *cis*-9, *trans*-11	1.012 ± 0.028	1.001 ± 0.034	1.031 ± 0.045	1.029 ± 0.040	0.920
C20:0	0.126 ± 0.002	0.126 ± 0.002	0.130 ± 0.003	0.127 ± 0.002	0.530
C20:1 *cis*-5	0.060 ± 0.001	0.060 ± 0.002	0.062 ± 0.002	0.060 ± 0.002	0.891
C20:1 *cis*-9	0.151 ± 0.002	0.152 ± 0.003	0.149 ± 0.004	0.151 ± 0.003	0.890
C20:1 *cis*-11	0.077 ± 0.001	0.075 ± 0.002	0.076 ± 0.002	0.073 ± 0.002	0.093
C20:3 *cis*-8,11,14	0.030 ± 0.001	0.031 ± 0.001	0.032 ± 0.001	0.029 ± 0.001	0.137
C20:4 *cis*-5,8,11,14	0.035 ± 0.001	0.035 ± 0.001	0.034 ± 0.001	0.034 ± 0.001	0.927
C22:0	0.064 ± 0.001	0.067 ± 0.002	0.066 ± 0.002	0.066 ± 0.002	0.519
C22:1 *trans*-13	0.066 ± 0.001	0.068 ± 0.002	0.070 ± 0.002	0.067 ± 0.002	*0.190*
C20:5 *cis*-5,8,11,14,17	0.088 ± 0.00	0.089 ± 0.001	0.089 ± 0.002	0.088 ± 0.002	0.936
C24:0	0.044 ± 0.001	0.046 ± 0.001	0.044 ± 0.001	0.045 ± 0.001	0.294
C22:5 *cis*-7,10,13,16,19	0.123 ± 0.002	0.122 ± 0.003	0.123 ± 0.004	0.123 ± 0.003	0.996
SCFA	4.030 ± 0.027	3.979 ± 0.033	3.989 ± 0.043	3.977 ± 0.038	0.451
MCFA	19.841 ± 0.148	19.661 ± 0.182	19.997 ± 0.239	19.353 ± 0.212	*0.127*
LCFA	48.492 ± 0.252	48.826 ± 0.310	48.201 ± 0.407	48.706 ± 0.360	0.587
Total C18:1	16.300 ± 0.168	16.383 ± 0.206	16.400 ± 0.271	16.654 ± 0.240	0.634
Total C18:2	2.911 ± 0.041	2.876 ± 0.051	2.969 ± 0.067	2.940 ± 0.059	0.664
Total C18:3	0.880 ± 0.011	0.849 ± 0.013	0.893 ± 0.017	0.873 ± 0.015	*0.123*
Omega 3	1.017 ± 0.011	0.987 ± 0.014	1.030 ± 0.018	1.011 ± 0.016	*0.173*
Omega 6	0.832 ± 0.008	0.815 ± 0.010	0.838 ± 0.013	0.836 ± 0.011	0.317
MUFA	19.792 ± 0.174	19.790 ± 0.214	19.858 ± 0.281	20.128 ± 0.249	0.657
PUFA	4.067 ± 0.044	4.002 ± 0.054	4.139 ± 0.071	4.088 ± 0.063	0.410
Branched FA	1.611 ± 0.014	1.593 ± 0.017	1.620 ± 0.022	1.605 ± 0.020	0.735
Total UFA	23.859 ± 0.206	23.792 ± 0.254	23.997 ± 0.334	24.216 ± 0.295	0.663
Total SFA	72.548 ± 0.225	72.643 ± 0.277	72.362 ± 0.364	72.210 ± 0.323	0.701
Unsaturated index	24.756 ± 0.218	24.678 ± 0.269	24.911 ± 0.353	25.119 ± 0.313	0.677
C10:1 index	8.252 ± 0.136	8.032 ± 0.167	7.972 ± 0.219	8.321 ± 0.194	0.410
C12:1 index	2.328 ± 0.034	2.239 ± 0.042	2.266 ± 0.055	2.256 ± 0.048	0.243
C14:1 index	7.424 ± 0.148	7.133 ± 0.183	7.087 ± 0.240	7.256 ± 0.213	0.427
C16:1 index	3.281 ± 0.052	3.200 ± 0.063	3.319 ± 0.083	3.338 ± 0.074	0.435
C18:1 index	65.594 ± 0.331	65.409 ± 0.407	65.514 ± 0.535	65.913 ± 0.474	0.855
CLA index	26.626 ± 0.269	26.339 ± 0.332	26.725 ± 0.436	26.868 ± 0.386	0.716

[1] SCFA- short-chain length fatty acid; MCFA—medium-chain length fatty acid; LCFA—long-chain length fatty acid; MUFA—monounsaturated fatty acid; PUFA—polyunsaturated fatty acid; UFA—unsaturated fatty acid; SFA—saturated fatty acid. [2] Predicted means and standard error of those means derived from general linear mixed-effects models (GLMM). Myostatin genotype (categorical variable), cow age (categorical), herd (categorical), and days in milk (continuous) were fitted to the model as fixed effects. Means within a row that do not share a superscript letter are separated at $p < 0.05$. [3] $0.05 < p < 0.2$ in italics, while $p < 0.05$ in bold.

4. Discussion

This is the first study demonstrating association of *MSTN* sequence variants with the component levels of two milk FAs in NZ HF × J cross cattle.

All seven of the nucleotide substitutions identified in this study have been previously reported in a study of NZ cattle breeds, which included; Hereford, Angus, Shorthorn, Charolais, Red Poll, South Devon, Simmental, Murray Grey, HF × J cross cattle, and some composite breeds [19]. The nucleotide substitutions c.373+751G/T, c.373+803T/G, c.373+877A/G, c.373+895G/C, and c.374−909C/T were identified in all the ten aforementioned breeds, while c.374−842G/C was found in all but four breeds (Red Poll, Shorthorn, Simmental, and Composites breeds) and the c.374−812A/G was only found in Shorthorn and HF × J cross breeds.

In a previous study, Smith et al. [22] identified a mutant "*mh*" allele of bovine *MSTN* with an 11-bp deletion in the exon 3 region. This was described in British South Devon cattle, and it was the same allele first identified in the Belgian Blue breed, causing the "double-muscled" phenotype. Also, in a later investigation of *MSTN* in 146 British South Devon cattle, Wiener et al. [23] showed that this same mutant "*mh*" allele, reduced the levels of total SFA and total MUFA in the muscle ($p < 0.05$). Their report also revealed an increase in the ratio of PUFA: SFA in total lipid to be greater in *mh/mh* individuals than in the other two genotype (*mh*/+ and +/+) classes ($p < 0.001$). This suggested that the "*mh*" allele is associated with reduced fat levels, particularly with the levels of SFA and MUFA. The increase in the ratio of PUFA to SFA would be expected, especially when the concentration of muscle fat decreases.

In a similar report on associations between the "*mh*" allele carrying the 11-bp deletion with intramuscular fatty acid composition in *MSTN* of Belgian Blue young bulls, Raes et al. [24] revealed that animals with the +/+ (normal) genotype showed a higher relative amount of the SFA; C14:0 and C16:0 and a higher relative amount of all MUFA C16:1; C17:1 and C18:1, but the relative proportion of PUFA in total fatty acids increased with increasing mutant "*mh*" alleles. In the current study, the genotype model revealed that cows carrying the *AA* genotype showed an increase in the amount of two medium-chain length SFAs (C12:0 and C12:1) relative to *AD* cows. While it may be difficult to conclude that the findings in the current study are similar to the work of Weiner et al. [23] and Raes et al. [24], especially because it investigated the intronic region of *MSTN* in non-doubled muscled cattle breed, it is important to note that variations in the intronic regions are equally capable of influencing gene expression and/or altering the functionality of a gene [25]. In a previous investigation, He et al. [25] transformed C2C12 cell-lines with a transgene construct that contained bovine *MSTN* promoter (pMD-MSTNPro) and a second construct that contained the first intron of bovine *MSTN* (pMD-Intron1). They observed an increase in the mean fluorescence intensity of green fluorescent protein (GFP) gene and the percentage of fluorescence positive cells, and concluded from this that the presence of intron 1 of bovine *MSTN* increased the expression of GFP in the transformed cells.

The findings in this study are in part in agreement with the findings of Buske et al. [26]. That study revealed that one copy of the so-called "wild-type + allele" of *MSTN* was associated with higher milk, protein, and fat yields in dual purpose Belgian Blue (DP-BB) cows, whereas a single copy of the mutant "*mh*" allele (associated with double muscling) was associated with a decrease in the SFA content of milk. Even though the current study did not involve the *MSTN* "*mh*" allele, it has perhaps suggested that antagonistic effects may exist between milk and meat production traits in the context of *MSTN*.

Since the *MSTN* variations in the current study occurred in a non-expressed region of the gene, it is perhaps less obvious how they might affect the structure or function of the MSTN protein. They might however influence the rate of transcription and/or translation of the gene, as was described by Liu et al. [27] in an investigation that involved both transgenic mice and in vitro studies. In that study, intron sequences in the transgene that encoded rat growth hormone were observed to stimulate transcription by promoting assembly of an ordered nucleosome array in the vicinity of the promoter. Additionally, while the observed intron variation does not yield any amino acid change, several studies have shown that synonymous nucleotide substitutions can affect the phenotypic characteristics of the protein product by altering mRNA structure, protein stability, the electrical charge of the resulting polypeptide, and codon usage during mRNA translation [28,29].

Previous reports on the differences in the composition of milk FAs between Holstein-Friesian and the Jersey breeds suggest that milk from Jersey cows tends to have higher concentrations of some short- and medium-chain length SFA, but lower concentrations of some UFA [16]. It is therefore interesting to note that the current investigation of NZ HF × J cross cows, revealed that variant *D* was associated with a decrease in the amount of two medium-chain length SFAs in milk. This discovery could be of benefit in terms of its potential applicability in cross-breeding and gene-marker development, particularly in selection for decreased SFA in milk.

5. Conclusions

The findings suggest that variation in *MSTN* affects two milk FA traits and this may be of value in breeding dairy cattle. Notably, variant *D* of *MSTN* is associated with a decrease in two medium-chain length SFA levels. Cows with the *AD* genotype might therefore produce a "preferred" FA profile in milk. However, because there were insufficient cattle with the homozygous genotypes *BB*, *CC*, and *DD*, or the heterozygous genotypes (*BC*, *BD*, *CE*, and *CD*) in the samples investigated, further investigation involving much larger sample sizes across different farms and breeds of cattle is needed to validate this claim.

Author Contributions: Conceptualization, J.G.H.H. and H.Z.; methodology, I.L.H., U.J.E., and H.A.-N.; software, Y.L., and H.A.-N.; validation, J.G.H.H., and H.Z.; formal analysis, Y.L., H.A.-N.; investigation, I.L.H. and U.J.E.; resources, J.G.H.H. and H.Z.; data curation, H.Z.; writing—original draft preparation, I.L.H.; writing—review and editing, I.L.H., J.G.H.H., and H.Z.; visualization, H.Z.; supervision, J.G.H.H.; project administration, J.G.H.H. All authors have read and agreed to the published version of the manuscript.

Funding: This research received no external funding.

Acknowledgments: The authors would like to thank Freeman Fang for technical support.

Conflicts of Interest: The authors declare no conflict of interest.

References

1. McPherron, A.C.; Lawler, A.M.; Lee, S.J. Regulation of skeletal muscle mass in mice by a new TGF-b superfamily member. *Nature* **1997**, *387*, 83–90. [CrossRef] [PubMed]
2. Mendias, C.L.; Bakhurin, K.I.; Faulkner, J.A. Tendons of myostatin-deficient mice are small, brittle, and hypocellular. *Proc. Natl. Acad. Sci. USA* **2008**, *105*, 388–393. [CrossRef] [PubMed]
3. Whittemore, L.A.; Song, K.; Li, X.; Aghajanian, J.; Davies, M.; Girgenrath, S.; Hill, J.J.; Jalenak, M.; Kelley, P.; Knight, A.; et al. Inhibition of myostatin in adult mice increases skeletal muscle mass and strength. *Biochem. Biophys. Res. Commun.* **2003**, *300*, 965–971. [CrossRef]
4. Kambadur, R.; Sharma, M.; Smith, T.P.; Bass, J.J. Mutations in myostatin (GDF8) in double-muscled Belgian Blue and Piedmontese cattle. *Genome Res.* **1997**, *7*, 910–916. [CrossRef]
5. Hickford, J.G.H.; Forrest, R.H.; Zhou, H.; Fang, Q.; Han, J.; Frampton, C.M.; Horrell, A.L. Polymorphisms in the ovine myostatin gene (*MSTN*) and their association with growth and carcass traits in New Zealand Romney sheep. *Anim. Genet.* **2010**, *41*, 64–72. [CrossRef]
6. McPherron, A.C.; Lee, S.J. Suppression of body fat accumulation in myostatin-deficient mice. *J. Clin. Investig.* **2002**, *109*, 595–601. [CrossRef]
7. Grummer, R.R. Effects of feed on the composition of milk fat. *J. Dairy Sci.* **1991**, *74*, 3244–3257. [CrossRef]
8. Palmquist, D.L.; Beaulieu, A.D.; Barbano, D.M. Feed and animal factors influencing milk fat composition. *J. Dairy Sci.* **1993**, *76*, 1753–1771. [CrossRef]
9. Dewhurst, R.J.; Shingfield, K.J.; Lee, M.R.F.; Scollan, N.D. Increasing the concentrations of beneficial polyunsaturated fatty acids in milk produced by dairy cows in high-forage systems. *Anim. Feed Sci. Technol.* **2006**, *131*, 168–206. [CrossRef]

10. Soyeurt, H.; Dardenne, P.F.; Dehareng, G.; Lognay, G.; Veselko, D.; Marlier, M.; Bertozzi, C.; Mayeres, P.; Gengler, N. Estimating fatty acid content in cow milk using mid-infrared spectrometry. *J. Dairy Sci.* **2006**, *89*, 3690–3695. [CrossRef]
11. Stoop, W.M.; van Arendonk, J.A.M.; Heck, J.M.L.; van Valenberg, H.J.F.; Bovenhuis, H. Genetic parameters for major milk fatty acids and milk production traits of Dutch Holstein Friesians. *J. Dairy Sci.* **2008**, *91*, 385–394. [CrossRef] [PubMed]
12. Palladino, R.A.; Buckley, F.; Prendiville, R.; Murphy, J.J.; Callan, J.; Kenny, D.A. A comparison between Holstein-Friesian and Jersey dairy cows and their f1 hybrid on milk fatty acid composition under grazing conditions. *J. Dairy Sci.* **2010**, *93*, 2176–2184. [CrossRef] [PubMed]
13. Maurice-Van Eijndhoven, M.H.T.; Hiemstra, S.J.; Calus, M.P.L. Short communication: Milk fat composition of 4 cattle breeds in the Netherlands. *J. Dairy Sci.* **2011**, *94*, 1021–1025. [CrossRef] [PubMed]
14. Strucken, E.M.; Laurenson, Y.C.S.M.; Brockmann, G.A. Go with the flow-biology and genetics of the lactation cycle. *Front. Genet.* **2015**, *6*, 118. [CrossRef] [PubMed]
15. Bauman, D.E.; Bruce Currie, W. Partitioning of nutrients during pregnancy and lactation: A review of mechanisms involving homeostasis and homeorhesis. *J. Dairy Sci.* **1980**, *63*, 1514–1529. [CrossRef]
16. Arnould, V.M.R.; Soyeurt, H. Genetic variability of milk fatty acids. *J. Appl. Genet.* **2009**, *209*, 29–39. [CrossRef]
17. Li, Y.; Zhou, H.; Cheng, L.; Edwards, G.R.; Hickford, G.H.J. Effect of DGAT1 variant (K232A) on milk traits and milk fat composition in outdoor pasture-grazed dairy cattle. NZ. *J. Agric. Res.* **2019**. [CrossRef]
18. Zhou, H.; Hickford, G.H.J.; Fang, Q. A two-step procedure for extracting genomic DNA from dried blood spots on filter paper for polymerase chain reaction amplification. *Anal. Biochem.* **2006**, *354*, 159–161. [CrossRef]
19. Haruna, I.L.; Ekegbu, U.J.; Ullah, F.; Amirpour-Najafabadi, H.; Zhou, H.; Hickford, J.G.H. Genetic variation and haplotypic diversity in the myostatin gene of New Zealand cattle breeds. *Gene* **2020**, *740*. [CrossRef]
20. Byun, S.O.; Fang, Q.; Zhou, H.; Hickford, J.G.H. An effective method for silver-staining DNA in large numbers of polyacrylamide gels. *Anal. Biochem.* **2009**, *385*, 174–175. [CrossRef]
21. Gong, H.; Zhou, H.; Hickford, J.G.H. Diversity of the glycine/tyrosine-rich keratin-associated protein 6 gene (KAP6) family in sheep. *Mol. Biol. Rep.* **2011**, *38*, 31–35. [CrossRef] [PubMed]
22. Smith, J.A.; Lewis, A.M.; Wiener, P.; Williams, J.L. Genetic variation in the bovine myostatin gene in UK beef cattle: Allele frequencies and haplotype analysis in the South Devon. *Anim. Genet.* **2000**, *31*, 306–309. [CrossRef] [PubMed]
23. Wiener, P.; Woolliams, J.A.; Frank-Lawale, A.; Ryan, M.; Richardson, R.I.; Nute, G.R.; Wood, J.D.; Homer, D.; Williams, J.L. The effects of a mutation in the myostatin gene on meat and carcass quality. *Meat Sci.* **2009**, *83*, 127–134. [CrossRef] [PubMed]
24. Raes, K.; de Smet, S.; Demeyer, D. Effect of double muscling in Belgian Blue young bulls on the intramuscular fatty acid composition with emphasis on conjugated linoleic acid and polyunsaturated fatty acids. *Anim. Sci.* **2001**, *73*, 253–260. [CrossRef]
25. He, Y.L.; Wu, Y.H.; Quan, F.S.; Liu, Y.G.; Zhang, Y. Molecular analysis of the first intron in the bovine myostatin gene. *Mol. Biol. Rep.* **2010**, *38*, 4643–4649. [CrossRef]
26. Buske, B.; Gengler, N.; Soyeurt, H. Short-communication: Influence of the muscle hypertrophy mutation of the myostatin gene on milk production traits and milk fatty acid composition in dual-purpose Belgian Blue dairy cattle. *J. Dairy Sci.* **2011**, *94*, 3687–3692. [CrossRef]
27. Liu, K.; Sandgren, E.P.; Palmiter, R.D.; Stein, A. Rat growth hormone gene introns stimulate nucleosome alignment in vitro and in transgenic mice. *Proc. Natl. Acad. Sci. USA* **1995**, *92*, 7724–7728. [CrossRef]
28. Sauna, Z.E.; Kimchi-Sarfaty, C. Understanding the contribution of synonymous mutations to human disease. *Nat. Rev. Genet.* **2011**, *12*, 683–691. [CrossRef]
29. Hunt, R.C.; Simhadri, V.L.; Iandoli, M.; Sauna, Z.E.; Kimchi-Sarfaty, C. Exposing synonymous mutations. *Trends Genet.* **2014**, *30*, 308–321. [CrossRef]

animals

MDPI

Review

Amino Acids Supplementation for the Milk and Milk Protein Production of Dairy Cows

Jung-Eun Kim and Hong-Gu Lee *

Department of Animal Science and Technology, Sanghuh College of Life Sciences, Konkuk University, Seoul 05029, Korea; sumzzzing@gmail.com
* Correspondence: hglee66@konkuk.ac.kr; Tel.: +82-2-450-0523

Simple Summary: The composition of milk not only has nutritional implications, but is also directly related to the income of dairy producers. As regards milk's composition, concerns around milk protein have emerged from the increased consumption of casein products. The synthesis of proteins in milk is a highly complex and high-cost process, because the conversion efficiency of dietary protein to milk protein is very low in dairy cows. Thus, some studies have increased milk protein by using protein supplements or a single amino acid (AA) supply. AAs are the building blocks of protein, and can also stimulate the protein synthetic pathway. This review mainly concerns the use of AAs for producing milk protein in high-producing dairy cows, particularly with methionine, lysine, and histidine. Understanding the mechanisms of AAs will help to promote milk protein synthesis in the dairy industry.

Abstract: As the preference of consumers for casein products has increased, the protein content of milk from dairy cows is drawing more attention. Protein synthesis in the milk of dairy cows requires a proper supply of dietary protein. High protein supplementation may help to produce more milk protein, but residues in feces and urine cause environmental pollution and increase production costs. As such, previous studies have focused on protein supplements and amino acid (AA) supply. This review concerns AA nutrition for enhancing milk protein in dairy cows, and mainly focuses on three AAs: methionine, lysine, and histidine. AA supplementation for promoting protein synthesis is related to the mammalian target of rapamycin (mTOR) complex and its downstream pathways. Each AA has different stimulating effects on the mTOR translation initiation pathway, and thus manifests different milk protein yields. This review will expand our understanding of AA nutrition and the involved pathways in relation to the synthesis of milk protein in dairy cows.

Keywords: protein metabolism; amino acids; milk protein; dairy cows

Citation: Kim, J.-E.; Lee, H.-G. Amino Acids Supplementation for the Milk and Milk Protein Production of Dairy Cows. *Animals* **2021**, *11*, 2118. https://doi.org/10.3390/ani11072118

Academic Editor: João Simões

Received: 28 June 2021
Accepted: 13 July 2021
Published: 16 July 2021

1. Introduction

Milk's composition relates to its nutritional value, but it is also directly linked to the income of dairy farms. Milk's fat content has been scrutinized due to its nutritional value; however, as the preference of consumers has shifted towards protein products, studies on milk protein production have emerged.

Milk protein synthesis in dairy cows requires adequate supplies of energy and dietary crude protein (CP), specifically individual amino acids (AAs). Cows exhibit very low efficiency in converting nitrogen (N) into protein in its body or milk: the utilization efficiency of N is 25–35% [1]. According to National Research Council (NRC) (2001), the maximum yield of milk and milk protein is attributed to 22% of the dietary CP. However, due to the high cost of dietary protein sources, controlling single AAs may be a cost-effective strategy in high-producing dairy cows.

Although review papers on protein and amino acid metabolism in dairy cows have been introduced [2], this article focuses on AA nutrition for milk and milk protein production in dairy cows, with a specific focus on protein metabolism in the animal's body, the importance of feeding with AAs, studies on AA supply, and the protein-synthetic pathway. As regards the AAs used, we mainly discuss methionine (Met), lysine (Lys), and histidine (His), which have been extensively studied previously.

2. Protein Metabolism in Ruminants

Ruminants obtain nitrogen (N) sources that are available for metabolism from dietary intake, microbial proteins, and endogenous N (Figure 1). Dietary protein is divided into rumen-degradable protein (RDP) and rumen-undegradable protein (RUP). RDP is degraded by rumen microbes, then synthesized into microbial protein or partly bypassed, whereas RUP directly bypasses the rumen. The digesta, escaping the stomach and reaching the small intestine, is disassembled and absorbed into the blood stream, and undigested proteins are excreted as feces. The absorbed AAs pass through the liver to the kidneys, or flow into the blood stream. The AAs arriving at the mammary glands are synthesized into milk protein and subsequently secreted into milk.

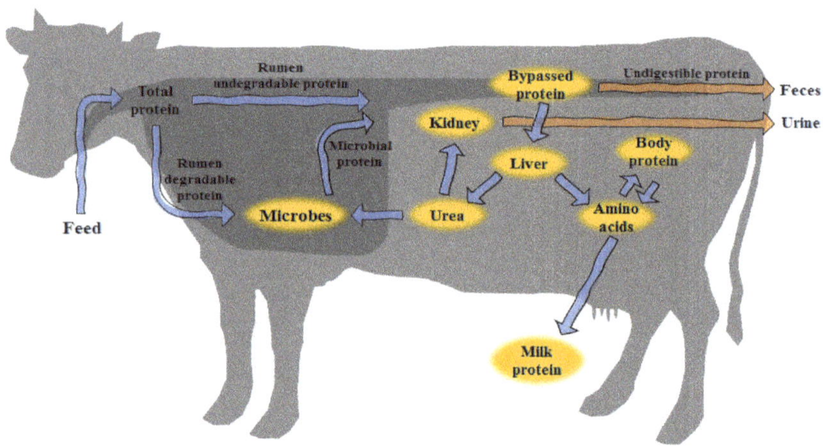

Figure 1. Protein metabolism of dairy cows.

3. Concept of AA Supplementation

The productive results in Section 3 are summarized in Supplementary Table S1.

Many studies have reported on oral administration and post-ruminal infusion as sources of protein for dairy cows. Soybean meal (SBM) and casein are typical protein supplements.

The effect of gradually increasing the level of SBM was studied [3,4]. The medium-protein group (formaldehyde-treated SBM inclusion; 15.4% of CP) showed a 1.4 kg/d increase in milk yield and a 48 g/d increase in milk protein yield compared to the low-protein group (no supplemented SBM; 11.3% of CP) when Holstein–Friesian cows received silage and three levels of protein concentration (ratio of 40:60) [4]. However, the high-protein group showed a similar milk yield to the low-protein group, and a similar milk protein yield to the medium-protein group. A linear increase in milk yield (26.6 to 28.0 kg/d) and milk protein yield (940 to 969 g/d) was observed when Swedish Red cows were fed grass silage-based diets (15.3% of CP), with various levels of SBM (17.3%, 19.0%, and 21.0% of CP) in their early lactation [3]. The supplementation of SBM to early-lactating Finnish Ayrshire, fed grass silage-based diets (17.0% of CP), significantly increased their milk yields and components by 3.0 kg/d, as compared to the control group [5]. In particular, the milk protein yield increased 113 g/d, and the milk protein concentration increased 0.14%, as compared to the control.

In a study on the abomasal infusion of sodium caseinate or enzymatically hydrolyzed casein into mid-lactating Holstein cows fed diets containing SBM (14.2% of CP) or SBM plus corn gluten meal (13.8% of CP), the sodium caseinate group showed increasing trends of milk yield (1.1 kg/d when SBM-fed and 1.2 kg/d when SBM plus corn gluten meal fed) and significant increases in milk protein yield (50 g/d), as compared to the saline-infused group [6]. However, the hydrolyzed casein group showed similar results to the saline group. The duodenal infusion of calcium caseinate produced more milk yield (1.01 kg/12 h), milk true protein concentration (0.90 g/kg), and milk true protein yield (36 g/12 h) in early-lactating Holstein cows as compared to those of the control group [7]. However, in Holstein cows fed basal diets containing 15.6% CP, the abomasal infusion of sodium caseinate did not affect lactational performance (numerical increases in milk yield and milk protein yield only) [8].

High-CP diets based on protein supplements may increase milk productivity. However, the problem is that N efficiency is also reduced [9]. Decreased N utilization efficiency causes huge losses to dairy farmers. Furthermore, N excreted in feces and urine (Figure 1) causes environmental pollution [10,11]. The price of protein supplements is relatively high when compared to other feed ingredients. In particular, the preference of SBM is high because it increases the productivity of cows, as described above. However, due to the high price of SBM, research on other protein feeds that can replace SBM has been reported. Some authors reported canola meal inclusion in dairy cow diets as a replacement for SBM [12,13]. When SBM or canola meal (CM) was included in diets, the CM group showed an increase [12] or no change [13] to milk and milk protein yield, as compared to those of the SBM group. In addition, studies on various other by-products or protein supplements that are used to replace SBM have been reported. Several protein products may be a good substitute for SBM. However, due to problems such as price and environmental pollution, the supplementation of excess protein sources should be limited. Therefore, it is necessary to carefully examine the supply of individual AAs, instead of surplus protein.

3.1. Balancing AA

When increasing milk protein, the level of a single AA supply is important, but it is also necessary to evaluate whether the supplied AAs are balanced. Concerns with balancing AAs have been increasing. Some authors have pointed out that increased milk production may be related to the increased or balanced AA supply that is a result of protein supplementation [3,5]. A decreased production of milk protein in dairy goats, due to deficiencies or imbalances of AAs, was observed in treatments employing abomasal AA mixture infusions with the deletion of Lys, arginine (Arg), Met, or His [14]. Similar results were reported when Holstein dairy cows received abomasal infusions of AA mixtures (the ratio of milk protein) with Met, Lys, or His eliminated [15]. A deficiency or imbalance in His is induced by the dietary inclusion of feather meal (12.6%) [16]. Feather meal treatment, in the context of a standard protein diet, reduced both milk yield and milk protein yield by 7.4 kg/d and 331 g/d, respectively [16].

It is important to supplement deficient AAs for balancing the body's AA pool well. Therefore, we must also assess which AAs are deficient, or risk producing an imbalanced environment.

3.2. Limiting AA

Essential AAs (EAAs) are synthesized by the animal itself at lower levels or not at all, and thus must be provided in the diet. For cows, the EAAs are the following: Arg, His, isoleucine (Ile), Leu, Lys, Met, phenylalanine (Phe), threonine (Thr), tryptophan (Trp), and valine (Val) [17]. Insufficient AAs, referred to as limiting AAs, are the AAs in the animal's body pool that interrupt protein synthesis when they are deficient. Generally, Met and Lys are known as the co-limiting factors in corn and alfalfa silage-based diets [17,18], and His is known as a limiting AA when cows are fed grass silage-based diets [19–22]. Therefore, these three AAs have been widely reported on.

The first experiment on the intravenous infusion of Met, Lys, and His was reported in 1972 [23]. The infusion of 11.2 g/d Met increased milk protein yield by 30 g/d, as compared to controls, in early-lactating Holstein dairy cows fed corn silage-based diets. However, the infusion of His reduced the milk protein content, and the infusion of Lys did not affect milk protein. Met and Lys were shown to be co-limiting factors of milk protein synthesis when Holstein cows were fed corn-based diets (14.5% of CP) and receiving various combinations of 10 EAAs via abomasal infusion [17]. Another study suggested that Met and Lys were the two most limiting AAs when Holstein dairy cows in four different stages of lactation were duodenally infused with DL-Met, L-Lys, DL-Met plus L-Lys, or casein [18]. The infusion of both DL-Met and L-Lys increased milk protein content in early, mid, and late lactation, and milk protein yield in peak, early, and mid lactation, as compared to the DL-Met or L-Lys treatments. Met and Lys are the first-limiting AAs in corn and alfalfa silage-based diets in the United States because these AAs are laced in feed ingredients. Corn and SBM, typical feed ingredients, contain low levels of Lys and Met, respectively [1]. Hence, Met and Lys have been suggested as co-limiting AAs in United States diets.

Compared to those receiving the intravenous infusion of a mixture of four AAs (Met, Lys, His, and Trp), reduced milk yields (−3.2 kg/d) and milk protein yields (−159 g/d) were observed in the His-deprived group when feather meal, barley, and grass silage was offered to Friesian cows [19]. The continuous intravenous infusion to Frisian cows fed grass silage-based diets was conducted with a treatment of EAA (a composition of ten AAs in casein), three AAs (Met, Lys, and His), or His alone [20]. The results showed that all treatments increased the milk and milk protein yields when compared to the control, and no changes were observed between treatments. The abomasal infusion of His (6.5 g/d) increased milk yield (0.7 kg/d) and milk protein yield (26 g/d) in mid-lactating Finnish Ayrshire fed grass silage diets [22]. However, similar results were verified in infusion groups of His and Lys, His and Met, and three AAs, as compared to the His alone infusion group. Accordingly, His was considered a first-limiting AA when cows were fed grass silage diets. This is due to the low content of His in barley and feather meal, as well as in the rumen microbes [24].

Studies on other AAs, such as Ile, Leu, and Val, have also been performed. It is well known that branched-chain AAs (BCAAs; Ile, Leu and Val) are involved in protein metabolism [25]. The abomasal infusion of 150 g/d of BCAAs mixture did not affect milk or milk protein yields, as compared to the control group, when Holstein dairy cows were fed a total mixed ration (TMR) consisting mainly of corn and alfalfa hay (16.2% of CP) [26]. As regards diets based on cereal and grass silage, lactating Finnish Ayrshires showed no changes when infused abomasally with either a mixture of BCAAs and His or a mixture with one of BCAAs removed [27]. The abomasal infusion of His increased 0.8 kg/d of milk yield and 24 g/d of milk protein yield; however, no effect was observed when infusing Leu with His, as compared to the His-infused group, in early-lactating Finnish Ayrshire fed with grass silage [28]. The jugular infusion of Met and Lys or 2AAs plus BCAAs was reported when early-lactating Holstein cows were fed corn-based diets (16.1% of CP) [29]. Both groups presented increased milk protein content and increasing tendencies in milk protein yield; however, BCAA had no effect. An increase in milk yield (2.2 kg/d) and milk protein yield (40 g/d) was reported following the jugular infusion of Ile and Leu in Holstein cows fed corn-based diets (15.2% of CP) [30]. However, further experiments on the effects of BCAA on milk protein in dairy cows are still required.

3.3. Rumen-Protected AA

Artificially infused EAAs are a better choice for delivering AAs post-ruminally; however, this process is not applicable at the herd level. Thus, various rumen-protected (RP) AA products have been developed for supplying EAAs to the small intestine without being degraded by rumen microbes.

The RP-AAs, Met, Lys, or both, have been widely studied and used, with diverse results. Two trials with RP-Met were performed with Holstein dairy cows fed alfalfa and corn silage-based diets [31]. In trial 1, treatments of 17.3% of CP plus 5 g of RP-Met and 16.1% of CP plus 10 g of RP-Met significantly increased the milk yield (+1.9 kg), as compared to 18.6% of CP with no supplementation group; however, the milk protein yield was unchanged. The supplementation of diets containing 17.3% or 16.1% CP with 10 g of RP-Met resulted in similar milk yields and composition as compared to the no RP-Met group. In early-lactating dairy cows, fed 14.5% of CP, RP-Met (0.03% of diets) tended to decrease milk protein content due to the numerical decrease in milk protein yield [32]. RP-Lys, supplemented at approximately 94.4 g/d/cow, increased milk and milk protein yields (2.03 kg/d and 80 g/d, respectively) in early-lactating Holstein cows fed alfalfa- and corn-based diets (17.0% of CP) [33]. In the same study with mid-lactating cows, RP-Lys supply tended to increase milk and milk protein yields (0.82 kg/d and 20 g/d, respectively). However, similar milk production levels were also reported as being between 0 or 60 g/d of RP-Lys-supplemented groups, when early-lactating Holstein cows were fed corn silage-based diets (16.4% of CP) with 10 or 20% of dried distillers grains plus solubles [34].

The effects of RP-Met and RP-Met plus RP-Lys were tested in early-lactating Holstein dairy cows fed alfalfa hay and heated whole soybean-based diets (19.5% of CP) [35]. A linear increase was observed in both milk protein content and yield when 5.25 and 10.5 g/d of RP-Met were supplemented. However, the supplementation of RP-Met plus RP-Lys (11.5 and 14.7 g/d, respectively) did not present changes as compared to the 10.5 g/d RP-Met group [35]. Contrary to the experiment mentioned above, the supplementation of 40 g/d of RP-Met and Lys product increased both milk and milk protein yields, as compared to the 15 g/d RP-Met group, when Holstein dairy cows were fed corn-based diets containing 16.0% or 18.5% of dietary CP [36]. The tablet forms of RP-Met (18.2 g/d supply), RP-Lys (11.7 g/d), or both were supplied to early-lactating Holstein cows fed maize silage and cereal-based diets (14.5% of CP) [37]. RP-Met or RP-Lys alone caused no changes in milk and milk protein yields; however, RP-Met plus RP-Lys increased the milk protein yield.

For estimating the effects of RP-Met on milk production and milk protein, a meta-analysis was conducted on papers that used two widely studied products: Mepron (Evonik Industries, Hanau, Germany) and Smartamine M (Adisseo, Antony, France) [38]. The analysis predicted that RP-Met supply would lead to 0.07% more produce and 27 g of true milk protein content and yield (true milk protein: milk CP × 0.94). In the same paper, the authors also analyzed whether AA deficiency and forage sources would affect true milk protein, and the results suggested that both true milk protein content and yield would be increased, regardless of the state of the AAs (adequate or deficient Met, Met + Lys, or Met + Lys + one other AA) and the main forage source (alfalfa, corn silage, grass, or grass + corn silage). Overall, the aforementioned studies and meta-analysis indicate that supplementation with RP-Met has the potential to increase the milk protein yield of dairy cows in various conditions.

After considering His, as a potential AA, to be a first-limiting AA, the effects of RP-His with or without RP-Met and RP-Lys on milk production were investigated. Gradually increasing amounts of RP-His (0, 82, 164, and 246 g/d) were supplemented to mid-lactating Holstein dairy cows fed corn silage-based diets (15.1% of CP, containing 11 g/d of RP-Met) [39]. A tendency to increase was observed in both milk yield (linearly) and milk protein yield (quadratically), and an actual increase in milk protein yield (+50 g/d) was presented when 256 g/d of RP-His was supplied. Morris and Kononoff (2020) reported the effects of RP-Lys, RP-His, or both in Jersey dairy cows fed corn silage-based diets (17.1% of CP) with hydrolyzed feather meal (His-deficient source). Supplementation with 70 g/d of RP-Lys did not change milk production; however, a supply of RP-His (32 g/d) caused a significant increase in milk yield and a tendency towards an increase in milk protein yield [40]. In some studies, the effect of His was tested when metabolizable protein (MP)-deficient diets (MPDs) were offered to cows [41–43]. A supply of MPD (13.6% of CP; corn silage and alfalfa haylage-based diets) with RP-Met (30 g/d/cow) plus RP-Lys (100 g/d/cow) with or without RP-His (50 g/d/cow) increased milk protein yield as compared to the MPD with no supplementation group, and the resulting yield was similar, as compared to the MP-adequate (MPA; 15.7% of CP) diet group. However, compared to MPD with RP-Met and RP-Lys, the effect of RP-His was not verified [43]. The effects of MPD, slow-release urea, RP-Met (30 g/d), and RP-His (50 g/d) on the productivity of dairy cows were estimated in Holstein dairy cows fed corn silage-based diets [42]. Milk protein production was not affected by MPD (14.8% of CP), MPD plus urea (15.8% of CP), or RP-Met, but it was affected by RP-His. Moreover, the supplementation of MPD with RP-His plus urea and RP-Met increased the milk protein yield (90 g/d) as compared to the MPA group (16.7% of CP). The same author studied MPD, RP-Met, Lys, and His [41]. The milk protein yield showed only numerical changes in each of the Met, Lys, and His alone groups as compared to MPD (average 14.5% of CP); however, supplied with mixture of three AAs to MPD, milk protein yield was increased 100 g/d. Due to the diversity of the effects and the limitations of these studies, further research on the effects of RP-His is still required. Research on other AAs, besides the three mentioned, has also been carried out (RP-Leu [44], RP-Phe [45–47], RP-Trp [48], RP-AAs (mixture of Lys, Ile, Val, and His in RP form) [49], etc.). Further studies on RP-AAs in various feeding types and under various animal conditions are required.

4. Effects of AA Supplementation on Mammary Translational Expression

For protein synthesis, the pivotal substances are as follows: AAs (building blocks), glucose (energy source), and insulin. Protein is synthesized by the mammalian target of rapamycin (mTOR) complex and its downstream pathway [50,51]. Due to the high-energy demands of protein synthesis, mTOR translational regulation is performed by the energy source, usually glucose [52,53]. When glucose is recognized by the AMP-activated protein kinase (AMPK), its activation inhibits mTOR. Insulin is essential to the cellular metabolism. Phosphoinositide 3-kinase (PI3K) is stimulated by insulin, and further stimulates protein kinase B (Akt). The AAs are not only building blocks of protein; they are also potential regulators of mTOR translational regulation [50,52,53]. In particular, leucine (Leu) is known to stimulate the mTOR pathway either indirectly or directly [54,55]. Once the mTOR complex is phosphorylated, the activity of the ribosomal protein S6 kinase 1 (S6K1) is blocked, and the activated ribosomal protein S6 (RPS6) enters for further translation processes. Meanwhile, mTOR phosphorylates the eukaryotic initiation factor (eIF) 4E binding protein 1 (4EBP1). Then, through the translation initiation and elongation stage, involving eIF and eukaryotic translation elongation factor 2 (eEF2), milk proteins are produced. An illustration of the basic mTOR pathway, supplemented with our recent in vitro results, is presented in Figure 2 [56–58].

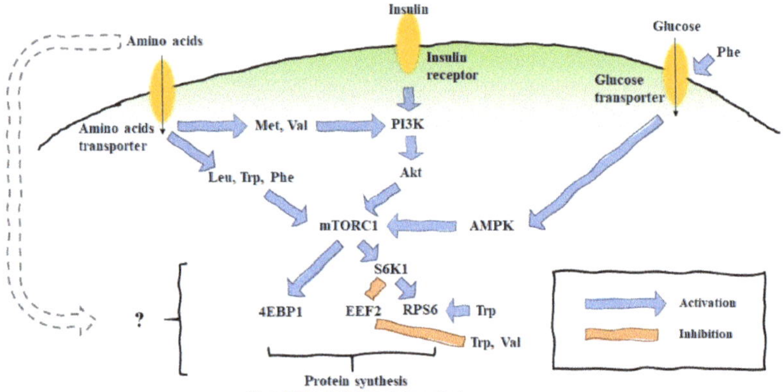

Figure 2. Mammalian target of rapamycin pathway.

The effects of AAs on the protein synthetic pathway have been assessed via the in vitro incubation of mammary cells and mammary biopsies taken from cows supplemented with AA. Due to the lack of experimental animals, many studies on milk protein synthesis with the mTOR pathway have been conducted in vitro, using immortalized or primary bovine mammary epithelial cells [56–61].

In our previous studies, AA treatments of immortalized bovine mammary epithelial cells (MAC-T cells) proceeded through different pathways [56–58]. The supplementation of 0.6 mM of L-Met in MAC-T cells was shown to increase β-casein expression and stimulate the PI3K pathway; however, an effect on mTOR was not observed [57]. eEFs were detected when 0.9 mM of L-Phe and 0.6 mM of L-Val were supplied to MAC-T cells; however, neither the mTOR nor the β-casein were altered in these treatments [58]. When 0.9 mM of L-Trp was added to the MAC-T cell medium, significant increases were observed in the mRNA expressions of mTOR, RPS6, and β-casein [56]. The inclusion of 0.15 mM of L-His increased β-casein expression, as compared to the control, when MAC-T cells were cultured in nutrient-restricted medium [60]. The effects of His, Lys, Met, and Leu on the mTOR pathway and β-casein were reported in Chinese Holstein mammary epithelial cells (CMEC-H cells) [59]. The expressions of mTOR, S6K1, RPS6, and β-casein increased, as compared to the control, following supplementation with 0.15 mM of His, 0.5 mM of Lys, 0.12 mM of Met, and 0.45 mM of Leu. In that study, His also increased the expression of 4EBP1, while Met decreased it [59]. The supplementation of Arg to primary mammary epithelial cells (pMEC) from Chinese Holstein cows increased the mRNA levels of four types of casein, mTOR, and S6K1, but decreased 4EBP1 [61].

The effects of EAAs and glucose infused by the jugular vein were reported when early-lactating Holstein cows experienced 22 h of nutrient deprivation [62]. Mammary biopsy samples taken after the treatment were analyzed for gene expression, and the results showed phosphorylated S6K1 (pS6K1) as a result of Leu, and phosphorylated RPS6 as a result of Leu and Met plus Lys. Although changes were observed in mTOR-pathway-related genes, the infusion of His, Leu, and Met plus Lys presented only a numerical increase in milk protein yield. Two studies were reported that investigated the effects of the abomasal infusion of EAAs (ratio of casein) on the milk protein yield and mRNA translation of mammary samples [63,64]. The elimination of His, Phe, BCAAs, Leu, and Lys from 10 EAAs significantly decreased the milk protein yield as compared to the 10-EAAs-infused group [63,64]. In one study, the phosphorylation of S6K1 (pS6K1/total S6K1) was enhanced following the removal of His as compared to supplementation with 10 EAAs [63]; however, no changes were observed in another study [64]. The jugular infusion of Arg with Lys, Met, Phe, and Ile (ratio of casein) increased milk yield (+2.71 kg/d), milk protein yield (+120 g/d), α-casein (+5.54 g/L) and κ-casein (+1.00 g/L) in Chinese Holstein dairy cows fed diets containing 14.08% of CP [65]. However, a significant increase in casein gene expression was only observed for αs_1-casein and αs_2-casein. In the case of mTOR and S6K1 expression following mammary biopsy, increases of more than 18-fold in mTOR and 5-fold in S6K1 were observed, as compared to the control; however, 4EBP1 expression was decreased.

Each of the AAs influence protein synthesis via different pathways and genes. Studies on mammary translational regulation via AAs in dairy cows are still required, as these will expand our understanding of milk protein production in dairy cows.

5. Summary and Conclusions

The supplementation of AAs to dairy cows is a possible method for increasing milk and milk protein yield. Supplying protein feed seems to be a more certain method by which to increase the cow's productivity, as compared to other methods (see Table S1). However, high-protein feeds in dairy cow diets have several problems: high cost, low utilization due to the cow's low efficiency, high N excretion, and environmental contamination such as eutrophication. Therefore, it is of great concern to control individual AAs, rather than supplying surplus protein only. The advantage of intravenous infusion is that we can incorporate AAs into the blood stream of the animal's body, as we choose to target. Abomasal infusion can also supplement AAs according to the needs of the researcher; however, it is a relatively indirect method to affect productivity, when compared to intravenous infusion, because it passes the digestive tract. Both infusion methods are useful in experiments; nevertheless, they are not applicable in an industrial environment. The final way to supplement AAs to cows is to use RP-AAs. RP-AAs can mitigate insufficient single AAs in the body's AA pool, but not excess protein. Therefore, it is believed that there is a possibility to maintain or increase the productivity of animals while lowering the CP content in their diets. However, as shown in Table S1, RP-AAs provide inconsistent results. Overall, although supplying RPAAs appears to be cost effective, further testing and development will be needed for actual applications with dairy cows. For dairy cows, the most-limiting amino acids are considered to be Met, Lys, and His. It mainly depends on the AA composition in their diets; thus, Met and Lys are the co-limiting factors in corn- and alfalfa-based diets and His is the first-limiting AA in grain and grass silage-based diets. Milk proteins are synthesized through the mTOR translation initiation pathway. Studies on milk protein stimulation using AAs, with analysis of mammary samples, have been conducted; however, remarkable results are still limited. Future research on AA nutrition and milk protein synthesis are required, and will provide more field-applicable understanding.

Supplementary Materials: The following are available online at https://www.mdpi.com/article/10.3390/ani11072118/s1, Figure S1: Summarized productive results according to various feeding methods as described in Section 3.

Author Contributions: Conceptualization, J.-E.K. and H.-G.L.; writing—original draft preparation, J.-E.K.; writing—review and editing, H.-G.L. All authors have read and agreed to the published version of the manuscript.

Funding: This research was supported by Konkuk University in 2019.

Institutional Review Board Statement: Not applicable.

Informed Consent Statement: Not applicable.

Data Availability Statement: No new data were created or analyzed in this study. Data sharing is not applicable to this article.

Conflicts of Interest: The authors declare no conflict of interest. The funders had no role in the design of the study.

References

1. NRC. *Nutrient Requirements of Dairy Cattle: 2001*; National Academies Press: Washington, DC, USA, 2001.
2. Schwab, C.G.; Broderick, G.A. A 100-Year Review: Protein and amino acid nutrition in dairy cows. *J. Dairy Sci.* **2017**, *100*, 10094–10112. [CrossRef]
3. Gidlund, H.; Hetta, M.; Krizsan, S.J.; Lemosquet, S.; Huhtanen, P. Effects of soybean meal or canola meal on milk production and methane emissions in lactating dairy cows fed grass silage-based diets. *J. Dairy Sci.* **2015**, *98*, 8093–8106. [CrossRef]
4. Metcalf, J.A.; Wray-Cahen, D.; Chettle, E.E.; Sutton, J.D.; Beever, D.E.; Crompton, L.A.; MacRae, J.C.; Bequette, B.J.; Backwell, F.R. The effect of dietary crude protein as protected soybean meal on mammary metabolism in the lactating dairy cow. *J. Dairy Sci.* **1996**, *79*, 603–611. [CrossRef]
5. Korhonen, M.; Vanhatalo, A.; Huhtanen, P. Effect of protein source on amino acid supply, milk production, and metabolism of plasma nutrients in dairy cows fed grass silage. *J. Dairy Sci.* **2002**, *85*, 3336–3351. [CrossRef]
6. Seymour, W.M.; Polan, C.E.; Herbein, J.H. Effects of dietary protein degradability and casein or amino acid infusions on production and plasma amino acids in dairy cows. *J. Dairy Sci.* **1990**, *73*, 735–748. [CrossRef]
7. Raggio, G.; Lemosquet, S.; Lobley, G.E.; Rulquin, H.; Lapierre, H. Effect of casein and propionate supply on mammary protein metabolism in lactating dairy cows. *J. Dairy Sci.* **2006**, *89*, 4340–4351. [CrossRef]
8. Danes, M.A.C.; Hanigan, M.D.; Arriola Apelo, S.I.; Dias, J.D.L.; Wattiaux, M.A.; Broderick, G.A. Post-ruminal supplies of glucose and casein, but not acetate, stimulate milk protein synthesis in dairy cows through differential effects on mammary metabolism. *J. Dairy Sci.* **2020**, *103*, 6218–6232. [CrossRef] [PubMed]
9. Colmenero, J.J.; Broderick, G.A. Effect of dietary crude protein concentration on milk production and nitrogen utilization in lactating dairy cows. *J. Dairy Sci.* **2006**, *89*, 1704–1712. [CrossRef]
10. Castillo, A.; Kebreab, E.; Beever, D.; France, J. A review of efficiency of nitrogen utilisation in lactating dairy cows and its relationship with environmental pollution. *J. Anim. Feed. Sci.* **2000**, *9*, 1–32. [CrossRef]
11. Kebreab, E.; France, J.; Beever, D.; Castillo, A. Nitrogen pollution by dairy cows and its mitigation by dietary manipulation. *Nutr. Cycl. Agroecosyst.* **2001**, *60*, 275–285. [CrossRef]

12. Broderick, G.A.; Faciola, A.P.; Armentano, L.E. Replacing dietary soybean meal with canola meal improves production and efficiency of lactating dairy cows. *J. Dairy Sci.* **2015**, *98*, 5672–5687. [CrossRef]
13. Lage, C.F.A.; Raisanen, S.E.; Stefenoni, H.; Melgar, A.; Chen, X.; Oh, J.; Fetter, M.E.; Kniffen, D.M.; Fabin, R.A.; Hristov, A.N. Lactational performance, enteric gas emissions, and plasma amino acid profile of dairy cows fed diets with soybean or canola meals included on an equal protein basis. *J. Dairy Sci.* **2021**, *104*, 3052–3066. [CrossRef]
14. Ying, F.; Lin, X.; Ma, W.; Chi, H.; Yan, Z.; Song, Y.; Wang, Z. Metabolic responses to the deficiency of Lys, Arg, Met, or His in the mammary gland of lactating goats. *Small Rumin. Res.* **2013**, *113*, 219–230. [CrossRef]
15. Weekes, T.L.; Luimes, P.H.; Cant, J.P. Responses to amino acid imbalances and deficiencies in lactating dairy cows. *J. Dairy Sci.* **2006**, *89*, 2177–2187. [CrossRef]
16. Stahel, P.; Purdie, N.; Cant, J. Use of dietary feather meal to induce histidine deficiency or imbalance in dairy cows and effects on milk composition. *J. Dairy Sci.* **2014**, *97*, 439–445. [CrossRef] [PubMed]
17. Schwab, C.G.; Satter, L.; Clay, A. Response of lactating dairy cows to abomasal infusion of amino acids. *J. Dairy Sci.* **1976**, *59*, 1254–1270. [CrossRef]
18. Schwab, C.G.; Bozak, C.K.; Whitehouse, N.L.; Mesbah, M.M. Amino acid limitation and flow to duodenum at four stages of lactation. 1. Sequence of lysine and methionine limitation. *J. Dairy Sci.* **1992**, *75*, 3486–3502. [CrossRef]
19. Kim, C.H.; Choung, J.J.; Chamberlain, D.G. Determination of the first-limiting amino acid for milk production in dairy cows consuming a diet of grass silage and a cereal-based supplement containing feather meal. *J. Sci. Food Agric.* **1999**, *79*, 1703–1708. [CrossRef]
20. Kim, C.H.; Choung, J.J.; Chamberlain, D.G. Variability in the ranking of the three most-limiting amino acids for milk protein production in dairy cows consuming grass silage and a cereal-based supplement containing feather meal. *J. Sci. Food Agric.* **2000**, *80*, 1386–1392. [CrossRef]
21. Korhonen, M.; Vanhatalo, A.; Varvikko, T.; Huhtanen, P. Responses to graded postruminal doses of histidine in dairy cows fed grass silage diets. *J. Dairy Sci.* **2000**, *83*, 2596–2608. [CrossRef]
22. Vanhatalo, A.; Huhtanen, P.; Toivonen, V.; Varvikko, T. Response of dairy cows fed grass silage diets to abomasal infusions of histidine alone or in combinations with methionine and lysine. *J. Dairy Sci.* **1999**, *82*, 2674–2685. [CrossRef]
23. Fisher, L. Response of lactating cows to the intravenous infusion of amino acids. *Can. J. Anim. Sci.* **1972**, *52*, 377–384. [CrossRef]
24. Sok, M.; Ouellet, D.R.; Firkins, J.L.; Pellerin, D.; Lapierre, H. Amino acid composition of rumen bacteria and protozoa in cattle. *J. Dairy Sci.* **2017**, *100*, 5241–5249. [CrossRef] [PubMed]
25. Zhang, S.; Zeng, X.; Ren, M.; Mao, X.; Qiao, S. Novel metabolic and physiological functions of branched chain amino acids: A review. *J. Anim. Sci. Biotechnol.* **2017**, *8*, 10. [CrossRef] [PubMed]
26. Mackle, T.R.; Dwyer, D.A.; Bauman, D.E. Effects of branched-chain amino acids and sodium caseinate on milk protein concentration and yield from dairy cows. *J. Dairy Sci.* **1999**, *82*, 161–171. [CrossRef]
27. Korhonen, M.; Vanhatalo, A.; Huhtanen, P. Evaluation of isoleucine, leucine, and valine as a second-limiting amino acid for milk production in dairy cows fed grass silage diet. *J. Dairy Sci.* **2002**, *85*, 1533–1545. [CrossRef]
28. Huhtanen, P.; Vanhatalo, A.; Varvikko, T. Effects of abomasal infusions of histidine, glucose, and leucine on milk production and plasma metabolites of dairy cows fed grass silage diets. *J. Dairy Sci.* **2002**, *85*, 204–216. [CrossRef]
29. Appuhamy, J.; Knapp, J.; Becvar, O.; Escobar, J.; Hanigan, M. Effects of jugular-infused lysine, methionine, and branched-chain amino acids on milk protein synthesis in high-producing dairy cows. *J. Dairy Sci.* **2011**, *94*, 1952–1960. [CrossRef]
30. Yoder, P.S.; Huang, X.; Teixeira, I.A.; Cant, J.P.; Hanigan, M.D. Effects of jugular infused methionine, lysine, and histidine as a group or leucine and isoleucine as a group on production and metabolism in lactating dairy cows. *J. Dairy Sci.* **2020**, *103*, 2387–2404. [CrossRef]
31. Broderick, G.A.; Stevenson, M.J.; Patton, R.A.; Lobos, N.E.; Olmos Colmenero, J.J. Effect of supplementing rumen-protected methionine on production and nitrogen excretion in lactating dairy cows. *J. Dairy Sci.* **2008**, *91*, 1092–1102. [CrossRef]
32. Tamura, T.; Inoue, K.; Nishiki, H.; Sakata, M.; Seki, M.; Koga, T.; Ookubo, Y.; Akutsu, K.; Sato, S.; Saitou, K.; et al. Effects of rumen-protected methionine on milk production in early lactation dairy cattle fed with a diet containing 14.5% crude protein. *Anim. Sci. J.* **2019**, *90*, 62–70. [CrossRef] [PubMed]
33. Robinson, P.; Swanepoel, N.; Shinzato, I.; Juchem, S. Productive responses of lactating dairy cattle to supplementing high levels of ruminally protected lysine using a rumen protection technology. *Anim. Feed. Sci. Technol.* **2011**, *168*, 30–41. [CrossRef]
34. Paz, H.A.; de Veth, M.J.; Ordway, R.S.; Kononoff, P.J. Evaluation of rumen-protected lysine supplementation to lactating dairy cows consuming increasing amounts of distillers dried grains with solubles. *J. Dairy Sci.* **2013**, *96*, 7210–7222. [CrossRef] [PubMed]
35. Armentano, L.E.; Bertics, S.J.; Ducharme, G.A. Response of lactating cows to methionine or methionine plus lysine added to high protein diets based on alfalfa and heated soybeans. *J. Dairy Sci.* **1997**, *80*, 1194–1199. [CrossRef]
36. Socha, M.T.; Putnam, D.E.; Garthwaite, B.D.; Whitehouse, N.L.; Kierstead, N.A.; Schwab, C.G.; Ducharme, G.A.; Robert, J.C. Improving intestinal amino acid supply of pre- and postpartum dairy cows with rumen-protected methionine and lysine. *J. Dairy Sci.* **2005**, *88*, 1113–1126. [CrossRef]
37. Třináctý, J.; Křížová, L.; Richter, M.; Černý, V.; Říha, J. Effect of rumen-protected methionine, lysine or both on milk production and plasma amino acids of high-yielding dairy cows. *Czech J. Anim. Sci.* **2009**, *54*, 239–248. [CrossRef]
38. Patton, R.A. Effect of rumen-protected methionine on feed intake, milk production, true milk protein concentration, and true milk protein yield, and the factors that influence these effects: A meta-analysis. *J. Dairy Sci.* **2010**, *93*, 2105–2118. [CrossRef]

39. Zang, Y.; Silva, L.H.P.; Ghelichkhan, M.; Miura, M.; Whitehouse, N.L.; Chizzotti, M.L.; Brito, A.F. Incremental amounts of rumen-protected histidine increase plasma and muscle histidine concentrations and milk protein yield in dairy cows fed a metabolizable protein-deficient diet. *J. Dairy Sci.* **2019**, *102*, 4138–4154. [CrossRef]
40. Morris, D.L.; Kononoff, P.J. Effects of rumen-protected lysine and histidine on milk production and energy and nitrogen utilization in diets containing hydrolyzed feather meal fed to lactating Jersey cows. *J. Dairy Sci.* **2020**, *103*, 7110–7123. [CrossRef]
41. Giallongo, F.; Harper, M.T.; Oh, J.; Lopes, J.C.; Lapierre, H.; Patton, R.A.; Parys, C.; Shinzato, I.; Hristov, A.N. Effects of rumen-protected methionine, lysine, and histidine on lactation performance of dairy cows. *J. Dairy Sci.* **2016**, *99*, 4437–4452. [CrossRef]
42. Giallongo, F.; Hristov, A.N.; Oh, J.; Frederick, T.; Weeks, H.; Werner, J.; Lapierre, H.; Patton, R.A.; Gehman, A.; Parys, C. Effects of slow-release urea and rumen-protected methionine and histidine on performance of dairy cows. *J. Dairy Sci.* **2015**, *98*, 3292–3308. [CrossRef]
43. Lee, C.; Hristov, A.N.; Cassidy, T.W.; Heyler, K.S.; Lapierre, H.; Varga, G.A.; de Veth, M.J.; Patton, R.A.; Parys, C. Rumen-protected lysine, methionine, and histidine increase milk protein yield in dairy cows fed a metabolizable protein-deficient diet. *J. Dairy Sci.* **2012**, *95*, 6042–6056. [CrossRef] [PubMed]
44. Arriola Apelo, S.I.; Bell, A.L.; Estes, K.; Ropelewski, J.; de Veth, M.J.; Hanigan, M.D. Effects of reduced dietary protein and supplemental rumen-protected essential amino acids on the nitrogen efficiency of dairy cows. *J. Dairy Sci.* **2014**, *97*, 5688–5699. [CrossRef] [PubMed]
45. Swanepoel, N.; Robinson, P.; Erasmus, L.J. Effects of ruminally protected methionine and/or phenylalanine on performance of high producing Holstein cows fed rations with very high levels of canola meal. *Anim. Feed. Sci. Technol.* **2015**, *205*, 10–22. [CrossRef]
46. Swanepoel, N.; Robinson, P.; Erasmus, L.J. Impacts of adding ruminally protected phenylalanine to rations containing high levels of canola meal on performance of high producing Holstein cows. *Anim. Feed. Sci. Technol.* **2016**, *216*, 108–120. [CrossRef]
47. Swanepoel, N.; Robinson, P.; Erasmus, L.J. Production responses of high producing Holstein cows to ruminally protected phenylalanine and tyrosine supplemented to diets containing high levels of canola meal. *Anim. Feed. Sci. Technol.* **2018**, *243*, 90–101. [CrossRef]
48. Kollmann, M.T.; Locher, M.; Hirche, F.; Eder, K.; Meyer, H.H.; Bruckmaier, R.M. Effects of tryptophan supplementation on plasma tryptophan and related hormone levels in heifers and dairy cows. *Domest. Anim. Endocrinol.* **2008**, *34*, 14–24. [CrossRef]
49. Robinson, P.; Swanepoel, N.; Evans, E. Effects of feeding a ruminally protected lysine product, with or without isoleucine, valine and histidine, to lactating dairy cows on their productive performance and plasma amino acid profiles. *Anim. Feed. Sci. Technol.* **2010**, *161*, 75–84. [CrossRef]
50. Proud, C.G. mTOR-mediated regulation of translation factors by amino acids. *Biochem. Biophys. Res. Commun.* **2004**, *313*, 429–436. [CrossRef] [PubMed]
51. Wang, X.; Proud, C.G. The mTOR pathway in the control of protein synthesis. *Physiology* **2006**, *21*, 362–369. [CrossRef]
52. Kim, S.G.; Buel, G.R.; Blenis, J. Nutrient regulation of the mTOR complex 1 signaling pathway. *Mol. Cells* **2013**, *35*, 463–473. [CrossRef] [PubMed]
53. Tokunaga, C.; Yoshino, K.; Yonezawa, K. mTOR integrates amino acid- and energy-sensing pathways. *Biochem. Biophys. Res. Commun.* **2004**, *313*, 443–446. [CrossRef] [PubMed]
54. Kimball, S.R.; Jefferson, L.S. Signaling pathways and molecular mechanisms through which branched-chain amino acids mediate translational control of protein synthesis. *J. Nutr.* **2006**, *136*, 227S–231S. [CrossRef] [PubMed]
55. Stipanuk, M.H. Leucine and protein synthesis: mTOR and beyond. *Nutr. Rev.* **2007**, *65*, 122–129. [CrossRef] [PubMed]
56. Conejos, J.R.V.; Ghassemi Nejad, J.; Kim, J.-E.; Moon, J.-O.; Lee, J.-S.; Lee, H.-G. Supplementing with L-Tryptophan Increases Medium Protein and Alters Expression of Genes and Proteins Involved in Milk Protein Synthesis and Energy Metabolism in Bovine Mammary Cells. *Int. J. Mol. Sci.* **2021**, *22*, 2751. [CrossRef]
57. Jeon, S.W.; Conejos, J.R.; Kim, J.; Kim, M.J.; Lee, J.E.; Lee, B.S.; Park, J.S.; Moon, J.O.; Lee, J.S.; Lee, H.G. Supplementing conjugated and non-conjugated L-methionine and acetate alters expression patterns of CSN2, proteins and metabolites related to protein synthesis in bovine mammary cells. *J. Dairy Res.* **2020**, *87*, 70–77. [CrossRef]
58. Kim, J.; Lee, J.-E.; Lee, J.-S.; Park, J.-S.; Moon, J.-O.; Lee, H.-G. Phenylalanine and valine differentially stimulate milk protein synthetic and energy-mediated pathway in immortalized bovine mammary epithelial cells. *J. Anim. Sci. Technol.* **2020**, *62*, 263–275. [CrossRef]
59. Gao, H.; Zhao, S.; Zheng, N.; Zhang, Y.; Wang, S.; Zhou, X.; Wang, J. Combination of histidine, lysine, methionine, and leucine promotes β-casein synthesis via the mechanistic target of rapamycin signaling pathway in bovine mammary epithelial cells. *J. Dairy Sci.* **2017**, *100*, 7696–7709. [CrossRef]
60. Kim, J.; Lee, H.G. Effects of L-Histidine and Sodium Acetate on beta-Casein Expression in Nutrient-Restricted Bovine Mammary Epithelial Cells. *Animals* **2021**, *11*, 1444. [CrossRef]
61. Wang, M.; Xu, B.; Wang, H.; Bu, D.; Wang, J.; Loor, J.J. Effects of Arginine concentration on the in vitro expression of Casein and mTOR pathway related genes in mammary epithelial cells from dairy cattle. *PLoS ONE* **2014**, *9*, e95985. [CrossRef]
62. Toerien, C.A.; Trout, D.R.; Cant, J.P. Nutritional stimulation of milk protein yield of cows is associated with changes in phosphorylation of mammary eukaryotic initiation factor 2 and ribosomal s6 kinase 1. *J. Nutr.* **2010**, *140*, 285–292. [CrossRef]

63. Doelman, J.; Curtis, R.; Carson, M.; Kim, J.; Metcalf, J.; Cant, J. Essential amino acid infusions stimulate mammary expression of eukaryotic initiation factor 2Bε but milk protein yield is not increased during an imbalance. *J. Dairy Sci.* **2015**, *98*, 4499–4508. [CrossRef] [PubMed]
64. Doelman, J.; Kim, J.J.; Carson, M.; Metcalf, J.A.; Cant, J.P. Branched-chain amino acid and lysine deficiencies exert different effects on mammary translational regulation. *J. Dairy Sci.* **2015**, *98*, 7846–7855. [CrossRef] [PubMed]
65. Zhang, X.; Wang, Y.; Wang, M.; Zhou, G.; Chen, L.; Ding, L.; Bu, D.; Loor, J. Arginine Supply Impacts the Expression of Candidate microRNA Controlling Milk Casein Yield in Bovine Mammary Tissue. *Animals* **2020**, *10*, 797. [CrossRef] [PubMed]

animals

MDPI

Article

Effects of *Saccharomyces cerevisiae* Culture on Ruminal Fermentation, Blood Metabolism, and Performance of High-Yield Dairy Cows

Xiaoge Sun [1], Yue Wang [2], Erdan Wang [1], Shu Zhang [1], Qianqian Wang [1], Yan Zhang [1], Yajing Wang [1], Zhijun Cao [1], Hongjian Yang [1], Wei Wang [1,*] and Shengli Li [1,*]

1 State Key Laboratory of Animal Nutrition, Beijing Engineering Technology Research Center of Raw Milk Quality and Safety Control, College of Animal Science and Technology, China Agricultural University, No. 2 Yuanmingyuan West Road, Haidian, Beijing 100193, China; xiaogesun@163.com (X.S.); wangerdan@cau.edu.cn (E.W.); zhangshu200666@126.com (S.Z.); qq_221w@163.com (Q.W.); dxzxyan8@163.com (Y.Z.); yajingwang@cau.edu.cn (Y.W.); caozhijun@cau.edu.cn (Z.C.); yang_hongjian@cau.edu.cn (H.Y.)

2 Animal Production Systems Group, Wageningen University & Research, 6708 PB Wageningen, The Netherlands; yue3.wang@wur.nl

* Correspondence: wei.wang@cau.edu.cn (W.W.); lisheng0677@163.com (S.L.)

Simple Summary: Nowadays, the lifetime milk production of dairy cows, as well as the fat and protein contents of milk, has reached an unprecedented high. These improvements pose threats and challenges to animal health and welfare due to metabolic stress. The cows, during the high-yielding period, are especially susceptible to metabolic diseases such as digestive alterations, rumen acidosis, and lameness. This study assessed the effects of *Saccharomyces cerevisiae* culture (SC), a food supplement, on ruminal pH, volatile fatty acid (VFA), inflammatory cytokines, and the performance of high-yield dairy cows. The results show that supplementing high-yield lactating cows with the SC of 100 g/d increases milk yield, milk fat content, and milk lactose content, but does not affect protein content. SC supplementation affects overall ruminal VFA concentration and induces a significantly greater ruminal pH. It has the potential to enhance the rumen microbial growth and decrease the inflammation response. Our research suggests that SC supplementation has a positive effect on the productivity and health of dairy cows.

Citation: Sun, X.; Wang, Y.; Wang, E.; Zhang, S.; Wang, Q.; Zhang, Y.; Wang, Y.; Cao, Z.; Yang, H.; Wang, W.; et al. Effects of *Saccharomyces cerevisiae* Culture on Ruminal Fermentation, Blood Metabolism, and Performance of High-Yield Dairy Cows. *Animals* **2021**, *11*, 2401. https://doi.org/10.3390/ani11082401

Academic Editor: João Simões

Received: 7 July 2021
Accepted: 11 August 2021
Published: 13 August 2021

Abstract: High-yield dairy cows with high-concentrate diets are more prone to experiencing health problems associated with rumen microbial imbalance. This study assessed the effects of *Saccharomyces cerevisiae* culture (SC), a food supplement, on ruminal pH, volatile fatty acid (VFA), inflammatory cytokines, and performance of high-yield dairy cows. Forty Holstein cows with similar characteristics (e.g., milk yield, days of milk, and parity) were randomly divided into two groups: an experimental group fed the basal ration supplemented with the SC of 100 g of SC per cow per day (hour, SC group), and a control group fed the same basal ration diet without SC (i.e., CON group). On average, the supplementation of SC started at 73 days of lactation. The experimental period lasted approximately 70 days (from 18 January to 27 March 2020), including 10 days for dietary adaptation. Milk yield was recorded daily. Rumen fluid and milk samples were collected after 2 h of feeding in the morning of day 0, 15, 30, and 60. The data showed that rumen pH increased ($p < 0.05$) when cows were provided with SC. On average, the cows in the SC group produced 1.36 kg ($p < 0.05$) more milk per day than those in the CON group. Milk fat content of cows in the SC group was also higher (4.11% vs. 3.96%) ($p < 0.05$). Compared with the CON group, the concentration of acetic acid in the rumen fluid of dairy cows in the SC group was significantly higher ($p < 0.05$). There were no differences ($p > 0.05$) found in milk protein content and propionic acid between groups. The SC group had a tendency increase in butyric acid ($p = 0.062$) and total VFA ($p = 0.058$). The result showed that SC supplementation also enhanced the ratio between acetic and propionic. Most of the mean inflammatory cytokine (IL-2, IL-6, γ-IFN, and TNF-α) concentrations ($p < 0.05$) of the SC group were lower than CON group. This study demonstrated that high-yield cows receiving supplemental SC could produce more milk with higher fat content, have higher rumen acetate, and potentially less inflammatory cytokines.

Keywords: *Saccharomyces cerevisiae*; high-yield cows; pH; VFA; inflammatory cytokines

1. Introduction

With the improvement of genetics, nutrition, and farm management, both the milk yield of dairy cows and the nutritional value of milk (e.g., protein and fat content) have been increased [1,2]. These improvements, however, pose threats to animal health and welfare due to metabolic stress. The cows, during the high-yielding period, are especially susceptible to metabolic and infectious diseases, such as digestive alterations, mastitis, rumen acidosis, and lameness [3]. To cope with the challenges, some additives were proposed to be added to the diet to enhance the productivity and health of animals, such as Boswellia sacra resin [4], rumen-protected fat [5], and *Saccharomyces cerevisiae* culture (SC) [6]. SC is a concentrated and dried product of *Saccharomyces cerevisiae* strain after solid or fluid fermentation. It contains fermentation substrates, bacterial proteins, yeast metabolites, yeast cell walls, and other beneficial substances, and it has the functions of balancing animal intestinal flora, improving immunity, relieving stress, and improving productivity [7]. Obeidat et al. [8] reported that adding SC to the diet increased the dry matter intake (DMI), nutrient digestibility, and rumen fluid pH of ruminants. It could also improve the milk quality of dairy cows [9]. Ogunade and McCoun [10] showed that the addition of SC could effectively stabilize the internal environment of rumen fermentation. A study revealed that SC increased the degradation of dry matter (DM) and neutral detergent fiber (NDF) of forage, which could be explained by the increased amount of total bacteria, fungi, protozoa, and lactate-utilizing bacteria in the rumen, and the decreased amount of starch-degrading and lactate-producing bacteria [11]. SC stabilized rumen pH, under normal conditions, tended to decrease ruminal lipopolysaccharide (LPS) after feeding, enhanced the milk fat depression during subacute ruminal acidosis (SARA), and reduced the extent of SARA-associated inflammatory cytokines [12].

In the previous studies, most of the SC experimental data were obtained under in vitro conditions, which mainly focus on ruminal microorganism fermentation [13–16], while some studies under in vivo situations paid more attention to milk yield instead of liver function and inflammatory cytokines [17–19]. Thus far, there is limited understanding of how SC would affect the performance of the high-yield cows associated with serum inflammatory cytokines, enzymes, metabolism, and rumen fermentation function. Therefore, this study systematically observed the effects of SC on the production performance, rumen pH, volatile fatty acid (VFA), serum biochemical, and immune indicators of high-yield dairy cows.

2. Materials and Methods

2.1. Animals, Diets, and Experimental Design

Forty healthy high-yield lactation Holstein cows (milk production > 30 kg/d) were selected and randomly divided into two homogeneous groups: parity (2.9 ± 0.7 vs. 2.7 ± 0.7), days of milk (73.2 ± 9 vs. 72.7 ± 6.8), weight (655.7 ± 38.1 kg vs. 656.1 ± 41.1 kg), body condition score (2.80 ± 0.1 vs. 2.86 ± 0.2), somatic cell counter (SCC, 23.6 ± 7.3 vs. 25.2 ± 8.8), fat percentage (4.02 ± 0.2 vs. 4.06 ± 0.3), protein percentage (3.26 ± 0.3 vs. 3.27 ± 0.3), lactose percentage (5.34 ± 0.1 vs. 5.37 ± 0.1), and daily milk production (40.8 ± 6.3 kg vs. 40.7 ± 5.2 kg). Body condition score (1–5 grading scale) was evaluated independently by 3 individual observers, and the average value was adopted for each cow, according to the method of Paul et al. [20]. The experimental group received a basal diet supplemented with the SC of 100 g/head/d (i.e., SC group) and control group fed the same basal diet without SC (i.e., CON group). The commercial name of SC is Naijiaoyi (lot number 201901050103 and patent number ZL2016200553625) provided by Xi'an Xinhanbao Biological Technology Company (Shanxi, China). The product contains the nutrients of the yeast cells (>1.5 × 10^{11}/g), the metabolites formed after fermentation, and the denatured medium needed for the growth of lactic acid bacteria. The DM content of the product was 93%, and the content of crude protein, crude ash, and mannan was 16.9%, 3.9%, and 2.3%, respectively (as DM basis). The composition and nutrient levels of the diet are shown in Table 1. The nutritional requirements of dairy cows met the standard in NRC (2001) [21]. The free-stall barn housing system was adopted: feeding and milking three times a day, drinking water ad libitum. The experimental period lasted approximately 70 days (from 18 January to 27 March 2020), including 10 days for dietary adaptation. The trial was carried out in Beijing Shounong Animal Husbandry Development Company (39°30′ N, 116°33′ E) in northern China.

Table 1. Composition and nutrient levels of the basal diet (DM basis).

Items	Basal Diet (%)
Ingredients	
Whole corn silage	26.8
Alfalfa hay	12.5
Oat grass	1.6
Steamflaked corn	11.6
Extruded soybean meal	1.6
High yield concentrate [1]	39.1
Soybean hull	1.2
Fat power	2.1
Beet pulp	1.1
Sunflower meal	0.8
Molasses cane	1.6
Total	100
Nutrient levels	
NE_L/(Mcal/kg) [2]	1.8
NDF	30.7
ADF	20.4
CP	17.8
EE	5.9
Starch	27.3
Calcium	1.0
Phosphorus	0.5

[1] high yield concentrate is a concentrate supplement provided by Beijing Capital Agribusiness Group, including (DM basis): CP 24.75%, NDF 21.11%, Starch 36.26%, Fat 3.41%, Ash 7.54%, Calcium 1.04%, Phosphorus 0.53%, NaCl 1.0%, Fe 105 mg/kg, Zn 65 mg/kg, Mn 24 mg/kg, Cu 7 mg/kg, Mg 2 g/kg, K 10 g/kg, VA 20,000 IU/kg, VD 2300 IU/kg, and VE 88 IU/kg. [2] NE_L: Net energy of lactation, a calculated value according to NRC (2001) [21], while the other nutrient levels were measured values.

2.2. *Data and Sample Collection*

2.2.1. Milk Yield and Milk Profile

Daily milk yield was recorded using an automatic milking system (ALPROTM by DeLaval©, Tumba, Sweden). Milk samples were collected after 2 h of feeding in the mornings of day 0, 15, 30, and 60 of the experiment period. Individual milk samples (500 mL/each) were collected according to the ratio of 4:3:3 for morning, afternoon, and evening, respectively. The obtained milk samples were added to the potassium dichromate preservative and stored at −4 °C refrigeration. The milk samples were sent to Beijing Dairy Cow Center for composition analysis, including milk protein, milk fat, lactose, SCC, and DM content. A near-infrared reflectance spectroscopy analyzer (Seris300 Combi-FOSS; Foss Electric, Hillerød, Denmark) was used for determination, which is a seamless integration of MilkoScan^{RM} (Hillerød, Denmark) and Fossomatic^{TM} (Flow Cytometry, Hillerød, Denmark).

2.2.2. Blood Sample Collection and Chemical Composition Determination

Blood samples were obtained from each animal at day 0, 15, 30, and 60 of the formal trial period. The samples were collected from the jugular vein in vacuum tubes without anticoagulants. Serum was obtained by centrifugation (3000× g for 15 min at 4 °C). An aliquot of each sample was frozen and stored at −20 °C for chemical analysis. The following biochemical blood components were measured by an auto-analyzer (CLS880, ZECEN Biotech Co., Ltd., Qingdao, Shandong, China): total bilirubin (TBIL), total protein (TP), albumin (ALB), alkaline phosphatase (ALP), alanine aminotransferase (ALT), glucose (GLU) and triglyceride (TG), in which TBIL, TP, ALB, ALP, and ALT were tested by kits from ZECEN Biotech Co., Ltd. (Jiangsu, China), and GLU and TG were tested by kits from Jiancheng Bioengineering Institute (Nanjing, China). β-hydroxybutyrate (BHB) and non-esterified fatty acid (NEFA) levels were measured using a spectrophotometer (Model 722, Gaomi Caihong Analytical Instrument Co., Weifang, Shandong, China) with kits supported by Jiancheng Bioengineering Institute (Nanjing, China). The levels of cytokines in the samples were measured by ELISA kit (ELISA, Thermo Multiskan Ascent, US), including interleukin-1β (IL-1β), interleukin-2 (IL-2), interleukin-6 (IL-6), interleukin-10 (IL-10), γ-IFN (γ- interferon), and tumor necrosis factor-α (TNF-α). The lowest detection level of the marker that the antibody pair used in the ELISA kit was < 5 pg/mL. The coefficients of variation of inter-assay and the intra-assay were 4.8% and 4.2%, respectively.

2.2.3. Rumen Fluid Collection and Analysis

At day 0, 15, 30, and 60 of the formal trial period, 200 mL rumen fluid from each cow was collected through various stomach tubes, a method described by Shen et al. [22]. The rumen fluid was immediately filtered with four layers of gauze, and the filtrate was divided into two 50 mL centrifuge tubes. Once collected, one of the aliquots was used to measure pH with the use of sophisticated handheld pH meters (Starter 300; Ohaus Instruments Co. Ltd., Shanghai, China). Another one was stored at a −20 °C refrigerator and analyzed for VFA (acetic acid, propionic acid, butyric acid, and total VFA). The concentrations of VFA in rumen fluid were determined using a gas chromatograph (6890N; Agilent technologies, Avondale, PA, USA) equipped with a capillary column (HP-INNOWax 19091N-213, Agilent). More details regarding the method were described in Cao et al. [23].

2.3. Statistical Analysis

The SAS 9.4 statistical software (SAS institute, Carry, NC, USA) was used for statistical analysis. Milk yield and milk index data were analyzed by split-plot in time ANOVA; for repeated measures refer to Tesfaye and Hailu [24].

Fermentation data, including pH, acetate, propionate, butyrate, total VFA, lactate, the ratio between acetate and propionate (A:P), and all of the blood parameters data, were analyzed by using repeated measures data of MIXED procedure with model (1).

$$\text{Yijk} = \mu + \text{Di} + \text{Tj} + (\text{DT})\text{ij} + \text{Aik} + \varepsilon\text{ijk}, \tag{1}$$

where μ is the overall mean. Di is the fixed effect of treatment (i = 1–2). Tj is the fixed effect of sample collecting time (j = 1–4). (DT)ij is the fixed interaction effect of Dj and Tj. Aik is the random effect of the animal within Di, and εijk is the random error. Time was used as a repeated measure.

The results were expressed as least squares mean and standard error of mean. $p < 0.05$ indicates a significant difference. $0.05 < p < 0.1$ means there is a tendency difference.

3. Results

3.1. Effects of SC on Milk Yield and Profile

Table 2 showed that the milk yield and 3.5% fat corrected milk (3.5% FCM) of the SC group were increased by 1.36 kg/d ($p < 0.01$) and 2.63 kg/d ($p < 0.001$), respectively, when compared with the CON group. The addition of SC significantly increased milk fat percentage ($p < 0.01$), lactose percentage ($p < 0.05$), and DM content ($p < 0.01$). However, supplementary SC supplementation did not significantly change the milk protein in milk ($p > 0.05$). The SCC in the SC group was lower than that of the CON group ($p < 0.01$).

Table 2. Effects of SC on dry matter intake milk yield and milk composition of lactating dairy cows.

Items	CON [3]	SC [4]	*p*-Value
Milk yield (kg/d)	46.58 ± 0.36	47.94 ± 0.29	0.003
3.5% FCM (kg/d) [1]	50.03 ± 0.57	52.66 ± 0.52	<0.001
Milk composition			
Milk fat (%)	3.96 ± 0.16	4.11 ± 0.15	0.005
Milk protein (%)	3.29 ± 0.0.06	3.23 ± 0.05	0.54
Milk lactose (%)	5.19 ± 0.03	5.29 ± 0.03	0.046
DM content (%)	13.14 ± 0.16	13.84 ± 0.19	0.006
SCC/($\times 10^4$/mL) [2]	32.4 ± 1.37	23.4 ± 0.98	0.005

[1] 3.5% FCM: 3.5% fat corrected milk = (kg milk × 0.432) + (kg fat × 16.216) [25]. [2] SCC: Somatic cell count. [3] CON: Control group. [4] SC: Treated group.

3.2. Effects of SC on Ruminal pH and VFA

As shown in Figure 1a,b, the SC group had a greater pH ($p < 0.05$) and a higher concentration of acetic acid in the rumen fluid ($p < 0.05$), compared with those of the CON group. The addition of SC did not affect propionic acid ($p > 0.05$) (Figure 1c). The SC group tended to have more butyric acid ($p = 0.062$) and total VFA ($p = 0.058$) (Figure 1d,e). The concentration of individual VFAs ($p < 0.05$) and total VFA ($p < 0.05$) were affected by the sampling time. Figure 1f showed that SC supplementation also enhanced A:P, but the gap between the CON and SC group became smaller over time. Meanwhile, the A:P of the SC (from 2.39 to 2.66) and CON (from 1.94 to 2.64) group was also affected by the time of sampling ($p < 0.05$). There was no interaction effect found between the treatment and the time for the individual VFAs ($p > 0.05$), total VFA concentration ($p > 0.05$) and A:P ($p > 0.05$).

3.3. Effects of SC on Hepatic Function and Energy Metabolism

The serum TBIL ($p < 0.01$) and ALT ($p < 0.01$) decreased over time, but the values between the two groups did not differ ($p > 0.05$) (Table 3). The concentration of serum total protein (TP) ($p < 0.01$) was lower in SC group, compared with the control group (Table 3). It was not affected ($p > 0.05$) by the interaction of sampling time and treatment group (Table 3).

Table 3. Effects of SC on hepatic function.

Item [1]	CON [2]	SC [3]	SEM (±) [4]	*p*-Value		
				Treatment	Time [5]	Interaction [6]
TBIL (umol/L)	7.75	7.18	1.68	0.185	<0.001	0.204
TP(g/L)	77.47	73.97	3.48	<0.001	0.088	0.575
ALB (g/L)	40.07	40.7	1.16	0.100	0.475	0.749
ALP (U/L)	50.75	55.95	13.88	0.225	0.227	0.167
ALT (U/L)	30.60	30.03	4.66	0.513	<0.001	0.014

[1] TBIL = total bilirubin, TP = total protein, ALB = albumin, ALP = Alkaline phosphatase, and ALT = alanine aminotransferase. [2] CON: Control group. [3] SC: Treated group.[4] SEM: Standard error of the mean. [5] Time: Sampling time effect. [6] Interaction: The interaction between sampling time and treatment group.

Table 4 shows that the sampling time affected the concentration of glucose (GLU) ($p = 0.001$). The treatment and interaction did not change the concentration of non-esterified fatty acid (NEFA) ($p > 0.05$), β-hydroxybutyrate (BHBA) ($p > 0.05$), GLU ($p > 0.05$), and triglyceride (TG) ($p > 0.05$).

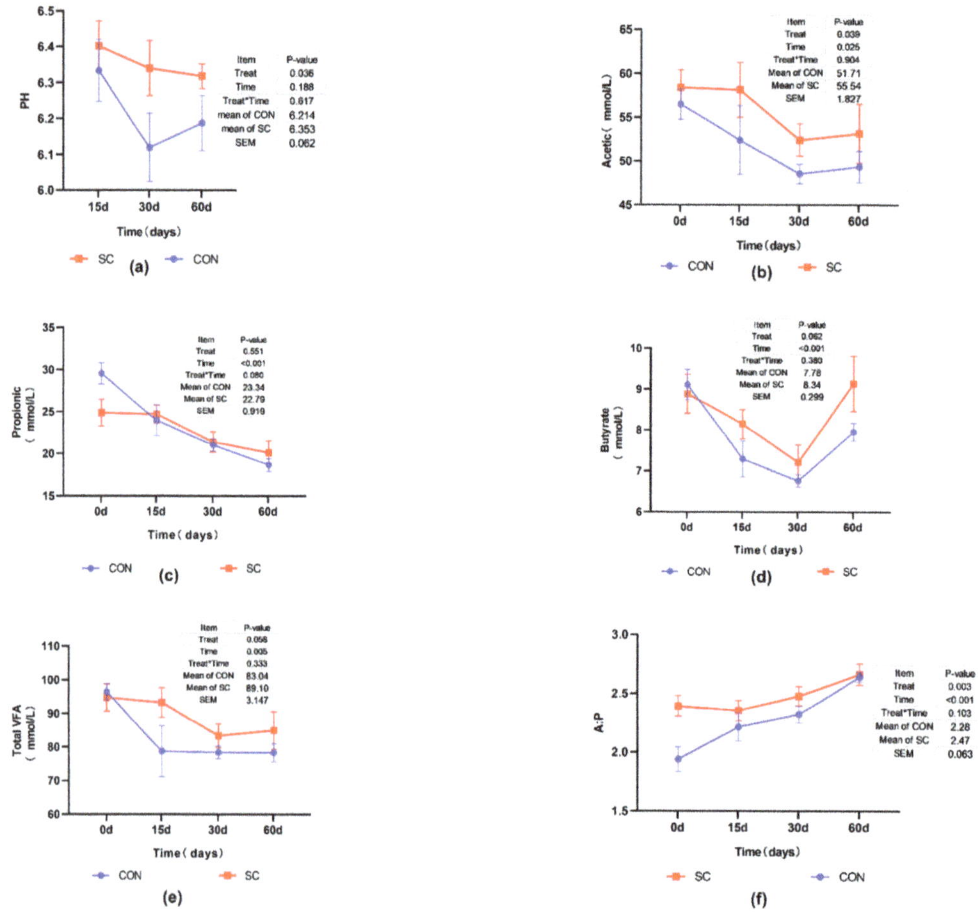

Figure 1. The effect of *Saccharomyces cerevisiae* culture (SC) on rumen VFA fermentation. SC is the experimental group diet supplemented with SC; CON is the control group without SC. (**a**–**f**) is the picture of pH, acetic, propionic, butyrate, total VFA concentrate, and acetate: propionate (A:P) with or without SC during the period, respectively. The data presented in the picture is mean ± standard error of mean.

Table 4. Effects of SC on energy metabolism of high-yield lactating cows.

Item [1]	CON [2]	SC [3]	SEM (±) [4]	*p*-Value		
				Treatment	Time [5]	Interaction [6]
NEFA (umol/L)	42.74	41.96	7.45	0.790	0.445	0.117
BHBA(mmol/L)	0.37	0.32	0.08	0.125	0.918	0.396
GLU (mmol/L)	4.25	4.34	0.50	0.493	0.001	0.568
TG (mmol/L)	0.25	0.27	0.04	0.253	0.772	0.296

[1] NEFA: Non-esterified fatty acid, BHBA = β-hydroxybutyrate, GLU = glucose, and TG = triglyceride. [2] CON: Control group. [3] SC: Treated group. [4] SEM: Standard error of the mean.[5] Time: Sampling time effect. [6] Interaction: The interaction between sampling time and treatment group.

3.4. Effects of SC on Inflammatory Cytokine

The effects of the treatment group on inflammatory cytokine concentrations are summarized in Table 5. Most of the mean inflammatory cytokine (IL-2, IL-6, γ-IFN, and TNF-α) concentrations ($p < 0.05$) in the SC group were lower than those in the CON group. The

IL-1β concentration had a tendency decrease (p = 0.061), but the IL-10 was unaffected (p > 0.05). The effect of time and the interaction of time and treatment did not affect these inflammatory cytokine concentrations in blood (p > 0.05).

Table 5. Effects of SC on inflammatory cytokine of high-yield lactating cows.

Item [1]	CON [2]	SC [3]	SEM (±) [4]	p-Value		
				Treatment	Time [5]	Interaction [6]
IL-1β (ng/L)	44.51	39.60	7.62	0.061	0.820	0.487
IL-2 (pg/mL)	205.73	180.30	33.17	0.025	0.928	0.978
IL-6 (ng/L)	519.06	454.81	72.67	0.009	0.994	0.854
IL-10 (ng/L)	214.84	201.77	26.69	0.161	0.896	0.901
γ-IFN (pg/mL)	87.93	77.69	14.03	0.028	0.703	0.684
TNF-α (ng/L)	370.49	324.79	57.49	0.018	0.953	0.772

[1] IL = interleukin, γ-IFN = γ- interferon, and TNF-α = tumor necrosis factor. [2] CON: Control group. [3] SC: Treated group. is the experimental group received a basal diet supplemented with 100g/head/day SC. [4] SEM: Standard error of the mean. [5] Time: Sampling time effect. [6] Interaction: The interaction between sampling time and treatment group.

4. Discussion

A random-effects meta-analysis showed that the increase in milk production estimated for cows supplemented with SC in peer-reviewed studies was 1.2 kg/d more milk, 1.6 kg/d more 3.5% FCM, or 1.7 kg/d more energy-corrected milk [9]. A recent study also revealed that milk fat percentage and milk production were significantly increased by feeding SC to dairy ruminants [26]. In our study, the addition of SC significantly increased the milk production (1.36 kg/d) and 3.5% FCM yield (2.63 kg/d) of high-yield dairy cows. Our study showed that the percentage of milk fat was increased by 0.15% in SC supplementation cows. The increases in both milk production and milk fat percentage result in a higher milk fat yield. This is in line with the results in Poppy et al. [9] and Ma et al. [26]. These effects could be probably explained by Li et al. [12], which showed the stabilizing effect of SC on rumen pH and fermentation.

The variation of ruminal pH is related to many factors. One of the most important is the dietary type, which has impacts on the rumen fermentation model. Nowadays, high-yield lactating dairy cows are always fed with high proportion of rapidly fermentable non-fiber carbohydrates, which affect the ruminal pH stabilization [27]. In our study, the high-yield cows supplemented with SC had significantly higher rumen pH. This might be explained by the increasing of lactate-utilizing bacteria [28], which reduced lactate concentrations in the rumen [29]. It suggested that this SC product stabilized rumen acidity. The stabilized rumen condition allows enhanced growth and activity of fiber-digesting bacteria [30], resulting in improved fiber digestion [31] and, subsequently, higher acetic acid production and A:P ratio in the rumen [32]. Similar results were found in our study, the SC supplementation cows have a higher acetic acid production, and A:P ratio. Milk fat concentration is highly influenced by nutrition and rumen fermentation. For example, studies show that milk fat concentration is positively associated with acetate [33,34], which was also observed in our study. In the present research, the milk protein and propionate did not differ between SC and control group, which is in line with the results in Desnoyers et al. [35] and Thrune et al. [36], who found SC had no influence on milk protein and propionate concentration. In addition, the VFA production decreased over time in both groups (Figure 1). This might be partly explained by the high-concentrate diet used during the entire experiment period, with a concentrate to forage ratio of 60:40. This was similar to the study in that long-term high-grain diet feeding gradually lowers VFA production in cattle [37]. Further studies are required to better understand those variations.

This study showed that there were no differences between treatments in concentrations of TBIL, ALB, ALP, ALT, NEFA, BHBA, GLU, and TG in blood serum, in which the TBIL, ALB, ALP, and ALT relate to the liver function of the dairy cows. Therefore, our study suggested that the SC supplements did not affect liver function in high-yield lactating cows. The NEFA, BHBA, GLU, and TG reflect energy metabolism. Hence, it indicates that the SC supplementation did not alter the energy metabolism of the high-yield cows. These results were inconsistent with some other studies, who concluded that the live yeast supplementation favorably influenced the metabolic status and might have a liver-protecting effect on the high-yield cows [38,39]. The non-significant results in our study could probably be explained by the fact that the cows used in this experiment were at approximately 70 days into lactation. The negative energy balance is less pronounced at this time, once the peak of ingestion was reached. However, the data obtained in our study showed that the lactose content of milk increased in the SC group. Energy balance in dairy cows is positively correlated with lactose percentage [40,41], especially for high-yield cows [42]. It implies the SC has the potential to affect energy metabolism.

The concentration of serum TP decreased when the SC was added to the diet. Elevated TP may indicate inflammation or infections. In agreement with this finding, we observed that the proinflammatory cytokines (IL-2, IL-6, γ-IFN, and TNF-α) of dairy cows were significantly lower in the SC group compared with the CON group. A previous study showed that being fed with high concentrate diets is simultaneous with activation of a non-specific acute phase reaction (APR) in cows [43,44], triggering the activation of a systemic APR due to the translocation of LPS into the blood of systemic circulation stimulates the release of proinflammatory cytokines, such as IL-1, IL-6, and TNF-α by liver macrophages [45]. As a strong negative relationship between rumen pH and high concentrate diets was well known [46], we inferred that the decrease in proinflammatory cytokines (IL-2, IL-6, γ-IFN, and TNF-α) in SC group cows was due to the higher pH, which reduced the LPS flow into the systemic circulation. Previous studies also showed that feeding SC to the dairy cows in early lactation could reduce plasma LPS concentration and relieve the symptoms of sub-acute ruminal acidosis [47,48]. A previous study showed a significant negative relationship between pre-feeding rumen pH and concentration of LPS in the rumen fluid [49], and rumen pH play a modulatory role in the accumulation and release of LPS attribute to its effects on metabolic processes, changes in the cell membrane of rumen bacteria, maintenance of bacterial ecological balances, and other physiological functions of the rumen [50]. Our results showed that SC has the potential to reduce the inflammatory cytokines, which partially explained the lower SCC level in the milk of SC supplementation cows. This effect of SC may be partly attributed to the stabilizing effect on rumen fermentation, as mentioned above.

5. Conclusions

This study showed that supplementing high-yield lactating cows with 100 g/d SC increased milk production, milk fat content, and milk lactose content, but did not affect protein content. SC supplementation affected overall ruminal VFA concentration and induced a significantly greater ruminal pH of dairy cows. It had the potential to enhance the rumen microbial growth and decreased the inflammatory cytokines, but it did not affect the blood parameters reflecting liver function and energy metabolism. It is concluded that the SC supplementation could increase the performance and health of high-yield dairy cows by stabilizing the rumen environment and decreasing the inflammatory cytokines.

Author Contributions: Conceptualization, S.L. and W.W.; methodology, Y.W. (Yajing Wang), Z.C., and H.Y.; software, X.S.; validation, S.L.; formal analysis, X.S.; investigation, X.S., E.W., and S.Z.; resources, Q.W. and Y.Z.; data curation, Y.W. (Yue Wang); writing—original draft preparation, X.S.; writing—review and editing, Y.W. (Yue Wang) and W.W.; visualization, X.S.; supervision, S.L.; project administration, W.W.; and funding acquisition, S.L. All authors have read and agreed to the published version of the manuscript.

Funding: This research was funded by National Key Research and Development Program of China, grant number 2018YFD0501600 and the 2115 Talent Development Program of China Agricultural University.

Institutional Review Board Statement: All experimental procedures were approved by the Ethical Committee of the College of Animal Science and Technology of China Agriculture University (Protocol number: 2020-5-LZ).

Informed Consent Statement: Not applicable.

Data Availability Statement: Not applicable.

Acknowledgments: The authors thank Xi'an Xinhanbao Biological Technology Company for providing the samples of SC and Beijing Shounong Animal Husbandry Development Company for providing the experimental animals.

Conflicts of Interest: The authors declare no conflict of interest.

References

1. Oltenacu, P.A.; Broom, D.M. The impact of genetic selection for increased milk yield on the welfare of dairy cows. *Anim. Welf.* **2010**, *19*, 39–49.
2. Walker, G.; Dunshea, F.; Doyle, P. Effects of nutrition and management on the production and composition of milk fat and protein: A review. *Aust. J. Agric. Res.* **2004**, *55*, 1009–1028. [CrossRef]
3. Opsomer, G. Interaction between metabolic challenges and productivity in high yielding dairy cows. *Jpn. J. Vet. Res.* **2015**, *63*, S1–S14.
4. Soltan, Y.; Morsy, A.; Hashem, N.; Sallam, S. Boswellia sacra resin as a phytogenic feed supplement to enhance ruminal fermentation, milk yield, and metabolic energy status of early lactating goats. *Anim. Feed Sci. Technol.* **2021**, *277*, 114963. [CrossRef]
5. Hashem, N.M.; El-Zarkouny, S.Z. Metabolic attributes, milk production and ovarian activity of ewes supplemented with a soluble sugar or a protected-fat as different energy sources during postpartum period. *Ann. Anim. Sci.* **2017**, *17*, 229. [CrossRef]
6. Olagaray, K.; Sivinski, S.; Saylor, B.; Mamedova, L.; Sauls-Hiesterman, J.; Yoon, I.; Bradford, B. Effect of *Saccharomyces cerevisiae* fermentation product on feed intake parameters, lactation performance, and metabolism of transition dairy cattle. *J. Dairy Sci.* **2019**, *102*, 8092–8107. [CrossRef]
7. Bruno, R.G.; Rutigliano, H.M.; Cerri, R.; Robinson, P.H.; Santos, J.E. Effect of feeding Saccharomyces cerevisiae on performance of dairy cows during summer heat stress. *Anim. Feed Sci. Technol.* **2009**, *150*, 175–186. [CrossRef]
8. Obeidat, B.S.; Mahmoud, K.Z.; Obeidat, M.D.; Ata, M.; Kridli, R.T.; Haddad, S.G.; Titi, H.H.; Jawasreh, K.I.; Altamimi, H.J.; Subih, H.S. The effects of *Saccharomyces cerevisiae* supplementation on intake, nutrient digestibility, and rumen fluid pH in Awassi female lambs. *Vet. World* **2018**, *11*, 1015. [CrossRef]
9. Poppy, G.; Rabiee, A.; Lean, I.; Sanchez, W.; Dorton, K.; Morley, P. A meta-analysis of the effects of feeding yeast culture produced by anaerobic fermentation of *Saccharomyces cerevisiae* on milk production of lactating dairy cows. *J. Dairy Sci.* **2012**, *95*, 6027–6041. [CrossRef]
10. Ogunade, I.M.; McCoun, M. Effects of adding live *Saccharomyces cerevisiae* and Aspergillus-based enzyme extracts on ruminal fermentation, plasma polyamine concentrations, and fiber digestibility in beef steers fed a high-forage diet. *Appl. Anim. Sci.* **2021**, *37*, 21–26. [CrossRef]
11. Ding, G.; Chang, Y.; Zhao, L.; Zhou, Z.; Ren, L.; Meng, Q. Effect of *Saccharomyces cerevisiae* on alfalfa nutrient degradation characteristics and rumen microbial populations of steers fed diets with different concentrate-to-forage ratios. *J. Anim. Sci. Biotechnol.* **2014**, *5*, 1–9. [CrossRef] [PubMed]
12. Li, S.; Yoon, I.; Scott, M.; Khafipour, E.; Plaizier, J. Impact of *Saccharomyces cerevisiae* fermentation product and subacute ruminal acidosis on production, inflammation, and fermentation in the rumen and hindgut of dairy cows. *Anim. Feed Sci. Technol.* **2016**, *211*, 50–60. [CrossRef]
13. Lynch, H.; Martin, S. Effects of *Saccharomyces cerevisiae* culture and *Saccharomyces cerevisiae* live cells on in vitro mixed ruminal microorganism fermentation. *J. Dairy Sci.* **2002**, *85*, 2603–2608. [CrossRef]
14. Lila, Z.; Mohammed, N.; Yasui, T.; Kurokawa, Y.; Kanda, S.; Itabashi, H. Effects of a twin strain of *Saccharomyces cerevisiae* live cells on mixed ruminal microorganism fermentation in vitro. *J. Anim. Sci.* **2004**, *82*, 1847–1854. [CrossRef] [PubMed]
15. Rossi, F.; Di Luccia, A.; Vincenti, D.; Cocconcelli, P.S. Effects of peptidic fractions from *Saccharomyces cerevisiae* culture on growth and metabolism of the ruminal bacteria Megasphaera elsdenii. *Anim. Res.* **2004**, *53*, 177–186. [CrossRef]
16. Cherdthong, A.; Prachumchai, R.; Supapong, C.; Khonkhaeng, B.; Wanapat, M.; Foiklang, S.; Milintawisamai, N.; Gunun, N.; Gunun, P.; Chanjula, P.; et al. Inclusion of yeast waste as a protein source to replace soybean meal in concentrate mixture on ruminal fermentation and gas kinetics using in vitro gas production technique. *Anim. Prod. Sci.* **2019**, *59*, 1682–1688. [CrossRef]
17. Kim, H.S.; Ahn, B.S.; Chung, S.G.; Moon, Y.H.; Ha, J.K.; Seo, I.J.; Ahn, B.H.; Lee, S.S. Effect of yeast culture, fungal fermentation extract and non-ionic surfactant on performance of Holstein cows during transition period. *Anim. Feed Sci. Technol.* **2005**, *126*, 23–29. [CrossRef]
18. Guedes, C.M.; Gonçalves, D.; Rodrigues, M.A.M.; Dias-da-Silva, A. Effects of a *Saccharomyces cerevisiae* yeast on ruminal fermentation and fibre degradation of maize silages in cows. *Anim. Feed Sci. Technol.* **2007**, *145*, 27–40. [CrossRef]

19. Carpinelli, N.A.; Halfen, J.; Trevisi, E.; Chapman, J.D.; Sharman, E.D.; Anderson, J.L.; Osorio, J.S. Effects of peripartal yeast culture supplementation on lactation performance, blood biomarkers, rumen fermentation, and rumen bacteria species in dairy cows. *J. Dairy Sci.* **2021**. [CrossRef]
20. Paul, A.; Bhakat, C.; Mondal, S.; Mandal, A. An observational study investigating uniformity of manual body condition scoring in dairy cows. *Ind. J. Dairy Sci.* **2020**, *73*, 77–80. [CrossRef]
21. NRC. *Nutrient Requirements of Dairy Cattle*, 7th rev. ed.; National Academies Press: Washington, DC, USA, 2001.
22. Shen, J.; Chai, Z.; Song, L.; Liu, J.; Wu, Y. Insertion depth of oral stomach tubes may affect the fermentation parameters of ruminal fluid collected in dairy cows. *J. Dairy Sci.* **2012**, *95*, 5978–5984. [CrossRef]
23. Cao, Z.; Li, S.; Xing, J.; Ma, M.; Wang, L. Effects of maize grain and lucerne particle size on ruminal fermentation, digestibility and performance of cows in midlactation. *J. Anim. Physiol. Anim. Nutr.* **2008**, *92*, 157–167. [CrossRef]
24. Tesfaye, A.; Hailu, Y. The effects of probiotics supplementation on milk yield and composition of lactating dairy cows. *J. Phytopharm.* **2019**, *8*, 12–17. [CrossRef]
25. Davidson, S.; Hopkins, B.; Odle, J.; Brownie, C.; Fellner, V.; Whitlow, L. Supplementing limited methionine diets with rumen-protected methionine, betaine, and choline in early lactation Holstein cows. *J. Dairy Sci.* **2008**, *91*, 1552–1559. [CrossRef] [PubMed]
26. Ma, Z.Z.; Cheng, Y.Y.; Wang, S.Q.; Ge, J.Z.; Shi, H.P.; Kou, J.C. Positive effects of dietary supplementation of three probiotics on milk yield, milk composition and intestinal flora in Sannan dairy goats varied in kind of probiotics. *J. Anim. Physiol. Anim. Nutr.* **2020**, *104*, 44–55. [CrossRef] [PubMed]
27. Krause, K.M.; Oetzel, G.R. Understanding and preventing subacute ruminal acidosis in dairy herds: A review. *Anim. Feed Sci. Technol.* **2006**, *126*, 215–236. [CrossRef]
28. Callaway, E.; Martin, S. Effects of a *Saccharomyces cerevisiae* culture on ruminal bacteria that utilize lactate and digest cellulose. *J. Dairy Sci.* **1997**, *80*, 2035–2044. [CrossRef]
29. Williams, P.; Tait, C.; Innes, G.; Newbold, C. Effects of the inclusion of yeast culture (*Saccharomyces cerevisiae* plus growth medium) in the diet of dairy cows on milk yield and forage degradation and fermentation patterns in the rumen of steers. *J. Anim. Sci.* **1991**, *69*, 3016–3026. [CrossRef] [PubMed]
30. Harrison, G.; Hemken, R.; Dawson, K.; Harmon, R.; Barker, K. Influence of addition of yeast culture supplement to diets of lactating cows on ruminal fermentation and microbial populations. *J. Dairy Sci.* **1988**, *71*, 2967–2975. [CrossRef]
31. Yoon, I.; Garrett, J.E. Yeast Culture and Processing Effects on 24-Hour in Situ Ruminal Degradation of Corn Silage. In Proceedings of the Proceeding of World Animal Production Conference, Seoul, Korea, 28 June–4 July 1998; pp. 322–323.
32. Russell, J. The importance of pH in the regulation of ruminal acetate to propionate ratio and methane production in vitro. *J. Dairy Sci.* **1998**, *81*, 3222–3230. [CrossRef]
33. Urrutia, N.L.; Harvatine, K.J. Acetate dose-dependently stimulates milk fat synthesis in lactating dairy cows. *J. Nutr.* **2017**, *147*, 763–769. [CrossRef]
34. Urrutia, N.; Bomberger, R.; Matamoros, C.; Harvatine, K. Effect of dietary supplementation of sodium acetate and calcium butyrate on milk fat synthesis in lactating dairy cows. *J. Dairy Sci.* **2019**, *102*, 5172–5181. [CrossRef] [PubMed]
35. Desnoyers, M.; Giger-Reverdin, S.; Bertin, G.; Duvaux-Ponter, C.; Sauvant, D. Meta-analysis of the influence of *Saccharomyces cerevisiae* supplementation on ruminal parameters and milk production of ruminants. *J. Dairy Sci.* **2009**, *92*, 1620–1632. [CrossRef]
36. Thrune, M.; Bach, A.; Ruiz-Moreno, M.; Stern, M.; Linn, J. Effects of *Saccharomyces cerevisiae* on ruminal pH and microbial fermentation in dairy cows: Yeast supplementation on rumen fermentation. *Livest. Sci.* **2009**, *124*, 261–265. [CrossRef]
37. Ogata, T.; Makino, H.; Ishizuka, N.; Iwamoto, E.; Masaki, T.; Ikuta, K.; Kim, Y.-H.; Sato, S. Long-term high-grain diet altered the ruminal pH, fermentation, and composition and functions of the rumen bacterial community, leading to enhanced lactic acid production in Japanese Black beef cattle during fattening. *PLoS ONE* **2019**, *14*, e0225448. [CrossRef]
38. Kumprechtová, D.; Illek, J.; Julien, C.; Homolka, P.; Jančík, F.; Auclair, E. Effect of live yeast (*Saccharomyces cerevisiae*) supplementation on rumen fermentation and metabolic profile of dairy cows in early lactation. *J. Anim. Physiol. Anim. Nutr.* **2019**, *103*, 447–455. [CrossRef]
39. Burdick Sanchez, N.C.; Broadway, P.R.; Carroll, J.A. Influence of yeast products on modulating metabolism and immunity in cattle and swine. *Animals* **2021**, *11*, 371. [CrossRef]
40. Reist, M.; Erdin, D.; Von Euw, D.; Tschuemperlin, K.; Leuenberger, H.; Chilliard, Y.; Hammon, H.; Morel, C.; Philipona, C.; Zbinden, Y. Estimation of energy balance at the individual and herd level using blood and milk traits in high-yielding dairy cows. *J. Dairy Sci.* **2002**, *85*, 3314–3327. [CrossRef]
41. Larsen, T.; Moyes, K. Are free glucose and glucose-6-phosphate in milk indicators of specific physiological states in the cow? *Animal* **2015**, *9*, 86–93. [CrossRef]
42. Lemosquet, S.; Delamaire, E.; Lapierre, H.; Blum, J.; Peyraud, J.-L. Effects of glucose, propionic acid, and nonessential amino acids on glucose metabolism and milk yield in Holstein dairy cows. *J. Dairy Sci.* **2009**, *92*, 3244–3257. [CrossRef] [PubMed]
43. Khafipour, E.; Krause, D.; Plaizier, J. A grain-based subacute ruminal acidosis challenge causes translocation of lipopolysaccharide and triggers inflammation. *J. Dairy Sci.* **2009**, *92*, 1060–1070. [CrossRef] [PubMed]
44. Emmanuel, D.; Madsen, K.; Churchill, T.; Dunn, S.; Ametaj, B. Acidosis and lipopolysaccharide from Escherichia coli B: 055 cause hyperpermeability of rumen and colon tissues. *J. Dairy Sci.* **2007**, *90*, 5552–5557. [CrossRef] [PubMed]

45. Gozho, G.; Plaizier, J.; Krause, D.; Kennedy, A.; Wittenberg, K. Subacute ruminal acidosis induces ruminal lipopolysaccharide endotoxin release and triggers an inflammatory response. *J. Dairy Sci.* **2005**, *88*, 1399–1403. [CrossRef]
46. Zhang, R.; Zhu, W.; Jiang, L.; Mao, S. Comparative metabolome analysis of ruminal changes in Holstein dairy cows fed low-or high-concentrate diets. *Metabolomics* **2017**, *13*, 1–15. [CrossRef]
47. Anzhi, L.; Changming, Q.; Hualin, C.; Lin, B.; Xiuqin, W.; Wencheng, W.; Qing, Y. Influence of Yikang XP on Plasma Endotoxin Concentration and Additional Indexes in Lactating Cows. *China Dairy Cattle* **2005**, *2*, 12–15.
48. Li, S.; Khafipour, E.; Krause, D.; Kroeker, A.; Rodriguez-Lecompte, J.; Gozho, G.; Plaizier, J. Effects of subacute ruminal acidosis challenges on fermentation and endotoxins in the rumen and hindgut of dairy cows. *J. Dairy Sci.* **2012**, *95*, 294–303. [CrossRef]
49. Ametaj, B.N.; Zebeli, Q.; Iqbal, S. Nutrition, microbiota, and endotoxin-related diseases in dairy cows. *Rev. Bras. Zootec.* **2010**, *39*, 433–444. [CrossRef]
50. Russell, J.B.; Rychlik, J.L. Factors that alter rumen microbial ecology. *Science* **2001**, *292*, 1119–1122. [CrossRef]

animals MDPI

Article

The Effect of Feeding Management and Culling of Cows on the Lactation Curves and Milk Production of Primiparous Dairy Cows

Jolanta Różańska-Zawieja [1], Stanisław Winnicki [2], Joanna Zyprych-Walczak [3], Alicja Szabelska-Beręsewicz [3], Idzi Siatkowski [3], Włodzimierz Nowak [4], Barbara Stefańska [4,5], Ryszard Kujawiak [6] and Zbigniew Sobek [1,*]

1 Department of Genetics and Animal Breeding, Poznan University of Life Sciences, ul. Wołyńska 33, 60-637 Poznań, Poland; jolanta.rozanska-zawieja@up.poznan.pl
2 Institute of Technology and Life Sciences, Poznań Branch, ul. Biskupińska 67, 60-463 Poznań, Poland; st-winnicki@wp.pl
3 Department of Mathematical and Statistical Methods, Poznan University of Life Sciences, ul. Wojska Polskiego 28, 60-637 Poznań, Poland; joanna.zyprych-walczak@up.poznan.pl (J.Z.-W.); alicja.szabelska-beresewicz@up.poznan.pl (A.S.-B.); idzi.siatkowski@up.poznan.pl (I.S.)
4 Department of Animal Nutrition, Poznań University of Life Sciences, ul. Wołyńska 33, 60-637 Poznań, Poland; wlodzimierz.nowak@up.poznan.pl (W.N.); barbara.stefanska@up.poznan.pl (B.S.)
5 Department of Grassland and Natural Landscape Sciences, Poznań University of Life Sciences, ul. Dojazd 11, 60-632 Poznań, Poland
6 Sano-Modern Animal Feeding, Lubiń, ul. Lipowa 10, 64-541 Sękowo, Poland; rku@sano.pl
* Correspondence: zbigniew.sobek@up.poznan.pl

Citation: Różańska-Zawieja, J.; Winnicki, S.; Zyprych-Walczak, J.; Szabelska-Beręsewicz, A.; Siatkowski, I.; Nowak, W.; Stefańska, B.; Kujawiak, R.; Sobek, Z. The Effect of Feeding Management and Culling of Cows on the Lactation Curves and Milk Production of Primiparous Dairy Cows. *Animals* **2021**, *11*, 1959. https://doi.org/10.3390/ani11071959

Academic Editor: João Simões

Received: 5 May 2021
Accepted: 27 June 2021
Published: 30 June 2021

Simple Summary: Curves for milk yield and milk constituents were plotted according to Wood's model. The curves were compared for four herds with a high milk yield (over 10,000 kg). The farms used different feeding systems. Differences were noted in the number of days in milk at which the herds reached the peak of lactation. The herd that reached peak production latest had the highest peak milk yield as well as the highest average milk yield for the entire lactation. The effect of early culling of primiparous cows after 30, 60 and 90 days in milk was analysed in herd T1. Such early culling of primiparous cows prevents their negative impact on the average yield of the entire herd. This increases the chance of improving yield in the herd by quickly introducing new, genetically more valuable cows.

Abstract: The study attempted to estimate the lactation curves of primiparous dairy cows in relation to their feeding management. Therefore, the first aim of the study was to determine and compare the lactation curves of primiparous dairy cows using Wood's model and to estimate the association between the lactation curves and feeding management. The second objective was to investigate the effect of the culling rate on improvement in the milk yield of primiparous dairy herds. The study was conducted on four commercial dairy farms of Polish Holstein–Friesian cows using different feeding systems (TMR—total mixed ration and PMR—partial mixed ration) and management (T1—one TMR throughout lactation; P1—one PMR throughout lactation; T2 and T3—three feed periods such as FRESH, TMR I and TMR II according to days in milk). The data used for the study were obtained from monthly milk performance evaluations of 1662 primiparous cows conducted by the Polish Federation of Cattle Breeding and Dairy Farmers throughout the year 2015. Wood's lactation model was used to plot curves for milk yield, fat and protein content, lactose content, and milk urea contents. The highest milk yield for the whole lactation and in the peak lactation phase was recorded for cows in herd T1. This herd reached peak lactation on day 105 of milking, with an average milk yield of 42.1 kg, which was about 5 kg more milk than in the other herds. The study showed that the culling of primiparous cows in herd T1 after 30, 60 and 90 days of lactation prevented a significant reduction in milk yield in a 305-day lactation. It also increased average milk production by 1586.9 kg per primiparous dairy cow.

Keywords: primiparous cows; lactation curves; feeding system; herd management

1. Introduction

Lactation curves are graphical representations of the course of lactation, taking into account all factors that affect milk production. The lactation curve illustrates the occurrence of changes in milk during lactation, involving a decrease in milk yield accompanied by an increase in fat and protein content [1]. The standard lactation curves of dairy cows show a maximum daily milk yield between 4 to 8 weeks after calving, followed by a daily decrease in yield until the dry period [1]. It is important for dairy cow farmers to know what level of milk yield the cow will reach the peak of lactation and how long after calving this peak will be observed.

There are currently many models and types of lactation curves in the literature. Common lactation models include the Ali–Schaeffer model [2], the Wilmink model [3], and the Wood model [4], which is the most popular. Analysis of lactation curves suggests that a flattened curve resulting from uniform lactation, during which milk yield remains at the same level for an extended period, is the most desirable. This is closely linked to lactation persistence, i.e., the rate at which milk production drops after the peak of lactation; the slower drop in production, the higher the persistence, which is also economically beneficial. Cows with higher lactation persistence ensure greater economic profit. It is assumed that milk production should decrease by 8–10% per month, with a 5% more rapid decrease in primiparous cows than in multiparous cows [5,6]. Moreover, higher milk yield could be associated with fertility, a 56-day increase in the calving-to-first-service interval [7]. Numerous studies also demonstrate that the lactation curve of primiparous cows has a lower peak and is more persistent than that of multiparous cows [8]. After the peak of lactation, milk production gradually decreases, but the dynamics of this process depend on many factors, nutrition, breed, age, diseases, frequency of milking, and length of dry period [9–12]. The productive performance of a herd is influenced by the level of culling of cows in successive months of lactation. Juszczak et al. (2003) reported that the cost of milk produced by cows used only in the first three lactations is equal to its market value, and only when used for a longer time do they begin to generate profit [13]. Therefore, early culling of cows will prevent them from reaching their production peak [14]. However, there are still limited scientific data on how lactation curves of primiparous dairy cows are influenced by feeding management and culling rates. Therefore, the first aim of this study was to determine and compare the lactation curves of primiparous dairy cows depending on the feeding management in field conditions using Wood's model. The second objective was to investigate the effect of the culling rate on the improvement in the milk yield of dairy herds.

2. Materials and Methods

2.1. Farms

The study was conducted on four commercial dairy farms of Polish Holstein–Friesian cows (Table 1) using different feeding systems (TMR—total mixed ration and PMR—partial mixed ration) and management (T1—one TMR throughout lactation; P1—one PMR throughout lactation; T2 and T3—three feed periods as FRESH, TMR I and TMR II according to days in milk). The farms were selected according to milk yield (about 11,000 kg/305 days lactation per cow), the size of the farm (more than 100 lactating dairy cows), housing of cows (only free-stall barns), and milking system (automated). The data used in the study included milk yield and chemical composition during 305 day of lactation from primiparous cows that calved in 2015. Milk production performance data from morning and afternoon milking during monthly milk performance evaluations were used. The evaluations were conducted by the Polish Federation of Cattle Breeding and Dairy Farmers.

Table 1. General information on the herd and the composition and nutritional value of the diets.

	Herds [1]							
	T1	P1		T2		T3		
General information								
Number of cows in herd	1262	369		749			335	
Number of primiparous cows/farm	779	142		459			282	
Milk production (kg)/farm/per cow	12,778	10,934		11,505			11,346	
Milk fat (%)/farm/per cow	3.41	3.79		3.83			3.65	
Milk protein (%)/farm/per cow	3.25	3.36		3.38			3.23	
Calving interval (days)	409	397		418			417	
Milking system	Rotary	Automatic (robot)		Fish bone			Fish bone	
Diet during lactation	TMR	PMR [2]	FRESH [3]	TMR I [4]	TMR II [5]	FRESH	TMR I	TMR II
Average dry matter intake (kg, DM)	23.4	22.2	21.4	23.6	22.5	20.4	22.6	21.5
Ingredient, (% DM)								
Maize silage	30.9	32.7	28.5	33	39.2	26.5	28.3	42.3
Alfalfa silage		23.8	15.9	8.4	8.8	16	13.4	16.7
Sugar beet pulp, ensiled		8.0		4.2	5.3	8.1	6.9	9.0
Grass silage	11.8		9.5			8.2		6.3
Brewer's grain silage								
Hay				2.4	2.2			
Straw	3.9	9.6	9.3	1.7	1.8	8.2	6.2	
Maize grain, ensiled		5.5	4.5	8.1	7.2	5.8	9.7	
Maize grain			4.7	6.8	3.5	5.9		
Maize husks							6	
Barley grain		4.5	5.3	5.1	5.3	4.1	9.7	13.1
Wheat grain		4.5		5	5.2			
Triticale grain								
Rapeseed meal		5.5	4.2	9.2	8.5	3.3	6.4	5
Soybean meal		1.9	7.9	9.1	8.1	6.1	6.1	4.9
Sugar beet pulp, dry	10.7			1.7				
Mineral and vitamin mix [6]		4.0	5.7	2.3	2.4	4.4	3.3	2.7
Molasses	3.9		4.5		2.1	3.4	2.6	
Inert fat				1.9			1.1	
Calcium carbonate				0.6			0.3	
Rock salt				0.3	0.2			
Lactasan [7]	38.8							
Urea				0.2	0.2			42.3

Table 1. *Cont.*

	Herds [1]							
	T1	**P1**		**T2**			**T3**	
	Nutrient composition (%)							
DM—dry matter	44.8	40.4	49.3	49.2	45.3	49.3	42.0	42.4
ME (Mcal/kg) [8]	2,91	2.61	2.60	2.90	2.60	2.60	2.85	2.60
CP—crude protein	16.5	15.5	15.2	17.1	17.1	15.2	15.9	16.6
NDF—neutral detergent fibre	29.5	39.2	31.3	28.1	30.1	31.3	28.8	33.2
ADF—acid detergent fibre	17.3	24.2	20.2	16.3	17.8	20.2	17.8	19.2
NFC—non-fibre carbohydrates	44.9	33.9	41.3	42.9	43.3	41.3	44.1	40.7
Starch	24.7	17.3	20.7	24.7	24.8	20.7	25	24.3
Ether extract	3.3	2.8	2.2	2.9	2.4	2.2	3.3	2.4
Ca	0.97	0.97	0.97	1.04	0.87	0.97	0.74	0.75
P	0.42	0.39	0.42	0.42	0.41	0.42	0.43	0.42
K	1.37	1.71	1.83	1.54	1.60	1.83	1.60	1.70
DCAD meq/100g SMDietary cation–anion difference	28.21	25.97	30.50	20.60	22.42	30.50	28.21	28.28

[1] Herds—T1—one TMR throughout lactation; P1—one PMR throughout lactation; T2 and T3—three feeding periods: FRESH, TMR I and TMR II, according to days in milk; [2] PMR—partial mixed ration; ground concentrates were supplied in the feed bunk of the milking robots: FRESH concentrate (0–40 day) containing 21.8% CP, 26.7% starch and 12.8% NDF; lactation period concentrate (41–305 day) consisting of a mixture of two concentrates in a 3:1 ratio, containing 19.5% CP, 31.6% starch and 13.5% NDF and 23.1% CP, 22.9% starch and 11.3% NDF, respectively; [3] FRESH—feeding period from 7 day before calving to 30 day of lactation; [4] TMR I—total mixed ration I; feeding period from 31 day to 100 day of lactation; [5] TMR II—total mixed ration II; feeding period from 101 day to 305 day of lactation; [6] Mineral and vitamin mix: composition—21.5% Ca, 4.0% P, 6.5% Na, 5.5% Mg, 1200 mg/kg Cu, 4000 mg/kg Mn, 15 mg/kg Co, 10,000 mg/kg Zn, 60 mg/kg Se, 1,200,000 jm/kg vitamin A, 180,000 jm/kg vitamin D, 6000 jm/kg vitamin E; [7] Lactasan—supplementary feed containing 27% CP, 23% starch and 6.0 % crude fibre and vitamins and macro- and microelements. [8] Calculated according to NRC (2001).

Analysis of the effect of early culling on lactation yield was performed only for herd T1, because it was the largest and had the highest milk yield (kg). The herd consisted of 779 primiparous cows (Table 1), but only 628 cows, whose lactation lasted 1–305 days, were included in the analysis. The remaining 151 cows were not included in the calculations because they were milked for longer than 305 days.

2.2. Chemical Composition of Diets

The diets were balanced based on analysed content of nutrients using Dairy Max System software (Cargill, Kiszkowo, Poland) formulated according to National Research Council guidelines [15]. On all farms the cows were fed diets based on maize silage; alfalfa silage; beet pulp; grass silage or wilted, ensiled high-moisture maize grain; barley, wheat, triticale, and maize grain; rapeseed and soybean meals; and mineral and vitamin supplements. Using wet chemistry analysis, monthly forage, concentrate, TMR, and PMR samples were tested for dry matter (DM, method No. 6496), crude protein (CP, method No. 976.05), neutral detergent fibre (NDF, method No. 942.05), acid detergent fibre (ADF, method No. 973.18), non-fibre carbohydrates (NFC, method No. 64.785), starch (method No. 64.785), ether extract (EE, method No. 989.05), calcium (Ca, method No. 6869), phosphorus (P, method No. 6491), and potassium (K, method No. 6869) according to Procedures of the Association of Official Analytical Chemists [16]. Crude fibre was determined with an Ankom 220 fibre analyzer (ANKOM Technology, Macedon, NY, USA). TMR and PMR diets were administered at 9 a.m. and 2.30 p.m. throughout the experimental period (Table 1).

2.3. Statistical Analysis

Estimation of lactation curves of primiparous cows based on test-day milk yield

The calculations were carried out using the R-package [17], with the following procedures: agricolae [18], easynls [19], ggplot2 [20], and openxlsx [21].

The lactation curves of primiparous cows for each trait (milk yield, fat and protein content, lactose content, and urea content) were determined according to Wood's function [4], using the following model:

$$y = a \times x^{b} \times \exp^{-cx}$$

where:

y—a dependent variable determining the milk traits on the test day (day x)
x—number of days after calving
a—parameter determining the average milk traits on the test day
b—parameter determining the slope of the increasing part of the lactation function
c—parameter determining the slope of the decreasing part of the lactation function
exp—exponential function. For detailed interpretations of Woods function parameters [22,23].

For the lactation curves determined based on Wood's function, milk yields were calculated using the definite integral (1) from calving to 305 days of milking; (2) from calving to lactation peak; (3) from calving to day of culling; (4) from calving to 305 days of milking for the curve estimated on the basis of shortened lactations. The coordinates x and y for the maximum value of y for this curve were calculated, as well as the areas under the curves, which correspond to the milk yields in the corresponding periods.

Since the data on milk yield curves were not subject to normal distribution (checked using the Shapiro–Wilk test), the Kruskal–Wallis test was used to compare the lactation curves instead of analysis of variance (ANOVA). A detailed comparison was made using Dunn's non-parametric multiple comparison test with the Holma–Šidák correction [24–26].

To determine how the early culling of cows with a yield below the adopted criterion (24 kg milk) affects the total productivity of the herd, predicted yield was calculated for a situation without this culling, taking into account the numbers of cows in each period. If culled cows had been milked up to 305 days, without culling of those with the lowest production, the total 305 d yield for the primiparous cows in herd T1 would have been

9798.13 kg of milk. By eliminating cows whose yield was too low, the average yield in the herd was increased by 1586.87 kg of milk per cow.

To test for differences between feeding strategies (or farms) regarding peak milk yields or specific milk charts (up to 30, 60, 90, etc. days), we used the definite integrals for Wood's function.

3. Results

The shape of lactation curves determined on the basis of average daily milk yields in all herds was consistent with the pattern generally accepted as normal. The flattest curve was obtained for the T3 herd (Figure 1). The lactation peak was noted on day 98 of lactation, and the average milk yield at that time was 37.8 kg. Similar milk yields (37.4 and 37.3 kg) at the peak of lactation were obtained for the cows in herds T2 and P1. The most rapid decrease in the lactation curve was observed in the P1 herd, in which the lactation peak was noted on day 75. The highest milk yield for the whole lactation and in the peak lactation phase was recorded for cows in herd T1. For the latest peak (105 days), this portion accounted for only 32.8% of the total lactation. Cows reached peak lactation in this herd on day 105 of milking, with an average milk yield of 42.2 kg, which was about 5 kg more milk than in the other herds.

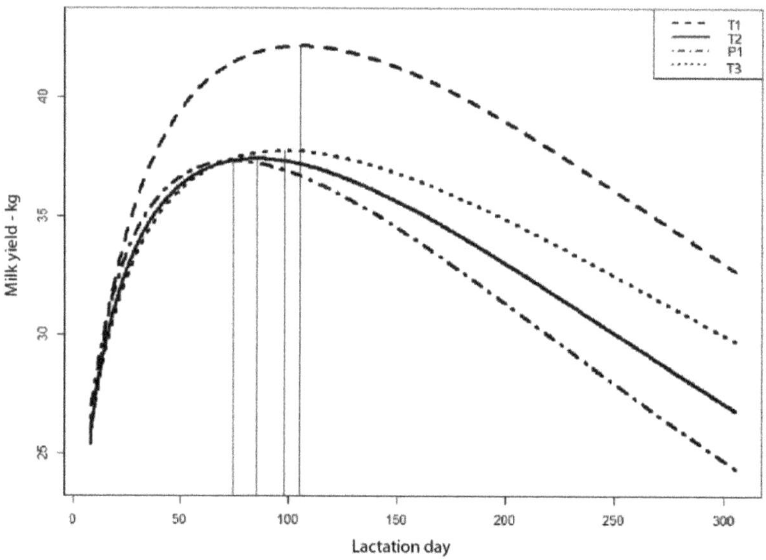

Figure 1. Lactation curves of primiparous cows with different feeding management.

Herds: T1—one TMR throughout lactation; P1—one PMR throughout lactation; T2 and T3—three feeding periods: FRESH, TMR I and TMR II, according to days in milk (kg).

The results show that the time to reach the peak of lactation was different in herds with different feeding management (Figure 1). Thus, the highest lactation yield was recorded in herd T1, in which the peak of lactation was reached the latest (on the 105th day in milk). Interestingly, a later peak of lactation was associated with a smaller decrease in productivity following the peak. The share of this fraction increases with earlier peaks in lactation. For peak on the 75th day, the share of this fraction increased to 36% in the P1 herd.

A decrease in the protein content of milk up to 50–60 days in milk was observed in herds T1 and P1, after which the value of this parameter significantly improved. This curve was completely different for herd T2, in which there was no such decrease; the protein content in the milk was very low at the beginning of lactation and subsequently showed a marked increase. Additionally, analysis of the fat content of milk revealed a decline up to about day 100. A much slower decrease in the percentage content of fat was noted in the T1 system, only until about day 20–30 of milking. After this, it was relatively stable, with a slight upward trend. The curves for lactose content in milk for herds P1 and T3 were almost identical. The T1 herd had a much more stable lactose content, but at the lowest level. The T2 herd initially had the highest lactose content, but then it decreased the most sharply, reaching 4.75% at the end of lactation. In the P1 and T3 systems, the milk urea (MU) concentration increased significantly from the beginning of lactation and exceeded 350 mg/l at the end of lactation. In the T1 and T2 systems, the values were more stable, oscillating around 250 mg/L. In the case of herd T1, the curve for MU was similar to the curve for milk yield, with a maximum of about 100 days of lactation for both curves (Figure 2).

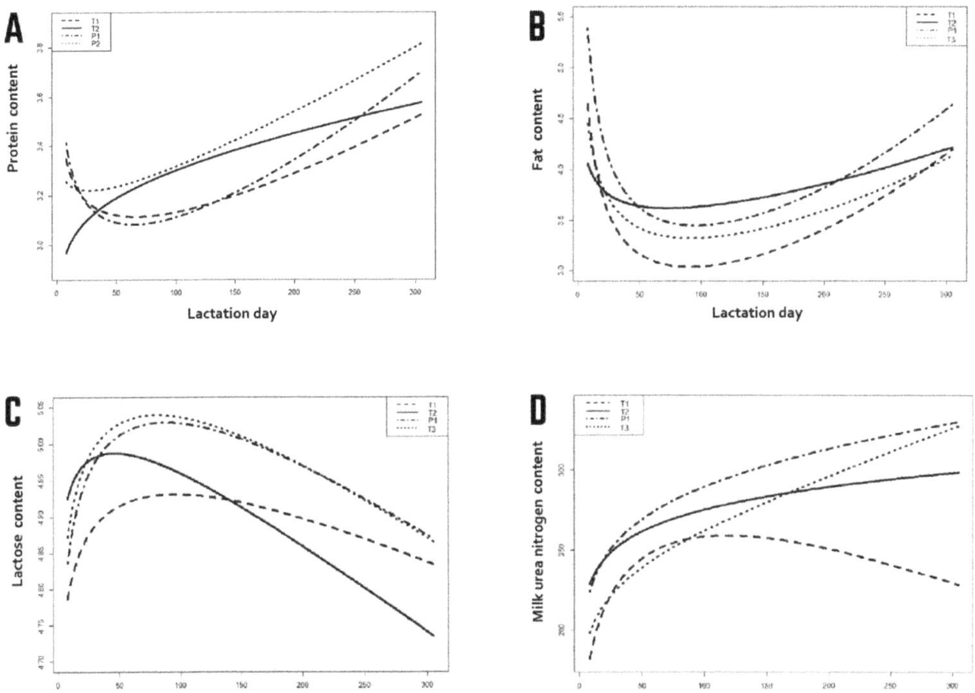

Figure 2. Presents the results for other milk traits in herds with different feeding management systems. Curves for protein (**A**), fat (**B**), lactose (**C**), milk urea (**D**) contents Herds: T1—one TMR (total mixed ration).throughout lactation; P1—one PMR (partial mixed ration) throughout lactation; T2 and T3—three feeding periods: FRESH, TMR I and TMR II, according to days in milk (kg).

Analysis of the areas under the lactation curves in herd T1, determined by their intersection with the lines denoting 100, 200 and 305 days of lactation, revealed the highest percentage of the total area for the period from 100–200 days of lactation, whose share in the total lactation was 35.6–36%. This share of phase 2 in the total lactation yield indicates highly uniform milk production over the course of lactation.

The impact of culled primiparous cows on milk production was examined in herd T1. Predicted lactation yields were estimated on the basis of Wood's function, increasing the number of days based on which this function was determined. The functions were used to estimate the expected 305 day yield. Knowing the actual milk yield of the T1 herd (11,385 kg) and the estimation of this yield based on initial data from the lactation, it can be concluded that the accuracy of the prediction increases with the number of days in milk (Table 2.)

Table 2. Herd T1 Wood function parameters and actual milk yields per x days of milking with predicted yield for 305 days.

	Parameters			Actual Yield	Predicted Yield
x–d	a	b	c	Area under curve (0–x day)	Area under curve (0–305 day)
30	15.0170	0.2425	−0.0013	833.222	18,036.89
60	12.5080	0.3321	0.0026	1999.796	12,258.44
90	11.0229	0.3903	0.0047	3243.969	10,430.16
120	11.9886	0.3561	0.0038	4513.761	10,997.52
150	12.1962	0.3491	0.0036	5771.006	11,146.57
180	12.9338	0.3273	0.0032	6986.659	11,309.45
210	13.3877	0.3152	0.003	8172.895	11,386.48
240	13.1579	0.321	0.0031	9271.689	11,333.51
270	13.4694	0.3132	0.003	10,337.96	11,347.22
305	13.7833	0.3056	0.0029	11,367.83	11,367.83

For herd T1 with very strict culling, curves were created for milk yields for actual days of milking of cows culled up to 30, 60, 90, etc. days. Figure 3 clearly indicates that for the shortened curves that reached the peak of lactation, the peak was very early and had a much lower value than for the rest of the herd, milked for 305 days.

If primiparous cows had not been culled due to low yields, this would clearly have reduced the lactation yield of the whole herd. To determine the losses that would have been caused by not culling these cows, a yield prediction using the Wood function was prepared (Figure 3). Curves for shortened lactations were compared with the total yield curve of the herd composed of previously culled cows and those that were milked until the end of lactation (thick black line). Additionally, a curve (Figure 3) was plotted (dashed line) for all cows except those culled before day 305 of lactation.

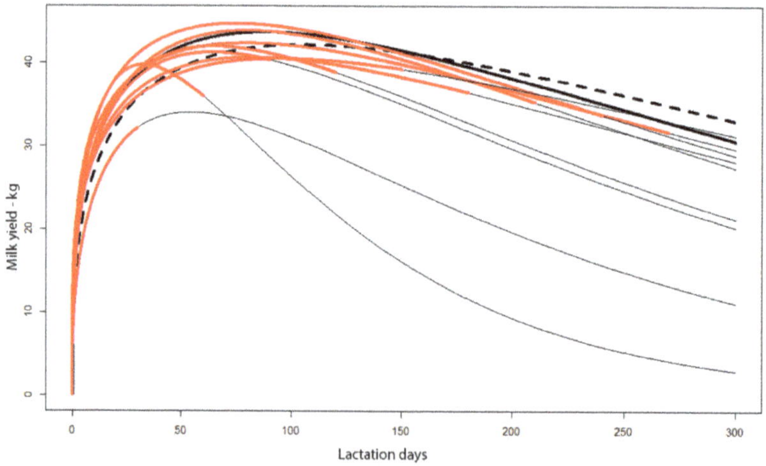

Figure 3. Milk yield curves for cows culled after day x of milking (red lines), extended based on predictions made using the Wood function (thin black lines). Total curve for cows milked for 305 days and those with shortened lactation (dashed line). Curve for cows milked at least 305 days (thick black line).

Using the curves determined by the Wood function, the integrals from day 0 to day x of milking and milk yield predict for 305 days were calculated for each successive function. The values of these integrals correspond to the milk yield over the corresponding periods (Table 3). These calculations are confirmed by the graphic (Figure 4), suggesting that culling of cows after 30, 60 and 90 days of lactation prevented a significant reduction in the milk yield of the whole herd, to 7109.765, 5610.461 and 9717.556 kg of milk, respectively, for 305 days of milking.

Table 3. Milk yields (kg) depending on the number of milking days—actual data ($0–x_i$) and predicted data ($0–x_{305}$).

Milking Period ($0–x_i$ [1])	Number of Milkings	Number of Cows	Yield in Period $0–x_i$	Yield in Period $0–x_{305}$
0–305	910	91	11,462.84	11,462.84
0–270	513	57	10,236.95	11,156.88
0–240	352	44	9602.265	11,478.32
0–210	434	62	8303.107	11,116.52
0–180	390	65	6779.897	10,653.34
0–150	255	51	5639.378	10,953.72
0–120	164	41	4622.313	9985.867
0–90	195	65	3356.83	9717.556
0–60	124	62	2138.613	5610.461
0–30	90	90	759.2602	7109.765

[1] x_i–final day of milking.

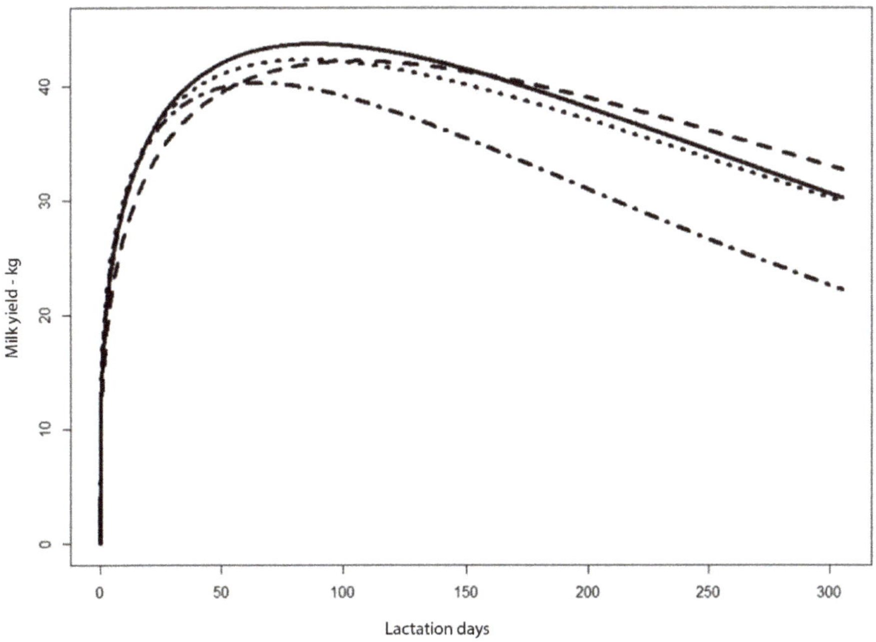

Figure 4. Predicted milk yield curves for culled primiparous cows. Legend: Dashed line—data for all cows milked in herd T1 up to day 305 (including culled cows). Continuous line—data only for cows milked up to day 305. Dotted line—data as for the dashed line, but excluding cows culled up to day 90. Dashed dotted line—data for average values from Figure 3.

The milk yield for the period of 0–305 days of milking was calculated for the curves from Figure 4. These values are similar and range from 11,348.95 to 11,614.94.

To determine how the early culling of cows with a yield below the adopted criterion (24 kg milk) affects the total production of the herd, predicted yield was calculated for a situation without this culling, taking into account the numbers of cows in each period. If culled cows had been milked up to 305 days, without culling of those with the lowest production, the total 305 d yield for the primiparous cows in herd T1 would have been 9798.13 kg of milk. By eliminating cows whose yield was too low, an additional 1586.87 kg of milk per cow was obtained in this herd.

The result of the Kruskal–Wallis test ($p = 0.003$) indicated that the curves differed statistically significantly. Dunn's non-parametric test of simultaneous multiple comparisons with the Holma–Šidák correction was used for detailed comparisons. The results indicate that charts y_0 and y_{90-305} do not differ significantly statistically ($p = 0.71$), charts y_0 and y_{305} differ statistically significantly ($p = 0.05$), and charts y_{90-305} and y_{305} differ statistically significantly ($p = 0.001$).

4. Discussion

Comparing the curves of milk yield and its components during lactation for different feeding systems is a good way to check whether the diet used has met expectations, with the cow as a biological test, in field conditions. Having access to feed components and very precise computer software for balancing feed rations, many farmers use well-balanced diets (Bach, 2020) [24]. However, it is not always possible to achieve the same results in different herds. The daily amount of concentrated feed and the organization of its use are also relevant.

Comparison of herds with well-balanced diets and very high milk yields (about 11,000 kg of milk for each herd) indicated that the T1 system resulted in by far the highest milk yield. The cows fed the same TMR diet throughout lactation produced more milk on average (42.2 vs. 37.7 kg), and their curves had a higher peak compared with animals fed PMR diets. Moreover, significant differences for peak milk yield were found between feeding management systems (P1 vs. T1) with no division into feeding groups according to days in milk). On the other hand, with the increase in milk production and higher peaks, the contents and curves for protein, fat, and milk urea decreased. In the case of herd T1, the curve for milk urea was similar to the curve for milk yield, with a maximum at about 100 days of lactation for both curves. The similar genetic potential of the herds might suggest that the reasons for these differences can be found in the feeding management systems.

A similar conclusion was drawn by Sabbioni et al. (2012), who found a significant difference in milk yields between traditional and TMR feeding [26,27]. Cichocki et al. (2007) found that the use of TMR and PMR systems did not result in a statistical difference in the level of milk production [28]. Several studies have compared TMR with PMR and traditional feeding systems, in which the forage and concentrate components of the diet are offered to cows separately. Bargo et al. (2002) compared three feeding systems (pasture with concentrate, TMR, and PMR) and found that the TMR feeding system resulted in the highest total milk production: cows produced 6.1 kg/day more milk than the cows from the PMR treatment [29]. This finding was similar to the results of the present study, in which dairy cows fed the same TMR ration throughout lactation, with no division into feeding groups, produced 5 kg/day more milk than in the PMR treatment. Gordon et al. (1995) found that feeding a complete diet resulted in 3.04 kg/day more milk than feeding concentrate and silage separately, without altering the milk concentrations of fat and protein [30]. In the present study, an increase in milk production and the highest peak of the lactation curve were observed in the herd receiving the same TMR ration throughout lactation. In the PMR treatment, the contents of protein, fat, and milk urea in the milk were reduced. Lactation curves for milk and milk components in dairy cattle show variation in peak yield and persistency of yield, partially explained by dietary composition and feeding management (Caccamo et al. 2012) [31]. Cabrita et al. (2007) [32] observed that low-protein, low-starch diets decreased dry matter intake and milk production in mid-lactation cows, but milk production responded to increases in dietary CP, starch, or both. Based on the meta-analysis performed by Hristov et al. (2002) [33], starch (energy content) and forage quality significantly affected herd curve traits, whereas Oba and Allen (2003) [34] observed that cows in early lactation fed a high-starch diet (32%, DM) versus low-starch diets (21%, DM) produce more milk and protein. Ikonen et al. (2004) found that milk yield was negatively correlated with fat ($r = -0.25$) and protein ($r = -0.27$) percentages in milk [35]. Moreover, Bargo et al. (2002) concluded that milk yield and milk chemical composition curves were most likely to respond to TMR in studies involving high-yielding cows (>28 kg/day) in early lactation [29].

MU (milk urea: MU = MUN $\times$ 21.4) is used in Europe and milk urea nitrogen (MUN) in North America as a tool for monitoring diets (Siachos et al. 2017) [36]. Variance in MU has been shown to be related to the ratio of dietary CP to energy, extended CP degradation in the rumen and the amount of ammonia in excess of microbial N requirements, and protein or energy intake in relation to feeding standards (Nousiainen et al. 2004) [37]. After a meal, microbial degradation of dietary protein is likely to cause an increase in the concentration of rumen ammonia which, due to the transport of rumen ammonia to the blood and the subsequent conversion of blood ammonia into urea by the liver, is followed by rising in plasma urea nitrogen, and due to diffusion between blood and milk as MU [38]. However, diets with a high rapidly degradable carbohydrate fraction may result in a decrease of rumen ammonia immediately after feeding, because the degraded protein in the rumen to ammonia is utilized for microbial protein synthesis [38]. In the current study, in the case of herd T1, the curve for milk urea was similar to the curve for milk yield, with a maximum at about 100 days of lactation for both curves, which could probably mean that this feeding strategy, where cows fed the same TMR diet throughout lactation might have the best balance of CP to carbohydrate ratio in the diet and utilization of nitrogen. The above results are similar to those presented by Wood et al. (2003) [39], who reported an increase in MU concentration after peaking of lactation. However, Mucha et al. (2011) [40] showed that the lactation curves of MU did not seem to be affected by the milk yield curve and differences may have been due to management or nutrition strategies. Surprisingly in the current study of other feeding management systems such as P1, T2, and T3, the increase in milk production was not correlated with MU concentration. The results might suggest that after peaking of lactation, CP intake was too high or supply rapidly degradable carbohydrates was too less for the effective microbial protein synthesis, which had an effect on MU contents and curves.

Professional feeding of dairy cows is a very important target based on economic and animal health reasons. Feed intake is characterized as dry matter intake (DMI) to compare diets of variable moisture concentrations. DMI is affected by both animal and feed factors. The weight of cows, milk production, and the stage of lactation or gestation are the major animal factors. In practice, the grouping of cows in nutritional groups is based on milk production, reproductive cycle and the days in milking. The cows are usually grouped monthly, after milk recording. Metabolic processes increase if milk productivity increases in dairy cows. Metabolization of body energy reserves during the early lactation enables the cow to close the gap between the alimentary energy intake and its loss through milk production [41]. Since the alterations in energy reserves have a considerable influence upon the productivity, health, and reproduction of dairy cows [42], the monetarization of optimal management of energy reserves is obviously needed. Indicators, which characterize dairy cows metabolic processes is body condition score (BCS). The use of a body condition scoring system is more useful than the measurement of body weight (BW) in feeding. BCS has been widely recommended as a method of evaluating the nutritional management of dairy cows [43]. It is a management tool used to elucidate if rations meet animal needs or not. Feeding a cow according to its needs leads to optimal performance. Stefańska et al. (2016) defined BCS as an indicator of how well the animal maintains energy reserves, reflective of the relationship between nutrition and milk production in a herd. Additionally, the production of cows correlates with their body condition which is a wide and effective method to evaluate the nutritional management of dairy cows [44]. Optimal BCS of dairy cows is essential to obtain elite herd and quantity milk production because thin or fat cows may have a greater risk of lower milk yield and higher milk somatic cell counts [45]. Additionally, Atasever et al. (2017) found that cows with lower than 3.0 BCS at calving showed lower milk production [46]. The highest lactation curve, greatest peak and least persistency were obtained from cows calving at BCS 3.25. On the other hand, over-conditioned animals, especially at the end of lactation or under-conditioned animals especially at the beginning of lactation, would have a health problem. Agenas et al. (2003) reported that, at the peak of lactation, the energy needs exceed the energy supply, which generated negative energy balance (NEB) and in consequence decrease BCS and metabolic disease occurrence [47]. On the other hand, after the peak of lactation, the energy needs decreasing, which also is associated with decreasing the milk yield and lactation curve. In this case, if the diets are not well balancing (too high-energy diet level in compiling to the milk production) it might affect improving the BCS in late-lactation dairy cows and then during the dry-off period, which has a negative effect because excessive energy intake prepartum leads to metabolic problems and decreases DMI in the following lactation and loss in milk production [48].

The milk yield curves were influenced by the herd management systems, including the culling rate. The management method in the T1 herd increased average milk production by about 1586.87 kg per primiparous cow. According to Zając-Mazur (2007), an increase in the culling level contributes to progress in the herd through a faster change of generations [49]. Culling for economic reasons increases the utility and breeding value of the whole herd. A study by Rogers et al. (1988) on the American population demonstrated that the optimal percentage of herd culling for farms should be within the range of 25–35% to achieve an economic profit [50]. In Europe, the culling rate is lower; Bergk and Swalve (2011) report that more than 10% of primiparous cows are eliminated from the herd in Germany during the first 300 days and more than 20% within 450 days after calving [51]. However, many other studies demonstrate that in order to achieve satisfactory production results, a cow should be used as long as its productivity is sufficient to cover the expenses incurred for its rearing and maintenance. Juszczak et al. (2003) estimated that the cost of milk produced by cows used only for the first three lactations is equal to its market value, and only when used for a longer period do they begin to generate profit [13]. The authors also state that the main source of economic losses lies in the cost of milk production from primiparous cows used only up to 200 days, especially those culled during the first 100 days of lactation. This effect is exactly the opposite of that observed in our study, which showed that earlier removal of a low-yield animal from the herd and its replacement with an animal meeting the productivity criterion increases the production of the entire herd, with better economic results. This has also been confirmed by Borkowska and Januś (2009), who found that the milk performance of cows was affected by the yield in the first lactation [14]. By far the highest values, especially in the case of yield per day of milk use, were found for the most productive primiparous cows. At the same time, the lower yield of the primiparous cows was found to be associated with a higher percentage of milk fat and a lower percentage of protein.

5. Conclusions

To sum up, the best production results in combination with a favourable lactation curve were obtained in herd T1, in which cows were fed in the TMR system with no division into feeding groups according to days in milk. A high rate of culling of cows that does not meet the high production criterion in the first stage of lactation increases the annual milk yield of the herd.

Author Contributions: Conceptualization, S.W. and Z.S.; methodology, J.R.-Z. and Z.S.; software, I.S.; validation, J.Z.-W. and A.S.-B.; formal analysis, R.K.; writing—original draft preparation, J.R.-Z.;

writing—review and editing, W.N. and B.S.; project administration, Z.S. All authors have read and agreed to the published version of the manuscript.

Funding: This research received no external funding.

Institutional Review Board Statement: Not applicable.

Informed Consent Statement: Informed consent was obtained from all farm owners involved in the study.

Data Availability Statement: Data is contained within the article.

Conflicts of Interest: The authors declare no conflict of interest.

References

1. Macciotta, N.P.; Vicario, D.; Cappio-Borlino, A. Detection of different shapes of lactation curve for milk yield in dairy cattle by empirical mathematical models. *J. Dairy Sci.* **2005**, *88*, 1178–1191. [CrossRef]
2. Ali, T.E.; Schaeffer, L.R. Accounting for covariance's among test day milk yields in dairy cows. *Can. J. Anim. Sci.* **1987**, *67*, 637–644. [CrossRef]
3. Wilmink, J.B.M. Adjustment of test-day milk, fat and protein yield for age, season and stage of lactation. *Livest. Prod. Sci.* **1987**, *16*, 335–348. [CrossRef]
4. Wood, P. Algebraic model of the lactation curve in cattle. *Nature* **1967**, *216*, 164–165. [CrossRef]
5. Strabel, T.; Kopacki, W.; Szwaczkowski, T. Genetic conditioning of the persistence of cow lactation. In Proceedings of the Material of the XIV Conference of Polish Genetic Society, Poznań, Poland, 11–13 June 2001; pp. 224–225.
6. Mustafa, A. *Lactation Curve*; University McGill Publications; Dairy Production: Montréal, QC, Canada, 2003; pp. 342A–450A.
7. Bogucki, M. Analysis of functional traits in highly-milking cows. *Rocz. Nauk. Pol. Tow. Zootech.* **2008**, *4*, 9–14.
8. Sawa, A.; Neja, W.; Bogucki, M.; Rępuszewska, D. Course of lactation in first calvers and older cows as related to milk yield. *Rocz. Nauk. Pol. Tow. Zootech.* **2007**, *4*, 205–216.
9. Capuco, A.; Ellis, S.; Hale, S.; Long, E.; Erdman, R.; Zhao, X.; Paape, M. Lactation persistency: Insights from mammary cell proliferation studies. *J. Anim. Sci.* **2003**, *81*, 18–31. [CrossRef] [PubMed]
10. Borkowska, D.; Polski, R.; Januś, E.; Grzesik, K. Influence of the condition of cows during the period of dryness for persistence and course of lactation. In Proceedings of the Conference of the LXXI Meeting PTZ in Bydgoszcz, Bydgoszcz, Poland, 18–20 September 2006.
11. Dekkers, J.; TenHag, J.; Weersink, A. Economic aspects of persistency of lactation in dairy cattle. *Acta Sci. Pol. Zootech.* **2007**, *6*, 15–20. [CrossRef]
12. Topolski, P.; Choroszy, B.; Choroszy, Z. Effect of production level in Polish Holstein-Friesian cows of Black-and-White variety on lactation persistency and length of calving interval. *Rocz. Nauk. Zootech.* **2008**, *35*, 93–99.
13. Juszczak, J.; Hibner, A.; Ziemiński, R.; Tomaszewski, A. Causes and consequences of early culling of dairy cows. *Med. Wet.* **2003**, *59*, 432–435.
14. Borkowska, D.; Januś, E. Milk yield of primiparas and their lifetime performance. *Rocz. Nauk. Pol. Tow. Zootech.* **2009**, *5*, 87–92.
15. NRC. *Nutrient Requirements of Dairy Cattle: Seventh Revised Edition*; The National Academies Press: Washington, DC, USA, 2001.
16. AOAC. Agricultural chemicals. In *Official Methods of Analysis*; Association of Official Analytical Chemists: Gaithersburg, MD, USA, 2010; Volume 1.
17. R Core Team. *R: A Language and Environment for Statistical Computing*; R Foundation for Statistical Computing: Vienna, Austria, 2020. Available online: https://www.R-project.org/ (accessed on 12 December 2020).
18. Mendiburu, F. Agricolae: Statistical Procedures for Agricultural Research. R Package Version 1.3-3. 2020. Available online: https://CRAN.R-project.org/package=agricolae (accessed on 12 December 2020).
19. Arnhold, E. Easynls: Easy Nonlinear Model. R Package Version 5.0. 2017. Available online: https://CRAN.R-project.org/package=easynls (accessed on 12 December 2020).
20. Wickham, H. *ggplot2: Elegant Graphics for Data Analysis*; Springer: New York, NY, USA, 2016.
21. Schauberger, P.; Walker, A. *Openxlsx: Read, Write and Edit Xlsx Files*; R Package Version 4.2.3; R Foundation for Statistical Computing: Vienna, Austria, 2020.
22. Bouallegue, M.; M'Hamdi, N. Mathematical Modeling of Lactation Curves: A Review of Parametric Models. 2019. Available online: https://cdn.intechopen.com/pdfs/70375.pdf (accessed on 5 February 2020).
23. Macciotta, N.P.P.; Dimauro, C.; Rassu, S.P.G.; Steri, R.; Giuseppe, P. The mathematical description of lactation curves in dairy cattle. *Ital. J. Anim. Sci.* **2011**, *10*, e51. [CrossRef]
24. Bach, A.; Terré, M.; Vidal, M. Symposium review: Decomposing efficiency of milk production and maximizing profit. *J. Dairy Sci.* **2020**, *103*, 5709–5725. [CrossRef] [PubMed]
25. Dunn, O.J. Multiple comparisons using rank sums. *Technometrics* **1964**, *6*, 241–252. [CrossRef]
26. Holm, S. A simple sequentially rejective multiple test procedure. *Scand. J. Stat.* **1979**, *6*, 65–70.

27. Sabbioni, A.; Beretti, V.; Tardini, L.; Vezzali, S.; Paini, V.; Superchi, P. Milk production and lactation curves of bioaca Val Padana and Italian Friesian dairy cows in relations to the management system. *Ital. J. Anim. Sci.* **2012**, *26*, 140–145. [CrossRef]
28. Cichocki, M.; Wroński, M.; Szydłowski, R. Milk performance of cows fed a TMR and PMR. *Acta Sci. Pol. Zootech.* **2007**, *6*, 15–20.
29. Bargo, F.; Muller, L.D.; Delahoy, J.E.; Cassidy, T.W. Performance of high producing dairy cows with three different feeding systems combining pasture and total mixed rations. *J. Dairy Sci.* **2002**, *85*, 2948–2963. [CrossRef]
30. Gordon, F.J.; Patterson, D.C.; Yan, T.; Porter, M.G.; Mayne, C.S.; Umsworth, E.F. The influence of genetic index for milk production on the response to complete diet feeding and the utilization of energy and nitrogen. *Anim. Sci.* **1995**, *61*, 199–210. [CrossRef]
31. Caccamo, M.; Veerkamp, R.F.; Licitra, G.; Petriglieri, R.; La Terra, F.; Pozzebon, A.; Ferguson, J.D. Association of total-mixed-ration chemical composition with milk, fat, and protein yield lactation curves at the individual level. *J. Dairy Sci.* **2012**, *95*, 6171–6183. [CrossRef]
32. Cabrita, A.R.J.; Bessa, R.J.B.; Alves, S.P.; Dewhurst, R.J.; Fonseca, A.J.M. Effects of dietary protein and starch on intake, milk production, and milk fatty acid profiles of dairy cows fed corn silage-based diets. *J. Dairy Sci.* **2007**, *90*, 1429–1439. [CrossRef]
33. Hristov, A.N.; Price, W.J.; Shafii, B. Dietary factors influencing milk yield and milk protein yield in dairy cows. In Proceedings of the Pacific Northwest Animal Nutrition Conference, Vancouver, BC, Canada, 3–5 October 2002; CR Press: Portland, OR, USA, 2002; pp. 147–166.
34. Oba, M.; Allen, M.S. Effects of corn grain conservation method on feeding behavior and productivity of lactating dairy cows at two dietary starch concentrations. *J. Dairy Sci.* **2003**, *86*, 174–183. [CrossRef]
35. Ikonen, T.; Morri, S.; Tyriseva, A.M.; Ruottinen, O.; Ojala, M. Genetic and phenotypic correlations between milk coagulation properties, milk production traits, somatic cell count, casein content, and pH of milk. *J. Dairy Sci.* **2004**, *87*, 458–467. [CrossRef]
36. Siachos, N.; Panousis, N.; Arsenos, G.; Valergakis, G. Investigation of milk urea nitrogen concentration and factors affecting its variation in Greek Holstein herds. *J. Hell. Vet. Med. Soc.* **2017**, *68*, 423–432. [CrossRef]
37. Nousiainen, J.; Shingfield, K.J.; Huhtanen, P. Evaluation of Milk Urea Nitrogen as a Diagnostic of Protein Feeding. *J. Dairy Sci.* **2004**, *87*, 386–398. [CrossRef]
38. Spek, J.W.; Dijkstra, J.; Van Duinkerken, G.; Bannink, A. A review of factors influencing milk urea concentration and its relationship with urinary urea excretion in lactating dairy cattle. *J. Agric. Sci.* **2013**, *151*, 407–423. [CrossRef]
39. Wood, G.M.; Boettcher, P.J.; Jamrozik, J.; Jansen, G.B.; Kelton, D.F. Estimation of genetic parameters for concentrations of milk urea nitrogen. *J. Dairy Sci.* **2003**, *86*, 2462–2469. [CrossRef]
40. Mucha, S.; Strandberg, E. Genetic analysis of milk urea nitrogen and relationships with yield and fertility across lactation. *J. Dairy Sci.* **2011**, *94*, 5665–5672. [CrossRef]
41. Bünemann, K.; Von Soosten, D.; Frahm, J.; Kersten, S.; Meyer, U.; Hummel, J.; Zeyner, A.; Dänicke, S. Effects of Body Condition and Concentrate Proportion of the Ration on Mobilization of Fat Depots and Energetic Condition in Dairy Cows during Early Lactation Based on Ultrasonic Measurements. *Animals* **2019**, *9*, 131. [CrossRef]
42. Harder, I.; Stamer, E.; Junge, W.; Thaller, G. Lactation curves and model evaluation for feed intake and energy balance in dairy cows. *J. Dairy Sci.* **2019**, *102*, 7204–7216. [CrossRef]
43. Van Hoeij, R.J.; Dijkstra, J.; Bruckmaier, R.M.; Gross, J.J.; Lam, T.J.G.M.; Remmelink, G.J.; Kemp, B.; Van Knegsel, A.T.M. Consequences of dietary energy source and energy level on energy balance, lactogenic hormones, and lactation curve characteristics of cows after a short or omitted dry period. *J. Dairy Sci.* **2017**, *100*, 8544–8564. [CrossRef] [PubMed]
44. Stefańska, B.; Nowak, W.ł.; Pruszyńska-Oszmałek, E.; Mikuła, R.; Stanisławski, D.; Kasprowicz-Potocka, M.; Frankiewicz, A.; Maćkowiak, P. The effect of body condition score on the biochemical blood indices and reproductive performance of dairy cows. *Ann. Anim. Sci.* **2016**, *16*, 129–143. [CrossRef]
45. Roche, J.R.; Friggens, N.C.; Kay, J.K.; Fisher, M.W.; Stafford, K.J.; Berry, D.P. Invited review: Body condition score and its association with dairy cow productivity, health, and welfare. *J. Dairy Sci.* **2009**, *92*, 5769–5801. [CrossRef] [PubMed]
46. Atasever Szilágyi, S.; Gräff, M.; Jónás, E.M. Effect of body conditio at calving on lactation curve of Holstein cows. In Proceedings of the 11th International Symposium Modern Trends in Livestock Production, Belgrade, Serbia, 11–13 October 2017; pp. 396–405.
47. Agenas, S.; Burstedt, E.; Holtenius, K. Effects of feeding intensity during the dry period. 1. Feed intake, body weight, and milk production. *J. Dairy Sci.* **2003**, *86*, 870–882. [CrossRef]
48. Stefańska, B.; Poźniak, A.; Nowak, W.Ł. Relationship between the pre- and postpartum body condition scores and per parturient indices and fertility in high-yielding dairy cows. *J. Vet. Res.* **2016**, *60*, 81–90. [CrossRef]
49. Zając-Mazur, M. Culling level of dairy cows as a measure of breeding work efficiency in the herd. *Wiadomości. Zootech.* **2007**, *3*, 1–15.
50. Rogers, G.W.; Van Arendork, J.A.M.; Mc Daniel, B.T. Influence of Production and Prices on Optimum Culling Rates and Annualized Net Revenue. *J. Dairy Sci.* **1988**, *12*, 3453–3462. [CrossRef]
51. Bergk, N.; Swalve, H. Der Fett-Eiweiß-Quotient in der Frühlaktation als Indikator für den Verbleib von Erstkalbinnen in der Milchkuhherde. *Züchtungskunde* **2011**, *83*, 89–103. (In Deutsch)

animals
MDPI

Article

Transcriptome Functional Analysis of Mammary Gland of Cows in Heat Stress and Thermoneutral Condition

Shuangming Yue, Zhisheng Wang, Lizhi Wang, Quanhui Peng and Bai Xue *

Animal Nutrition Institute, Sichuan Agricultural University, Chengdu 611130, China; yueshuangming@stu.sicau.edu.cn (S.Y.); wangzs@sicau.edu.cn (Z.W.); 12825@sicau.edu.cn (L.W.); 14101@sicau.edu.cn (Q.P.)

* Correspondence: xuebai@sicau.edu.cn

Received: 15 May 2020; Accepted: 8 June 2020; Published: 10 June 2020

Simple Summary: The current study employed RNA-seq technology to analyze the impact of heat stress on the whole transcript sequencing profile in the mammary glands of lactating Holstein dairy cows. In the findings of the current study, heat stress downregulated the expression of casein genes, which resulted in a decrease in milk production. Moreover, heat stress upregulated the gene expression of *HSPA1A* and *HSP90B1*, while it downregulated the expression of immune response-related genes that resulted in a reduction in milk yield. Furthermore, there was an increased synthesis of heat shock proteins and unfolded proteins that could reduce the availability of circulating amino acids for milk protein synthesis. The findings of the current experiment may help to explore the impact of heat stress on immune function, milk production, and milk protein synthesis in cows.

Abstract: Heat stress (HS) exerts significant effects on the production of dairy animals through impairing health and biological functions. However, the molecular mechanisms related to the effect of HS on dairy cow milk production are still largely unknown. The present study employed an RNA-sequencing approach to explore the molecular mechanisms associated with a decline in milk production by the functional analysis of differentially expressed genes (DEGs) in mammary glands of cows exposed to HS and non-heat-stressed cows. The results of the current study reveal that HS increases the rectal temperature and respiratory rate. Cows under HS result in decreased bodyweight, dry matter intake (DMI), and milk yield. In the current study, a total of 213 genes in experimental cow mammary glands was identified as being differentially expressed by DEGs analysis. Among identified genes, 89 were upregulated, and 124 were downregulated. Gene Ontology functional analysis found that biological processes, such as immune response, chaperone-dependent refolding of protein, and heat shock protein binding activity, were notably affected by HS. The Kyoto Encyclopedia of Genes and Genomes enrichment analysis found that almost all of the top-affected pathways were related to immune response. Under HS, the expression of heat shock protein 90 kDa beta I (*HSP90B1*) and heat shock 70 kDa protein 1A was upregulated, while the expression of bovine lymphocyte antigen (*BoLA*) and histocompatibility complex, class II, DRB3 (*BoLA-DRB3*) was downregulated. We further explored the effects of HS on lactation-related genes and pathways and found that HS significantly downregulated the casein genes. Furthermore, HS increased the expression of phosphorylation of mammalian target of rapamycin, cytosolic arginine sensor for mTORC1 subunit 2 (*CASTOR2*), and cytosolic arginine sensor for mTORC1 subunit 1 (*CASTOR1*), but decreased the phosphorylation of Janus kinase-2, a signal transducer and activator of transcription factor-5. Based on the findings of DMI, milk yield, casein gene expression, and the genes and pathways identified by functional annotation analysis, it is concluded that HS adversely affects the immune function of dairy cows. These results will be beneficial to understand the underlying mechanism of reduced milk yield in HS cows.

Keywords: heat stress; dairy cow; whole transcript sequencing; immune response; stress response

1. Introduction

In livestock production, heat stress (HS) negatively affects livestock health and production [1]. However, the effects of HS on livestock production are different in different livestock breeds [2,3]. For example, HS in beef breeds is generally believed less critical as compare to dairy breeds because beef breeds have lower metabolic rates and lower body heat production [2,3]. Therefore, most of the studies explore the impact of HS on dairy animals' health and production.

In dairy animals, the Holstein breed is one of the standard commercial dairy cattle breeds and is widely known due to its high milk yield. In hot and humid seasons, the ability of Holstein dairy cattle to dissipate body heat via skin evaporation is restrained due to its relatively low surface area to body weight ratio, dense body surface hair, and underdeveloped sweat glands [4]. Therefore, Holstein dairy cattle are at higher risk of facing even more severe heat stress [5]. Previous studies have reported that HS negatively influences the production of milk and protein contents of milk in Holstein dairy cows [6,7]. It is traditionally believed that HS results in lower dry matter intake (DMI), which reduces the production of milk yield and protein contents of milk in dairy cows [8,9]. However, the utilization of pair-fed thermal neutral (TN) controls in recent studies have demonstrated that decreased DMI only partially (about 35–50%) explains the decrease in productivity [6,7]. It has also been reported that lower DMI in HS cows is not only one of the main reasons for the decline in milk production [6,7]. Furthermore, it has also been reported that HS influences the cellular response that is responsible for the decline in milk production and milk quality [10]. For example, Cowley et al. [7] found that HS results in a reduction in milk protein as a result of downregulation of bovine mammary epithelial cell (BMEC) activity [7]. Moreover, at the cellular level, heat stress adversely affects the function and gene expression of casein in BMECs [11,12], especially the Alpha casein S1 (*CSN1S1*) [13]. The in-vitro studies explored that high ambient temperature downregulates the gene expression involved in cell structure, biosynthesis, and transport, whereas it upregulates the gene expression involved in protein repair and degradation in BMECs [7,14].

However, the effect of HS on the whole transcript sequencing in mammary tissue of dairy cows in-vivo is still unknown. Therefore, the current experiment was designed to investigate the global expression profile of dairy cows' mammary gland tissue during normal and HS state and to identify the molecular pathways regulated in heat-stressed dairy cows by using RNA sequencing (RNA-Seq). The objective of the current study is to explore the effect of HS on the molecular mechanism that reduces the performance of dairy cows.

2. Materials and methods

2.1. Animals, Management and Experimental Treatments

All experimental protocols in the current study were conducted following the guidelines of the Animal Care and Use Committee, constituted by Sichuan Agricultural University (Chengdu, China). The current experiment was conducted at Qingbaijiang Dairy Farm of New Hope Dairy Co., Ltd. (Chengdu, China). A total of twenty Holstein cows, with healthy and symmetrical udders, were selected for this study. All the experimental cows were reared in a closed-type cowshed to avoid seasonal variation and photoperiods on metabolism. Individual pens were assigned to each cow in such a way that each cow had free access to fresh drinking water. The managemental regimes were the same for all the experimental cows. A group of ten cows was considered for HS experimental treatment, while the second group of 10 cows was considered for TN experimental treatment. In both experimental groups, each cow was considered as a replicate. The basic information of experimental cows is shown in Table 1.

Table 1. Characteristics of the experimental cows.

Parameter	TN (Thermal Neutral)	HS (Heat Stress)	*p*-Value
Number of cows	10	10	
Parity	2.1 ± 1.0	2.2 ± 1.3	0.89
Lactation days	130.5 ± 15	123.4 ± 20	0.92
305-day milk yield (kg)	8993.7 ± 767.5	8933.2 ± 757.1	0.95
Average bodyweight (kg)	605.8 ± 58.1	603.5 ± 45.3	0.97

The samples of HS treatment were collected in the summer season (from mid-July to late August in 2018) when the environmental temperature–humidity index (THI) was enhanced from 72.5 to 86.9 over one month and stabled at 80.5 for one week. The samples of TN experimental treatments were collected in the spring season (from mid-March to late April in 2018) when environmental THI steadily increased from 52.1 to 65.2 over a one-month period. Experimental cows were on the same diet throughout the trial period. The duration of the adaptation period was 15 days, while the duration of the experimental data collection was 30 days.

Animals were fed the diet in the form of a total mixed ration (TMR). The delivery of TMR was at 07:00, 14:00, and 20:00 h of the day throughout the experimental period. The diets of experimental animals were formulated according to the recommendation of National Research Council (NRC 2001) for dairy cows [15]. The experimental feed ingredients profile and chemical composition are shown in Table 2.

Table 2. Ingredients and chemical composition of the basal diet.

Ingredients (% of DM)	Content
Corn	14.94
Soybean meal	4.98
Cottonseed meal	1.30
Rapeseed meal	0.42
Extruded full-fat soybean	2.21
Corn gluten meal	0.83
Dried distillers grains	1.11
Oat hay	5.03
Alfalfa hay	10.75
Whole cottonseed	3.42
Whole corn silage	38.87
Beet pulp	2.52
Molasses	3.42
Fatty power	0.57
Brewer's grains	6.86
Limestone	0.77
Dicalcium phosphate	0.42
Vitamin–mineral premix [1]	0.48
Sodium bicarbonate	0.70
MgO	0.14
NaCL	0.28
Chemical composition (%)	
NE_L (Mcal/Kg) [2]	1.70
Crude protein	16.6
Ether extract	5.50
NDF [3]	30.89
ADF [4]	19.27
Ca	0.78
P	0.43
Ash	7.50

[1] Provided TMR/kg: a min. of 10,000 IU of vitamin A.; 1850 IU of vit. D; 50 IU. of vit. E; 16.73 mg of niacin; 54 mg of Zn; 12.5 mg of Cu; 0.45 mg of Se; 20.5 mg of Mn; 0.54 mg of Co; 0.945 mg of I. [2] Calculated. following NRC (2001) recommendation and was, based on actual DMI.. [3] Neutral detergent fiber. [4] Acid detergent fiber.

In the current experiment, relative humidity and temperatures were recorded on a daily basis. The time of recording of temperatures and relative humidity was 07:00, 14:00, and 20:00 h of the day. A hygrometer and a thermometer on one instrument panel (Jiangsu JingChuang Electric Co. Ltd., Nanjing, China) were used for the determination of relative humidity and temperatures. The temperature-humidity index was calculated according to NRC.1971.

$$\text{THI} = 1.8 \times \text{T} + 32 - 0.55 \times (1 - \text{RH}) \times (\text{T} \times 1.8 + 32 - 58) \quad (1)$$

where T.was the ambient temperature determined by the dry bulb in °C. and RH was the relative humidity in %.

Bodyweight (BW), rectum temperature (RT), respiratory rate (RR), and milk quality indexes were recorded or determined for all experimental cows. Respiratory rate and RT were measured three times a day at 07:00, 14:00, and 20:00 h. Respiratory rate was calculated by counting the total number of flank movements/min for 120 s. Rectal temperature was obtained with a GLA 525/550 digital thermometer.

2.2. Sample Collection

On the last day of the experimental period, blood samples (duplicate samples) were taken from the median coccygeal vein of each cow before first feeding (morning). Blood was collected into evacuated tubes with and without anticoagulant. The samples of blood were kept at a cool place until they were centrifuged for 15 min at 3000× *g* (4 °C) to separate serum or plasma. Obtained serum and plasma were stored at −80 °C for further analysis. The concentrations of heat shock protein 70 (HSP70) and lipopolysaccharide (LPS) in plasma were determined using an ELISA kit (Beyotime Biochemical Reagent Co., Shanghai, China). Furthermore, the serum obtained was used to analyze nonesterified fatty acid (NEFA) and glucose concentration by automatic biochemical analyzer 7600 (Hitachi, Tokyo, Japan).

After a 3-week rearing of experimental cows in either HS or TN group, three cows with the same average milk yield were chosen for taking mammary gland samples. The biopsy procedure was carried out according to established methods, as described in the previous study [16]. Biopsies were operated after approximately 6 h of milk accumulation. To carry out the biopsy procedures, experimental cows were properly restrained, and an intravenous injection of xylazine hydrochloride (35–45 μg/mg of BW, romazine 2%, Healton Animal Health, Neijiang, China) was applied. A 10-cm^2 area of udder skin on the right rear quarter was clipped, cleaned, and sterilized. The area for biopsy was anesthetized by injection (subcutaneous) of 3 mL of lignocaine hydrochloride (20 mg/mL. of lopaine, Healton Animal Health, Neijiang, China). A 1–2 cm incision was made through the skin and gland capsule. The incision was made in such a way to avoiding any large subcutaneous blood vessels.

The biopsy instrument (Wuhan Anscitech Farming Technology, Wuhan, China) was used to cut a core (70 × 4 mm. in diameter) of mammary tissue. To control bleeding, we inserted a 3 × 5 cm surgical plug (Healton Animal Health, Neijiang, China) into the wound. After that, Michel suture clips were used to close the skin incision. Antibiotic powder was also applied onto the wound (terramycin powder oxytetracycline hydrochloride (2% wt/wt), North China Pharmaceutical Group Corporation Veterinary, Shijiazhuang, China). A single intramuscular dose of penicillin and streptomycin (4 mL/1000 kg of BW, North China Pharmaceutical Group Corporation Veterinary, Shijiazhuang, China) was also given instantly after the biopsy. After the biopsy, the cows were machine milked. To remove intramammary blood clots, hand-stripping was used. Furthermore, cows were hand-stripped as required at each milking for the next 4–7 days until all blood clots were removed entirely. Michel suture clips were removed 7–10 days after the biopsy. In the subsequent duration of the experiment, after the first milking, both rear glands received a prophylactic dose of intramammary antibiotic (200 mg of sodium cloxacillin, North China Pharmaceutical Group Corporation Veterinary, Shijiazhuang, China). The same intramammary antibiotic dosage was repeated after every two days. Representative tissues of the mammary gland were sampled, weighed, washed by cold phosphate-buffered saline, and then kept in liquid nitrogen until further analysis.

2.3. RNA Isolation and Library Preparation

Total RNA was extracted from the mammary gland tissue of cows of the HS group and the TN group by utilizing trizol reagent (Invitrogen, South San Francisco, CA, USA). The manufacturer's protocol was strictly followed to obtain total RNA. A NanoDrop 2000 spectrophotometer (Thermo Scientific Scientific, Inc., Waltham, MA, USA) was used to evaluate RNA purity and quantification. Furthermore, to evaluate RNA integrity, an Agilent 2100 Bioanalyzer (Agilent Technologies, Santa Clara, CA, USA) was used. Samples with an RNA Integrity Number (RIN) >7 were further subjected for analysis. The libraries were constructed by employing the TruSeq Stranded mRNA LTSample Prep kit (Illumina, San Diego, CA, USA) by following the manufacturer's protocol. Then, these libraries were sequenced on an Illumina HiSeq X Ten platform (OE Biotech Co., Ltd, Shanghai, China), and 150-bp paired-end reads were generated.

2.4. Quality Control and Mapping

Raw reads were generated from the images by using Base Calling, and the quality of the raw reads was checked by using Trimmomatic (San Diego, CA, USA). The low-quality reads and those containing poly-N were removed [17]. Then, the clean reads were mapped to the cow genome (*Bos taurus*) by using Hisat2 [18].

2.5. RNA-Seq Data Analysis

The FPKM (fragments per kb per million reads) value of each gene was measured by using cufflinks [19]. Then, HTSeq-Count was used to obtain the read counts of each gene. Differentially expressed genes (DEGs) were analyzed using the DESeq R package [20]. The corrected $p < 0.05$ and fold change (FC) > 1.5 or fold change (FC) < 0.67 was set as the thresholds for significantly differential expression. A hierarchical cluster analysis of DEGs was performed to examine gene expression patterns. The DEGs were annotated by Gene ontology (GO) functional enrichment and Kyoto Encyclopedia of Genes and Genomes (KEGG) pathway enrichment using the R programming language (3.5 versition, http://www.r-project.org/), based on the hypergeometric distribution.

2.6. Quantitative Real-Time PCR Analysis (qRT-PCR).

To verify the expression of DEGs identified by the RNA-seq approach, four DEGs, including *HSP70*, *HSP90B1*, bovine lymphocyte antigen (*BoLA*), and major histocompatibility complex, class II, DRB3 (*BoLA-DRB3*), were randomly picked for real-time PCR analysis. Moreover, 10 genes involved in lactation were selected for RT-PCR analysis, including *CSN1S1*, *CSN2*, *CSN3*, *PRLR*, STAT5A, *STAT5B*, *CASTOR1*, *CASTOR2*, *mTOR*, and *JAK2*. The PrimeScript™ RT reagent Kit (TaKaRa, Kyoto, Japan) was used to synthesis cDNAs in qRT-PCR from the same RNA extractions used for the RNA-seq. Furthermore, cDNAs were diluted 1:5. Glyceraldehyde-3 phosphate dehydrogenase (*GAPDH*) was selected as the control gene. qRT-PCRs were carried out in an ABI 7500 real-time thermocycler (Applied Biosystems, Foster City, CA, USA). The primers used are listed in Supplementary Materials Table S1. The comparative cycle threshold ($2^{-\Delta\Delta Ct}$) method was used to determine the relative gene expression. Data are expressed as mean.±.standard error of the mean and were analyzed using SAS 9.2 software (SAS Institute Inc, Cary, NC, USA).

2.7. Western Blot Analysis

Mammary gland tissues were solubilized in RIPA Lysis and Extraction Buffer (Invitrogen; Thermo Fisher Scientific, Inc., Waltham, MA, USA) to extract total protein. After boiling for 5–10 min, the protein samples extracted from different cell suspensions were separated by SDS-PAGE and then transferred onto a nitrocellulose membrane. The membrane was blocked with 5% nonfat dry milk prepared in Tris-buffer and incubated with the primary antibodies (Santa Cruz Biotechnology, Inc., Santa Cruz, CA, USA) for 1 h at room temperature. The membrane was then incubated with a horseradish peroxidase-conjugated anti-rabbit

IgG secondary antibody (Santa Cruz Biotechnology, Inc., Santa Cruz, CA, USA), for 4 h at room temperature. The blot was developed using the ECL™ Western Blotting Detection Reagent (GE Healthcare, Piscataway, NJ, USA), and the proteins were visualized by enhanced chemiluminescence (Amersham Biosciences, Piscataway, NJ, USA).

2.8. Statistical Analysis

Data for DMI, milk yield, milk composition, somatic cell count (SCC), HSP70, LPS, glucose, NEFA, RR, and RT were analyzed by the *t*-test (SAS 2003, ver. 9.2, Inst. Inc., Cary, NC, USA). Each individual cow was considered as an experimental unit. Statistical significance was set at $p < 0.05$.

3. Results

3.1. Temperature–Humidity Index and Physiological Index

The mean daily THI ranged from 55.6 to 63.2 during the spring period (averaged 59.8 in the whole period). The mean daily THI was from 79.0 to 83.2 during the summer period (averaged 79.5 in the whole period). The details of THI in both experimental treatments are presented in Figure 1. Respiratory rate and RT were higher ($p < 0.01$) in the HS experimental group than those in the TN experimental group (Table 3).

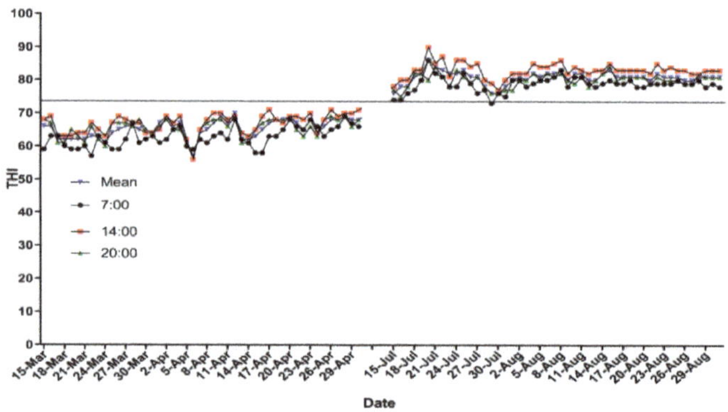

Figure 1. Temperature–humidity index at 07:00, 14:00, and 20:00 h, and mean.

Table 3. Respiratory rate and rectal temperature of lactating Holstein cows in TN and HS groups.

Table	Treatment		SEM	*p*-Value
	TN	HS		
Respiration rate (breath/min)				
0700	35.6	72.4	2.51	0.03
1400	39.3	89.6	3.31	<0.01
2000	41.3	84.3	4.32	<0.01
Rectal temperature (°C)				
0700	38.2	39.2	0.08	<0.01
1400	38.5	39.5	0.09	<0.01
2000	38.4	39.3	0.07	<0.01

3.2. Effect of Exposure of Heat Stress on Milking Performance and Blood Biochemical Indexes

The results of milk production, milk composition, and blood biochemical indexes are presented in Table 4. Milk production decreased in animals of the HS group ($p < 0.01$) as compared to the TN group animals. Results also revealed a dramatic loss of BW and DMI in the animals of the HS group ($p < 0.01$). Both protein and fat in milk were decreased in the HS group animals ($p < 0.05$). In addition, the concentrations of plasma LPS, HSP70, and NEFA were enhanced, whereas the level of glucose was reduced in the HS group ($p < 0.05$).

Table 4. Milk yield, dry matter intake, milk composition, and somatic cell count of cows in the TN and HS groups.

Items	Treatment		SEM	p-Value
	TN	HS		
Milk yield (kg/d)	42.5	35.6	2.03	<0.01
DMI (kg/d)	23.5	21.4	0.10	<0.01
Milk Protein %	3.2	2.8	0.07	0.03
Milk protien yield (kg/d)	1.3	1.0	0.06	0.12
Milk Fat %	4.3	3.8	0.14	0.04
Milk Lactose %	4.6	4.7	0.03	<0.01
UN of milk (mg/dL)	13.9	14.6	0.86	0.51
LPS (EU/L)	691.4	948.1	72.81	0.01
HSP 70 (ng/mL)	7.8	14.0	1.31	<0.01
Glucose (mm/L)	2.8	2.2	0.22	0.03
NEFA (μm/L)	167.0	238.2	18.51	0.02
SCC 1000/ML	266.0	307.1	132.46	0.76

3.3. HS Group Changed the Transcriptome in Mammary Gland Tissues

An average of 49.30 million raw reads was generated from TN and HS libraries. After filtering out low-quality sequences, we had 47.4 to 48.76 million total tag numbers per library. High-quality ($Q > 30$) reads accounted for 95.27–96.04% of the reads and the GC percentage of each library ranged from 46.90% to 49.24% (Supplementary Materials Table S2). In order to evaluate the transcript profiling data, principal component analysis (PCA) was applied to capture the overall variance among the samples in two dimensions (Figure 2B). The principal component analysis illustrated that a significant difference existed between the TN and HS groups. Two hundred and thirteen DEGs were identified, among which 89 were upregulated and 124 downregulated (Figure 2E). The DEG profiles in six samples are shown in Figure 2C,D.

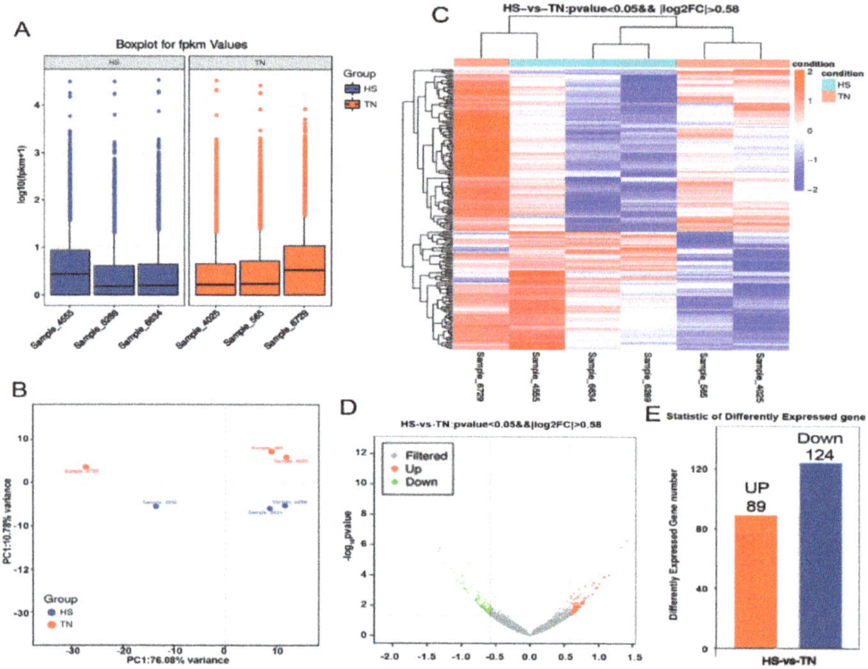

Figure 2. Effect of HS on transcriptomes in mammary gland tissues. (**A**) The distribution of FPKM (fragments per kb per million reads) values in HS and TN groups. (**B**) Principal component analysis was utilized to determine the reliability of the data, simplify the complexity of RNA-seq, and reduce the effective dimension of gene expression space. Heatmap (**C**) and volcano map (**D**) show the differentially expressed genes between HS and TN groups. (**E**) A total of 213 differentially expressed genes (DEGs) were identified, among which 89 were upregulated, and 124 downregulated between the groups studied.

3.4. Gene Ontology Enrichment Analysis

GO analysis was carried out to determine the biological function of DEGs in dairy cow mammary glands, which were enriched by 913 GO terms ($p < 0.05$). The top ten terms in each category, including cellular component, biological process, as well as molecular function, are listed in the column charts (Figure 3). The result showed that the most affected functional category was related to immune response, including antigen processing (GO: 0002476), T-cell-mediated cytotoxicity regulation (GO: 0001916), acute-phase response (GO: 0006953), and immune response (GO: 0006955). Most of those immunology-related terms in which HS downregulated the expression of genes were also the top terms in the category. On the other hand, most of the terms in which heat stress upregulated the expression of genes were involved in heat shock response, such as chaperone-dependent refolding of protein (GO: 0051085), inflammatory response (GO: 006954), and heat shock protein (HSP) binding activity (GO: 0031072).

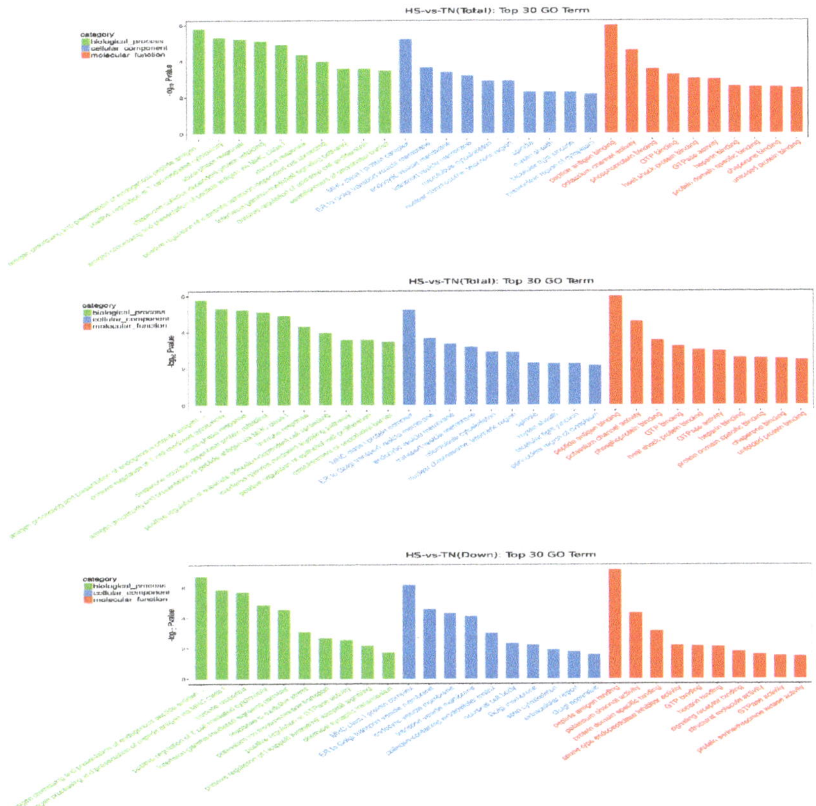

Figure 3. Gene ontology enrichment analysis. The genes from hierarchical clustering were further analyzed with the Database for Annotation Visualization and Integrated Discovery. The categories of the most affected genes were shown as the most upregulated (up), downregulated (down), and overall (total) effected. The top-10 terms in each category, including cellular component, biological process, as well as molecular function, are listed in the column charts.

3.5. Biological Pathway Affected by HS

To get a better insight into biological function, the GO information of the DEGs were further analyzed with the help of the KEGG database. The evaluation and comparison of the 20 different most-affected pathways were filtered out via *p*-value significance (Figure 4 and Supplementary Materials Figure S1). Nearly all these top-affected pathways could be related to the immune and heat shock response mentioned in Table 5. Six pathways are associate with exogenous pathogen invasion, including viral myocarditis, Epstein-Barr virus infection, *Herpes simplex* infection, *Staphylococcus aureus* infection, leishmaniasis, and antigen processing and presentation. In addition, four of them have a very close relationship to an autoimmune disorder, including rheumatoid arthritis, autoimmune thyroid disease, asthma, type I diabetes mellitus asthma, and all of them are associated with cellular responses by heat stress, including antigen processing and presentation, protein processing in the endoplasmic reticulum, and spliceosome. Therefore, all the results are directly or indirectly linked with immune regulation.

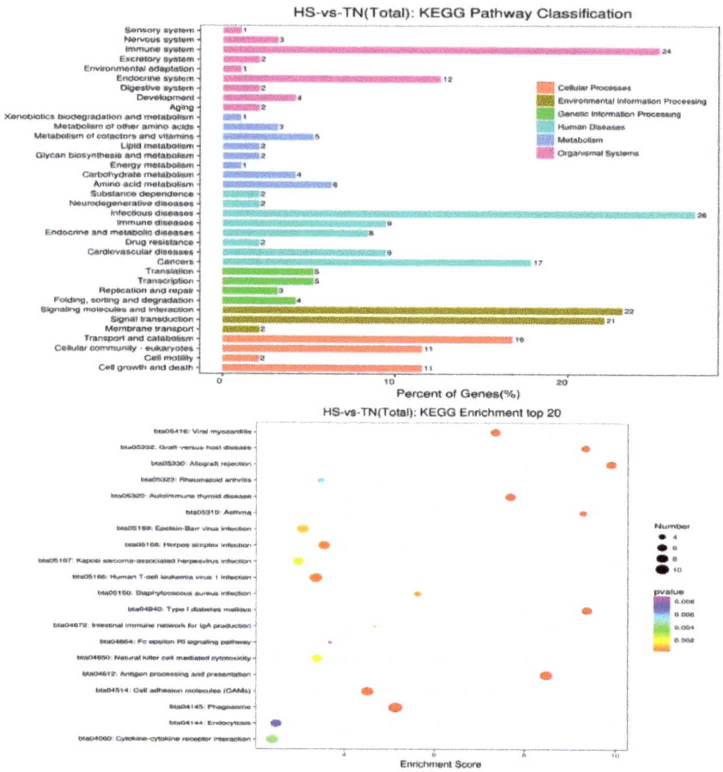

Figure 4. The top-20 pathways of HS-vs.-TN (total) of Kyoto Encyclopedia of Genes and Genomes enrichment.

Table 5. Significantly enriched KEGG pathways of downregulated and upregulated genes.

Term	*p*-Value	Genes	Gene Expression
bta05416, Viral myocarditis	<0.01	*BoLA; BOLA; BOLA-DRB3; BOLA-DYA; LOC512672; LOC524810*	Down
bta05169, Epstein-Barr virus infection	<0.01	*BoLA; BOLA; BOLA-DRB3; BOLA-DYA; LOC512672; LOC524810*	Down
bta05168, Herpes simplex infection	<0.01	*BoLA; BOLA; BOLA-DRB3; BOLA-DYA; LOC512672; SP100*	Down
bta05150, Staphylococcus aureus infection	<0.01	*BOLA-DRB3; BOLA-DYA; LOC524810*	Down
bta5140, Leishmaniasis		*BOLA-DRB3; BOLA-DYA; LOC524810*	Down
bta04612, Antigen processing and presentation	0.001	*BoLA; BOLA; BOLA-DRB3; BOLA-DYA; LOC512672*	Down
bta05323, Rheumatoid arthritis	0.003	*BOLA-DRB3; BOLA-DYA; LOC524810*	Down
bta05320, Autoimmune thyroid disease	<0.01	*BoLA; BOLA; BOLA-DRB3; BOLA-DYA; LOC512672; LOC524810*	Down
bta05310, Asthma	<0.01	*BOLA-DRB3; BOLA-DYA; LOC524810; MS4A2*	Down
bta04940, Type I diabetes mellitus	<0.01	*BoLA; BOLA; BOLA-DRB3; BOLA-DYA; CPE; LOC512672*	Down
bta04612, Antigen processing and presentation	0.001	*HSPA1A; HSPA8; LOC618733, HSP90B1*	Up
bta04141, Protein processing in endoplasmic reticulum	0.002	*DNAJA1; HSPA1A; HSPA8; HSPH1, HSP90B1*	Up
bta03040, Spliceosome	<0.01	*HSPA1A; HSPA8; PHF5A; SNRPG*	Up

3.6. *Identification of Differentially Expressed Genes in Response to HS Group*

The results of qRT-PCR are presented in Figure 5. Results revealed that fourteen genes were significantly affected by HS. RNA-seq results showed that *HSP70* and *HSP90B1* were upregulated while BoLA and BoLA-DRB3 were downregulated in the HS group. Moreover, HS decreased the gene expression of *CSN1S1*, *CSN2*, *CSN3*, *PRLR*, *STAT5A*, *STAT5B*, and *JAK2*, while HS increased the gene expression of *CASTOR1*, *CASTOR2*, and *mTOR*. The change was also consistent with that of their total protein products (Figure 6). The phosphorylation of *mTOR* was increased while the phosphorylation of *JAK2* and *STAT5* was decreased in the HS group.

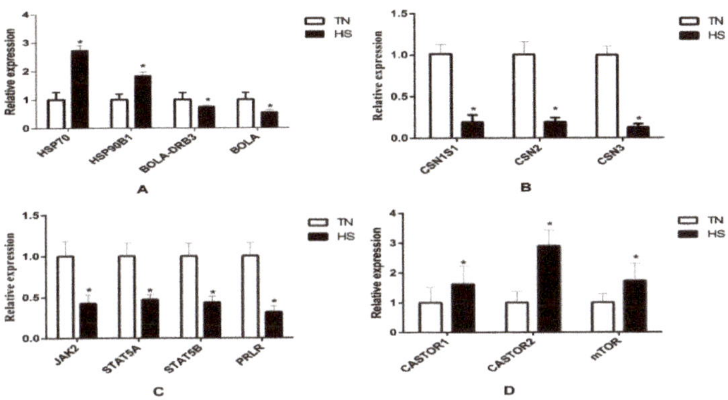

Figure 5. The relative expression of heat shock and immune response genes (**A**), casein genes (**B**), *JAK/STAT5* pathway genes (**C**), and *mTOR* pathway genes between both TN and HS groups (**D**). * $p < 0.05$ vs. TN group.

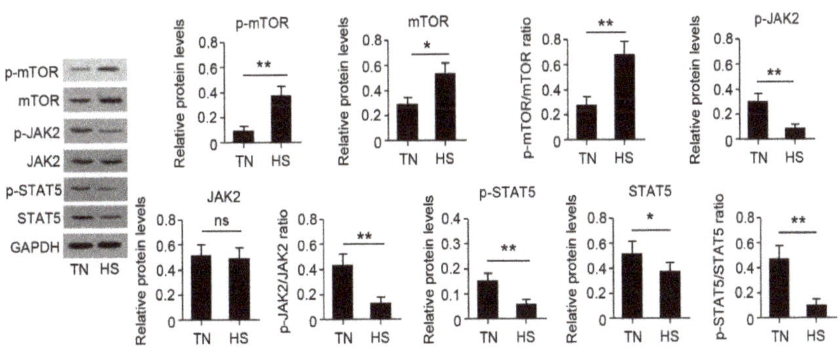

Figure 6. Identification of differentially expressed protein by Western blot assay (WBA). The WBA was carried out to find the protein and phosphorylation levels. * $p < 0.05$ or ** $p < 0.01$ vs. TN group. Ns: no significance.

4. Discussion

In intensive management systems, THI has been widely used as vital indicator to assess HS in dairy production, and it is generally recognized that when THI is above 68, dairy cows experienced HS [21]. In the current study, it was observed that the averaged THI in the HS group was ranged from 79.0 to 83.2 (Figure 1), which was much higher than the minimum THI (THI = 68) needed to induce HS in cows. Therefore, it could be assumed that cows in the HS group suffered from heat stress. It has been reported that milk yield drops sharply when THI is above 69 [22]. In the present study, the reduction of 16% milk yield could be related to higher THI in HS group cows. Respiratory rate and RT are the most imperative physiological indexes for evaluating the occurrence of HS in lactating cows [23,24]. It has been reported that heat stress rapidly enhances the lactating dairy cows' RT and respiration rate [25]. Moreover, it has been reported that higher respiration rates and RTs are signs of HS for lactating dairy cows [26]. In the current study, higher respiration rates and RTs in the HS group further confirmed that the HS group cows experienced heat stress in the current experiment. Several study reports have suggested that heat stress persuades cell apoptosis [13], disturbs the normal biological activity of BMECs [12,13], and provoked intracellular thermotolerance responses of BMECs [10]. Furthermore, HS inhibited the transcription and translation of RNA in BMECs [14,27], especially the gene expression of *CSN1S1* [12,13]. In the current experiment, casein-coding genes were downregulated in the animals of the HS experimental treatment group as compared to the animals of the TN experiment treatment group. Findings of the current experiment are in line with the findings of researchers who have reported that at the cellular level, heat stress adversely affects the function and gene expression of casein in BMECs [10–12], especially the Alpha casein S1 (*CSN1S1*) [13]. Downregulation of casein-coding genes in the current study could be another reason for decreased milk production in HS group cows.

HSPs function as molecular chaperones and are the most recognized cellular responses by heat stress. HSP70 is an abundant and sensitive acute protein during heat stress periods [28] and known as a reliable indicator of harsh environmental stress [29,30]. HSP70 also plays an important role in conferring thermo-adaptability and high levels of heat resistance of cells [31]. In the present experiment, an enhanced concentration of HSP70 in plasma was found in cows from the HS group. Furthermore, gene expression of *HSP90B1* and *HSPA1A* (*HSP70*) in mammary gland tissue were also upregulated in cows of the HS group. These results are similar to the results of previous researchers who have reported that the gene expression and synthesis of HSP (such as *HSPA1A*, *HSP90*, *HSP27*) were elevated when the cows suffered heat stress [11,12,32]. Moreover, GO enrichment analysis showed that *HSP90B1* and *HSPA1A* (*HSP70*) genes were involved in chaperone-dependent refolding of proteins (GO: 0051085) and HSP binding activity (GO: 0031072). It has been reported that the endogenous protein stores might have alternative fates during heat stress, and the synthesis of HSP70 during harsh weather may reduce the presence of circulating amino acids for synthesis of milk protein [33]. Therefore, when dairy cows experience heat stress, their cells produce a large number of HSPs, improperly folded protein, and unfold protein, which could be one of the factors contributing to the decrease in synthesis of milk protein. The results of blood glucose in the present study are in line with previous reports that have revealed that plasma glucose concentration was decreased by heat stress [34,35]. Heat stress could attenuate the response of the immune system significantly through effectively suppressing the blood glucose in dairy cows [36]. In vitro studies have suggested that heat stress weakens the functionality of immune cells isolated from thermal-neutral dairy cattle, including the ability to migrate and proliferate, phagocytose, and kill [37,38]. In our study, a GO enrichment analysis result showed that most of the affected functional category were related to immune response, such as antigen processing (GO: 0002476), T-cell-mediated cytotoxicity regulation (GO: 0001916), acute-phase response (GO: 0006953), and immune response (GO: 0006955). These results indicate a possible severe compromise in the immune system after the dairy cows experience heat stress.

The KEGG pathway analysis found that almost all of the top-affected pathways were related to immune response. Many genes are involved in those pathways, including *HSPA1A* (*HSP70*), *HSP90B1*, *BoLA*, *BoLA-DYA*, *LOC512672*, *LOC524810*, and *BoLA-DRB3*. A unique and most vital observation of this study is that the antigen processing and presentation pathway (bta04612) was significantly affected in HS group cows (Supplementary Materials Figure S2). In this pathway, the expression of the major histocompatibility complex (MHC) genes (*BoLA*, *BoLA-DYA*, and *BoLA-DRB3*) was downregulated while the gene expression of *HSPA1A* and *HSP90B1* was elevated. Previous studies have also reported that in hot environmental conditions, *HSP90* mRNA expression was significantly higher in Sahiwal, Tharparkar, and Murrah buffalo [39], Frieswal cattle [40] and goats [41]. It has been reported that HSP90B1 is the most available glycoprotein in the cell endoplasmic reticulum [42], and it is involved in the transport and processing of excreted proteins [42,43]. *HSP90B1* has been associated as an essential immune chaperone to regulate both adaptive and innate immunity [44]. It has also been reported that *HSP90B1* is necessary for the start of the innate immune response in animals [45]. Thus, it could be speculated that the enhanced expression of *HSP90B1* in heat stressed cows of the current study is related to the innate immune response.

The MHC genes like BoLA have received attention because of their relationship with the induction and regulation of immune responses of dairy cows [46]. Their function is to present foreign antigens (such as *Staphylococcus aureus*), after intracellular processing, to T-cells for a successful immune response [47]. It was reported that class II positive leukocytes specifically respond to local infusion of *Streptococcus uberis* in the bovine mammary gland [48–50], which could be related to clinical mastitis of dairy cows [51]. In the current experiment, we found that the expression of *BoLA* and *BoLA-DRB3* was downregulated in the pathway of antigen processing and presentation and the pathway of *Staphylococcus aureus* infection (bta05150), suggesting that the infection rate of cows was elevated in HS group due to exposure of heat stress. These findings are similar to the reports of earlier researchers who reported that dairy cows had a higher infection rate of *Staphylococcus aureus* and *Corynebacterium pseudotuberculosis*, a higher rate of clinical mastitis, and higher milk SCC when exposed to hot environments [51,52].

Both the acquired and innate immune responses can identify parts of pathogens called pathogen-associated molecular patterns, such as peptidoglycan bacterial DNA and LPS [53]. We observed that the cows in the HS experimental treatment had an expressively high level of LPS compared to the cows in the TN experimental treatment. The higher blood level of LPS and LPS-binding protein in the HS group animals suggest that heat stress activates a systemic inflammatory response [54]. It was reported that infusion of LPS intravenously in cows activated the immune system and resulted in the consumption of more glucose within 12 h after infusion, and it was concluded that immune activation redirects available nutrients from production to the immune system [55]. Therefore, the decrease of plasma glucose concentration and the increase of LPS concentration, along with the upregulation of *HSP90B1* gene expression, and the downregulation of *BoLA* and *BoLA-DDR3* gene expression in HS cow mammary gland tissues of the current experiment, means that exposure to heat activates the inflammatory responses and immune system, which redirect the nutrients from production to the dairy cows' immune system.

It has been reported that prolactin (PRL) is correlated with hormones such as hydrocortisone and insulin during the lactation periods of cows [56] and known to turn on milk protein gene expression [57] via STAT5 phosphorylation by the PRLR and JAK2 [56,58]. The importance of STAT5 and PRLR in lactation has been shown in knockout mice in a previous study; it was reported that the removal of one of these proteins results in diminished mammary gland development and lactation [59]. It has also been reported that suppressors of cytokine signaling (SOCS) proteins and cytokine-inducible SH2-containing proteins (CIS) compose of a family of intracellular proteins [60] that are stimulated by PRL, act through feedback to inhibit cytokine signaling [61], and regulate the responses of immune cells to cytokines [60]. In particular, SOCS-1 and SOCS-3 have been shown to bind to cytokine receptors or to receptor-associated Janus-associated kinases to inhibit the activation of signal transducers and activators of transcription members, and ultimately interferon signaling [62]. Interestingly, in our study, the Jak/STAT5 pathway members, like JAK2, PRLR, STAT5A, and STATB, were also downregulated in the HS group, implying that exposure to heat could suppress the JAK2/STAT5 pathway to reduce milk production and suppress immune response. Therefore, it could be assumed that heat exposure to cows affects the immune function through changes in the PRL signaling pathway.

Previous studies on cows and mice have reported that amino acids, like leucine, isoleucine, methionine, and threonine, act as signaling molecules and positively regulate milk synthesis and lactation [63]. The milk synthesis and lactation regulation are generally believed to be conducted through amino acids sensors like vacuolar H^+-ATPase, SLC38A9, Sestrin2, CASTOR1 homodimer, and CASTOR1-CASTOR2, which serve as upstream activators of mTORC1 [64]. After being activated, the mTOR pathway regulates its targets to promote milk protein synthesis [65]. It has been reported that bovine mammary cells, which were incubated in high temperature (42 °C), reduced protein translation by reducing the mTOR downstream pathway activity [66,67]. In contrast to the findings of Kaufman et al. [66] and Salama et al. [67], we observed that heat exposure to cows enhanced the gene expression of *CASTOR1*, *CASTOR2* and phosphorylation of mTOR. However, findings of the present experiment are similar with the results of previous researchers, who stated that mTOR is negatively related to milk production [68]. It has been further reported that mTOR is one of the most classic autoimmune response regulators [69,70], and we also found that several autoimmune-related pathways were found to be significantly changed by HS exposure, which suggests a tight connection among the increased mTOR, reduced milk production, and immune response.

5. Conclusions

The current study employed the RNA-seq technology to analyze the impact of HS on the whole transcript sequencing profile in the mammary glands of lactating Holstein dairy cows. The results of the current study reveal that heat stress downregulates the expression of casein genes, which result in a decrease in milk production. Moreover, according to functional annotation analysis, heat stress upregulates the gene expression of *HSPA1A* and *HSP90B1*, while iy downregulates the expression of immune response-related genes (*BoLA* and *BoLA-DRB3*) that eventually affected the immune function of the dairy cows and resulted in a reduction in milk yields. Furthermore, the results also reveal that under heat stress, the synthesis of heat shock and unfolded proteins was increased, which could reduce the accessibility of circulating amino acids for milk protein synthesis. Findings of current experiment may help to explore the impact of heat stress on immune function, milk production, and milk protein synthesis in cows.

Supplementary Materials: The following are available online at http://www.mdpi.com/2076-2615/10/6/1015/s1. Figure S1: The top-20 pathway of the up and down pathway of KEGG Enrichment between HS and TN dairy cows, Figure S2: The pathway of antigen processing and presentation—*Bos taurus* (cow). The genes in the red box are significantly regulated by heat stress, Table S1: The primer information in the PCR assay, Table S2: The results of sequencing data quality preprocessing.

Author Contributions: S.Y. and B.X. designed the studies and prepared the manuscript, with comments from all authors. S.Y. performed all the experiments and analyzed the data. Z.W., L.W. and Q.P. revised the manuscript. All authors have read and agreed to the published version of the manuscript.

Funding: This research was funded by the Science and Technology Supporting Program of the National Science and Technology Ministry, China, grant number 2012BAD12B02.

Acknowledgments: The authors gratefully acknowledge grants from the experimental farm at Qingbaijiang Dairy Farm of New Hope Dairy Co., Ltd. (Chengdu, Sichuan province, China) for the provision of experimental animals and their assistance in animal handling throughout the experimental period.

Conflicts of Interest: The authors declare no conflict of interest.

References

1. Bernabucci, U.; Biffani, S.; Buggiotti, L.; Vitali, A.; Lacetera, N.; Nardone, A. The effects of heat stress in Italian Holstein dairy cattle. *J. Dairy Sci.* **2014**, *97*, 471–486. [CrossRef] [PubMed]
2. St-Pierre, N.; Cobanov, B.; Schnitkey, G. Economic losses from heat stress by US livestock industries. *J. Dairy Sci.* **2003**, *86*, E52–E77. [CrossRef]
3. Nardone, A.; Ronchi, B.; Lacetera, N.; Ranieri, M.; Bernabucci, U. Effects of climate changes on animal production and sustainability of livestock systems. *Livest. Sci.* **2010**, *130*, 57–69. [CrossRef]

4. Rhoads, M.L.; Kim, J.W.; Collier, R.J.; Crooker, B.A.; Boisclair, Y.R.; Baumgard, L.H.; Rhoads, R.P. Effects of heat stress and nutrition on lactating Holstein cows: II. Aspects of hepatic growth hormone responsiveness. *J. Dairy Sci.* **2010**, *93*, 170–179. [CrossRef] [PubMed]
5. Liu, J.; Li, L.; Chen, X.; Lu, Y.; Wang, D. Effects of heat stress on body temperature, milkproduction, and reproduction in dairy cows: A novel idea for monitoring and evaluation of heat stress—A review. *Asian Australas. J. Anim. Sci.* **2019**, *32*, 1332–1339. [CrossRef] [PubMed]
6. Wheeock, J. Effects of heat stress on energetic metabolism in lactating Holstein cows. *J. Dairy Sci.* **2010**, *93*, 644–655. [CrossRef] [PubMed]
7. Cowley, F.C.; Barber, D.G.; Houlihan, A.V.; Poppi, D.P. Immediate and residual effects of heat stress and restricted intake on milk protein and casein composition and energy metabolism. *J. Dairy Sci.* **2015**, *98*, 2356–2368. [CrossRef] [PubMed]
8. Fuquay, J.W. Heat stress as it affects animal production. *J. Anim. Sci.* **1981**, *52*, 164. [CrossRef] [PubMed]
9. West, J. Effects of heat-stress on production in dairy cattle. *J. Dairy Sci.* **2003**, *86*, 2131–2144. [CrossRef]
10. Collier, R.J.; Collier, J.L.; Rhoads, R.P.; Baumgard, L.H. Invited review: Genes involved in the bovine heat stress response. *J. Dairy Sci.* **2008**, *91*, 445–454. [CrossRef] [PubMed]
11. Collier, R.J.; Stiening, C.M.; Pollard, B.C.; Vanbaale, M.J.; Coussens, P.M. Use of gene expression microarrays for evaluating environmental stress tolerance at the cellular level in cattle. *J. Anim. Sci.* **2006**, *84* (Suppl. S13), E1–E13. [CrossRef] [PubMed]
12. Han, Z.Y.; Mu, T.; Yang, Z. Methionine protects against hyperthermia-induced cell injury in cultured bovine mammary epithelial cells. *Cell Stress Chaperon.* **2015**, *20*, 109–120. [CrossRef] [PubMed]
13. Hu, H.; Wang, J.; Gao, H.; Li, S.; Zhang, Y.; Zheng, N. Heat-induced apoptosis and gene expression in bovine mammary epithelial cells. *Anim. Prod. Sci.* **2016**, *56*, 918. [CrossRef]
14. Li, L.; Sun, Y.; Wu, J.; Li, X.; Luo, M.; Wang, G. The global effect of heat on gene expression in cultured bovine mammary epithelial cells. *Cell Stress Chaperon.* **2015**, *20*, 381–389. [CrossRef] [PubMed]
15. NRC. *Nutrient Requirements of Dairy Cattle*, 7th ed.; National Academy Press: Washington, DC, USA, 2001.
16. Farr, V.C.; Stelwagen, K.; Cate, L.R.; Molenaar, A.J.; Mcfadden, T.B.; Davis, S.R. An improved method for the routine biopsy of bovine mammary tissue. *J. Dairy Sci.* **1996**, *79*, 543–549. [CrossRef]
17. Simon, A.; Theodor, P.P.; Wolfgang, H. HTSeq-a Python framework to work with high-throughput sequencing data. *Bioinformatics* **2014**, *31*, 166–169.
18. Kim, D.; Langmead, B.; Salzberg, S.L. Hisat: A fast spliced aligner with low memory requirements. *Nat. Methods* **2015**, *12*, 357–360. [CrossRef] [PubMed]
19. Trapnell, C.; Roberts, A.; Goff, L.; Pertea, G.; Kim, D.; Kelley, D.R.; Pimentel, H.; Salzberg, S.L.; Rinn, J.L.; Pachter, L. Differential gene and transcript expression analysis of RNA-seq experiments with TopHat and Cufflinks. *Nat. Proto.* **2012**, *7*, 562–578. [CrossRef] [PubMed]
20. Wang, L.; Feng, Z.; Wang, X.; Wang, X.; Zhang, X. DEGseq: An R package for identifying differentially expressed genes from RNA-seq data. *Bioinformatics* **2010**, *26*, 136–138. [CrossRef] [PubMed]
21. Armstrong, D. Heat stress interaction with shade and cooling. *J. Dairy Sci.* **2002**, *77*, 2044–2050. [CrossRef]
22. Bouraoui, R.; Lahmar, M.; Majdoub, A.; Belyea, R. The relationship of temperature-humidity index with milk production of dairy cows in a Mediterranean climate. *Anim. Res.* **2002**, *51*, 479–491. [CrossRef]
23. Dikmen, S.; Alava, E.; Pontes, E.; Fear, J.M.; Dikmen, B.Y.; Olson, T.A.; Hansen, P.J. Differences in thermoregulatory ability between slick-haired and wild-type lactating holstein cows in response to acute heat stress. *J. Dairy Sci.* **2008**, *91*, 3395–3402. [CrossRef] [PubMed]
24. Dikmen, S.; Hansen, P. Is the temperature-humidity index the best indicator of heat stress in lactating dairy cows in a subtropical environment? *J. Dairy Sci.* **2009**, *92*, 109–116. [CrossRef] [PubMed]
25. Zimbelman, R.; Rhoads, R.; Rhoads, M.; Duff, G.; Baumgard, L.; Collier, R. A re-evaluation of the impact of temperature humidity index (THI) and black globe humidity index (BGHI) on milk production in high producing dairy cows. In Proceedings of the 24h Southwest Nutrition and management Conference, Savoy, IL, USA, 26–27 February 2009; pp. 158–169.
26. Kadzere, C.; Murphy, M.; Silanikove, N.; Maltz, E. Heat stress in lactating dairy cows: A review. *Livest. Prod. Sci.* **2002**, *77*, 59–91. [CrossRef]
27. Sonna, L.A.; Fujita, J.; Gaffin, S.L.; Lilly, C.M. Invited review: Effects of heat and cold stress on mammalian gene expression. *J. Appl. Physiol.* **2002**, *92*, 1725–1742. [CrossRef] [PubMed]

28. Trinklein, N.D.; Murray, J.I.; Hartman, S.J.; Botstein, D.; Myers, R.M. The role of heat shock transcription factor 1 in the genome-wide regulation of the mammalian heat shock response. *Mol. Biol. Cell* **2004**, *15*, 1254–1261. [CrossRef] [PubMed]
29. Gaughan, J.; Bonner, S.; Loxton, I.; Mader, T. Effects of chronic heat stress on plasma concentration of secreted heat shock protein 70 in growing feedlot cattle. *J. Dairy Sci.* **2013**, *91*, 120–129.
30. Romero, R.D.; Pardo, A.M.; Montaldo, H.H.; Rodríguez, A.D.; Cerón, J.H. Differences in body temperature, cell viability, and HSP-70 concentrations between Pelibuey and Suffolk sheep under heat stress. *Trop. Anim. Heal. Prod.* **2013**, *45*, 1691–1696. [CrossRef] [PubMed]
31. Patir, H.; Upadhyay, R. Purification, characterization and expression kinetics of heat shock protein 70 from Bubalus bubalis. *Res. Veter. Sci.* **2010**, *88*, 258–262. [CrossRef] [PubMed]
32. Weng, X.; Monteiro, A.P.A.; Guo, J.; Li, C.; Orellana, R.M.; Marins, T.N.; Bernard, K.; Tomlinson, D.J.; Defrain, J.M.; Wohgemuth, S.; et al. Effects of heat stress and dietary zinc source on performance and mammary epithelial integrity of lactating dairy cows. *J. Dairy Sci.* **2018**, *101*, 2617–2630. [CrossRef] [PubMed]
33. Gao, S.; Guo, J.; Quan, S.; Nan, X.; Fernandez, M.S.; Baumgard, L.; Bu, D. The effects of heat stress on protein metabolism in lactating Holstein cows. *J. Dairy Sci.* **2017**, *100*, 5040–5049. [CrossRef] [PubMed]
34. Guo, J.; Gao, S.; Quan, S.; Zhang, Y.; Bu, D.; Wang, J. Blood amino acids profile responding to heat stress in dairy cows. *Asian Australas. J. Anim. Sci.* **2018**, *31*, 47–53. [CrossRef] [PubMed]
35. De Rensis, F.; Scaramuzzi, R.J. Heat stress and seasonal effects on reproduction in the dairy cow—A review. *Theriogenology* **2003**, *60*, 1139–1151. [CrossRef]
36. Rhoads, M.; Rhoads, R.; VanBaale, M.; Collier, R.J.; Sanders, S.; Weber, W.J.; Crooker, B.A.; Baumgard, L. Effects of heat stress and plane of nutrition on lactating Holstein cows: I. Production, metabolism, and aspects of circulating somatotropin. *J. Dairy Sci.* **2009**, *92*, 1986–1997. [CrossRef] [PubMed]
37. Elvinger, F.; Hansen, P.J.; Natzke, R.P. Modulation of function of bovine polymorphonuclear leukocytes and lymphocytes by high temperature in vitro and in vivo. *Am. J. Vet. Res.* **1991**, *52*, 1692–1698. [PubMed]
38. Nardone, A. Heat stress elicits different responses in peripheral blood mononuclear cells from brown swiss and holstein cows. *J. Dairy Sci.* **2006**, *89*, 4606–4612.
39. Kumar, A.; Ashraf, S.; Goud, T.S.; Grewal, A.; Singh, S.V.; Yadav, B.R.; Upadhyay, R.C. Expression profiling of major heat shock protein genes during different seasons in cattle (Bos indicus) and buffalo (Bubalus bubalis) under tropical climatic condition. *J. Therm. Biol.* **2015**, *51*, 55–64. [CrossRef] [PubMed]
40. Deb, R.; Sajjanar, B.; Singh, U.; Kumar, S.; Singh, R.; Sengar, G.; Sharma, A. Effect of heat stress on the expression profile of Hsp90 among Sahiwal (Bos indicus) and Frieswal (Bos indicus×Bos taurus) breed of cattle: A comparative study. *Gene* **2014**, *536*, 435–440. [CrossRef] [PubMed]
41. Nagayach, R.; Gupta, U.D.; Prakash, A. Expression profiling of hsp70 gene during different seasons in goats (Capra hircus) under sub-tropical humid climatic conditions. *Small Rumin. Res.* **2017**, *147*, 41–47. [CrossRef]
42. Randow, F.; Seed, B. Endoplasmic reticulum chaperone gp96 is required for innate immunity but not cell viability. *Nat. Cell Biol.* **2001**, *3*, 891–896. [CrossRef]
43. Yang, Y.; Liu, B.; Dai, J.; Srivastava, P.K.; Zammit, D.J.; Lefrançois, L.; Li, Z. Heat shock protein gp96 is a master chaperone for toll-like receptors and is important in the innate function of macrophages. *Immunity* **2007**, *26*, 215–226. [CrossRef] [PubMed]
44. Schild, H.; Rammensee, H.G. gp96—the immune system's swiss army knife. *Nat. Immunol.* **2000**. [CrossRef] [PubMed]
45. Galea-Lauri, J.; Latchman, D.S.; Katz, D.R. The role of the 90-kDa heat shock protein in cell cycle control and differentiation of the monoblastoid cell line U937. *Exp. Cell Res.* **1996**, *226*, 243–254. [CrossRef] [PubMed]
46. Usinger, W.R.; Curie-Cohen, M.; Benforado, K.; Pringnitz, D.; Stone, W.H. The bovine major histocompatibility complex (BoLA): Close linkage of the genes controlling serologically defined antigens and mixed lymphocyte reactivity. *Immunogenetics* **1981**, *14*, 423–428. [CrossRef] [PubMed]
47. Soloski, M.J.; Szperka, M.E.; Davies, A.; Wooden, S.L. Host immune response to intracellular bacteria: A role for MHC-linked class-Ib antigen-presenting molecules. *Exp. Bio. Med.* **2000**, *224*, 231–239. [CrossRef] [PubMed]
48. Grant, R.G. The Role of the Bovine Mammary Gland in Immunological Responses to StreptococcusUberis. Ph.D. Thesis, University of Bristol, Bristol, UK, 1995.
49. Fitzpatrick, J.; Cripps, P.; Hill, A.; Bland, P.; Stokes, C. MHC class II expression in the bovine mammary gland. *Veter. Immunol. Immunopathol.* **1992**, *32*, 13–23. [CrossRef]

50. Lundén, A.; Sigurdardóttir, S.; Edfors-Lilja, I.; Danell, B.; Rendel, J.; Andersson, L. The effect of bovine MHC class. II polymorphism on bull breeding values for clinical mastitis and somatic cell counts in milk. In Proceedings of the 4th World Congress on Genetics applied to Livestock Production, Edinburgh, UK, 23–27 July 1990.
51. Cook, N.B.; Bennett, T.B.; Emery, K.M. Monitoring nonlactating cow intramammary infection dynamics using DHI somatic cell count data. *J. Dairy Sci.* **2002**, *85*, 1119–1126. [CrossRef]
52. Bertocchi, L.; Vitali, A.; Lacetera, N.; Nardone, A.; Varisco, G.; Bernabucci, U. Seasonal variations in the composition of Holstein cow's milk and temperature-humidity index relationship. *Animal* **2014**, *8*, 667–674. [CrossRef] [PubMed]
53. Hornef, M.W.; Frisan, T.; Vandewalle, A.; Normark, S.; Richterdahlfors, A. Toll-like receptor 4 resides in the golgi apparatus and colocalizes with internalized lipopolysaccharide in intestinal epithelial cells. *J. Exp. Med.* **2002**, *195*, 559–570. [CrossRef] [PubMed]
54. Bannerman, D.D.; Paape, M.J.; Hare, W.R.; Sohn, E.J. Increased levels of LPS-binding protein in bovine blood and milk following bacterial lipopolysaccharide challenge. *J. Dairy Sci.* **2003**, *86*, 3128–3137. [CrossRef]
55. Kvidera, S.K.; Horst, E.A.; Abuajamieh, M.; Mayorga, E.J.; Fernandez, M.V.S.; Baumgard, L.H. Glucose requirements of an activated immune system in lactating holstein cows. *J. Dairy Sci.* **2017**, *100*, 2360–2374. [CrossRef] [PubMed]
56. Yulee, L.; Luo, G.; Book, M.L.; Morris, S.M. Lactogenic hormone signal transduction. *Biol. Reprod.* **1998**, *58*, 295–301. [CrossRef] [PubMed]
57. Naylor, M.J.; Li, N.; Cheung, J.; Lowe, E.T.; Lambert, E.; Marlow, R.; Wang, P.; Schatzmann, F.; Wintermantel, T.; Schüetz, G.; et al. Ablation of β1 integrin in mammary epithelium reveals a key role for integrin in glandular morphogenesis and differentiation. *J. Cell Biol.* **2005**, *171*, 717–728. [CrossRef] [PubMed]
58. Groner, B. Transcription factor regulation in mammary epithelial cells. *Domest. Anim. Endocrinol.* **2002**, *23*, 25–32. [CrossRef]
59. Liu, X.; Robinson, G.W.; Wagner, K.U.; Garrett, L.; Wynshaw-Boris, A.; Hennighausen, L. Stat5a is mandatory for adult mammary gland development and lactogenesis. *Genes Dev.* **1997**, *11*, 179–186. [CrossRef] [PubMed]
60. Yoshimura, A.; Naka, T.; Kubo, M. SOCS proteins, cytokine signalling and immune regulation. *Nat. Rev. Immunol.* **2007**, *7*, 454–465. [CrossRef] [PubMed]
61. Wall, E.H.; Auchtungmontgomery, T.L.; Dahl, G.E.; Mcfadden, T.B. Short communication: Short-day photoperiod during the dry period decreases expression of suppressors of cytokine signaling in mammary gland of dairy cows. *J. Dairy Sci.* **2005**, *88*, 3145–3148. [CrossRef]
62. Sasaki, A.T.; Yasukawa, H.; Suzuki, A.; Kamizono, S.; Syoda, T.; Kinjyo, I.; Sasaki, M.; Johnston, J.A.; Yoshimura, A. Cytokine-inducible SH2 protein-3 (CIS3/SOCS3) inhibits Janus tyrosine kinase by binding through the N-terminal kinase inhibitory region as well as SH2 domain. *Genes Cells* **1999**, *4*, 339–351. [CrossRef] [PubMed]
63. Trottier, N.L. Linking our understanding of mammary gland metabolism to amino acid nutrition. *Amino Acids* **2014**, *46*, 2447–2462.
64. Stransky, L.A.; Forgac, M. Amino acid availability modulates vacuolar H+-ATPase assembly. *J. Biol. Chem.* **2015**, *290*, 27360–27369. [CrossRef] [PubMed]
65. Gao, H.N.; Hu, H.; Zheng, N.; Wang, J.Q. Leucine and histidine independently regulate milk protein synthesis in bovine mammary epithelial cells via mTOR signaling pathway. *J. Zhejiang Univ. Sci. B* **2015**, *16*, 560–572. [CrossRef] [PubMed]
66. Kaufman, J.D.; Kassube, K.R.; Almeida, R.A.; Ríus, A.G. Short communication: High incubation temperature in bovine mammary epithelial cells reduced the activity of the mTOR signaling pathway. *J. Dairy. Sci.* **2018**. [CrossRef] [PubMed]
67. Salama, A.A.K.; Duque, M.; Wang, L.; Shahzad, K.; Olivera, M.; Loor, J. Enhanced supply of methionine or arginine alter MTOR signaling proteins, mRNA, and microRNA abundance in heat-stressed bovine mammary epithelial cells in vitro. *J. Dairy. Sci.* **2019**, *102*, 2469–2480. [CrossRef] [PubMed]
68. Liu, J.; Wang, Y.; Li, D.; Wang, Y.; Li, M.; Chen, C.; Fang, X.; Chen, H.; Zhang, C. Milk protein synthesis is regulated by T1R1/T1R3, a G protein-coupled taste receptor, through the mTOR pathway in the mouse mammary gland. *Mol. Nutr. Food Res.* **2017**, *61*, 1601017. [CrossRef] [PubMed]

69. Yang, H.; Wang, X.; Zhang, Y.; Liu, H.; Liao, J.; Shao, K.; Chu, Y.; Liu, G. Modulation of TSC-mTOR signaling on immune cells in immunity and autoimmunity. *J. Cell Physiol.* **2013**, *229*, 17–26. [CrossRef] [PubMed]
70. Zhang, S.L.; Pruitt, M.; Tran, D.; Du Bois, W.; Zhang, K.; Patel, R.; Hoover, S.; Simpson, R.M.; Simmons, J.; Gary, J.; et al. B cell-specific deficiencies in mTOR limit humoral immune responses. *J. Immunol.* **2013**, *191*, 1692–1703. [CrossRef] [PubMed]

animals

MDPI

Article

The Effect of Energy Metabolism up to the Peak of Lactation on the Main Fractions of Fatty Acids in the Milk of Selected Dairy Cow Breeds

Krzysztof Młynek *, Agata Danielewicz and Ilona Strączek

Faculty of Agrobioengineering and Animal Sciences, University of Natural Sciences and Humanities in Siedlce, 08-110 Siedlce, Poland; agata.danielewicz@uph.edu.pl (A.D.); ilonastraczek1984@gmail.com (I.S.)
* Correspondence: krzysztof.mlynek@uph.edu.pl

Simple Summary: The metabolism of cows up to the peak of lactation significantly influences production parameters and the quality of the production cycle. The aim of this study was to analyze the energy metabolism of selected breed groups of cows and its variability in different stages of early lactation. The analysis was performed using data on the following parameters: body condition score (BCS), fatty acid (FA) fractions, basic milk constituents, and serum parameters (BHBA, glucose, and leptin). Holstein-Friesian (HF) cows and HF crossbreds with Black-and-White Lowland (BW) bulls generally had higher daily yields and reached the peak of lactation earlier. These cows, however, suffered the greatest loss in BCS, which led to higher levels of non-esterified fatty acid (NEFA) in the blood than in the other groups. Cows entering the peak of lactation with less intensive production were shown to have lower levels of leptin and higher glucose concentrations in the blood. The smaller loss of BCS in these cows did not lead to rapid lipolysis, and therefore the release of large amounts of non-esterified fatty acids and β-hydroxybutyrate into the blood was limited. The milk of breed groups that had somewhat lower yield and reached the peak of lactation about 11 days later (Simmental and Black-and-White Lowland) had a more beneficial fatty acid profile and casein content. The results indicate that one of the important factors influencing the intensity of lipolysis is leptin, which regulates appetite. The Simmental and Black-and-White Lowland groups had lower concentrations of leptin. This suggests the need for further research on how appetite and the prolongation of feed intake is linked to longer availability of nutrients, including glucose.

Abstract: During early lactation in dairy cows, metabolic processes are adopted to provide energy and nutrients for the synthesis of milk compounds. High milk production potential includes sudden changes in energy metabolism (negative energy balance (NEB)) that can induce uncontrolled lipomobilization and high blood free fatty acid (FFA) levels. Destabilization of cows' energy may interfere with endocrine homeostasis, such as the secretion of leptin, a co-regulator of the appetite center. Therefore, it is important to analyze the physiological aspects of the maintenance of energy homeostasis in various dairy breeds. Usually it is crucial for the health of cows, influences the production cycle and lifetime yield, and determines the profitability of production and milk quality. The aim of this study was to analyze the energy metabolism of selected breed groups of cows and its variability in different stages of early lactation. The analysis was performed using data on the following parameters: body condition score (BCS), fatty acid (FA) fractions, basic milk constituents, and serum parameters (BHBA, glucose, and leptin). These results were analyzed in relation to parameters of energy metabolism during the stage up to the peak of lactation. An earlier peak of lactation was shown to be conducive to an increase in the content of non-esterified fatty acids (NEFAs) and of casein and κ-casein. During the study period, parameters characterizing the maintenance of energy homeostasis were usually lower in the Simmental and Black-and-White Lowland cows. Compared to the group with the highest production, their yield was from 2.8 to 4.7 kg lower, but the milk had a more beneficial fatty acid profile and nutrient content, determining suitability for cheese making. At the same time, they had lower levels of NEFAs and β-hydroxybutyrate in the blood, which indicates less spontaneous lipolysis of fat reserves. Concentrations of the appetite regulator leptin in the blood were correlated negatively ($p \leq 0.05$) with the glucose concentration (–0.259) and

Citation: Młynek, K.; Danielewicz, A.; Strączek, I. The Effect of Energy Metabolism up to the Peak of Lactation on the Main Fractions of Fatty Acids in the Milk of Selected Dairy Cow Breeds. *Animals* **2021**, *11*, 112. https://doi.org/10.3390/ani11010112

Received: 25 November 2020
Accepted: 5 January 2021
Published: 7 January 2021

positively with NEFA (0.416). The level of NEFAs was at the same time positively correlated with the content of saturated fatty acids in the milk (0.282–0.652; $p \leq 0.05$). These results contribute to our knowledge of the effect of production intensity on the maintenance of homeostasis up to the peak of lactation in dairy breeds with differing production potential. In practice, this may increase the possibilities of improving milk quality and the profitability of production.

Keywords: cattle; peak of lactation; lipolysis; fatty acids; casein

1. Introduction

Research on improvement of the fatty acid (FA) profile of milk indicates that apart from the yield of cows, the FA profile is largely determined by the availability of substrates in the diet, fermentation processes in the rumen [1], changes in FAs in the mammary gland [2], and the seasonality of production [3,4]. A significant proportion of substrate transformations, about 85%, involve ester hydrolysis and biohydrogenation in the rumen [5]. The main product of these transformations is energy derived from acetate and β-hydroxybutyrate (BHBA) [1,6]. Up to the peak of lactation, the level of energy is usually insufficient. This induces spontaneous and excessive lipolysis of fat tissue [7,8] and is conducive to the production of non-esterified fatty acids (NEFAs) [9]. Synergy of endogenous factors (negative energy balance, glucose level, and leptin secretion) usually antagonizes feed intake and intensifies the excessive production of NEFAs and BHBA in the liver [10–12]. This mechanism involves the appearance of a negative energy balance (NEB) and is linked to the physiological demands of maintaining body functions and increased milk production. According to Drackley et al. [13], these factors can increase the demand for nutrients threefold. The energy deficit increases the mobilization of reserves of energy deposited in the adipocytes. This induces lipolysis of adipose tissue [8] and causes a sharp decline in BCS, which is conducive to the release of oleic acid from the adipocytes [9,11]. Research by Ducháček et al. [14] and Vernon [15] shows that this process need not adversely affect the nutritional value of milk, but it does affect the overall health of cows [14]. Excessive lipolysis of adipose tissue is conducive to the production of non-esterified fatty acids (NEFAs) [9], probably because the FA profile of milk is determined at many levels. Apart from bacterial metabolism in the rumen, FAs are also synthesized de novo in the udder. Lipogenesis here is mediated by stearoyl–CoA desaturase, which mainly affects FAs containing 14 to 18 carbon atoms [16]. Entering the mammary gland together with the blood, these acids affect the quality of milk fat [2]. After the peak of lactation, the energy generated promotes the synthesis of short- and medium-chain fatty acids [2,9]. These acids can undergo further modification in the udder. This process mainly requires β-hydroxybutyrate and acetate derived from fermentation in the rumen [17]. Previous research has focused primarily on environmental factors and the possibility of modifying the FA profile through diet and feed quality. There is little information, however, linking the energy metabolism and productive potential of dairy breeds with the nutritional properties of the milk they produce. In production practice, up to the peak of lactation, this information can be valuable, providing the possibility to improve dairy herds and increase the suitability of milk for processing. The aim of this study was thus to analyze the energy metabolism of selected breed groups of cows and its variability in different stages of early lactation. The analysis was performed using data on the following parameters: BCS, FA fractions, basic milk constituents, and serum parameters (BHBA, glucose, and leptin).

2. Materials and Methods

The study was carried out on four breed groups of dairy cows: Holstein-Friesian (HF) cows with more than 75% genes of the breed (20 cows); Simmental (SIM) (20 cows); Black-and-White Lowland (BW), a conserved breed (20 cows); and HF×BW crossbred cows (50:50%—20 cows). The average milk production was characterized by the mean (least-squares method—LSM) and standard error (SE) and age of the cows (LSM ± SE) in each group were as follows: for HF, 8450 ± 42 L and age 2.7 ± 0.21; for SIM, 7342 ± 42 L and age 2.8 ± 0.31; for BW, 6253 ± 35 L and age 3.1 ± 0.52; and for HF×BW, 7235 ± 38 L and age 2.6 ± 0.81. The cows were from four farms and constituted separate experimental groups. The animals were evaluated on average from the 6th day postpartum. The experiment was completed after 110 days of lactation on average.

2.1. Feeding and Housing of Cows

The HF and HF×BW cows were kept in pens with separated resting boxes, whereas the cows in the other groups (SIM and BW) were tethered. The cows had direct access to feed and water (open drinkers), and their living conditions met the requirements of good production practice.

Nutrient requirements were established based on information about feed quality [18], the approximate body weight of the breed groups, and forecast milk production for the analyzed stages of lactation (SLs): days 5 to 30 (SL I), days 31 to 60 (SL II) and days 61 to 99 (SL III). The nutritional value of the feed was determined by chemical analyses performed a few days prior to each stage of lactation. Percentages of particle sizes in the feed were determined at the same time. The body condition score (BCS) of the cows was also assessed before each stage of lactation, using a five-point scale. The average from three independent scores was calculated. The BCS was used as an indicator for NEB.

Nutrient requirements and balancing of the diet were determined according to feeding standards for ruminants [19] and INRAtion software. HF and HF×BW cows were fed in a total mixed ration (TMR) system, with the ingredients mixed in a feed wagon. The SIM and BW groups were fed in a partially mixed ration (PMR) system, in which the feed components were placed directly on the feed platform and mixed. In all groups the cows were fed three times a day (on average every 8 h), using feed pushing. The composition of the diet was similar in all herds (Table 1).

Table 1. Feed components and balancing of the diet of the genetic groups. Sample collection and analytical methods. HF—Holstein-Friesian, BW—Black-and-White Lowland, SIM—Simmental.

Forage/Parameters	Breed Group			
	HF×BW	HF	SIM	BW
Number of cows in the group (n)	(20)	(20)	(20)	(20)
Silage maize (kg)	8.73	3.98	5.04	4.86
Haylage (kg)	3.66	5.06	6.35	5.98
Ground cereals [1] (kg)	1.31	0.44	0.87	1.31
Supplementary feed Krowimix 18 DE (kg)	1.34	0.45	0.8	1.07
Ground rapeseeds (kg)	1.41	0.88	1.58	1.58
Straw (kg)	0.42	0.48	0.47	0.54
Hays (kg)	0.81	1.64	1.58	1.64
Mineral compound suplement (kg)	0.15	0.14	0.12	0.15
Limestone (kg)	0.10	0.12	0.12	0.11
Total protein (%)	14.2	13.7	13.2	13.8
Dry matter(%)	48.1	47.9	48.5	48.8
Acid detergent fiber (ADF) (%)	22.1	22.5	20.2	19.9
Neutral detergent fiber (NDF) (%)	34.8	34.4	33.5	32.4
Physically effective NDF (peNDF) (%)	14.1	13.9	13.4	13.8
Uneaten feed(%/day)	16.7	17.1	13.4	15.1
Energy (MJ NEL):				
Request	128.5	132.6	125.8	121.5
Intake	112.6	112.8	115.6	110.9
Balace	−15.9	−19.8	−10.2	−10.6
Dry matter intake (DMI) (kg/day)	15.7	15.1	16.2	15.9

[1] barley to 50%, triticale to 20%, oats to 20%, rye to 10% of the participation; NEL—net energy for lactation

The average amount of uneaten feed in each group was determined based on weighing five times a month. On this basis, the average dry matter intake (DMI) was calculated. The average percentages of particle sizes in the diet (PSPS sieves) were as follows: >19 mm (7%), 8–19 mm (52%), 4–8 mm (19%), and ≤4mm (22%). About two weeks before calving, the cows received a preparatory diet. The animals did not show any disease symptoms, e.g., perinatal diseases, ketosis, or acidosis, indicated by milk urea concentration or blood BHBA. To rule out the effect of differences in the diet and housing conditions (farms), a cluster analysis was performed, taking into account the actual energy intake and DMI, considered to be the main factors inducing a negative energy balance (NEB). The analysis showed a high similarity between the farms for these parameters: 0.682 to 0.921 in the case of housing and 0.589 to 0.729 for the diets used on the farms.

2.2. Sample Collection and Analyses

Milk samples of about 250 mL were collected using a calibrated milk meter (DeLaval) that simultaneously measured the amount of milk. Milk was collected on average on the 7th day after the start of each SL and 7 days before the end of each SL. In total, 480 milk samples were collected (2 samplings × 3 SLs × 20 cows × 4 breeds). Milk from morning and evening milking was mixed into one sample. The samples were stored in refrigerated conditions (4 °C ± 0.5 °C). The content of protein, fat, lactose, dry matter, and urea in the milk was determined using the Bentley Combi 150 (Bentley Instruments, Chaska, MN, USA). The analysis was performed in a laboratory accredited by the Polish Accreditation Centre.

The content of caseins was determined in skimmed milk (after centrifuging). Total casein was determined using the Kjeldahl method [20] after separating and purifying the precipitate (casein) in buffer with pH 4.6. The content of κ-casein was determined by HPLC (Varian Inc., Palo Alto, CA, USA). Preparation of the matrix additionally involved dissolving the purified casein precipitate in buffer with pH 8.0 and filtering it (nylon syringe filter; 0.25 μm/0.45 μm; Alchem, Poland). Casein proteins were separated on an XB-C18 column (250 × 4.6 mm; Aeris Widepore, Phenomenex, US: 708385-2 S.N.) using carrier phases A—TFA/Acetonitryl (0.1N) and B—TFA/H_2O (0.1N).

The FA (fatty acid) profile of the milk fat was determined using GLC;Agilent 6890N (Agilent Technologies., Wilmington, DE). The following FA groups were distinguished: short-chain fatty acids (SCFAs), long-chain fatty acids (LCFAs), saturated fatty acids (SFAs), monounsaturated fatty acids (MUFAs), polyunsaturated fatty acids (PUFAs), and unsaturated fatty acids (UFAs). The share of non-esterified FAs (NEFAs) was based on the sum of C16:0, C18:0, and C18:1 cis-9. Fat was extracted using the Röse-Gottlieb method [21]. Transmethylation of FAs to methyl esters (FAME) was at 70 °C ± 0.5 °C (Thermo heat block). GLC modules: autosampler, split/splitless injector (split 1:5), flame ionization detector (FID). Separation was carried out on a 100-m, 0.250-mm column (HP-88; SN:UST458414H, Agilent Technologies Inc., US). Temperature program: injector and FID −250 °C; furnace −95 °C (5 min), 120 °C (15 °C/min for 15 min), 210 °C (25 °C/min for 30 min), 250 (20° C/min for 5 min). Carrier gas flow (He): 5.0 mL/min. The identification of FAs and determination of their percentages were based on retention times (reference Supelco 37.No:47885-U; Sigma Aldrich) in Agilent Tech GC Chemstation A09.03 software (Agilent Technologies., Wilmington, DE). The content of oleic (C18:1) and palmitic (C16:0) acid were the basic markers of a negative energy balance (NEB).

Blood for analysis was drawn before morning feeding from the jugular vein. Due to the potential effect of stress on milk yield, blood was taken 24 h after milk was collected for analysis. Blood was collected twice in each stage of lactation, for a total of 480 blood samples (2 × 3 × 80). Test tubes with sodium fluoride and sodium heparin (Medlab-Products Ltd.) were refrigerated. Blood for glucose determination was placed in ice. Glucose content in the blood was measured using original Randox kits (Randox Laboratories Ltd., Crumlin, UK) and a UV-Vis spectrophotometer (Varian Inc., Palo Alto, CA, USA). Samples for determination of BHBA (β-hydroxybutyrate) were centrifuged at 1500× *g* at 4 °C for

20 min. The supernatant was collected and stored at −75 °C until BHBA analysis using original Randox kits (Randox Laboratories Ltd.) and a UV-Vis spectrophotometer (Varian Inc., Palo Alto, CA, USA). The plasma samples were analyzed for leptin levels using a bovine-specific ELISA kit (EIAab, Wuhan, China). All samples were measured in triplicate.

2.3. *Statistical Analysis*

Statistical analysis of the results was performed in Statistica 13.0 software. Analysis of variance was carried out in a general linear model (GLM) with repeated measures. The effect of diet and housing was verified by cluster analysis, using the k-means method. The model included account the actual energy intake and DMI of the groups. The results are presented as means (least-squares method - LSM) and standard error of the mean(SEM). The significance of differences between means was estimated by Duncan's test at $p \leq 0.05$. Correlations (r) between selected parameters were estimated using the Pearson correlation model ($p \leq 0.05$).

3. Results

HF and HF×BW cows reached the peak of lactation on average 11 days earlier than SIM and BW cows. At the same time, they achieved higher daily milk production (DMP), on average by 2.62 kg in the case of HF×BW and 4.65 kg for HF ($p \leq 0.05$; Table 2).

Table 2. Average yield at the peak of lactation and parameters of energy metabolism in the breed groups.

Parameters	Breed Group *				
	HF×BW	HF	SIM	BW	SEM
Number of samples (n)	120	120	120	120	
Peak yield (kg)	33.8 ^{a}	35.8 ^{b}	31.5^{c}	30.8 ^{c}	1.022
Day of lactation (day)	66 ^{b}	62 ^{b}	74 ^{a}	76 ^{a}	2.093
BCS 5th day (point)	2.98 ^{a}	2.93 ^{a}	3.04 ^{b}	3.11 ^{b}	0.04
BCS 99th day (point)	2.34 ^{b}	2.22 ^{b}	2.53 ^{a}	2.52 ^{a}	0.02
Loss of BCS (%)	21.5 ^{a}	24.2 ^{b}	16.8 ^{c}	16.7 ^{c}	0.43
Urea (mmol/L^{-1})	172 ^{a}	175 ^{a}	164 ^{b}	163 ^{b}	4.591
BHBA (mmol/L^{-1})	1.072 ^{a}	1.069 ^{a}	0.987 ^{b}	0.950 ^{b}	0.005
NEFA (μmol/L^{-1})	242.2 ^{a}	268.9 ^{b}	237.8 ^{c}	231.8 ^{c}	5.962
Leptin (ng/mL^{-1})	2.82 ^{a}	2.80 ^{a}	2.71 ^{b}	2.67 ^{b}	0.028
Glucose (mmol/L)	2.45 ^{b}	2.41 ^{b}	2.63 ^{a}	2.59 ^{a}	0.036

^{abc} $p \leq 0.05$; * number of cows in the groups—20; BHBA—β-hydroxybutyrate; NEFA—non-esterified fatty acids (FAs).

The table presents body condition scores (BCS) and BCS loss (%) as an indicator of the degree of mobilization of lipolysis. HF and HF×BW, with greater production potential, were found to have greater losses of body condition. In these groups the BCS loss was 21.5% and 24.2%. The differences in relation to groups SIM and BW ranged from 4.75% to 7.45% ($p \leq 0.05$). The milk of HF and HF×BW cows contained on average 10.5 mmol/L^{-1} more urea. The blood of these groups contains more 0.103 mmol/L^{-1} BHBA and 32.9 μmol/L^{-1} NEFAs ($p \leq 0.05$). The data (Table 2) show that the HF and HF×BW cows had a higher leptin content in the blood, higher on average by 0.12 ng/mL^{-1} ($p \leq 0.05$). Compared to SIM and BW cows, HF and HF BW blood contained less glucose, lower on average by 0.18 mmol/L ($p \leq 0.05$).

The milk of the breed groups did not differ substantially in terms of total protein content (TP), though it was slightly lower in HF (Table 3). Differences were noted, however, in the content of nutrients determining suitability for cheese making. The milk from the HF and HF×BW groups had less casein than that of SIM and BW, on average by 1.4 p.p. ($p \leq 0.05$), and from 0.014 p.p. to 0.023 p.p. less κ-casein ($p \leq 0.05$). Their milk usually had a lower content of SCFAs, on average by 0.23 p.p., and higher content of LCFAs, on average by 0.78 p.p. ($p \leq 0.05$). The content of palmitic acid (C16:0), which usually makes up the largest share of SFA, can to some extent function as one of the markers of NEB at the peak of lactation. The highest level of this acid was noted in the milk of HF cows (26.96%). In comparison with the other genetic groups, in which the content of C16:0 was similar, the differences ranged from 2.49 to 0.83 p.p. ($p \leq 0.05$). The total content of the SFAs was 0.56 p.p. lower ($p < 0.05$) in the milk of SIM and BW (Table 3).

Table 3. Content of nutrients and main fatty acid fractions in the milk of the breed groups.

Parameters	Breed Group *				
	HF×BW	HF	SIM	BW	SEM
Number of samples (n)	120	120	120	120	
Total protein (TP) (%)	3.41	3.37	3.45	3.42	0.454
Fat (%)	4.26	4.23	4.26	4.22	0.394
Casein (% of TP)	77.6 ^{b}	77.2 ^{b}	78.9 ^{a}	78.7 ^{a}	2.418
κ-casein (%)	0.369 ^{cb}	0.361 ^{c}	0.384 ^{a}	0.375 ^{b}	0.014
Lactose (%)	5.48	5.49	5.43	5.49	0.626
Dry matter (DM) (%)	13.19	12.97	13.14	13.14	0.939
SCFA (%)	9.52 ^{b}	9.61 ^{b}	9.82 ^{a}	9.77 ^{a}	0.051
LCFA (%)	59.71 ^{a}	59.91 ^{a}	58.78 ^{c}	59.28 ^{b}	0.127
Palmitic acid C16: 0 (%)	24.47 ^{b}	26.96 ^{a}	24.13 ^{b}	24.43 ^{b}	0.542
SFA (%)	69.22 ^{a}	69.52 ^{a}	68.56 ^{c}	69.05 ^{b}	0.198
MUFA (%)	28.18 ^{b}	27.85 ^{c}	28.67 ^{a}	28.29 ^{b}	1.258
Oleic acid C18: 1 (%)	2.57 ^{b}	2.86 ^{a}	2.39 ^{c}	2.33 ^{c}	0.154
PUFA (%)	2.59 ^{c}	2.61 ^{c}	2.77 ^{a}	2.66 ^{b}	0.033
PUFA n-3 (%)	0.42 ^{a}	0.41 ^{a}	0.45 ^{b}	0.42	0.008
PUFA n-6 (%)	2.17 ^{c}	2.20 ^{bc}	2.32 ^{a}	2.23 ^{b}	0.054
UFA (%)	30.77 ^{b}	30.47 ^{c}	31.44 ^{a}	30.94 ^{b}	1.133
SFA/UFA	2.25 ^{b}	2.29 ^{a}	2.18 ^{c}	2.23 ^{b}	0.678
MUFA/SFA	0.40	0.39	0.41	0.41	0.043
PUFA n-6/n-3	5.20	5.24	5.17	5.27	0.966
Cholesterol (mg/100mL)	20.53 ^{a}	20.32 ^{a}	19.61 ^{b}	19.89 ^{b}	1.319

^{abc} $p \leq 0.05$; * number of cows in the groups—20; LCFA—long-chain fatty acid; SCFA—Short-chain FA; SFA—saturated FA; MUFA—monounsaturated FA; PUFA—polyunsaturated FA; UFA—unsaturated FA.

In the case of MUFAs and PUFAs, the greatest differences were found between the SIM and HF groups ($p \leq 0.05$). The analysis of MUFA took into account oleic acid (C18:1), which is the main deposit of adipocytes of fat reserves. It is considered a good marker of NEB. Its content was highest in the milk of HF cows (2.86%) and lowest in the milk of BW and SIM, 0.5 p.p. lower ($p \leq 0.05$) on average. This may indicate that a greater reduction in BCS was accompanied by more intensive lipolysis. The percentages of UFA fractions were lower in the case of the milk from the SIM groups. Compared to the milk of HF cattle, their content was 0.97 p.p. higher on average ($p \leq 0.05$). The milk of HF cows had the highest SFA/UFA ratio (Table 3). The difference was greatest in comparison to the milk of SIM cows, at 0.11 ($p \leq 0.05$). The milk of SIM cows also had one of the lowest values for cholesterol content.

A significant factor influencing the metabolism of cows throughout lactation is the dynamics of milk secretion up to the peak of lactation, and the animals' ability to maintain energy homeostasis is an important aspect of lactation persistence. Up to the peak of lactation, an upward trend was observed in the content of urea in the milk (Table 4). The average difference between stages of lactation (SL) was 14.5 mmol/L^{-1} ($p \leq 0.05$). In the initial stage of lactation (up to SL II), the more intensive increase in daily milk production (DMP), amounting to 8.5 kg/day ($p \leq 0.05$), was accompanied by lower production of BHBA (0.362 mmol/L^{-1}; $p \leq 0.05$). However, somewhat less intensive lipolysis of fat reserves was noted during this stage, as indicated by the difference in the content of NEFAs in the blood, amounting to 50.4 μmol/L^{-1} ($p \leq 0.05$). The more intense lipolysis during this stage is confirmed by the body condition results. The differences noted between stages of lactation were greatest up to SL II. They averaged 0.39 BCS points and were 0.18 points greater than in the later stage ($p \leq 0.05$).

Table 4. Milk yield and parameters of energy metabolism in cows up to 100 days of lactation.

Parameters	Stage of Lactation (SL) *			
	I: 5–30	**II: 31–60**	**III: 61–99**	**SEM**
Number of samples (n)	160	160	160	
Daily milk production (DMP) (kg/day)	21.2 ^{b}	29.7 ^{a}	33.4 ^{a}	3.029
Urea (mmol/L^{-1})	152 ^{c}	168 ^{b}	181 ^{a}	5.324
BHBA (mmol/L^{-1})	0.613 ^{c}	0.975 ^{b}	1.503 ^{a}	0.008
NEFA (μmol/L^{-1})	196.9 ^{c}	247.3 ^{a}	288.6 ^{b}	3.634
Leptin (ng/mL^{-1})	2.58 ^{c}	2.80 ^{b}	2.87 ^{a}	0.026
Glucose (mmol/L)	2.61 ^{a}	2.43 ^{b}	2.39 ^{b}	0.029
Body condition score (BCS) (points)	3.02 ^{a}	2.63 ^{b}	2.42 ^{c}	0.124

^{abc} $p \leq 0.05$; * number of cows in the groups—80; BHBA—β-hydroxybutyrate; NEFA—non-esterified FA.

In the period approaching the peak of lactation, in which the rate of increase in DMP was lower (3.7 kg/day; $p \leq 0.05$), the level of BHBA in the blood was higher, on average by 0.528 mmol/L^{-1} ($p \leq 0.05$). Interestingly, the level of NEFAs in the blood during this stage decreased on average by 41.3 μmol/L^{-1} ($p \leq 0.05$). This was most likely due in part to the effect of leptin, an appetite regulator, of which the level in the blood increased more rapidly up to SL II (0.22 ng/mL^{-1}; $p \leq 0.05$). During this time the synergistic effect with the decreasing level of glucose in the blood, on average 0.18 mmol/L ($p \leq 0.05$), may have been conducive to lipolysis of fat reserves and production of NEFAs. After SL II, however, when the increase in NEFAs was less dynamic, the level of glucose as a co-regulator of energy homeostasis was relatively stable (0.05 mmol/L; $p \leq 0.05$).

Analysis of the values of the NEB markers used in the study revealed the greatest loss of BCS points (Table 4) up to SL II (0.39 points, $p \leq 0.05$). This was accompanied by a larger increase in the share of C16:0, on average by 2.02 p.p., and C18:1, on average by 1.27 p.p. In the next 3rd SL, the loss of BCS was smaller, on average by 0.21 points ($p \leq 0.05$), and the increase in the content of C16:0 and C18:1 was 0.73 and 0.83 p.p., respectively ($p \leq 0.05$; Table 5).

Up to the peak of lactation no significant changes were noted in the protein content in the milk, which was 3.42% on average (Table 5). From SL II, however, the percentage of casein was on average 1.4 p.p. lower ($p \leq 0.05$). A similar tendency was observed for fat, but the difference was much smaller (0.06 p.p.; $p \leq 0.05$). It should be noted that in the stage between SL II and III, there was also a reduction in leptin and glucose in the blood (Table 4). In the initial stage after calving (up to SL II; Table 5) there was mainly an increase in the level of LCFAs, by 2.74 p.p. on average ($p \leq 0.05$), during this stage, whereas level of SFAs increased on average by 2.51 p.p. ($p \leq 0.05$). The changes were accompanied by an increase in the amount of NEFAs and BHBA released into the blood (Table 4), which explains the higher content of SFAs in the milk obtained after SL II. Up to the second stage of lactation (SL II), the greatest change was noted in MUFAs, of which the content decreased on average by 2.41 p.p. ($p \leq 0.05$). In the case of PUFAs, although the trend was similar, the average differences amounted to just 0.1 p.p. ($p \leq 0.05$; Table 5). Differences between SLs were also shown for n-3 and n-6 PUFA content ($p \leq 0.05$; Table 5). However, no significant differences were shown for the n-6/n-3 ratio. The share of these acids decreased, although only slightly, with the approach of the peak of lactation. The results indicate that the SLs significantly influenced the reduction in UFA content. Their share was lowest in SLs II and III. Compared to SL I, it was on average 2.51 p.p. lower

($p \leq 0.05$; Table 5). The SLs were not shown to influence cholesterol content. The data in Table 5 also did not indicate any influence of SL on the ratio of SFA to UFA and MUFA.

Table 5. Content of nutrients and main fatty acid fractions in milk up to 100 days of lactation.

Parameters	Stage of Lactation (SL) *			
	I: 5–30	II: 31–60	III: 61–99	SEM
Number of samples (n)	160	160	160	
Total protein (%)	3.42	3.39	3.42	0.538
Fat (%)	4.28 ^{a}	4.23 ^{b}	4.22 ^{b}	0.428
Casein (% of TP)	79.2 ^{a}	77.9 ^{b}	77.7 ^{c}	1.232
κ-casein (%)	0.372	0.370	0.373	0.033
Lactose (%)	5.54	5.53	5.35	0.959
Dry matter (%)	13.26	13.17	12.99	0.879
SCFA (%)	9.84 ^{a}	9.80 ^{a}	9.41 ^{b}	0.082
LCFA (%)	57.59 ^{b}	59.87 ^{a}	60.79 ^{a}	0.191
Palmitic acid C16: 0 (%)	23.41 ^{b}	25.43 ^{a}	26.16 ^{a}	0.432
SFA sum (%)	67.43 ^{b}	69.67 ^{a}	70.20 ^{a}	0.174
MUFA (%)	29.85 ^{a}	27.70 ^{b}	27.19 ^{b}	1.255
Oleic acid C18:1 (%)	1.44 ^{c}	2.71 ^{b}	3.54 ^{a}	0.064
PUFA (%)	2.73 ^{a}	2.64 ^{b}	2.61 ^{b}	0.067
PUFA n-3 (%)	0.44 ^{a}	0.42 ^{b}	0.42 ^{b}	0.011
PUFA n-6 (%)	2.29 ^{a}	2.21 ^{b}	2.20 ^{b}	0.071
UFA (%)	32.58 ^{a}	30.34 ^{b}	29.80 ^{b}	1.121
SFA/UFA	2.07	2.30	2.35	0.653
MUFA/SFA	0.42	0.39	0.40	0.031
PUFA n-6/n-3	5.26	5.21	5.20	0.998
Cholesterol (mg/100mL)	19.58	19.46	20.01	1.289

^{abc} $p \leq 0.05$; * number of cows in the groups—80; LCFA—long-chain fatty acid; SCFA—Short-chain FA; SFA—saturated FA; MUFA—monounsaturated FA; PUFA—polyunsaturated FA; UFA—unsaturated FA.

The correlation coefficients (r) presented in Table 6 indicate that daily milk production (DMP) and day of lactation (DL) are positively correlated with indicators of the intensity of spontaneous lipolysis. Their values with NEFAs were 0.906 and 0.788, respectively ($p \leq 0.05$), and the correlations with BHBA were 0.777 and 0.885 ($p \leq 0.05$). The NEFA × leptin correlation (0.416; $p \leq 0.05$) indicates that a higher level of the appetite-reducing leptin accompanied a higher concentration of NEFAs released during the energy deficit. Moreover, there was also a negative correlation with glucose (−0.259; $p \leq 0.05$) and glucose with NEFA (−0.386; $p \leq 0.05$).

Table 6. Correlations ($p \leq 0.05$) between selected parameters.

Indicators	BCS (points)	Urea (mmol/L)	BHBA (mmol/L)	Fat (%)	DL (day)	DMP (kg/day)	NEFA (%)	Leptin (ng/mL^{-1})	Glucose (mmol/L)
BCS (points)	-	−0.364	−0.687	–	−0.715	−0.640	−0.653	−0.480	0.227
DL (day)	−0.715	0.519	0.885	–	–	0.785	0.788	−0.441	−0.341
DMP (kg/day)	−0.640	0.552	0.777	−0.233	0.799	–	0.906	0.417	−0.399
NEFA (%)	−0.654	–	–	–	–	–	–	0.416	−0.386
Leptin	0.480	–	–	–	–	–	–	–	−0.259
SCFA (%)	0.216	–	0.282	0.323	0.289	0.399	0.274	–	–
LCFA (%)	−0.567	0.442	0.595	0.478	0.604	0.593	0.689	–	−0.380
Palmitic acid (%)	−0.302	0.308	0.381	−0.234	0.367	0.632	0.682	−0.293	−0.289
SFA (%)	−0.567	0.486	0.652	0.398	0.714	0.746	0.582	–	−0.287
MUFA (%)	0.558	−0.487	−0.690	−0.299	−0.733	−0.758	−0.682	–	0.292
Oleic acid (%)	−0.605	0.541	0.877	−0.232	0.789	0.871	0.865	−0.428	−0.317
PUFA (%)	–	–	−0.231	–	−0.258	−0.271	–	–	–
PUFA n–3 (%)	–	–	−0.356	–	–	–	–	–	–
PUFA n–6 (%)	–	–	–	–	–	–	–	–	–
UFA (%)	0.568	−0.482	−0.708	−0.324	−0.754	−0.761	−0.687	−0.352	0.293

BHBA—β-hydroxybutyrate; NEFA—non-esterified FA; DMP—daily milk production; DL—day of lactation; LCFA—long-chain fatty acid; SCFA—short-chain fatty acid; SFA—saturated FA; MUFA—monounsaturated FA; PUFA—polyunsaturated FA; UFA—unsaturated FA.

The correlation coefficients obtained for SFAs with parameters of lipid metabolism indicate a negative effect of DMP and DL. The increasing milk production up to the peak of lactation was positively correlated with the content of SFAs, which was confirmed by r values from 0.289 to 0.746 ($p \leq 0.05$; Table 6). The content of SFAs was correlated positively with the NEFA concentration in the blood (0.274 to 0.689; $p \leq 0.05$) and was usually negatively correlated with that of glucose (−0.287 to −0.380; $p \leq 0.05$). In the case of UFAs, negative correlations were noted with DMP and DL, from −0.258 to −0.761 ($p \leq 0.05$). While the significant relationship between BCS and DMP and BHBA and NEFAs are confirmed by the negative correlation ranging from −0.640 to −0.686 (p ≤ 0.05).

UFAs were negatively correlated with NEFAs, from −0.145 to −0.687 ($p \leq 0.05$), and with BHBA, from −0.231 to −0.708 ($p \leq 0.05$). Negative correlations were also shown between UFAs in the milk and leptin in the blood: from −0.158 to −0.352 ($p \leq 0.05$). Palmitic acid (C16:0) and oleic acid (C18:1) were found to be correlated negatively with BCS and positively with BHBA (Table 6). However higher values for these correlations were found for C18:1.

4. Discussion

Selection for milk yield increases the demand for energy and intensifies energy metabolism in dairy cows. One important factor in this case is the changes in the negative energy balance (NEB) and content of nutrients in the milk. In this regard, concentrations of NEFAs esterified to triglycerides and β–hydroxybutyrate (BHBA), as well as the release of C18:1 from the adipocytes, remain the basic parameters of the metabolic profile of dairy cows. Their content is linked to maintenance of energy homeostasis, which is associated with the synthesis of FAs in the rumen and mammary gland [22]. In the early lactation stage, increasing production is conducive to the appearance of NEB. Adipocytes then release FAs that are usually unfavorable for the nutritional quality of milk fat. Ducháček et al. [14] showed that NEB increased the share of hypercholesterolemic FAs in the milk of HF cows by 1.86 p.p. on average. Their study showed that NEB mainly leads to an increase in MUFA, in which case the difference was 1.81 p.p. In the case of UFA the difference was smaller, at 0.33 p.p. A similar tendency was observed in our study, but the milk of HF cows had higher content of MUFA and lower content of PUFA than in the study cited. The milk of the HF breed also had the highest content of C16:0 and C18:1, which is the main deposit in the adipocytes. This may also suggest that diet did not directly influence the FA profile in the milk. Additionally, Sobótka et al. [23] showed a significant influence of lactation phase and breed on the content of FAs. However, significant correlations were found only in the case of SFAs and PUFAs. This did not fully coincide with our results. Sobótka et al. [23] the lowest content of C16:0 was found in the milk of HF cattle. More C16:0 has been found in the Jersey breed and its HF hybrids. However, there was no effect associated with lactation stage. However, it is interesting that the C16:1 released from adipocytes appeared in greater amounts in the later stages of lactation. However, this can be explained by the greater availability of energy for C16:0 synthesis in the mammary gland de novo [16]. Incomplete compliance of our results with the studies by Sobótka et al. [23] may be explained by the research of Poulsen et al. [24]. These researchers explained that environmental variation affects individual FAs of milk differently, based on both breed and feed quality. The dietary effect on FA composition is obvious, but as they suggest, an important factor is also the source of origin of the feed ingredients. Furthermore, Samkova et al. [5] suggested that the shaping of milk FAs may be affected to a large degree by cow individuality, but also by the stage of lactation. According these studies, breed is a factor of lower significance. Samkowa et al. [5] suggest that biochemical changes, especially biohydrogenation in the rumen, should be studied in greater detail. In light of these studies [5,23,24], the obtained FA profiles of milk in our experiment could have been shaped by other environmental factors, as well as the production potential of the studied breeds and their reaction to the formation of NEB. During early lactation it stimulates the content of non-esterified FAs (NEFAs) generated from lipid tissue increases. It is usually proportional to the size of the energy deficit [25,26]. This leads to an increase in long-chain FAs in the milk, which enter the mammary gland with the blood. Craninx et al. [2] reported a lower content of C16:0 in the milk of Holstein cows in a grazing trial. However, the diet used by the authors, containing hay and silage from grass and maize, did not affect the content of C18:1. This may suggest that the diet does not directly affect the share of C16:0, which is influenced by NEB. Craninx et al. [2] showed that in early lactation, the milk of cows with higher yields and with higher fat content had a lower share of FAs, with less than 15 carbon atoms. In our study, this was confirmed by the positive correlations between the level of long-chain FAs (LCFAs) and C16:0 and milk yield (0.593, 0.632; $p \leq 0.05$) and NEFA content (0.689 and 0.682; $p \leq 0.05$). A study by Vanbergue et al. [26] showed that spontaneous lipolysis was mainly dependent on the breed of cows and season of production. The study did not confirm the effect of the feeding intensity or its interaction with breed on the course of lipolysis. Although the authors found that the influence of breed on SFA and MUFA was greater, feeding intensity was also found to have a minor effect. This effect, however, varied depending on the production season. In cows with high postpartum milk production, we noted higher concentrations of BHBA and NEFAs in the blood. At the same time, the blood glucose levels were usually lower.

Lipomobilization in high-producing dairy cows up to the peak of lactation could reach a pathological range, disturbing the liver's morphological and functional efficiency. Diokovic et al. [12] demonstrated that in the early-lactation cows, there is a rapid increase of fatty acids in the liver. The authors observed growth the lipomobilisation markers, especially the serum β–hydroxybutyrate and free fatty acid concentrations. According to the authors, the liver steatosis was affected, which disturbed the synthesis of hepatocyte. Consequently, it led to weaker concentrations of glucose and an increase in triglycerides. This effect has induced some cellular lesions, as evidenced by significant increases in the serum albumin and bilirubin concentrations. Šamanc et al. [27] showed that an NEB is conducive to excess accumulation of fat in the hepatocytes. This situation may be a major cause of endocrine disorders, including secretion leptin. This may result in impaired gluconeogenesis and reduced glycaemia. In comparison with our results, Vargová et al. [28], who studied the hormone profile of Slovak Pied Cattle, reported higher blood glucose concentrations. However, up to nine weeks postpartum they observed a downward trend in NEFAs and BHBA, accompanied by a decrease in body condition score (BCS). The authors [28] noted a positive correlation between BCS and leptin (0.360; $p \leq 0.001$). Contrary to the studies by Vargová et al. [28], in our study we showed a negative correlation between BCS and leptin (−0.480; $p \leq 0.05$) and a positive correlation between daily milk production and leptin (0.417; $p \leq 0.05$). As demonstrated in our studies, the NEFA × leptin correlation (0.416; $p \leq 0.05$) indicated that a higher level of leptin led to a higher concentration of NEFAs released during the energy deficit. This may be related to decreased appetite. In consequence, reduced feed intake may result in a lower glucose concentration in the blood. This is explained by the negative leptin × glucose (−0.259; $p \leq 0.05$) and NEFA × glucose (−0.386; $p \leq 0.05$) correlations. This may indicate that the level of the appetite regulator leptin is linked to the production potential of the breed. In our study, a lower leptin concentration was usually noted just after calving, but it was higher in the groups reaching the peak of lactation with a higher yield. The obtained values of the correlation coefficient with leptin were relatively low, but they may suggest that the stronger appetite in individuals with lower leptin levels may prolong feed intake. In effect, this may be linked to better energy availability and to energy conversions that are beneficial for FAs. This indicates more rapid induction of lipolysis due to reduced appetite in cows with higher yields [27,28]. Reduced appetite and higher production potential may also explain the lower glucose level in the blood of animals during the peak of lactation and reaching this peak with a higher yield. According to authors such as Lock and Garnsworthy [16], a greater energy deficit associated with milk production may reduce direct absorption of cis–9–C18:1 in the small intestine and limit processes involving desaturase in the mammary gland. Liefers et al. [29] reported a lower leptin content in the blood of cows that produced more milk, as in our study. They also showed lower DMI in cows with a negative energy balance. Our results were not confirmed in this case, as the content of C18:1 as an indicator of NEB was usually lower in cows with lower leptin levels. This relationship is confirmed by the correlation of C18:1 × leptin (–0.428; $p \leq 0.05$). Buttchereit et al. [30] also showed that strong mobilization of energy from lipid tissue is not only conducive to more intensive production of long-chain FAs, but can also reduce the protein content in milk. This is consistent with the tendency observed by Pupel et al. [31] and with our results regarding levels of κ–casein. Our results for the content of the main FA fractions are in agreement with those reported by Petit and Côrtes [32]. They found the highest BHBA level (on average 748 μmol/L) in milk with a higher content of fat and SFAs. The exception was a group whose diet included ground flaxseed. The milk of this group had the highest content of BHBA (1512 μmol/L) and high levels of MUFAs and PUFAs. In this case, the effect was explained by the greater availability of energy reaching the mammary gland, in which a considerable portion of these fractions is generated de novo. In this group of cows, Petit and Côrtes [32] also showed the highest content of non-esterified FAs in the blood (337 μmol/L), although the production level in this case was one of the lowest (29.5 kg/d). The cows in this group, however, had the lowest glucose level in the blood, which in our study was negatively correlated with NEFA (−0.386, $p \leq 0.05$). Puppel et al. [31] showed very strong negative correlations between the BHBA level in the blood and the content of CLA–9 and –10. Our results are consistent with those reported by Vanbergue et al. [26], who studied the milk of Holstein and Normande cows up to the peak of lactation. The share of SFAs ranged from 61.8% to 73.5% and that of MUFAs from 23.0% to 34.7%. Vanbergue et al. [26] also noted more intensive lipolysis of fat reserves and 1.1% to 5.6% higher content of cis–9 in the milk of cows fed less intensively. Adamska et al. [4], in comparison to our study, reported a higher content of SCFAs (15.99 g/100g FAs) and a similar content of PUFAs (2.53 g/100g FAs) in the milk of Simmental and HF cows. However, the milk of the Simmental breed usually had a more beneficial composition, and also contained more branched-chain FAs. The milk of HF cows, on the other hand, had the most LCFAs (52.11 g/100g FAs) and MUFAs (26.32 g/100g FAs). The results of our study also correspond with those reported by Król et al. [33], in which the milk of Simmental cows contained on average 30.53% SFA, which was on average 4.1 p.p. less than in the milk of HF cows. In the milk of Simmental cows, Król et al. [33] showed higher levels of MUFAs and PUFAs, at 27.13%and 3.43%, respectively. Such large differences were not observed in our study, which can be explained by the use of a similar diet (TMR) in all breed groups.

5. Conclusions

The information gathered here suggests that the metabolic reactions of the breed groups were mainly linked to the level of production at the peak of lactation. Breed groups that reach the peak of lactation earlier and with greater daily milk production (DMP) had a greater loss of BCS, on average by 22.8% ($p > 0.05$), and more pronounced changes in NEB markers. These groups, in contrast with cows with lower DMP and a later lactation peak, had higher values for BHBA, on average by 0.102 mmol/L^{-1}, and for NEFA, on average by 21.15 μmol/L^{-1} ($p \leq 0.05$). These changes were accompanied by a higher content of FAs C18:1 and C:16:0, released during NEB from adipocyte deposits. The relationships linking BCS and DMP with NEB markers were confirmed by the correlation coefficients, which ranged from −0.302 to −0.640 ($p \leq 0.05$) and from 0.632 to 0.906 ($p \leq 0.05$), respectively. The significant influence of DMP is also indicated by changes in the NEB indicators in the stage up to the peak of lactation. Despite lower DMP in the first stage of lactation, the values for parameters characterizing NEB, including C16:0, were usually higher. This may indicate that the increases in DMP and NEB are more closely linked to the physiology of the breed groups and mainly result from their production potential. The tendencies observed indicate that factors that should be taken into account are leptin and glucose, co-regulators of appetite and NEB, which were negatively correlated (−0.259; $p \leq 0.05$). However, changes in these parameters in the analyzed groups and stages of lactation, as well as the correlations with BCS, DMP, and C18:1, indicate that they have a relatively strong influence on NEB and lipolysis. The knowledge obtained on the production potential of the breeds can be used to model production adjusted to the environmental conditions of farms.

Author Contributions: Conceptualization, K.M.; data curation, I.S.; formal analysis, K.M., A.D. and I.S.; investigation, K.M. and A.D.; methodology, K.M.; writing—original draft, A.D. All authors have read and agreed to the published version of the manuscript.

Funding: This research received no external funding.

Institutional Review Board Statement: The study was conducted under a research protocol approved by the III Local Ethical Committee in Warsaw, Poland, No. 70/2013.

Informed Consent Statement: Not applicable for studies not involving humans.

Data Availability Statement: The data presented in this study are available on request from the corresponding author. The data are not publicly available due to: The research was carried out in private farms. Due to the privacy of farmers, the data will be made available in the form of a database.

Conflicts of Interest: The authors declare no conflict of interest.

References

1. Jenkins, T.C.; Wallace, R.J.; Moate, P.J.; Mosley, E.E. Board–invited review: Recent advances in biohydrogenation of unsaturated fatty acids within the rumen microbial ecosystem. *J. Anim. Sci.* **2008**, *86*, 397–412. [CrossRef] [PubMed]
2. Crarinx, M.; Steen, A.; Van Laar, H.; Van Nespen, T.; Martin–Tereso, J.; De Baets, B.; Fievez, V. Effect of lactation stage on the odd– and branched–chain milk fatty acids of dairy cattle under grazing and indoor conditions. *J. Dairy Sci.* **2008**, *91*, 2662–2677. [CrossRef] [PubMed]
3. Schonfeldt, H.C.; Hall, N.G.; Smit, L.E. The need for country specific composition data on milk. *Food Res. Int.* **2012**, *47*, 207–209. [CrossRef]

4. Adamska, A.; Rutkowska, J.; Przybylski, W. Comparison of Fatty Acid Composition of Milk from Simmental and Polish Holstein–Friesian Cows in Different Production Seasons. *Ann. Anim. Sci.* **2016**, *16*, 1211–1225. [CrossRef]
5. Samkova, E.; Spicka, J.; Pesek, M.; Pelikanova, T.; Hanus, O. Animal factors affecting fatty acid composition of cow fat: A review. *S. Afr. J. Anim. Sci.* **2012**, *42*, 83–100. [CrossRef]
6. Roca–Fernandez, A.I.; Gonzalez–Rodriguez, A. Effect of dietary and animal factors on milk fatty acids composition of grazing dairy cows: Review. *Iran. J. Appl. Anim. Sci.* **2012**, *2*, 97–109.
7. Müller, F.; Huber, K.; Pfanakuche, H.; Aschenbach, J.R.; Breves, G.; Gabel, G. Transport of ketone bodies and lactate in the sheep ruminal epithelium by monocarboxylate transporter. *Am. J. Physiol.* **2002**, *283*, 1139–1146. [CrossRef]
8. Sasson, G.; Ben–Shabat, S.K.; Seroussi, E.; Doron–Faigenboim, A.; Shterzer, N.; Yaacoby, S.; Miller, M.E.B.; White, B.A.; Halperin, E.; Mizrahi, I. Heritable bovine rumen bacteria are phylogenetically related and correlated with the cow's capacity to harvest energy from its feed. *Am. Soc. Microbiol.* **2017**, *8*, 4. [CrossRef]
9. Kay, J.K.; Weber, W.J.; Moore, C.E.; Bauman, D.E.; Hansen, L.B.; Chester–Jones, H.; Crooker, B.A.; Baumgard, L.H. Effects of week of lactation and genetic selection for milk yield on fatty acid composition in Holstein cows. *J. Dairy Sci.* **2005**, *88*, 3886–3893. [CrossRef]
10. Ingvartsen, K.L.; Dewhurst, R.J.; Friggens, N.C. On the relationship between lactational performance and health: Is it yield or metabolic imbalance that cause production diseases in dairy cattle? A position paper. *Livest. Prod. Sci.* **2003**, *83*, 277–308. [CrossRef]
11. Akbar, H.; Grala, T.M.; VailatiRiboni, M.; Cardoso, F.C.; Verkerk, G.; McGowan, J.; Macdonald, K.; Webster, J.; Schutz, K.; Meier, S.; et al. Body condition score at calving affects systemic and hepatic transcriptome indicators of inflammation and nutrient metabolism in grazing dairy cows. *J. Dairy Sci.* **2015**, *98*, 1019–1032. [CrossRef] [PubMed]
12. Djokovic, R.; Ilic, Z.; Kurcubic, V.; Petrovic, M.; Doskovic, V. Functional and morphological state of the liver in Simmental dairy cows during transitional period. *Revue Méd. Vét.* **2011**, *162*, 574–579.
13. Drackley, J.K.; Overton, T.R.; Douglas, G.N. Adaptations of Glucose and Long–Chain Fatty Acid Metabolism in Liver of Dairy Cows During the Periparturient Period. *J. Dairy Sci.* **2001**, *84*. [CrossRef]
14. Ducháček, J.; Stádník, S.; Ptáček, M.; Beran, J.; Okrouhlá, M.; Gašparík, M. Negative Energy Balance Influences Nutritional Quality of Milk from Czech Fleckvieh Cows due Changes in Proportion of Fatty Acids. *Animals* **2020**, *10*, 563. [CrossRef]
15. Vernon, R.G. Lipid metabolism during lactation: A review of adipose tissue–liver interactions and the development of fatty liver. *J. Dairy Res.* **2005**, *72*, 460–469. [CrossRef] [PubMed]
16. Lock, A.L.; Garnsworthy, P.C. Seasonal variation in milk conjugated linoleic acid and Δ9–desaturase activity in dairy cows. *Livest. Prod. Sci.* **2003**, *79*, 47–59. [CrossRef]
17. Ingvartsen, K.L.; Andersen, J.B. Integration of Metabolism and Intake Regulation:A Review Focusing on Periparturient Animals. *J. Dairy Sci.* **2000**, *83*, 1573–1597. [CrossRef]
18. Intl, A. *Official Methods of Analysis of AOAC International*, 18th ed.; AOAC International: Gaithersburg, MD, USA, 2007.
19. Strzetelski, J.E.D. National Research Institute of Animal Production INRA Norm of Ruminant Nutrition. In *The Nutritional Value of French and National Feed fur Ruminants*; Instytut Zootechniki PIB: Kraków, Poland, 2009. (In Polish)
20. IDF. Milk and Milk Products. Determination of Nitrogen Content Routine Method by Combustion According to the Dumas Principle. Available online: https://www.iso.org/standard/25871.html (accessed on 28 December 2020).
21. Official Methods of Analysis. International Method IDF–ISO–AOAC; Nr 905.02.Gravimetric method (Röse–Gottlieb). 2000. Available online: https://www.aoac.org/official-methods-of-analysis-21st-edition-2019/ (accessed on 28 December 2020).
22. Palmquist, D.L. Milk fat: Origin of fatty acids and influence of nutritional factors. In *Advanced Dairy Chemistry*; Fox, P.F., McSweener, P.L.H., Eds.; Threon: New York, NY, USA, 2006; Volume 2, pp. 43–80. [CrossRef]
23. Sobotka, W.; Miciński, J.; Matusevičius, P.; Staniškienė, B.; Sobiech, M.; Zwierzchowski, G.; Fiedorowicz, E.; Pietrzak–Fiećko, R. The effect of cattle breed and lactation stage on nutrient concentrations in milk and the fatty acid profile of milk fat. *Vet. ir Zootech.* **2014**, *65*, 85–90.
24. Poulsen, N.A.; Gustavsson, F.; Glantz, M.; Paulsson, M.; Larsen, L.B.; Larsen, M.K. The influence of feed and herd on fatty acid composition in 3 dairy breeds (Danish Holstein, Danish Jersey and Swedish Red). *J. Dairy Sci.* **2012**, *95*, 6362–6371. [CrossRef]
25. van Knegsel, A.T.M.; Van den Brand, H.; Dijkstra, J.; Tamminga, S.; Kemp, B. Effect of dietary energy source on energy balance, production, metabolic disorders and reproduction in lactating dairy cattle. *Reprod. Nutr. Dev.* **2005**, *45*, 665–688. [CrossRef]
26. Vanbergue, E.; Delaby, L.; Peyraud, J.L.; Colette, S.; Gallard, Y.; Hurtaud, C. Effects of breed, feeding system, and lactation stage on milk fat characteristics and spontaneous lipolysis in dairy cows. *J. Dairy Sci.* **2017**, *100*, 4623–4636. [CrossRef] [PubMed]
27. Šamanc, H.; Gvozdić, D.; Fratrić, N.; Kirovski, D.; Djoković, R.; Sladojević, Ž.; Cincović, M. Body condition score loss, hepatic lipidosis and selected blood metabolites in Holstein cows during transition period. *Anim. Sci. Pap. Rep.* **2015**, *33*, 35–47.
28. Vargová, M.; Petrovič, V.; Konvičná, J.; Kadaši, M.; Zaleha, P.; Kováč, G. 2015 Hormonal profile and body condition scoring in dairy cows during *pre-partum* and *postpartum* periods. *Acta Vet. Brno* **2015**, *84*, 141–151. [CrossRef]
29. Liefers, S.C.; Veerkamp, R.F.; Pas, M.F.W.; Delavaud, C.; Chilliard, Y.; van der Lende, T. Leptin Concentrations in Relation to Energy Balance, Milk Yield, Intake, Live Weight, and Estrus in Dairy Cows. *J. Dairy Sci.* **2003**, *86*, 799–807. [CrossRef]
30. Buttchereit, N.; Stamer, E.; Junge, W.; Thaller, G. Evaluation of five lactation curve models fitted for fat:protein ratio of milk and daily energy balance. *J. Dairy. Sci.* **2010**, *93*, 1702–1712. [CrossRef] [PubMed]

31. Puppel, K.; Gołębiewski, M.; Solarczyk, P.; Grodkowski, G.; Slósarz, J.; Kunowska-Slósarz, M.; Balcerak, M.; Przysucha, T.; Kalińska, A.; Kuczyńska, B. The relationship between plasma β–hydroxybutyric acid and conjugated linoleic acid in milk as a biomarker for early diagnosis of ketosis in postpartum Polish Holstein–Friesian cows. *BMC Vet. Res.* **2019**, *15*, 367. [CrossRef]
32. Petit, H.V.; Côrtes, C. Milk production and composition, milk fatty acid profile, and blood composition of dairy cows fed whole or ground flaxseed in the first half of lactation. *Anim. Feed Sci. Tech.* **2010**, *158*, 36–43. [CrossRef]
33. Król, J.; Litwińczuk, Z.; Matwijczuk, A. Fatty acid profile and cholesterol content of the milk of cows raised in a low–input system, taking into account the production season. (In Polish). *Rocz. Nauk. Pol. Tow. Zootech.* **2015**, *11*, 95–102.

animals

MDPI

Review

The Physiological Roles of Vitamin E and Hypovitaminosis E in the Transition Period of High-Yielding Dairy Cows

Satoshi Haga [1,*], Hiroshi Ishizaki [1] and Sanggun Roh [2]

[1] Division of Grassland Farming, Institute of Livestock and Grassland Science, NARO, Nasushiobara, Tochigi 329-2793, Japan; hishizak@affrc.go.jp

[2] Graduate School of Agricultural Science, Tohoku University, Sendai 980-8572, Japan; sanggun.roh.a3@tohoku.ac.jp

* Correspondence: hagatiku@affrc.go.jp; Tel.: +81-287-37-7239

Simple Summary: In high-yield cows, most production diseases occur during transition periods. Alpha-tocopherol, the most biologically active form of vitamin E, declines in blood and reaches the lowest levels (hypovitaminosis E) around calving. Hypovitaminosis E is associated with the incidence of peripartum diseases. Therefore, many studies which have been published for more than 30 years have investigated the effects of α-tocopherol supplementation. This α-tocopherol deficiency was thought to be caused by complex factors. However, until recently, the physiological factors or pathways underlying hypovitaminosis E in the transition period have been poorly understood. In the last 10 years, the α-tocopherol-related genes expression, which regulate the metabolism, transportation, and tissue distribution of α-tocopherol in humans and rodents, has been reported in ruminant tissues. In this paper, we discuss at least six physiological phenomena that occur during the transition period and may be candidate factors predisposing to a decreased blood α-tocopherol level and hypovitaminosis E with changes in α-tocopherol-related genes expression.

Citation: Haga, S.; Ishizaki, H.; Roh, S. The Physiological Roles of Vitamin E and Hypovitaminosis E in the Transition Period of High-Yielding Dairy Cows. *Animals* **2021**, *11*, 1088. https://doi.org/10.3390/ani11041088

Academic Editor: João Simões

Received: 27 March 2021
Accepted: 9 April 2021
Published: 11 April 2021

Abstract: Levels of alpha-tocopherol (α-Toc) decline gradually in blood throughout prepartum, reaching lowest levels (hypovitaminosis E) around calving. Despite numerous reports about the disease risk in hypovitaminosis E and the effect of α-Toc supplementation on the health of transition dairy cows, its risk and supplemental effects are controversial. Here, we present some novel data about the disease risk of hypovitaminosis E and the effects of α-Toc supplementation in transition dairy cows. These data strongly demonstrate that hypovitaminosis E is a risk factor for the occurrence of peripartum disease. Furthermore, a study on the effectiveness of using serum vitamin levels as biomarkers to predict disease in dairy cows was reported, and a rapid field test for measuring vitamin levels was developed. By contrast, evidence for how hypovitaminosis E occurred during the transition period was scarce until the 2010s. Pioneering studies conducted with humans and rodents have identified and characterised some α-Toc-related proteins, molecular players involved in α-Toc regulation followed by a study in ruminants from the 2010s. Based on recent literature, the six physiological factors: (1) the decline in α-Toc intake from the close-up period; (2) changes in the digestive and absorptive functions of α-Toc; (3) the decline in plasma high-density lipoprotein as an α-Toc carrier; (4) increasing oxidative stress and consumption of α-Toc; (5) decreasing hepatic α-Toc transfer to circulation; and (6) increasing mammary α-Toc transfer from blood to colostrum, may be involved in α-Toc deficiency during the transition period. However, the mechanisms and pathways are poorly understood, and further studies are needed to understand the physiological role of α-Toc-related molecules in cattle. Understanding the molecular mechanisms underlying hypovitaminosis E will contribute to the prevention of peripartum disease and high performance in dairy cows.

Keywords: alpha-tocopherol/vitamin E-related gene; calving; colostrum; high-yield dairy cows; inflammation; health; lactation; liver; mammary gland; oxidative stress

1. Introduction

Calving is an unavoidable event in the production of milk and young cattle in dairy farms. However, during the transition period (from three weeks before calving to three weeks after calving), high-yield dairy cows experience severe energy and nutrient deficiencies [1,2], metabolic and endocrine changes [3,4], peripartum stress [2], and inflammation [5]. This leads to an imbalance between pro-oxidants and antioxidants, eventually resulting in oxidative stress [6] and immune dysfunction [7], which increases the risk of peripartum diseases. Indeed, most production diseases (mastitis, ketosis, digestive disorders, and laminitis) occur before and soon after calving [8,9].

Vitamin E (VE) has essential antioxidant functions and is an important nutrient for cows. It is well known that blood VE levels decline gradually throughout prepartum, reaching the lowest levels (hypovitaminosis E) after delivery [2,10–13]. Several studies have documented that lower blood VE concentrations are associated with the incidence of peripartum diseases such as mastitis [14], retained foetal membranes [15] and left displaced abomasum [16]. The effect of VE supplementation, as a practical measure to counter hypovitaminosis E and the high risk of peripartum diseases, have been often controversial in published literature for more than 30 years. To the best of our knowledge, many detailed reviews reported in the last 30 years, focused primarily on the disease risk and the effect of VE supplementation on the health of dairy cows and heifers in transition period (two databases, PubMed and Web of Science were searched using key words: vitamin E, dairy cows, tocopherol, 1990–2021, and reviews were picked up whose full-text is currently available online; 1990s: [17–22], 2000s: [1,23–27], 2010s: [28–32]). This VE deficiency may be caused by complex factors, such as changes in VE intake and its transfer into colostrum around calving [26]. However, the physiological factors underlying hypovitaminosis E in the transition period of high-yielding dairy cows were less well understood until recently. In the last 10 years, several studies have shown evidence that the ovine [33,34] and bovine [35] liver may play an important role in the regulation of VE disposition because of the high expression of VE-related molecules. In addition, other peripheral tissues showed unique expression patterns for VE-related molecules and VE accumulation properties in cattle [2,35–38]. These findings indicate that the expression of VE-related genes in the liver and non-hepatic tissues may be involved in the regulation of VE status in cows. Therefore, it is necessary to explore and discuss the physiological factors underlying hypovitaminosis E in dairy cows based on the latest reports on the expression of VE-related genes in bovine tissues. Especially during peripartum period in dairy cows, dramatic changes in lipid metabolism [39], physiological stress and inflammation may lead to hepatic injury, dysfunctions [40] and hepatocyte apoptosis [41]. Furthermore, during the onset of lactation, the mammary gland undergoes dramatic functional and metabolic changes during the transition period [42]. Novel knowledge about the changes in the expression of VE-related genes in the liver and mammary glands of transition dairy cows is compelling.

Thus, the current study aimed to understand the occurrence of hypovitaminosis E during the transition period, and contribute to the development of an effective feeding system for the health and high performance of dairy cows. Therefore, the aims of this review are: (1) to summarise the basic information about VE and VE-related molecules from the latest literature; (2) to re-evaluate the physiological roles of VE and the relationship between the risk of peripartum diseases and hypovitaminosis E in transition dairy cows; and (3) to discuss the physiological factors underlying hypovitaminosis E in relation to changes in the VE-related genes expression levels, especially in the liver and mammary gland from late pregnancy to early lactation.

2. Vitamin E and VE-Related Molecules

2.1. *Vitamin E (α-Tocopherol)*

In 1922, Evans and Bishop [43] discovered VE as an essential micronutrient for reproduction in rats. Subsequently, considerable research has been conducted on VE function and metabolism globally during the last one-hundred years. Vitamin E is considered the most effective, fat-soluble trace compound and chain-breaking antioxidant, which protects cell membranes from peroxidative damage [44]. It also plays specific roles beyond its antioxidant function, such as cellular signalling and regulation of gene expression [45,46]. The clinical importance of VE is significantly increasing to prevent various diseases in humans; VE deficiency can lead to neurological abnormalities, such as ataxia [47] and blindness [48]. The importance is also well recognised in animal production and medicine to maintain the health of livestock. Vitamin E cannot be synthesised in the mammalian body and must therefore be provided through food or supplementation. The VE family is composed of four tocopherols and four tocotrienols (α, β, γ and δ). Tocopherols have a saturated phytyl side chain, whereas tocotrienols have a three-fold unsaturated isoprenoid side chain [44]. Unlike other nutrients, the body cannot interconvert among these forms. In the plasma and tissues of humans and animals, one form, α-tocopherol (α-Toc), is the predominant congener and the most biologically active form of VE [44] because other tocopherols and tocotrienols are usually found at very low-levels compared to α-Toc. Alpha-tocopherol regulates key cellular events by mechanisms unrelated to its antioxidant properties, such as the inhibition of protein kinase C activity [44]. Furthermore, the expression of many genes has been found to be under the non-antioxidant control of α-Toc [45,49], suggesting the high availability of α-Toc in protection against disease in humans and animals.

Natural α-Toc (RRR-α-Toc (2,5,7,8-tetramethyl-2R-(4′R,8′R,12-trimethyltridecyl)-6-chromanol)) has the highest biological activity and is maintained at the highest level in plasma and tissues of humans and animals [50]. However, most of the α-Toc used for supplementation of food and feed is synthetic in origin, designated as *all-rac-*α-Toc (2,5,7,8-tetramethyl-2RS-(4′RS,8′RS,12-trimethyltridecyl)-6-chromanol). *All-rac-*α-Toc (*dl-*α-Toc) has three asymmetric carbons at positions 2, 4′ and 8′ and consists of an equimolar mixture of eight stereoisomers (RRR, RRS, RSS, RSR, SRR, SSR, SRS and SSS). The measurement of α-Toc activity in terms of IU was based on fertility enhancement by the prevention of spontaneous abortions in pregnant rats [51]; 1 IU of α-Toc is defined as 1 mg of *all-rac-*α-tocopheryl acetate, as 0.74 mg of RRR-α-tocopheryl acetate, and as 0.67 mg of RRR-α-Toc. Data from cows comparing the bioavailability of various Toc stereoisomers are contradictory, and insufficient consistent data are available to determine IU conversion factors for VE for ruminants [52]. However, Meglia et al. [53] showed that RRR-α-Toc was the most predominant stereoisomer, constituting more than 86%, whereas the remaining part of α-Toc was made up of three synthetic 2R isomers, while the 2S isomers contributed less than 1% of the total α-Toc in plasma and milk from dairy cows supplemented with *all-rac-*α-tocopheryl acetate. Jensen et al. [54] showed that after a single dose injection of *all-rac-*α-tocopheryl acetate, the RRR-α-Toc was retained in plasma for the longest time and secreted into milk at the highest concentration followed by RRS-, RSS-, and RSR-α-Toc leaving the Σ2S-α-Toc to be retained in plasma for the shortest time and secreted into milk at the lowest concentration. In dairy cows during early lactation, the serum concentrations (nM) of γ-Toc, α-, β-, γ- and δ-tocotrienol were far lower than those of α-Toc (approximately 1/56, 1/214, 1/3947, 1/5000 and 1/2500, respectively) [55]. Among the naturally occurring forms of the VE family, α-Toc only meets VE requirements and α-Toc stereoisomers have different bioactivities because α-Toc transfer protein (αTTP) has a different affinity for VE (described in Section 2.2.1) and play an important role in the circulation and disposition of α-Toc in cattle. Throughout this review, the terms "VE" and "α-Toc" are used interchangeably and the plasma/serum concentrations of α-Toc is unified and expressed in "μg/mL" (1 μM = 0.43 μg/mL).

2.2. *Alpha-Tocopherol Transfer Protein and Other α-Toc-Related Molecules*

Alpha-tocopherol shows tissue-specific distribution in animals [56], and this property may affect the α-Toc potencies for each tissue. However, until recently, the molecular mechanisms underlying this tissue distribution and the action of α-Toc were poorly understood in cattle. Pioneering studies from the 1990s conducted with humans and rodents have identified and characterised some α-Toc-related proteins, molecular players involved in α-Toc regulation, followed by a study in ruminants from the 2010s.

2.2.1. Alpha-Tocopherol Transfer Protein

A cytosolic protein that specifically binds to α-Toc was purified from rat and human liver [57], and the full-length cDNA sequence of the rat and human homolog has been reported [58,59]. The protein is called α-Toc transfer protein (αTTP) encoded by the *TTPA* gene, which is classified as a member of the Sec14 like protein family, which has a CRAL-TRIO lipid-binding domain. Humans carrying mutations in the *TTPA* gene revealed a low α-Toc level in plasma and neurological disorders associated with elevated oxidative stress termed as ataxia with VE deficiency (AVED), indicating the importance of αTTP in regulating plasma α-Toc levels [60,61]. In agreement with this fact, $\alpha TTP^{-/-}$ mice demonstrated AVED-like symptoms [48]. Arita et al. [62] reported that α-Toc secretion was markedly stimulated when αTTP was overexpressed in a cultured rat hepatocyte cell line. These reports indicate that hepatic αTTP regulates α-Toc secretion from the liver into the circulation. In intracellular transfer mechanisms, αTTP binds to α-Toc in the endosomal membrane and enables its transport to the plasma membrane, where the αTTP interacts with phosphatidylinositol-4,5-bisphosphate [PI(4,5)P2] or [PI(3,4)P2] for the release of α-Toc and its incorporation into the plasma membrane [63]. Mutations in the *TTPA* prevent its binding with the membrane PIPs and transfer α-Toc to the membrane [63–65]. Alpha-tocopherol transfer protein translocate to the hepatic endosomal compartment to transfer available α-Toc [66–68], and the intracellular localisation of αTTP in hepatocytes is dynamic and responds to the presence of α-Toc [67]. Moreover, the regulation of *TTPA* expression in tissues in response to α-Toc supplementation is unclear, with complicated reports on rodents genes. Concerning the effect on *TTPA* mRNA expression in the liver, there were conflicting reports that α-Toc supplementation in rats was up-regulated [69], down-regulated [70] and showed no [71] effective actions on gene expression. Recently, higher expression of *TTPA* mRNA in chicken liver, in response to dietary α-Toc content, was reported, and the result may suggest its crucial role in the transport of α-Toc in chicken liver [72]. The αTTP, which is mainly expressed in the liver, regulates α-Toc secretion from the liver into circulation, and is also observed in other peripheral tissues and/or cells, such as mouse uterus [73], human leukocytes [74], human placental trophoblast cells [75], mouse lung [76], chicken small intestine, intestinal mucosal layer and adipose tissues [72].

Molecular cloning and characterisation of the full-length cDNA of ovine and bovine *TTPA* genes were conducted [33,77], which contained 2740 nucleotides, and the open reading frame contained 846 bp encoding 282 amino acids with 88% identity with the human genes. Bovine and ovine αTTP have an additional five amino acids (GEEVT) at the C terminus, which goat, bison, deer, dolphin, and killer whale αTTPs contain, whereas human, mouse, and rat αTTPs do not have, suggesting that the C-terminal sequences of αTTP are specific to *Cetartiodactyla* animals [77]. Zuo et al. [78] demonstrated the expression of *TTPA* mRNA in non-hepatic ovine tissues, including the heart, spleen, lung, kidney, and muscle. Haga et al. [35] reported the distribution of *TTPA* mRNA expression in 20 major tissues in calves, including metabolic, reproductive, endocrine, immune, digestive, and absorptive tissues. Furthermore, the hepatic [2,55,79] and mammary [2] *TTPA* mRNA expression in dairy cows was also investigated (described in Section 4). These reports suggest that αTTP is expressed not only in the liver, but also in various non-hepatic tissues in cattle, and may also play a crucial role in regulating α-Toc circulation and local α-Toc status.

2.2.2. Afamin

Another potential candidate protein for α-Toc binding and transport to plasma and extravascular fluids is afamin (AFM) encoded by the *AFM* gene [80,81]. Afamin belongs to the albumin (*ALB*) gene superfamily, which comprises ALB, α-fetoprotein, and vitamin D-binding protein [82] and has multiple binding sites for both α-Toc and γ-Toc [80].

Afamin is primarily expressed in the liver [82] and secreted into circulation, where it is partially associated with apolipoprotein A1 (ApoA1), which contains high density lipoprotein (HDL) subfractions in human plasma [81]. In women, nearly a two-fold increase in serum AFM concentrations was observed during uncomplicated pregnancy [83] and the elevated serum concentrations were associated with the presence of metabolic syndrome [84]. These findings suggest that plasma/serum AFM concentrations have the potential to serve as predictive markers for various medical conditions. However, AFM and α-Toc concentrations are significantly correlated in follicular and cerebrospinal fluids, but not in plasma or serum [81,85,86]. The expression of AFM has also been confirmed in other organs such as the human kidney [83], and Kratzer et al. [87] reported that AFM is synthesised by brain capillary endothelial cells and mediates α-Toc transport into the central nervous system across the blood-brain barrier. According to these reports, AFM might be a specific binding/transport protein contributing to α-Toc circulation and its status in various local tissues.

Haga et al. [35] reported the distribution of *AFM* mRNA expression in the liver, renal cortex, testis, thymus, duodenum, and jejunum tissues of calves. Although hepatic and mammary *AFM* mRNA expression in dairy cows was investigated, the *AFM* transcript in mammary gland tissue was not detected [2]. The understanding of the physiological role of AFM in ruminants is minimal. However, in dairy cows, 3% of plasma α-Toc was not associated with the lipoprotein fractions in circulation [88]; thus, AFM might be involved in α-Toc transportation in plasma and extravascular fluids.

2.2.3. Tocopherol-Associated Protein

Studies have reported the identification of tocopherol-associated protein (TAP/*SEC14L2*) in the cytosol of bovine liver, and TAP has a sequence that is homologous to the proteins with the CRAL-TRIO structural motif in common with αTTP [89,90]. Recombinant human TAP could bind only to α-Toc but not to other tocopherols, as shown by ligand competition analysis and α-Toc-dependent nuclear translocation and transcriptional activation properties in transfected COS-7 cells [91]. This report suggested that TAP might be associated with intracellular metabolism, non-antioxidative function, and the regulation of gene expression of α-Toc [92]. However, TAP, expressed in mouse mast cells, was predominantly localised in the cytoplasm and its subcellular localisation was not changed by α-Toc [93]. These results suggest that the physiological role of TAP in mast cells is not α-Toc-related, while as an α-Toc binding protein, TAP can promote α-Toc retention and thus increase its concentration in breast cancer cells [94]. *SEC14L2* mRNA has also been observed in various human tissues [89,90]; however, the biological roles of TAP in each tissue are still poorly understood.

Haga et al. [35] reported the distribution of bovine *SEC14L2* mRNA expression in 20 major tissues of calves. Furthermore, the hepatic [2,55] and mammary [2] *SEC14L2* mRNA expression in dairy cows was also investigated (described in Section 4).

2.2.4. Scavenger Receptor Class B, Type I

Some lipoprotein receptors and transporters might also be important for the control of α-Toc distribution in tissues because HDL, low density lipoprotein (LDL), very low-density lipoprotein (VLDL), and chylomicron (CM) are the major carriers of α-Toc in the bloodstream because of the hydrophobic properties of α-Toc [95]. Several studies using knock-out mice and over-expressing cells [95], and specific antibodies and a chemical inhibitor in enterocytes [96], have suggested that selective cholesterol ester uptake from HDL by scavenger receptor class B, Type I (SRBI), rather than endocytosis (Ex. VLDLs and LDLs), are important factors for α-Toc delivery into cells. Scavenger receptor class B, Type I, encoded by the *SCARB1* gene, is a member of a multiligand family that plays a well-established role as an HDL receptor [97]. In SRBI-deficient mutant mice, there was a significant increase in plasma α-Toc that was mostly distributed in HDL-like particles and a significant decrease in the α-Toc concentrations in bile and several tissues, including ovary, testis, lung, and brain, but not in the liver, spleen, kidney, or white fat [95]. These reports suggest that SRBI plays an important role in transferring α-Toc from plasma lipoproteins to specific tissues and the nervous system [98]. In addition, SRBI was also shown to mediate α-Toc efflux from the cytosolic compartment of Caco-2 cells to the apical medium, suggesting a potential regulatory role in α-Toc absorption [96].

In cattle, HDL is the major lipoprotein in the plasma and follicular fluid [99,100], and α-Toc is mainly located in HDL among lipoproteins [37]. Rajapaksha et al. [101] sequenced bovine *SCARB1* cDNA, which contains 509 amino acids. The changes in *SCARB1* mRNA levels were evaluated in developing bovine ovarian cells; however, the relationship between the mRNA level and α-Toc concentration in follicular fluid is unknown [99,100]. By contrast, Higuchi et al. [37] clarified that the upregulation of *SCARB1* mRNA in neutrophils in cattle supplemented with α-Toc and the cellular α-Toc contents were decreased after anti-SRBI treatment. These results suggest that SRBI is a crucial receptor in bovine neutrophils for the uptake of HDL-associated α-Toc. A study investigating the distribution of *SCARB1* mRNA in six tissues from cows demonstrated that their levels were high in the adrenal cortex and corpus luteum [101] because these organs take up large amounts of cholesterol from the bloodstream HDL to synthesise steroid hormones. Haga et al. [35] also reported high α-Toc accumulation in the adrenal gland and testis, with the highest expression levels of *SCARB1* mRNA among the 20 tissues in calves. These results suggest that the high expression of SRBI in these tissues may take up some α-Toc along with HDL.

2.2.5. ATP-Binding Cassette Transporter A1

ATP-binding cassette transporter A1 (ABCA1/*ABCA1*) is a cholesterol efflux regulatory protein. It is known that ABCA1 is involved in the regulation of cholesterol efflux from cells, and mutations in ABCA1 genes cause HDL deficiency [102]. In hepatocytes, lipid-free apoA1 is secreted to ABCA1, which localises on the plasma membrane and into intracellular sites, and nascent HDL (preβ-HDL) particles are formed [103,104]. It was also reported that ABCA1 mediates cellular secretion of α-Toc because hepatic α-Toc secretion is suppressed by ABCA1-RNAi or probucol (inactivator of ABCA1) in a rat hepatoma cell line and C57BL/6Cr mice in vivo [105]. Kono and Arai [66] demonstrated that αTTP transports α-Toc to the plasma membrane, where it is picked up by ABCA1 and excreted from the hepatocyte. The expression profile of human *ABCA1* mRNA in different tissues has been previously reported [106,107]. In non-hepatic cells, forced expression of ABCA1 markedly stimulated α-Toc efflux in baby hamster kidney cells [108]. Therefore, the regulation of the transportation and distribution of α-Toc must be closely linked to the complex mechanisms of cholesterol, lipoprotein and especially HDL metabolism via ABCA1.

Sequence analysis of bovine *ABCA1* cDNA revealed that the open reading frame of this gene consists of 6786 bases and encodes a protein of 2261 AA with a predicted molecular weight of 254 kDa [109]. Haga et al. [35] reported that the *ABCA1* mRNA level in Japanese Black beef calves was the highest in the liver, followed by heart muscle, lung, adipose, spleen and adrenal gland, which did not agree with the report of Farke et al. [109], who detected the highest mRNA level in the lungs of an adult lactating Holstein–Friesian cow. This discrepancy might be attributable to the differences in the bovine breeds and the life stages of the animals because of the different lipid metabolisms. In particular, the differences in cholesterol status may affect the distribution of *ABCA1* mRNA expression in tissues because *ABCA1* transcript activity is reportedly regulated by the liver X receptor and sterol regulatory element-binding protein (SREBP) 2, which are key proteins in cholesterol metabolism [102,110]. Hepatic gene expression in transition dairy cows has been reported in recent studies [2,111,112]. Furthermore, it is notable that the expression and localisation of ABCA1 in the bovine mammary epithelial cells, mammary gland, and milk fat globules have an important role in cholesterol homeostasis and milk fat synthesis [36,112–116]. The mammary expression of ABCA1 may be involved in the regulation and mechanism of α-Toc transfer into colostrum and milk.

2.2.6. Cytochrome P450 Family 4, Subfamily F, Polypeptide 2

Cytochrome P450 family 4, subfamily F, polypeptide 2 (CYP4F2/*CYP4F2*), is a member of the CYP4F subfamily ω-hydroxylate leukotriene B_4 [117–119]. In addition, tocopherols and tocotrienols are also metabolised by side chain degradation initiated by CYP4F2-catalyzed ω-hydroxylation, followed by β-oxidation, mainly in the liver [120]. The resulting water-soluble metabolites, carboxyethyl hydroxychromans (CEHC), are excreted in the urine [121,122]. Cytochrome P450 family 4, subfamily F, polypeptide 2 exhibited markedly higher catalytic activities for γ-Toc than α-Toc, resulting in preferential physiological retention of α-Toc and elimination of γ-Toc [120]. In particular, sesamin potently inhibited tocopherol-ω-hydroxylase activity exhibited by CYP4F2 [120], and dietary sesame seeds elevated α-Toc concentrations in the brain, liver and serum, and lowered the oxidative stress marker, thiobarbituric acid reactive substance (TBARS), in the brain of rats [123]. These data also emphasise the importance of CYP4F2 in α-Toc metabolism. The α-Toc levels in the body may be influenced by changes in the mRNA expression and enzyme activity of CYP4F2. In fact, the reduced expression of hepatic *TTPA*, *AFM* and *CYP4F2* genes probably leads to decreased plasma α-Toc levels and elevated α-Toc levels in the liver of streptozotocin-induced type 1 diabetes rat models [124]. Sterol regulatory element-binding proteins can transactivate *CYP4F2* transcription in hepatocytes [125], and decrease SREBP-1 proteins expression, resulting in reduced expression of CYP4F2, which slows the breakdown of α-Toc in experimental non-alcoholic fatty liver disease model mice [126].

In calves, the *CYP4F2* mRNA level was the highest in the liver, followed by the testis, adrenal gland, duodenum, and jejunum, which have high α-Toc accumulation [35]. The *CYP4F2* mRNA in lactating Holstein cows was significantly higher in the kidney than in the liver, lung, mammary gland, heart, skeletal muscle, spleen and uterus [38]. Furthermore, the hepatic [2,55] and mammary [2] *CYP4F2* mRNA expression in transition dairy cows was investigated (described in Section 4). The contribution of CYP4F2 to circulating α-Toc concentrations in transition dairy cows is not yet well understood. However, it is believed that the evidence of α-Toc metabolism by CYP4F2, similar to the metabolism of polyunsaturated fatty acids, can provide information regarding the physiological factors underlying hypovitaminosis E and the importance of CYP4F2 in the maintenance of dairy cow health [127].

As described above, in bovine species, the expression of *TTPA*, *AFM*, *SCARB1*, *ABCA1*, *SEC14L2* and *CYP4F2* genes in various tissues may play important roles in the regulation of α-Toc disposition (metabolism, transportation, and tissue distribution) (Figure 1). These genes may not be the whole explanation of α-Toc disposition mechanism [128–130]; however, the evidence of the expression of these α-Toc-related genes has the potential to help understand the physiological factors underlying hypovitaminosis E in the transition period of high-yielding dairy cows.

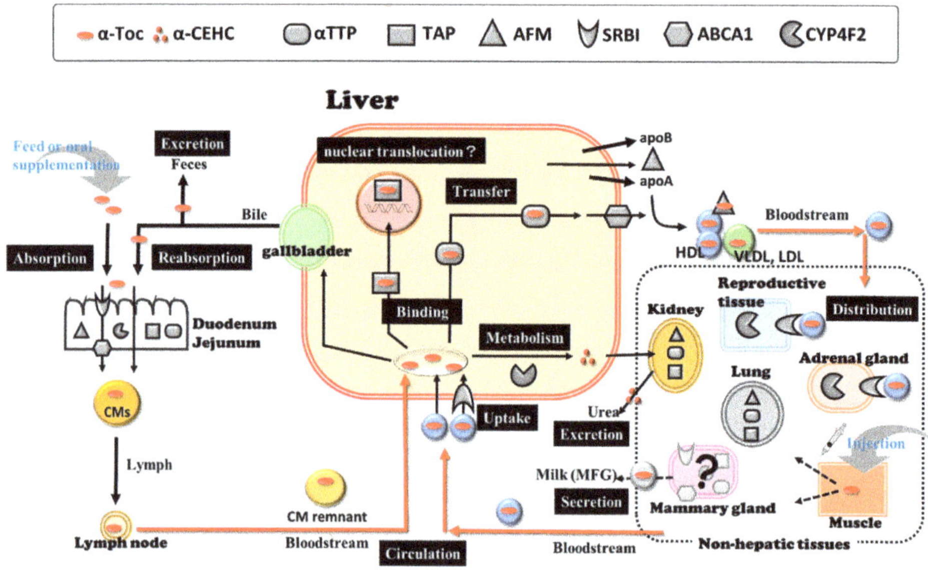

Figure 1. The possible mechanism of α-Toc disposition (metabolism, transportation, and tissue distribution) with the expression of α-Toc-related genes in cattle. Abbreviations: α-Toc, α-tocopherol; α-CEHC, α-carboxyethyl hydroxychromans; αTTP, α-tocopherol transfer protein; TAP, tocopherol associated protein; AFM, afamin; SRBI, scavenger receptor class B, Type I; ABCA1, ATP-binding cassette transporter A1; CYP4F2, cytochrome P450 family 4, subfamily F, polypeptide 2; CM, chylomicron; HDL, high density lipoprotein; VLDL, very low-density lipoprotein; LDL, low density lipoprotein; MFG, milk fat globules.

3. Hypovitaminosis E in Transition High-Yield Dairy Cows

3.1. Changes in α-Tocopherol Status in Transition Dairy Cows

It is well known that the plasma/serum concentrations of α-Toc in high-yield dairy cows gradually decrease throughout prepartum, starting from several weeks before calving, reaching a nadir at calving, and remaining at lower levels during the puerperal period (about 3–7 d), and increasing thereafter [2,10–13]. According to the fundamental research conducted by Weiss in 1990s [11,12,21,131–133], others [10,134] and NRC [52], based on disease risk and immune function in dairy cows, plasma/serum concentrations of α-Toc should be more than approximately 3 μg/mL in peripartum period; below this cut-off level is an α-Toc deficiency namely hypovitaminosis E. From 2000s to 2020s, the occurrence of hypovitaminosis E around calving has been still reported [2,13,15,16,79,135–138]. According to the NRC [52] recommendation to maintain this cut-off value, dry cows and heifers fed stored forages during the last 60 d of gestation require approximately 1.6 IU of supplemental α-Toc/kg BW (1120 IU/d for cows weighing 700 kg BW or approximately 80 IU/kg of DMI). However, the effect of supplementation and optimal dose may be far from certain. In cows supplemented with 1000 IU α-tocopheryl acetate per day (approximately 108 IU/kg of DMI) from 30 d prepartum to two weeks postpartum, the occurrence of hypovitaminosis E around calving was observed but with a lower decrease in plasma α-Toc concentration. However, when compared to the group with no α-Toc supplementation, the plasma α-Toc concentration was regained earlier after calving [135]. Hypovitaminosis E after calving also occurred in cows that were administered 1000 IU α-tocopheryl acetate per day (approximately 110 IU/kg of DMI) from 60 d prepartum to calving [139]. In agreement with these reports, the serum α-Toc concentration in high-yield dairy cows fed approximately 97 mg of α-toc/kg of DM from five weeks prepartum to calving, was less than 3 μg/mL during the prepartum period and approximately 1.5 μg/mL after calving [2].

By contrast, high-dose VE supplementation (3000 IU/day/cow) from eight weeks before the predicted calving date was sufficiently high to prevent hypovitaminosis E [140].

3.2. Disease Risk in Hypovitaminosis E and the Effects of α-Toc Supplemantetion in Transition Dairy Cows

To our knowledge, numerous detailed reviews about the disease risk of hypovitaminosis E and the α-Toc supplementation effect on the health of transition dairy cows and heifers have been published in the last 30 years (two databases, PubMed and Web of Science were searched using key words: vitamin E, dairy cows, tocopherol, 1990–2021, and reviewed full text is currently available online, 1990s: [17–22], 2000s: [1,23–27], 2010s: [28–32]). Based on three recent reviews published in the last 10 years [30–32], α-Toc supplementation has the potential to affect the incidence of mastitis, including milk somatic cell count (SCC) values, and retained foetal membranes (RFM). Some studies suggest that α-Toc supplementation at the level 1000 to 4000 IU/day/cow during the dry period can reduce the frequency of intramammary infection and the occurrence of clinical mastitis, as well as the levels of SCC in milk [17,133,141], suggesting that α-Toc deficiency may be a critical risk factor for the increased frequency of infection and duration in mammary glands during the transition period. Many studies have provided evidence suggesting that α-Toc supplementation can mitigate the immune dysfunction that occurs during the transition period [21,142–146]. The overproduction of reactive oxygen species (ROS) may contribute to several metabolic disturbances, resulting in the appearance of RFM [147]. The meta-analysis [148], performed to consolidate the results of studies that have evaluated the effect of α-Toc supplementation during the dry period on the incidence of RFM and found that α-Toc supplementation was associated with a decrease in the incidence of RFM. However, there are insufficient studies which have evaluated the effect of α-Toc alone on RFM, and much of the apparent significant benefit of α-Toc may be mediated by Se. The results of published studies were equivocal, and further work is required to assess α-Toc supplementation.

Next, we focus on relevant papers published in the last 10 years and present some novel data about the disease risk of hypovitaminosis E and α-Toc supplementation effects in transition dairy cows.

3.2.1. Left Displaced Abomasum

Left displaced abomasum (LDA) occurs in multiparous cows during the first month of lactation as part of the peripartal disease complex. Hasanpour et al. [149] reported that cows with LDA had approximately 42% lower serum α-Toc concentrations than healthy cows as control (2.7 vs. 4.7 μg/mL, respectively), and suggested use of supplementary α-Toc with Se for the LDA cattle. Qu et al. [16] investigated a retrospective case-control study to determine whether the lower serum α-Toc concentrations precede or remain after LDA in the transition period. Seven multiparous Holstein cows diagnosed with LDA between days 6 and 32 postpartum and 10 healthy cows from the same herd were analysed. Before calving, all 17 cows were visually healthy. These cows fed a TMR contained supplemental *all-rac*-α-tocopheryl acetate at 167 and 24.5 IU/kg DM before and after calving, respectively. Each cow in LDA had other diseases before LDA diagnosis (five cows had ketosis, four cows had metritis, two cows had milk fever, and one cow each retained placenta after twins, mastitis, or laminitis). Most cows had diseases after LDA correction (until day 49 postpartum: four cows had ketosis, one cow had mastitis, and one cow died from an intestinal ulcer 2 d after LDA diagnosis at day 34 postpartum). Serum α-Toc concentrations decreased dramatically in the first week postpartum in all cows; however, the nadir α-Toc concentration at day 7 postpartum in LDA cows was lower than that in controls, and the level (about 2.2 μg/mL) was diagnosed as hypovitaminosis E. Furthermore, cows with LDA during the first month postpartum (for 28 days) had, on average, lower serum α-Toc concentrations than the cut-off value. In control cows, α-Toc concentrations returned to prepartal concentrations (approximately 6.0 μg/mL) by four weeks postpartum, while serum α-Toc in cows with LDA remained lower than controls during the entire postpartum sampling period (seven weeks postpartum). A lower DMI might be a probable causative factor for the lower serum α-Toc concentrations in LDA cows, however, there was an absence of DMI data in the study [16]. The authors suggested that these findings indicate that lower serum α-Toc concentrations are a potential early indicator for the development of LDA in multiparous cows.

3.2.2. Retained Foetal Membranes, Stillbirth, and Reproductive Performance

Retained foetal membranes are an established risk factor for other peripartum diseases and fertility. Multiple physical, endocrine, and cellular factors are involved in RFM, and the immune and antioxidant potential before calving are important predisposing factors. Qu et al. [15] investigated the risk indicators of RFM using a nested case-control design and compared multiparous dairy cows that developed RFM with cows that remained healthy or cows that developed other diseases (metritis, mastitis, ketosis, or laminitis) in early lactation (each n = 32). These cows fed a TMR contained supplemental VE at 167 and 24.5 IU/kg DM as *all-rac*-α-tocopheryl acetate before and after calving, respectively. During the three weeks pre-calving, RFM cows had lower serum α-Toc concentrations than healthy cows. In addition, RFM cows tended to have lower serum α-Toc concentrations between three and two weeks prepartum than cows that developed other diseases. After calving, the α-Toc concentrations in RFM and other diseased cows were lower than the cut-off values (<3 μg/mL). Pontes et al. [150] evaluated the effects of injectable α-Toc supplementation during the last three weeks prepartum on the incidence of RFM and stillbirth, and reproductive performance in cows fed limited amounts of dietary α-Toc. During the prepartum period, cows were fed with less than 500 IU of supplemental *dl*-α-tocopherol per day. Cows were randomly assigned to remain as untreated controls (n = 441) or to receive three intramuscular injections of 1000 IU each of *dl*-α-tocopherol administered at three, two, and one week before calving (VitE group, n = 449). The serum α-Toc concentration at three weeks prepartum was similar between the control (n = 75) and VitE (n = 66), with an average of 2.97 μg/mL. The results showed that treatment with injectable α-Toc decreased the RFM rate from 20.1% to 13.5% and, decreased the incidence of stillbirth from 14.9% to 6.8%. In addition, cortisol is known to suppress leukocyte function in cattle, and its serum levels in cows receiving α-Toc were lower than those in control cows at 1 week before calving. The authors suggested that improved immune-cell function, through either antioxidant effects or other cellular signalling pathways activated by α-Toc, is likely to underlie the reduction in the incidence of RFM. Cows with VitE also tended to have improved pregnancy per insemination at first AI (36.7 vs. 30.1%) because of decreased pregnancy loss than control cows. Despite a similar insemination rate, VitE cows had a 22% higher pregnancy rate than control cows. The authors concluded that feeding dairy cows with less than 500 IU of VE per day during the last weeks of gestation may result in an inadequate level of plasma α-Toc that leads to compromised-peripartum health and subsequent reproduction.

3.2.3. Udder Health (Mastitis and SCC Values) and Milk Yield

Politis et al. [14] investigated the relationship between the incidence of clinical mastitis and blood α-Toc levels during dry off and calving. All cows (n = 146) were supplemented with *all-rac*-α-Toc at a rate of 3000 and 50 IU/cow per day during the dry period and lactation, respectively. According to the blood α-Toc concentrations, three groups at dry off were created: high (>6.25 μg/mL), medium (4.25–6.25 μg/mL), and low (<4.25 μg/mL). In addition, three groups at calving were created: high (>3 μg/mL), medium (2–3 μg/mL), and low (<2 μg/mL). No differences were observed in the incidence of mastitis between the three α-Toc groups during the dry-off period, however, the incidence of mastitis was four times lower in the high- (>3 μg/mL) and medium- (2–3 μg/mL) α-Toc groups than that in the low-α-Toc group (<2 μg/mL) at calving. In fact, cows with mastitis had lower concentrations of α-Toc (1.9 μg/mL) than healthy cows (2.74 μg/mL) at calving. These results also suggest that supplementation with *all-rac*-α-Toc at a rate of 3000 IU/cow per day during the dry period cannot always prevent hypovitaminosis E around calving.

Several other factors affect α-Toc availability and its physiological functions in cows, including the source of the α-Toc active substance, other fat-soluble nutrients in feeds, timing and period of supplementation, inclusion of Se, α-Toc content of the basal feeds, and method of administration (e.g., ruminal pellets or premixes in diet, or iv, sc and im injection). Moghimi-Kandelousi et al. [151] evaluated the effects of α-Toc supplementation on the serum level, milk yield, and SCC values of transition cows by considering a large set of variables that might influence the responses to α-Toc supplementation. To conduct a comprehensive meta-analysis of α-Toc supplementation effects during the transition period, after a broad search in journals and databases with keywords related to transition cows supplemented with VE and appropriate filtering of the results, 36 papers including 53 trials were selected from 528 publications (from 1979 to 2018), and their data were extracted into a database. Overall, 22 studies were conducted on Holstein cows, and the rest used other breeds. In 10 studies, primiparous and multiparous cows were studied, whereas in the remainder, treatments were applied only to multiparous cows. Six papers (12 independent trials) were used in the meta-analysis of milk yield, in which eight trials reported positive effects of VE supplementation. Furthermore, meta-regression showed that breed, Se supplementation, number of days treated prepartum, parity, and method of administration did not alter the effect of VE administration on milk yield in the first month of lactation. By contrast, the overall results of the meta-analysis showed non-significant changes in milk SCC with VE supplementation. The meta-analysis and meta-regression also showed that VE supplementation improved the reproductive performance of transition cows, such as shorter days open, reduced the number of services per conception, and decreased the odds of RFM. In conclusion, the authors suggested that up to 3600 IU/day of VE as an oral supplement during the transition period affects the milk production and reproduction performance of cows with Se supplementation.

As mentioned above, relevant papers published in the last 10 years with novel data indicated that hypovitaminosis E in the transition period is a risk factor for peripartum disease and lower performance in dairy cows. Alpha-tocopherol supplementation is an important method for the effective prevention of peripartum disease in high-yield dairy cows, although supplementation during the dry period cannot always prevent hypovitaminosis E around calving.

4. Physiological Factors Underlying Decreased Blood α-Toc Level and Hypovitaminosis E in Transition Period

The α-Toc deficiency may be caused by complex factors such as changes in the amount of α-Toc intake, increased oxidative stress and lipid peroxidation, and transfer of α-Toc into colostrum around calving. However, until recently, the physiological factors or pathways underlying hypovitaminosis E in the transition period of high-yielding dairy cows have been less well understood. Dramatic changes in lipid metabolism [39], endocrine status [3,4], physiological stress [2] and inflammation [5] occur and may damage the hepatic functions [40,41], playing a crucial role in the metabolism and disposition of α-Toc [33–35] during peripartum in dairy cows. Therefore, we discuss the candidate physiological factors underlying decreased blood α-Toc levels and hypovitaminosis E during late pregnancy to early lactation period.

4.1. The Decline in α-Toc Intake by Decreasing DMI from Close-Up Period to Calving

In dairy cows, daily intake of DM gradually decreases during the dry period, especially during the close-up period, and the DMI drops sharply to a nadir level at calving and increases thereafter toward early lactation [53,152,153]. It is believed that the major cause of the decline in DMI may be reduced rumen volume and capacity beyond the space requirement of developing foetuses in the close-up period, and physical and physiological stress-induced parturition. Decline in DMI is an unavoidable physiological phenomenon in dairy cows. If the α-Toc content in the feed is not different during this period and no supplementation is provided, its intake amount decreases with the decline in DMI [2]. Thus, to determine whether the decrease in serum α-Toc concentrations reflects the decline in α-Toc intake due to decreasing DMI around calving, Haga et al. [2] compared the rate of change between the serum α-Toc concentrations and its intake using monitoring data from high-yield Holstein cows (n = 28). During the close-up period, the α-Toc intake gradually decreased, declining at calving. After calving, DMI and α-Toc intake progressively recovered and increased with time. The serum concentrations of α-Toc decreased during the close-up period, reaching a nadir after parturition until 0.5 week after parturition. A comparison of the changes in α-Toc intake levels and serum α-Toc concentrations around the calving period (−2 to 2 weeks relative to parturition) revealed that, in prepartum, the decreasing α-Toc rates were similar between intake and blood; however, after calving, increasing serum α-Toc levels were significantly delayed compared to the recovery of α-Toc intake. These results suggest that decreased α-Toc intake levels may be one of the causes that strongly influence the decrease in serum α-Toc concentrations until pre-calving, but not the only influencer during post-calving.

4.2. Changes in the Digestive and Absorptive Functions of α-Toc with Change in the Expression of α-Toc-Related Genes

It was reported that *all-rac*-α-tocopheryl acetate was stable in the rumen of high-yielding dairy cows [154]. However, these cows are at the risk of subacute ruminal acidosis (SARA; diagnosed when reticulo-ruminal pH is <5.6 for more than three hours per day) during the periparturient period [155]. The stability of α-Toc under SARA conditions is not well known.

Blood was sampled through a trial from sheep with a ligated pylorus that received α-Toc, suggesting that no significant amount of α-Toc was absorbed from the preintestine region to the blood stream in ruminants [156]. This finding was indirectly supported by Haga et al. [35], who investigated the expression of α-Toc-related genes and α-Toc accumulation in weaned calves with and without oral administration of α-Toc. These results indicated that, in the gastrointestinal (GI) tract (rumen to the colon), the jejunum and duodenum had high α-Toc content and expressed high levels of *SRB1*, *AFM*, *TAP* and *CYP4F2* mRNA. The functions of αTTP and TAP in the GI tract have not been clarified, however, these results suggest that the expression of α-Toc-related genes is involved in the regulation of absorbed dietary α-Toc in the GI tract. Thus, these small intestine regions may be the major oral α-Toc absorption sites in cattle. However, to the best of our knowledge, there has been no detailed investigation of the changes in the digestive and absorptive function of α-Toc in the GI tract of high-yield dairy cows during the transition period, although there may be a possibility of changing the function with changes in the expression of α-Toc-related genes. Although challenging, further studies investigating in detail about the changes in the digestive and absorptive function of α-Toc, with α-Toc-related genes expression, in the GI tract of high-yield dairy cows are needed.

4.3. The Decline of Plasma HDL Level as an α-Toc Carrier from Close-Up Period to Calving

Alpha-tocopherol, a fat-soluble vitamin, requires a carrier system for bloodstream transportation. Herdt and Smith [88] investigated the distribution of α-Toc and cholesterol among the various lipoprotein density fractions in the blood of lactating Holstein cows; the percentage of total plasma α-Toc and cholesterol were VLDL (2% and 2%), LDL (17% and 22%), and HDL (77% and 72%, respectively). In addition, the α-Toc:cholesterol ratios were not significantly different among the lipoprotein fractions. These results indicated that α-Toc and cholesterol were distributed in equal proportions among lipoprotein fractions and HDL is a major lipoprotein carrier of α-Toc in the plasma of dairy cows. These results were supported by Higuchi et al. [37]. During the transition period, plasma HDL, VLDL, LDL, and cholesterol levels in dairy cows gradually decreased throughout prepartum, reaching a nadir at calving, and increasing thereafter [2,112,157]. This decline in lipoproteins may be caused by (1) the reduced DMI intake [2,53,152,153]; (2) changes in lipid metabolism and increased plasma NEFA and BHBA [39,158]; (3) the impaired hepatic export mechanism with reduced secretion of ApoB100 (decreased mRNA) and apoA1 [159], and (4) enhanced transfer of these compounds into the fat rich-colostrum [112]. Based on the changes monitored in blood α-Toc/HDL ratio, which assesses the rate-limiting levels of HDL concentration as an α-Toc carrier, from −2 to 2 weeks relative to parturition, Haga et al. [2] confirmed that the α-Toc/HDL ratio was significantly lower during the post-calving period than during the pre-calving period. These results suggest that remaining at lower serum α-Toc concentrations after calving might not result in lower HDL concentrations. It appeared that other causes might determine the lower serum α-Toc concentrations during the first week after calving in addition to changes in the levels of α-Toc intake and plasma HDL concentrations.

4.4. Increasing Systemic Oxidative Stress and Consumption of α-Toc as Antioxidant around Calving

Oxidative stress in living organisms is generated when free radical production exceeds the capacity of antioxidant mechanisms. Considerable evidence [2,14,160–163] and reviews [6,164] suggest that high-yield dairy cows experience severe oxidative stress around calving and during the onset of lactation. Bernabucci et al. [165] reported that dairy cows with higher BCS and higher body condition losses are more prone to oxidative stress during the periparturient period. It was confirmed that the increase in systemic oxidative stress around calving roughly coincided with a decrease in blood α-Toc concentration [2]. Furthermore, mRNA expression of the major antioxidant enzymes in the liver was markedly downregulated at calving [2]. Since α-Toc is considered an important antioxidant, these results suggest that both systemic and hepatic antioxidative/oxidative balance may be lower, and the consumption of α-Toc increases around calving. In fact, supplementation with α-Toc could reduce the markers of oxidative damage, serum/plasma malondialdehyde (the product of lipid peroxidation) [139,166] and heat shock protein 70 [139].

4.5. Decreasing Hepatic α-Toc Transfer to Circulation with Change in the Expression of α-Toc-Related Genes

The study on bovine tissues distribution [35] demonstrated that the liver may play a central role in the regulation of α-Toc disposition, as inferred by the high hepatic expression of six α-Toc-related genes (see Section 2.2 about the information and references). However, high-yield dairy cows experience physiological stress [2], systemic inflammation [5], oxidative stress [139,160,167], hepatic endoplasmic reticulum (ER) stress [2,168], hepatocyte apoptosis [41], hepatic injury (necrosis-like cell death) [79] and development of fatty liver resulting in severe negative energy balance (NEB) because of high milk production immediately after calving [41,159,169]. Thus, liver function will be changed and substantially inhibited around calving.

Gessner et al. [168] showed the upregulation of ER stress-induced genes of the unfolded protein response (UPR) markers in the liver at one week postpartum compared to three weeks prepartum. The expression levels of these genes decreased from 1 week postpartum to later lactation. Sadri et al. [55] reported that the hepatic *TTPA* and *TAP* mRNA in dairy cows during the transition period tended to be lower than those during the peak lactation period (105 d relative to parturition), although the changes in these gene expressions were not observed during the transition period (day −21, 1 and 21 d postpartum). However, using consecutive liver tissue biopsies in the peripartum period (−4, −1, 0, 1 and 4 weeks postpartum) [2] demonstrated that *TTPA*, *AFM* and *TAP* mRNA expression were strongly downregulated immediately after calving. In the experiment, the expression of *ALB* mRNA, a negative acute-phase protein that plays the most basal hepatic function, was also downregulated, and hepatic ER stress-induced UPR and acute-phase response occurred at calving. After the first week postpartum, when the mRNA expression of *TTPA*, *AFM*, *TAP* and *ALB* recovered, the elevated UPR markers and haptoglobin mRNA expression decreased. These results suggest that α-Toc transfer from the liver into the bloodstream may be suppressed in the first days after calving because of temporal downregulation of *TTPA* and *AFM*. Ongoing research [79] showed that the hepatic mRNA expression levels of *TTPA* and *ALB* were continuously downregulated at least during the

3 d after calving. These changes in the hepatic expression of α-Toc-related genes might be associated with the maintenance of lower serum α-Toc concentrations during the first week after calving. However, there is insufficient knowledge about the hepatic expression of α-Toc-related genes and proteins in transition high-yield dairy cows. More studies are needed to delineate the relationship between hepatic α-Toc transfer and metabolism and the occurrence of hypovitaminosis E around calving.

4.6. Increasing Mammary α-Toc Transfer from Blood to Colostrum with Change in the Expression of α-Toc-Related Genes at Calving

It is well known that in multiparous dairy cows the α-Toc concentration in colostrum is approximately 5- to 8-fold higher than that in mature milk [13,170]. Alpha-tocopherol is a lipid-soluble micronutrient and its concentration is strongly affected by the level of milk fat; however, the α-Toc concentrations in colostrum, which were normalised by the milk fat value, were also approximately five- to eight-fold greater than those in mature milk [2,137]. The calculated α-Toc efflux with milk (concentration × milk yield) was highest in colostrum and declined in transition milk (2–3 d relative to parturition), reaching nadir levels in mature milk after one week relative to parturition [2]. In addition, the estimated mammary extraction ratios [137,171,172] of α-Toc after calving (colostrum) and at 6 weeks lactation (mature milk) were at 1.3 and 0.05%, respectively [2]. These estimations suggest that α-Toc uptake during colostrum production might be more than 20-fold greater than that in mature milk production. Furthermore, α-Toc did not accumulate in precolostrum (at one week before parturition) [2]. These results indicate that high α-Toc concentration in colostrum might be caused by the presence of a mechanism that temporarily augments a specific α-Toc transfer from the blood to colostrum across the mammary gland at calving, which might be a mechanism contributing to a lower serum α-Toc concentration at calving.

To test the possibility of this mechanism, Haga et al. [2] measured the mRNA expression levels of α-Toc-related genes in biopsied mammary gland tissues. SRBI and ABCA1 play pivotal roles in cholesterol transport, milk-fat globule synthesis and these secretions from the mammary gland [36,114,115,173]. These genes might also contribute to blood α-Toc transfer into colostrum or mature milk. However, *SCARB1* mRNA expression in mammary gland tissues was downregulated after calving. In various mouse tissues, there are SRBI-dependent and -independent pathways for tissue α-Toc uptake [95]. These observations suggest that further studies are needed to investigate the expression of other receptors, such as LDL-R and Niemann-Pick C1-Like 1 [129,173], which may be involved in α-Toc uptake in bovine mammary glands. The *ABCA1* mRNA [113,114] and protein [115] levels in bovine mammary glands during the dry-off period were higher than those during lactation. In agreement with these findings, in the transition period, *ABCA1* mRNA levels declined after calving [2]. By contrast, Mani et al. [115] demonstrated that the subcellular distribution of ABCA1 changed throughout the pregnancy-lactation cycle, and ABCA1 was present in milk-fat globule membranes isolated from fresh mature milk. It has been reported that key acceptors of cholesterol and α-Toc efflux by ABCA1, apoA1, were present in milk-fat globule membranes and the protein level was significantly higher in milk fat globule membranes prepared from colostrum than in mature milk [174]. These results suggested that ABCA1 proteins could share a function in the regulation of α-Toc transfer through localisation in the basal, apical membranes, and cytoplasm of mammary epithelial cells. The expression of *TTPA*, *SEC14L2* and *CYP4F2* mRNA in bovine mammary gland tissue and the changes in these gene expression levels during peripartum were observed [2]. The functions of αTTP and TAP in mammary epithelial cells have not been clarified, however, these results suggest that α-Toc-related genes expressed in mammary gland tissues may play an important role in the transfer of α-Toc from blood to colostrum. Further investigation to explore the physiological function of these genes in mammary glands during peripartum is needed.

A comparison of blood α-Toc profiles between mastectomized and intact dairy cows can determine the cause of decreasing blood α-Toc because the metabolic demands of colostrum production and lactation are eliminated, with only the calving effect remaining.

Goff et al. [152] demonstrated that mastectomies reduced but did not eliminate loss of plasma α-Toc around calving, and strengthening the mammary transfer to colostrum does not exclusively affect the blood α-Toc concentration.

Based on recent literature, six physiological factors may be involved in α-Toc deficiency and hypovitaminosis E during the transition period of high-yielding dairy cows (Figure 2). However, the mechanisms and pathways are less well understood, and further studies are needed to understand the physiological role of α-Toc-related molecules in the GI tract and mammary gland.

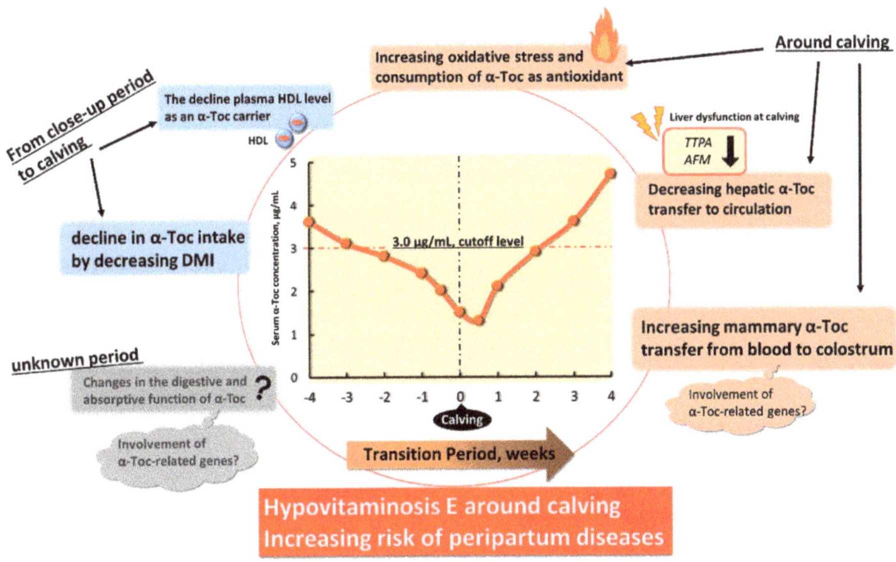

Figure 2. Six candidate physiological factors causing decreased blood α-tocopherol (α-Toc) level and hypovitaminosis E in transition high-yield dairy cows. Abbreviations: DMI, dry matter intake; *TTPA*, α-tocopherol transfer protein gene; *AFM*, afamin gene; HDL, high density lipoprotein.

5. Foresight

Many reviews on α-Toc and dairy cows, until Politis [30], focus primarily on the effects of supplementation on health and performance, and the utility of α-Toc as a biomarker for periparturient cow diseases has not yet been thoroughly considered. However, Qu et al. [16] indicated that lower serum α-Toc concentration is a potential early indicator for the development of LDA in multiparous cows. The same group [15] also implied that the best predictive indicators for disease were lower serum α-Toc concentrations and higher NEFA and BHBA concentrations during the prepartum period. Most recent studies on biomarkers for disease risk and milk production in periparturient dairy cows [138] performed a longitudinal, herd-based epidemiologic investigation of serum β-carotene, retinol, and α-Toc concentrations in dairy cattle on five commercial farms at three specific time points through the non-lactating and early lactation periods. The serum α-Toc concentrations, from dry-off to close-up, decreased ($p < 0.01$; LSM ± SE, 4.69 ± 1.09 to 3.00 ± 1.09 μg/mL) and then further decreased from close-up to early lactation ($p < 0.01$; 3.00 ± 1.09 to 1.44 ± 1.09 μg/mL). Higher α-Toc concentrations were associated with greater ME305 (305-d mature-equivalent milk yield), especially among cows in parity 1. Higher α-Toc concentrations were associated with decreased odds of disease among cows in parity 1, but were associated with increased odds of disease among cows in parity 2. The mechanism behind this finding is unknown, but may be associated with the increased stress cows' parity 1 experience in the peripartum period. No vitamins were significantly associated with lameness or mastitis in multivariable models. The authors suggested that future studies should further investigate the association between serum concentrations of lipid vitamins and periparturient cow diseases to establish serum ranges at which these biomarkers indicate increased disease risk. According to a recent study [175], whole blood samples can be directly used, and the measurement of lipid vitamin levels can be performed effortlessly in less than 5 min, even at the cow-side, using a field-portable fluorometer/spectrophotometer (iCheck) without further sample preparation. Based on this new development, the concentrations of VE, β-carotene, and vitamin A in the blood can be used as nutritional biomarkers to directly optimise nutritional interventions at the farm, together with stakeholders such as veterinarians, farmers, nutritional advisors, and feed consultants. Thus, the wave of the future, using serum vitamin concentrations as biomarkers for disease risk during the periparturient period, in addition to monitoring dietary supplementation, may prove to be an effective tool for improving animal health. Furthermore, a better understanding of the physiological factors underlying the cause of hypovitaminosis E in the transition period will improve the utility of α-Toc as a biomarker for periparturient high-yield dairy cow diseases.

6. Conclusions

Numerous studies and reviews about the disease risk of hypovitaminosis E (<3 μg/mL) and the effect of α-Toc supplementation on the health and performance of transition dairy cows and heifers have been published in the last 30 years. However, the risk and supplemental effects are controversial because several factors affect the availability of α-Toc and its physiological functions in cows. In the current review, we focused on relevant papers published in the last 10 years and presented some novel data about the disease risk of hypovitaminosis E and the effects of α-Toc supplementation in transition dairy cows. These data strongly demonstrate that hypovitaminosis E in the transition period is a risk factor for the occurrence of peripartum disease and lower performance in dairy cows. Alpha-tocopherol supplementation of more than 3000 IU/day during the prepartum period can be important for the effective prevention of peripartum disease in high-yield dairy cows. Furthermore, a study on the effectiveness of using serum vitamin levels as biomarkers to predict disease in dairy cows was reported, and a rapid field test (cow-side assay) for measuring vitamin levels using whole blood was developed. By contrast, evidence for how hypovitaminosis E occurred during the transition period was scarce until the 2010s. Pioneering studies conducted with humans and rodents have identified and characterised some α-Toc-related proteins, molecular players involved in α-Toc regulation, from the 1990s, followed by a study in ruminants from the 2010s. Based on the recent literature, six physiological factors may be involved in α-Toc deficiency and hypovitaminosis E during the transition period of high-yielding dairy cows. However, the mechanisms and pathways are less well understood, and further studies are needed to understand the physiological role of α-Toc-related molecules in cattle. In the future, understanding the molecular mechanisms underlying hypovitaminosis E will contribute to the prevention of peripartum disease and high performance in dairy cows.

Author Contributions: The paper was mainly designed by S.H. and S.R. Research articles were collected by S.H. and H.I. The manuscript was mainly written by S.H. and edited by S.R. and H.I. All authors have read and agreed to the published version of the manuscript.

Funding: This research received no external funding.

Institutional Review Board Statement: Not applicable.

Data Availability Statement: Not applicable.

Acknowledgments: This work was supported in part by JSPS KAKENHI grant number 25871102 and 15K21620.

Conflicts of Interest: The authors declare no conflict of interest.

References

1. Spears, J.W.; Weiss, W.P. Role of antioxidants and trace elements in health and immunity of transition dairy cows. *Vet. J.* **2008**, *176*, 70–76. [CrossRef] [PubMed]
2. Haga, S.; Miyaji, M.; Nakano, M.; Ishizaki, H.; Matsuyama, H.; Katoh, K.; Roh, S. Changes in the expression of α-tocopherol–related genes in liver and mammary gland biopsy specimens of peripartum dairy cows. *J. Dairy Sci.* **2018**, *101*, 5277–5293. [CrossRef]
3. Ohtani, Y.; Takahashi, T.; Sato, K.; Ardiyanti, A.; Song, S.H.; Sato, R.; Onda, K.; Wada, Y.; Obara, Y.; Suzuki, K.; et al. Changes in circulating adiponectin and metabolic hormone concentrations during periparturient and lactation periods in Holstein dairy cows. *Anim. Sci. J.* **2012**, *83*, 788–795. [CrossRef] [PubMed]
4. Nakano, M.; Suzuki, Y.; Haga, S.; Yamauchi, E.; Kim, D.; Nishihara, K.; Nakajima, K.; Gotoh, T.; Park, S.; Baik, M.; et al. Downregulated angiopoietin-like protein 8 production at calving related to changes in lipid metabolism in dairy cows. *J. Anim. Sci.* **2018**, *96*, 2646–2658. [CrossRef] [PubMed]
5. Bradford, B.J.; Yuan, K.; Farney, J.K.; Mamedova, L.K.; Carpenter, A.J. Invited review: Inflammation during the transition to lactation: New adventures with an old flame. *J. Dairy Sci.* **2015**, *98*, 6631–6650. [CrossRef] [PubMed]
6. Abuelo, A.; Hernández, J.; Benedito, J.L.; Castillo, C. Redox Biology in Transition Periods of Dairy Cattle: Role in the Health of Periparturient and Neonatal Animals. *Antioxidants* **2019**, *8*, 20. [CrossRef] [PubMed]
7. Sordillo, L.M. Nutritional strategies to optimize dairy cattle immunity. *J. Dairy Sci.* **2016**, *99*, 4967–4982. [CrossRef]
8. Ingvartsen, K.L.; Dewhurst, R.J.; Friggens, N.C. On the relationship between lactational performance and health: Is it yield or metabolic imbalance that cause production diseases in dairy cattle? A position paper. *Livest. Prod. Sci.* **2003**, *83*, 277–308. [CrossRef]
9. Mulligan, F.J.; Doherty, M.L. Production diseases of the transition cow. *Vet. J.* **2008**, *176*, 3–9. [CrossRef]
10. Goff, J.P.; Stabel, J.R. Decreased plasma retinol, α-tocopherol, and zinc concentration during the periparturient period: Effect of milk fever. *J. Dairy Sci.* **1990**, *73*, 3195–3199. [CrossRef]
11. Weiss, W.P.; Todhunter, D.A.; Hogan, J.S.; Smith, K.L. Effect of duration of supplementation of selenium and vitamin E on periparturient dairy cows. *J. Dairy Sci.* **1990**, *73*, 3187–3194. [CrossRef]
12. Weiss, W.P.; Hogan, J.S.; Smith, K.L.; Hoble, K.H. Relationships among selenium, vitamin E, and mammary gland health in commercial dairy herds. *J. Dairy Sci.* **1990**, *73*, 381–390. [CrossRef]
13. Kumagai, H.; Chaipan, Y. Changes of vitamin E status of periparturient dairy cows and newborn calves. *Anim. Sci. J.* **2004**, *75*, 541–547. [CrossRef]
14. Politis, I.; Theodorou, G.; Lampidonis, A.D.; Kominakis, A.; Baldi, A. Short communication: Oxidative status and incidence of mastitis relative to blood α-tocopherol concentrations in the postpartum period in dairy cows. *J. Dairy Sci.* **2012**, *95*, 7331–7335. [CrossRef] [PubMed]
15. Qu, Y.; Fadden, A.N.; Traber, M.G.; Bobe, G. Potential risk indicators of retained placenta and other diseases in multiparous cows. *J. Dairy Sci.* **2014**, *97*, 4151–4165. [CrossRef] [PubMed]
16. Qu, Y.; Lytle, K.; Traber, M.G.; Bobe, G. Depleted serum vitamin E concentrations precede left displaced abomasum in early-lactation dairy cows. *J. Dairy Sci.* **2013**, *96*, 3012–3022. [CrossRef]
17. Hogan, J.S.; Weiss, W.P.; Smith, K.L. Role of vitamin E and selenium in host defense against mastitis. *J. Dairy Sci.* **1993**, *76*, 2795–2803. [CrossRef]
18. Miller, J.K.; Brzezinska-Slebodzinska, E.; Madsen, F.C. Oxidative stress, antioxidants, and animal function. *J. Dairy Sci.* **1993**, *76*, 2812–2823. [CrossRef]
19. Smith, K.L.; Hogan, J.S.; Weiss, W.P. Dietary vitamin E and selenium affect mastitis and milk quality. *J. Anim. Sci.* **1997**, *75*, 1659–1665. [CrossRef] [PubMed]
20. Sordillo, L.M.; Shafer-Weaver, K.; DeRosa, D. Immunobiology of the mammary gland. *J. Dairy Sci.* **1997**, *80*, 1851–1865. [CrossRef]
21. Weiss, W.P. Requirements of fat-soluble vitamins for dairy cows: A review. *J. Dairy Sci.* **1998**, *81*, 2493–2501. [CrossRef]
22. Hemingway, R.G. The influences of dietary selenium and vitamin E intakes on milk somatic cell counts and mastitis in cows. *Vet. Res. Commun.* **1999**, *23*, 481–499. [CrossRef] [PubMed]
23. Allison, R.D.; Laven, R.A. Effect of vitamin E supplementation on the health and fertility of dairy cows: A review. *Vet. Rec.* **2000**, *147*, 703–708. [PubMed]
24. Spears, J.W. Micronutrients and immune function in cattle. *Proc. Nutr. Soc.* **2000**, *59*, 587–594. [CrossRef] [PubMed]
25. Hemingway, R.G. The influences of dietary intakes and supplementation with selenium and vitamin E on reproduction diseases and reproductive efficiency in cattle and sheep. *Vet. Res. Commun.* **2003**, *27*, 159–174. [CrossRef] [PubMed]
26. Baldi, A. Vitamin E in dairy cows. *Livest. Prod. Sci.* **2005**, *98*, 117–122. [CrossRef]
27. Wilde, D. Influence of macro and micro minerals in the peri-parturient period on fertility in dairy cattle. *Anim. Reprod. Sci.* **2006**, *96*, 240–249. [CrossRef]
28. Vagni, S.; Saccone, F.; Pinotti, L.; Baldi, A. Vitamin E Bioavailability: Past and Present Insights. *Food Sci. Nutr.* **2011**, *2*, 1088–1096. [CrossRef]
29. Yang, F.L.; Li, X.S.; He, B.X. Effects of vitamins and trace-elements supplementation on milk production in dairy cows: A review. *Afr. J. Biotechnol.* **2011**, *10*, 2574–2578.

30. Politis, I. Reevaluation of vitamin E supplementation of dairy cows: Bioavailability, animal health and milk quality. *Animals* **2012**, *6*, 1427–1434. [CrossRef]
31. Abuelo, A.; Hernández, J.; Benedito, J.L.; Castillo, C. The importance of the oxidative status of dairy cattle in the periparturient period: Revisiting antioxidant supplementation. *J. Anim. Physiol. Anim. Nutr. (Berl.)* **2015**, *99*, 1003–1016. [CrossRef] [PubMed]
32. Pinotti, L.; Manoni, M.; Fumagalli, F.; Rovere, N.; Tretola, M.; Baldi, A. The role of micronutrients in high-yielding dairy ruminants: Choline and vitamin E. *Ankara Üniv. Vet. Fak. Derg.* **2020**, *67*, 209–214. [CrossRef]
33. Liu, K.; Luo, H.L.; Yue, D.B.; Ge, S.Y.; Yuan, F.; Yan, L.Y.; Jia, H.N. Molecular cloning and characterization of the sheep α-TTP gene and its expression in response to different vitamin E status. *Gene* **2012**, *494*, 225–230. [CrossRef] [PubMed]
34. Liu, K.; Luo, H.L.; Zuo, Z.Y.; Jia, H.N.; Zhang, Y.W.; Chang, Y.F.; Jiao, L.J. Regulation of sheep α-TTP by dietary vitamin E and preparation of monoclonal antibody for sheep α-TTP. *Gene* **2014**, *540*, 110–116. [CrossRef] [PubMed]
35. Haga, S.; Nakano, M.; Ishizaki, H.; Roh, S.; Katoh, K. Expression of α-tocopherol-associated genes and α-tocopherol accumulation in Japanese Black (*Wagyu*) calves with and without α-tocopherol supplementation. *J. Anim. Sci.* **2015**, *93*, 4048–4057. [CrossRef]
36. Mani, O.; Sorensen, M.T.; Sejrsen, K.; Bruckmaier, R.M.; Albrecht, C. Differential expression and localization of lipid transporters in the bovine mammary gland during the pregnancy-lactation cycle. *J. Dairy Sci.* **2009**, *92*, 3744–3756. [CrossRef]
37. Higuchi, H.; Ito, E.; Iwano, H.; Oikawa, S.; Nagahata, H. Effects of vitamin E supplementation on cellular α-tocopherol concentrations of neutrophils in Holstein calves. *Can. J. Vet. Res.* **2013**, *77*, 120–125.
38. Kuhn, M.J.; Putman, A.K.; Sordillo, L.M. Widespread basal cytochrome P450 expression in extrahepatic bovine tissues and isolated cells. *J. Dairy Sci.* **2020**, *103*, 625–637. [CrossRef]
39. Drackley, J.K. Biology of dairy cows during the transition period: The final frontier? *J. Dairy Sci.* **1999**, *82*, 2259–2273. [CrossRef]
40. Trevisi, E.; Amadori, M.; Cogrossi, S.; Razzuoli, E.; Bertoni, G. Metabolic stress and inflammatory response in high-yielding, periparturient dairy cows. *Res. Vet. Sci.* **2012**, *93*, 695–704. [CrossRef]
41. Tharwat, M.; Takamizawa, A.; Hosaka, Y.Z.; Endoh, D.; Oikawa, S. Hepatocyte apoptosis in dairy cattle during the transition period. *Can. J. Vet. Res.* **2012**, *76*, 241–247.
42. Finucane, K.A.; McFadden, T.B.; Bond, J.P.; Kennelly, J.J.; Zhao, F.Q. Onset of lactation in the bovine mammary gland: Gene expression profiling indicates a strong inhibition of gene expression in cell proliferation. *Funct. Integr. Genomics* **2008**, *8*, 251–264. [CrossRef]
43. Evans, H.M.; Bishop, K.S. On the existence of a hitherto unrecognized dietary factor essential for reproduction. *Science* **1922**, *56*, 650–651. [CrossRef] [PubMed]
44. Brigelius-Flohe, R.; Traber, M.G. Vitamin E: Function and metabolism. *FASEB J.* **1999**, *13*, 1145–1155. [CrossRef] [PubMed]
45. Barella, L.; Muller, P.Y.; Schlachter, M.; Hunziker, W.; Stöcklin, E.; Spitzer, V.; Meier, N.; de Pascual-Teresa, S.; Minihane, A.M.; Rimbach, G. Identification of hepatic molecular mechanisms of action of alpha-tocopherol using global gene expression profile analysis in rats. *Biochim. Biophys. Acta* **2004**, *1689*, 66–74. [CrossRef]
46. Rimbach, G.; Moehring, J.; Huebbe, P.; Lodge, J.K. Gene-Regulatory Activity of α-Tocopherol. *Molecules* **2010**, *15*, 1746–1761. [CrossRef] [PubMed]
47. Aparicio, J.M.; Bélanger-Quintana, A.; Suárez, L.; Mayo, D.; Benítez, J.; Díaz, M.; Escobar, H. Ataxia with isolated vitamin E deficiency: Case report and review of the literature. *J. Pediatr. Gastroenterol. Nutr.* **2001**, *33*, 206–210. [CrossRef]
48. Yokota, T.; Igarashi, K.; Uchihara, T.; Jishage, K.; Tomita, H.; Inaba, A.; Li, Y.; Arita, M.; Suzuki, H.; Mizusawa, H.; et al. Delayed-onset ataxia in mice lacking α-tocopherol transfer protein: Model for neuronal degeneration caused by chronic oxidative stress. *Proc. Natl. Acad. Sci. USA* **2001**, *98*, 15185–15190. [CrossRef]
49. Azzi, A.; Ricciarelli, R.; Zingg, J.M. Non-antioxidant molecular functions of alpha-tocopherol (vitamin E). *FEBS Lett.* **2002**, *519*, 8–10. [CrossRef]
50. Jensen, S.K.; Nørgaard, J.V.; Lauridsen, C. Bioavailability of alpha-tocopherol stereoisomers in rats depends on dietary doses of *all-rac-* or RRR-alpha-tocopheryl acetate. *Br. J. Nutr.* **2006**, *95*, 477–487. [CrossRef]
51. Harris, P.L.; Ludwig, M.I. Relative vitamin E potency of natural and of synthetic alpha-tocopherol. *J. Biol. Chem.* **1949**, *179*, 1111–1115. [CrossRef]
52. National Academy Press. *Nutrient Requirements of Dairy Cattle;* National Academy Press: Washington, WA, USA, 2001.
53. Meglia, G.E.; Jensen, S.K.; Lauridsen, C.; Waller, K.P. Alpha-tocopherol concentration and stereoisomer composition in plasma and milk from dairy cows fed natural or synthetic vitamin E around calving. *J. Dairy Res.* **2006**, *73*, 227–234. [CrossRef] [PubMed]
54. Jensen, S.K.; Lashkari, S.; Kristensen, N.B. Pharmacokinetics of α-tocopherol stereoisomers in plasma and milk of cows following a single dose injection of *all-rac-*α-tocopheryl acetate. *Food Chem.* **2020**, *310*, 125931. [CrossRef]
55. Sadri, H.; Dänicke, S.; Meyer, U.; Rehage, J.; Frank, J.; Sauerwein, H. Tocopherols and tocotrienols in serum and liver of dairy cows receiving conjugated linoleic acids or a control fat supplement during early lactation. *J. Dairy Sci.* **2015**, *98*, 7034–7043. [CrossRef] [PubMed]
56. Hidiroglou, N.; Laflamme, L.F.; McDowell, L.R. Blood plasma and tissue concentrations of vitamin E in beef cattle as influenced by supplementation of various tocopherol compounds. *J. Anim. Sci.* **1988**, *66*, 3227–3234. [CrossRef]
57. Sato, Y.; Hagiwara, K.; Arai, H.; Inoue, K. Purification and characterization of the α-tocopherol transfer protein from rat liver. *FEBS Lett.* **1991**, *288*, 41–45. [CrossRef]
58. Sato, Y.; Arai, H.; Miyata, A.; Tokita, S.; Yamamoto, K.; Tanabe, T.; Inoue, K. Primary structure of alpha-tocopherol transfer protein from rat liver. Homology with cellular retinaldehyde-binding protein. *J. Biol. Chem.* **1993**, *268*, 17705–17710. [CrossRef]

59. Arita, M.; Sato, Y.; Miyata, A.; Tanabe, T.; Takahashi, E.; Kayden, H.J.; Arai, H.; Inoue, K. Human α-tocopherol transfer protein: cDNA cloning, expression and chromosomal localization. *Biochem. J.* **1995**, *306*, 437–443. [CrossRef]
60. Ouahchi, K.; Arita, M.; Kayden, H.; Hentati, F.; Hamida, M.B.; Sokol, R.; Arai, H.; Inoue, K.; Mandel, J.L.; Koenig, M. Ataxia with isolated vitamin E deficiency is caused by mutations in the α-tocopherol transfer protein. *Nat. Genet.* **1995**, *9*, 141–145. [CrossRef]
61. Hentati, A.; Deng, H.X.; Hung, W.Y.; Nayer, M.; Ahmed, M.S.; He, X.; Tim, R.; Stumpf, D.A.; Siddique, T. Human α-tocopherol transfer protein: Gene structure and mutations in familial vitamin E deficiency. *Ann. Neurol.* **1996**, *39*, 295–300. [CrossRef]
62. Arita, M.; Nomura, K.; Arai, H.; Inoue, K. α-Tocopherol transfer protein stimulates the secretion of α-tocopherol from a cultured liver cell line through a brefeldin A-insensitive pathway. *Proc. Natl. Acad. Sci. USA* **1997**, *94*, 12437–12441. [CrossRef] [PubMed]
63. Kono, N.; Ohto, U.; Hiramatsu, T.; Urabe, M.; Uchida, Y.; Satow, Y.; Arai, H. Impaired α-TTP-PIPs interaction underlies familial vitamin E deficiency. *Science* **2013**, *340*, 1106–1110. [CrossRef] [PubMed]
64. Qian, J.; Morley, S.; Wilson, K.; Nava, P.; Atkinson, J.; Manor, D. Intracellular trafficking of vitamin E in hepatocytes: The role of tocopherol transfer protein. *J. Lipid Res.* **2005**, *46*, 2072–2082. [CrossRef]
65. Qian, J.; Atkinson, J.; Manor, D. Biochemical consequences of heritable mutations in the alpha-tocopherol transfer protein. *Biochemistry* **2006**, *45*, 8236–8242. [CrossRef] [PubMed]
66. Kono, N.; Arai, H. Intracellular Transport of Fat-Soluble Vitamins A and E. *Traffic* **2015**, *16*, 19–34. [CrossRef]
67. Chung, S.; Ghelfi, M.; Atkinson, J.; Parker, R.; Qian, J.; Carlin, C.; Manor, D. Vitamin E and phosphoinositides regulate the intracellular localization of the hepatic α-tocopherol transfer protein. *J. Biol. Chem.* **2016**, *291*, 17028–17039. [CrossRef]
68. Irías-Mata, A.; Sus, N.; Flory, S.; Stock, D.; Woerner, D.; Podszun, M.; Frank, J. α-Tocopherol transfer protein does not regulate the cellular uptake and intracellular distribution of α- and γ-tocopherols and -tocotrienols in cultured liver cells. *Redox Biol.* **2018**, *19*, 28–36. [CrossRef] [PubMed]
69. Fechner, H.; Schlame, M.; Guthmann, F.; Stevens, P.A.; Rustow, B. α- and δ-tocopherol induce expression of hepatic α-tocopherol-transfer-protein mRNA. *Biochem. J.* **1998**, *331*, 577–581. [CrossRef]
70. Kim, H.S.; Arai, H.; Arita, M.; Sato, Y.; Ogihara, T.; Inoue, K.; Mino, M.; Tamai, H. Effect of α-tocopherol status on α-tocopherol transfer protein expression and its messenger RNA level in rat liver. *Free Radic. Res.* **1998**, *28*, 87–92. [CrossRef]
71. Shaw, H.M.; Huang, C.J. Liver α-tocopherol transfer protein and its mRNA are differentially altered by dietary vitamin E deficiency and protein insufficiency in rats. *J. Nutr.* **1998**, *128*, 2348–2354. [CrossRef]
72. Rengaraj, D.; Truong, A.D.; Hong, Y.; Pitargue, F.M.; Kim, J.H.; Hong, Y.H.; Han, J.Y.; Kil, D.Y. Identification and expression analysis of alpha tocopherol transfer protein in chickens fed diets containing different concentrations of alpha-tocopherol. *Res. Vet. Sci.* **2019**, *123*, 99–110. [CrossRef]
73. Kaempf-Rotzoll, D.E.; Igarashi, K.; Aoki, J.; Jishage, K.; Suzuki, H.; Tamai, H.; Linderkamp, O.; Arai, H. α-tocopherol transfer protein is specifically localized at the implantation site of pregnant mouse uterus. *Biol. Reprod.* **2002**, *67*, 599–604. [CrossRef] [PubMed]
74. Misaki, K.; Takitani, K.; Ogihara, T.; Inoue, A.; Kawakami, C.; Kuno, T.; Kawamura, N.; Miyake, M.; Nakagawa, T.; Tamai, H. Alpha-tocopherol content and α-tocopherol transfer protein expression in leukocytes of children with acute leukemia. *Free Radic. Res.* **2003**, *37*, 1037–1042. [CrossRef] [PubMed]
75. Kaempf-Rotzoll, D.E.; Horiguchi, M.; Hashiguchi, K.; Aoki, J.; Tamai, H.; Linderkamp, O.; Arai, H. Human placental trophoblast cells express α-tocopherol transfer protein. *Placenta* **2003**, *24*, 439–444. [CrossRef] [PubMed]
76. Valacchi, G.; Vasu, V.T.; Yokohama, W.; Corbacho, A.M.; Phung, A.; Lim, Y.; Aung, H.H.; Cross, C.E.; Davis, P.A. Lung vitamin E transport processes are affected by both age and environmental oxidants in mice. *Toxicol. Appl. Pharmacol.* **2007**, *222*, 227–234. [CrossRef]
77. Taguchi, Y.; Komatsu-Tanaka, M.; Hirose, N.; Kaji, Y.; Saeki, K. Molecular Cloning and Expression Analysis of Bovine Alpha-tocopherol Transfer Protein (α-TTP). *Annu. Res. Rev.* **2017**, *12*, 1–7. [CrossRef]
78. Zuo, Z.; Zuo, Z.Y.; Luo, H.L.; Liu, K.; Jia, H.N.; Zhang, Y.W.; Jiao, L.J.; Chang, Y.F. Dietary vitamin E affects α-TTP mRNA levels in different tissues of the Tan sheep. *Gene* **2014**, *541*, 1–7. [CrossRef]
79. Haga, S.; Ishizaki, H.; Suzuki, T. Effects of orally administered *Rhizopus oryzae* aqueous extracts in the close-up dry period on postpartum serum liver injury markers and hepatic gene expression levels in dairy cows. *J. Farm Anim. Infect. Dis.* **2020**, *9*, 69–81. (In Japanese)
80. Voegele, A.F.; Jerković, L.; Wellenzohn, B.; Eller, P.; Kronenberg, F.; Liedl, K.R.; Dieplinger, H. Characterization of the vitamin E-binding properties of human plasma afamin. *Biochemistry* **2002**, *41*, 14532–14538. [CrossRef]
81. Jerkovic, L.; Voegele, A.F.; Chwatal, S.; Kronenberg, F.; Radcliffe, C.M.; Wormald, M.R.; Lobentanz, E.M.; Ezeh, B.; Eller, P.; Dejori, N.; et al. Afamin is a novel human vitamin E-binding glycoprotein characterization and in vitro expression. *J. Proteome Res.* **2005**, *4*, 889–899. [CrossRef]
82. Lichenstein, H.S.; Lyons, D.E.; Wurfel, M.M.; Johnson, D.A.; McGinley, M.D.; Leidli, J.C.; Trollinger, D.B.; Mayer, J.P.; Wright, S.D.; Zukowski, M.M. Afamin is a new member of the albumin, alpha-fetoprotein, and vitamin D-binding protein gene family. *J. Biol. Chem.* **1994**, *269*, 18149–18154. [CrossRef]
83. Hubalek, M.; Buchner, H.; Mörtl, M.G.; Schlembach, D.; Huppertz, B.; Firulovic, B.; Köhler, W.; Hafner, E.; Dieplinger, B.; Wildt, L.; et al. The vitamin E-binding protein afamin increases in maternal serum during pregnancy. *Clin. Chim. Acta* **2014**, *434*, 41–47. [CrossRef]

84. Seeber, B.; Morandell, E.; Lunger, F.; Wildt, L.; Dieplinger, H. Afamin serum concentrations are associated with insulin resistance and metabolic syndrome in polycystic ovary syndrome. *Reprod. Biol. Endocrinol.* **2014**, *12*, 88. [CrossRef]
85. Varga, V.E.; Lőrincz, H.; Szentpéteri, A.; Juhász, L.; Seres, I.; Paragh, G., Jr.; Balla, J.; Paragh, G.; Harangi, M. Changes in serum afamin and vitamin E levels after selective LDL apheresis. *J. Clin. Apher.* **2018**, *33*, 569–575. [CrossRef] [PubMed]
86. Erol, S.A.; Tanacan, A.; Anuk, A.T.; Tokalioglu, E.O.; Biriken, D.; Keskin, H.L.; Moraloglu, O.T.; Yazihan, N.; Sahin, D. Evaluation of maternal serum afamin and vitamin E levels in pregnant women with COVID-19 and its association with composite adverse perinatal outcomes. *J. Med. Virol.* **2021**, *93*, 2350–2358. [CrossRef] [PubMed]
87. Kratzer, I.; Bernhart, E.; Wintersperger, A.; Hammer, A.; Waltl, S.; Malle, E.; Sperk, G.; Wietzorrek, G.; Dieplinger, H.; Sattler, W. Afamin is synthesized by cerebrovascular endothelial cells and mediates α-tocopherol transport across an in vitro model of the blood–brain barrier. *J. Neurochem.* **2009**, *108*, 707–718. [CrossRef]
88. Herdt, T.H.; Smith, J.C. Blood-lipid and lactation-stage factors affecting serum vitamin E concentrations and vitamin E cholesterol ratios in dairy cattle. *J. Vet. Diagn. Investig.* **1996**, *2*, 228–232. [CrossRef] [PubMed]
89. Stocker, A.; Zimmer, S.; Spycher, S.E.; Azzi, A. Identification of a novel cytosolic tocopherol-binding protein: Structure, specificity, and tissue distribution. *IUBMB Life* **1999**, *48*, 49–55. [CrossRef] [PubMed]
90. Zimmer, S.; Stocker, A.; Sarbolouki, M.N.; Spycher, S.E.; Sassoon, J.; Azzi, A. A novel human tocopherol-associated protein: Cloning, in vitro expression, and characterization. *J. Biol. Chem.* **2000**, *275*, 25672–25680. [CrossRef]
91. Yamauchi, J.; Iwamoto, T.; Kida, S.; Masushige, S.; Yamada, K.; Esashi, T. Tocopherol-associated protein is a ligand-dependent transcriptional activator. *Biochem. Biophys. Res. Commun.* **2001**, *285*, 295–299. [CrossRef]
92. Porter, T.D. Supernatant protein factor and tocopherol-associated protein: An unexpected link between cholesterol synthesis and vitamin E (review). *J. Nutr. Biochem.* **2003**, *14*, 3–6. [CrossRef]
93. Ikeda, T.; Murakami, M.; Funaba, M. Expression of tocopherol-associated protein in mast cells. *Clin. Diagn. Lab. Immunol.* **2004**, *11*, 1189–1191. [CrossRef]
94. Wang, X.; Ring, B.Z.; Seitz, R.S.; Ross, D.T.; Woolf, K.; Beck, R.A.; Hicks, D.G.; Yeh, S. Expression of a-Tocopherol-Associated protein (TAP) is associated with clinical outcome in breast cancer patients. *BMC Clin. Pathol.* **2015**, *15*, 21. [CrossRef] [PubMed]
95. Mardones, P.; Strobel, P.; Miranda, S.; Leighton, F.; Quiñones, V.; Amigo, L.; Rozowski, J.; Krieger, M.; Rigotti, A. Alpha-tocopherol metabolism is abnormal in scavenger receptor class B type I (SR-BI)-deficient mice. *J. Nutr.* **2002**, *132*, 443–449. [CrossRef] [PubMed]
96. Reboul, E.; Klein, A.; Bietrix, F.; Gleize, B.; Malezet-Desmoulins, C.; Schneider, M.; Margotat, A.; Lagrost, L.; Collet, X.; Borel, P. Scavenger receptor class B type I (SR-BI) is involved in vitamin E transport across the enterocyte. *J. Biol. Chem.* **2006**, *281*, 4739–4745. [CrossRef] [PubMed]
97. Acton, S.; Rigotti, A.; Landschulz, K.T.; Xu, S.; Hobbs, H.H.; Krieger, M. Identification of scavenger receptor SR-BI as a high density lipoprotein receptor. *Science* **1996**, *271*, 518–520. [CrossRef] [PubMed]
98. Lee, P.; Ulatowski, L.M. Vitamin E: Mechanism of transport and regulation in the CNS. *IUBMB Life* **2019**, *71*, 424–429. [CrossRef] [PubMed]
99. Argov, N.; Moallem, U.; Sklan, D. Lipid transport in the developing bovine follicle: Messenger RNA expression increases for selective uptake receptors and decreases for endocytosis receptors. *Biol. Reprod.* **2004**, *71*, 479–485. [CrossRef]
100. Argov, N.; Moallem, U.; Sklan, D. Summer heat stress alters the mRNA expression of selective-uptake and endocytotic receptors in bovine ovarian cells. *Theriogenology* **2005**, *64*, 1475–1489. [CrossRef]
101. Rajapaksha, W.R.; McBride, M.; Robertson, L.; O'Shaughnessy, P.J. Sequence of the bovine HDL-receptor (SR-BI) cDNA and changes in receptor mRNA expression during granulosa cell luteinization in vivo and in vitro. *Mol. Cell. Endocrinol.* **1997**, *134*, 59–67. [CrossRef]
102. Knight, B.L. ATP-binding cassette transporter A1: Regulation of cholesterol efflux. *Biochem. Soc. Trans.* **2004**, *32*, 124–127. [CrossRef] [PubMed]
103. Schmitz, G.; Langmann, T. Transcriptional regulatory networks in lipid metabolism control ABCA1 expression. *Biochim. Biophys. Acta* **2005**, *1735*, 1–19. [CrossRef]
104. Liu, M.; Chung, S.; Shelness, G.S.; Parks, J.S. Hepatic ABCA1 and VLDL triglyceride production. *Biochim. Biophys. Acta* **2012**, *1821*, 770–777. [CrossRef] [PubMed]
105. Shichiri, M.; Takanezawa, Y.; Rotzoll, D.E.; Yoshida, Y.; Kokubu, T.; Ueda, K.; Tamai, H.; Arai, H. ATP-binding cassette transporter A1 is involved in hepatic alpha-tocopherol secretion. *J. Nutr. Biochem.* **2010**, *21*, 451–456. [CrossRef] [PubMed]
106. Kielar, D.; Dietmaier, W.; Langmann, T.; Aslanidis, C.; Probst, M.; Naruszewicz, M.; Schmitz, G. Rapid quantification of human ABCA1 mRNA in various cell types and tissues by real-time reverse transcription-PCR. *Clin. Chem.* **2001**, *47*, 2089–2097. [CrossRef]
107. Langmann, T.; Mauerer, R.; Zahn, A.; Moehle, C.; Probst, M.; Stremmel, W.; Schmitz, G. Real-time reverse transcription-PCR expression profiling of the complete human ATP-binding cassette transporter superfamily in various tissues. *Clin. Chem.* **2003**, *49*, 230–238. [CrossRef] [PubMed]
108. Oram, J.F.; Vaughan, A.M.; Stocker, R. ATP-binding cassette transporter A1 mediates cellular secretion of α-tocopherol. *J. Biol. Chem.* **2001**, *276*, 39898–39902. [CrossRef]
109. Farke, C.; Viturro, E.; Meyer, H.H.D.; Albrecht, C. Identification of the bovine cholesterol efflux regulatory protein ABCA1 and its expression in various tissues. *J. Anim. Sci.* **2006**, *84*, 2887–2894. [CrossRef]

110. Wang, J.; Xiao, C.; Wei, Z.; Wang, Y.; Zhang, X.; Fu, Y. Activation of liver X receptors inhibit LPS-induced inflammatory response in primary bovine mammary epithelial cells. *Vet. Immunol. Immunopathol.* **2018**, *197*, 87–92. [CrossRef]
111. Schlegel, G.; Ringseis, R.; Keller, J.; Schwarz, F.J.; Eder, K. Changes in the expression of hepatic genes involved in cholesterol homeostasis in dairy cows in the transition period and at different stages of lactation. *J. Dairy Sci.* **2012**, *95*, 3826–3836. [CrossRef]
112. Kessler, E.C.; Gross, J.J.; Bruckmaier, R.M.; Albrecht, C. Cholesterol metabolism, transport, and hepatic regulation in dairy cows during transition and early lactation. *J. Dairy Sci.* **2014**, *97*, 5481–5490. [CrossRef] [PubMed]
113. Farke, C.; Meyer, H.H.; Bruckmaier, R.M.; Albrecht, C. Differential expression of ABC transporters and their regulatory genes during lactation and dry period in bovine mammary tissue. *J. Dairy Res.* **2008**, *75*, 406–414. [CrossRef] [PubMed]
114. Mani, O.; Körner, M.; Sorensen, M.T.; Sejrsen, K.; Wotzkow, C.; Ontsouka, C.E.; Friis, R.R.; Bruckmaier, R.M.; Albrecht, C. Expression, localization, and functional model of cholesterol transporters in lactating and nonlactating mammary tissues of murine, bovine, and human origin. *Am. J. Physiol. Regul. Integr. Comp. Physiol.* **2010**, *299*, 642–654. [CrossRef]
115. Mani, O.; Körner, M.; Ontsouka, C.E.; Sorensen, M.T.; Sejrsen, K.; Bruckmaier, R.M.; Albrecht, C. Identification of ABCA1 and ABCG1 in milk fat globules and mammary cells—implications for milk cholesterol secretion. *J. Dairy Sci.* **2011**, *94*, 1265–1276. [CrossRef] [PubMed]
116. Chen, Z.; Chu, S.; Wang, X.; Fan, Y.; Zhan, T.; Arbab, A.A.I.; Li, M.; Zhang, H.; Mao, Y.; Loor, J.J.; et al. MicroRNA-106b Regulates Milk Fat Metabolism via ATP Binding Cassette Subfamily A Member 1 (ABCA1) in Bovine Mammary Epithelial Cells. *J. Agric. Food Chem.* **2019**, *67*, 3981–3990. [CrossRef]
117. Kikuta, Y.; Kusunose, E.; Kondo, T.; Yamamoto, S.; Kinoshita, H.; Kusunose, M. Cloning and expression of a novel form of leukotriene B4 omega-hydroxylase from human liver. *FEBS Lett.* **1994**, *348*, 70–74. [CrossRef]
118. Kikuta, Y.; Miyauchi, Y.; Kusunose, E.; Kusunose, M. Expression and molecular cloning of human liver leukotriene B4 omega-hydroxylase (CYP4F2) gene. *DNA Cell Biol.* **1999**, *18*, 723–730. [CrossRef]
119. Kikuta, Y.; Kusunose, E.; Kusunose, M. Characterization of human liver leukotriene B(4) omega-hydroxylase P450 (CYP4F2). *J. Biochem.* **2000**, *127*, 1047–1052. [CrossRef]
120. Sontag, T.J.; Parker, R.S. Cytochrome P450 omega-hydroxylase pathway of tocopherol catabolism. Novel mechanism of regulation of vitamin E status. *J. Biol. Chem.* **2002**, *277*, 25290–25296. [CrossRef]
121. Schultz, M.; Leist, M.; Petrzika, M.; Gassmann, B.; Brigelius-Flohé, R. Novel urinary metabolite of alpha-tocopherol, 2,5,7,8-tetramethyl-2(2′-carboxyethyl)-6-hydroxychroman, as an indicator of an adequate vitamin E supply? *Am. J. Clin. Nutr.* **1995**, *62*, 1527S–1534S. [CrossRef]
122. Brigelius-Flohe, R. Vitamin E and drug metabolism. *Biochem. Biophys. Res. Commun.* **2003**, *305*, 737–740. [CrossRef]
123. Abe, C.; Ikeda, S.; Yamashita, K. Dietary sesame seeds elevate alpha-tocopherol concentration in rat brain. *J. Nutr. Sci. Vitaminol. (Tokyo)* **2005**, *51*, 223–230. [CrossRef]
124. Takitani, K.; Inoue, K.; Koh, M.; Miyazaki, H.; Kishi, K.; Inoue, A.; Tamai, H. Alpha-Tocopherol status and altered expression of alpha-tocopherol-related proteins in streptozotocin-induced type 1 diabetes in rat models. *J. Nutr. Sci. Vitaminol. (Tokyo)* **2014**, *60*, 380–386. [CrossRef]
125. Hsu, M.H.; Savas, U.; Griffin, K.J.; Johnson, E.F. Regulation of human cytochrome P450 4F2 expression by sterol regulatory element-binding protein and lovastatin. *J. Biol. Chem.* **2007**, *282*, 5225–5236. [CrossRef]
126. Bartolini, D.; Torquato, P.; Barola, C.; Russo, A.; Rychlicki, C.; Giusepponi, D.; Bellezza, G.; Sidoni, A.; Galarini, R.; Svegliati-Baroni, G.; et al. Nonalcoholic fatty liver disease impairs the cytochrome P-450-dependent metabolism of alpha-tocopherol (vitamin E). *J. Nutr. Biochem.* **2017**, *47*, 120–131. [CrossRef]
127. Kuhn, M.J.; Mavangira, V.; Sordillo, L.M. Invited review: Cytochrome P450 enzyme involvement in health and inflammatory-based diseases of dairy cattle. *J. Dairy Sci.* **2021**, *104*, 1276–1290. [CrossRef] [PubMed]
128. Traber, M.G. Vitamin E regulatory mechanisms. *Annu. Rev. Nutr.* **2007**, *27*, 347–362. [CrossRef] [PubMed]
129. Narushima, K.; Takada, T.; Yamanashi, Y.; Suzuki, H. Niemann-pick C1-like 1 mediates alpha-tocopherol transport. *Mol. Pharmacol.* **2008**, *74*, 42–49. [CrossRef] [PubMed]
130. Reboul, E. Vitamin E intestinal absorption: Regulation of membrane transport across the enterocyte. *IUBMB Life* **2019**, *71*, 416–423. [CrossRef]
131. Weiss, W.P.; Hogan, J.S.; Smith, K.L.; Todhunter, D.A.; Williams, S.N. Effect of supplementing periparturient cows with vitamin E on distribution of alpha-tocopherol in blood. *J. Dairy Sci.* **1992**, *75*, 3479–3485. [CrossRef]
132. Weiss, W.P.; Hogan, J.S.; Smith, K.L.; Williams, S.N. Effect of dietary fat and vitamin E on alpha-tocopherol and beta-carotene in blood of peripartum cows. *J. Dairy Sci.* **1994**, *77*, 1422–1429. [CrossRef]
133. Weiss, W.P.; Hogan, J.S.; Todhunter, D.A.; Smith, K.L. Effect of vitamin E supplementation in diets with a low concentration of selenium on mammary gland health of dairy cows. *J. Dairy Sci.* **1997**, *80*, 1728–1737. [CrossRef]
134. Hidiroglou, M.; Batra, T.R.; Zhao, X. Bioavailability of vitamin E compounds and the effect of supplementation on release of superoxide and hydrogen peroxide by bovine neutrophils. *J. Dairy Sci.* **1997**, *80*, 187–193. [CrossRef]
135. Chawla, R.; Kaur, H. Plasma antioxidant vitamin status of periparturient cows supplemented with α-tocopherol and β-carotene. *Anim. Feed Sci. Technol.* **2004**, *114*, 279–285. [CrossRef]
136. LeBlanc, S.J.; Herdt, T.H.; Seymour, W.M.; Duffield, T.F.; Leslie, K.E. Peripartum serum vitamin E, retinol, and beta-carotene in dairy cattle and their associations with disease. *J. Dairy Sci.* **2004**, *87*, 609–619. [CrossRef]

137. Calderon, F.; Chauveau-Duriot, B.; Martin, B.; Graulet, B.; Doreau, M.; Nozière, P. Variations in carotenoids, vitamins A and E, and color in cow's plasma and milk during late pregnancy and the first three months of lactation. *J. Dairy Sci.* **2007**, *90*, 2335–2346. [CrossRef]
138. Strickland, J.M.; Wisnieski, L.; Herdt, T.H.; Sordillo, L.M. Serum retinol, beta-carotene, and alpha-tocopherol as biomarkers for disease risk and milk production in periparturient dairy cows. *J. Dairy Sci.* **2021**, *104*, 915–927. [CrossRef] [PubMed]
139. Aggarwal, A.; Ashutosh; Chandra, G.; Singh, A.K. Heat shock protein 70, oxidative stress, and antioxidant status in periparturient crossbred cows supplemented with alpha-tocopherol acetate. *Trop. Anim. Health Prod.* **2013**, *45*, 239–245. [CrossRef] [PubMed]
140. Bouwstra, R.J.; Goselink, R.M.; Dobbelaar, P.; Nielen, M.; Newbold, J.R.; van Werven, T. The relationship between oxidative damage and vitamin E concentration in blood, milk, and liver tissue from vitamin E supplemented and nonsupplemented periparturient heifers. *J. Dairy Sci.* **2008**, *91*, 977–987. [CrossRef] [PubMed]
141. Bourne, N.; Wathes, D.C.; Lawrence, K.E.; McGowan, M.; Laven, R.A. The effect of parenteral supplementation of vitamin E with selenium on the health and productivity of dairy cattle in the UK. *Vet. J.* **2008**, *177*, 381–387. [CrossRef]
142. Weiss, W.P.; Hogan, J.S.; Wyatt, D.J. Relative bioavailability of *all-rac* and RRR vitamin E based on neutrophil function and total alpha-tocopherol and isomer concentrations in periparturient dairy cows and their calves. *J. Dairy Sci.* **2009**, *92*, 720–731. [CrossRef] [PubMed]
143. Politis, I.; Hidiroglou, M.; Batra, T.R.; Gilmore, J.A.; Gorewit, R.C.; Scherf, H. Effects of vitamin E on immune function of dairy cows. *Am. J. Vet. Res.* **1995**, *56*, 179–184. [PubMed]
144. Politis, I.; Hidiroglou, N.; White, J.H.; Gilmore, J.A.; Williams, S.N.; Scherf, H.; Frigg, M. Effects of vitamin E on mammary and blood leukocyte function, with emphasis on chemotaxis, in periparturient dairy cows. *Am. J. Vet. Res.* **1996**, *57*, 468–471.
145. Politis, I.; Hidiroglou, N.; Cheli, F.; Baldi, A. Effects of vitamin E on urokinase-plasminogen activator receptor expression by bovine neutrophils. *Am. J. Vet. Res.* **2001**, *62*, 1934–1938. [CrossRef]
146. Politis, I.; Bizelis, I.; Tsiaras, A.; Baldi, A. Effect of vitamin E supplementation on neutrophil function, milk composition and plasmin activity in dairy cows in a commercial herd. *J. Dairy Res.* **2004**, *71*, 273–278. [CrossRef]
147. McNaughton, A.P.; Murray, R.D. Structure and function of the bovine fetomaternal unit in relation to the causes of retained fetal membranes. *Vet. Rec.* **2009**, *165*, 615–622. [CrossRef]
148. Bourne, N.; Laven, R.; Wathes, D.C.; Martinez, T.; McGowan, M. A meta-analysis of the effects of Vitamin E supplementation on the incidence of retained foetal membranes in dairy cows. *Theriogenology* **2007**, *67*, 494–501. [CrossRef] [PubMed]
149. Hasanpour, A.; Saranjam, N.; Amuoghli Tabrizi, B. Antioxidant Concentration Status in the Serum of Cows with Left Displacement Abomasom. *Glob. Vet.* **2011**, *7*, 478–481.
150. Pontes, G.C.; Monteiro, P.L., Jr.; Prata, A.B.; Guardieiro, M.M.; Pinto, D.A.; Fernandes, G.O.; Wiltbank, M.C.; Santos, J.E.; Sartori, R. Effect of injectable vitamin E on incidence of retained fetal membranes and reproductive performance of dairy cows. *J. Dairy Sci.* **2015**, *98*, 2437–2449. [CrossRef]
151. Moghimi-Kandelousi, M.; Alamouti, A.A.; Imani, M.; Zebeli, Q. A meta-analysis and meta-regression of the effects of vitamin E supplementation on serum enrichment, udder health, milk yield, and reproductive performance of transition cows. *J. Dairy Sci.* **2020**, *103*, 6157–6166. [CrossRef]
152. Goff, J.P.; Kimura, K.; Horst, R.L. Effect of mastectomy on milk fever, energy, and vitamins A, E, and beta-carotene status at parturition. *J. Dairy Sci.* **2002**, *85*, 1427–1436. [CrossRef]
153. Miyaji, M.; Haga, S.; Matsuyama, H.; Hosoda, K. Effect of feeding brown rice instead of corn on lactation performance and blood metabolites in periparturient dairy cows. *Anim. Feed Sci. Technol.* **2016**, *219*, 234–240. [CrossRef]
154. Hymøller, L.; Jensen, S.K. Stability in the rumen and effect on plasma status of single oral doses of vitamin D and vitamin E in high-yielding dairy cows. *J. Dairy Sci.* **2010**, *93*, 5748–5757. [CrossRef]
155. Tsuchiya, Y.; Ozai, R.; Sugino, T.; Kawashima, K.; Kushibiki, S.; Kim, Y.H.; Sato, S. Changes in peripheral blood oxidative stress markers and hepatic gene expression related to oxidative stress in Holstein cows with and without subacute ruminal acidosis during the periparturient period. *J. Vet. Med. Sci.* **2020**, *82*, 1529–1536. [CrossRef] [PubMed]
156. Alderson, N.E.; Mitchell, G.E., Jr.; Little, C.O.; Warner, R.E.; Tucker, R.E. Preintestinal disappearance of vitamin E in ruminants. *J. Nutr.* **1971**, *101*, 655–659. [CrossRef]
157. Bernabucci, U.; Ronchi, B.; Basiricò, L.; Pirazzi, D.; Rueca, F.; Lacetera, N.; Nardone, A. Abundance of mRNA of apolipoprotein B100, apolipoprotein E, and microsomal triglyceride transfer protein in liver from periparturient dairy cows. *J. Dairy Sci.* **2004**, *87*, 2881–2888. [CrossRef]
158. Pilotto, A.; Savoini, G.; Baldi, A.; Invernizzi, G.; De Vecchi, C.; Theodorou, G.; Koutsouli, P.; Politis, I. Short communication: Associations between blood fatty acids, beta-hydroxybutyrate, and alpha-tocopherol in the periparturient period in dairy cows: An observational study. *J. Dairy Sci.* **2016**, *99*, 8121–8126. [CrossRef]
159. Katoh, N. Relevance of apolipoproteins in the development of fatty liver and fatty liver-related peripartum diseases in dairy cows. *J. Vet. Med. Sci.* **2002**, *64*, 293–307. [CrossRef]
160. Castillo, C.; Hernandez, J.; Bravo, A.; Lopez-Alonso, M.; Pereira, V.; Benedito, J.L. Oxidative status during late pregnancy and early lactation in dairy cows. *Vet. J.* **2005**, *169*, 286–292. [CrossRef] [PubMed]
161. Sordillo, L.M.; O'Boyle, N.; Gandy, J.C.; Corl, C.M.; Hamilton, E. Shifts in thioredoxin reductase activity and oxidant status in mononuclear cells obtained from transition dairy cattle. *J. Dairy Sci.* **2007**, *90*, 1186–1192. [CrossRef]

162. Konvičná, J.; Vargová, M.; Paulíková, I.; Kováč, G.; Kostecká, Z. Oxidative stress and antioxidant status in dairy cows during prepartal and postpartal periods. *Acta Vet. Brno* **2015**, *84*, 133–140. [CrossRef]
163. Abuelo, A.; Hernández, J.; Benedito, J.L.; Castillo, C. Oxidative stress index (OSi) as a new tool to assess redox status in dairy cattle during the transition period. *Animals* **2013**, *7*, 1374–1378. [CrossRef] [PubMed]
164. Sordillo, L.M.; Aitken, S.L. Impact of oxidative stress on the health and immune function of dairy cattle. *Vet. Immunol. Immunopathol.* **2009**, *128*, 104–109. [CrossRef] [PubMed]
165. Bernabucci, U.; Ronchi, B.; Lacetera, N.; Nardone, A. Influence of body condition score on relationships between metabolic status and oxidative stress in periparturient dairy cows. *J. Dairy Sci.* **2005**, *88*, 2017–2026. [CrossRef]
166. Rahayu, S.; Prasdini, W.A.; Djati, M.S.; Prasetyawan, S.; Ciptad, G. Malondialdehyde (MDA) level and protein profile of serum after calving towards the provision of selenium-vitamin E™ on dairy cow frisian holstein (FH). *J. Phys. Conf. Ser.* **2019**, *1146*. [CrossRef]
167. Khatti, A.; Mehrotra, S.; Patel, P.K.; Singh, G.; Maurya, V.P.; Mahla, A.S.; Chaudhari, R.K.; Das, G.K.; Singh, M.; Sarkar, M.; et al. Supplementation of vitamin E, selenium and increased energy allowance mitigates the transition stress and improves postpartum reproductive performance in the crossbred cow. *Theriogenology* **2017**, *104*, 142–148. [CrossRef]
168. Gessner, D.K.; Schlegel, G.; Ringseis, R.; Schwarz, F.J.; Eder, K. Up-regulation of endoplasmic reticulum stress induced genes of the unfolded protein response in the liver of periparturient dairy cows. *BMC Vet. Res.* **2014**, *10*, 46. [CrossRef]
169. Bobe, G.; Young, J.W.; Beitz, D.C. Invited review: Pathology, etiology, prevention, and treatment of fatty liver in dairy cows. *J. Dairy Sci.* **2004**, *87*, 3105–3124. [CrossRef]
170. Hidiroglou, M. Mammary transfer of vitamin E in dairy cows. *J. Dairy Sci.* **1989**, *72*, 1067–1071. [CrossRef]
171. Linzell, J.L. Mammary blood flow and methods of identifying and measuring precursors of milk. In *Lactation: A Comprehensive Treatise*; Larson, B.L., Smith, V.R., Eds.; Academic Press: New York, NY, USA, 1974; Volume 1, pp. 143–225.
172. Delamaire, E.; Guinard-Flament, J. Longer milking intervals alter mammary epithelial permeability and the udder's ability to extract nutrients. *J. Dairy Sci.* **2006**, *89*, 2007–2016. [CrossRef]
173. Ontsouka, E.C.; Albrecht, C. Cholesterol transport and regulation in the mammary gland. *J. Mammary Gland Biol. Neoplasia* **2014**, *19*, 43–58. [CrossRef] [PubMed]
174. Reinhardt, T.A.; Lippolis, J.D. Developmental changes in the milk fat globule membrane proteome during the transition from colostrum to milk. *J. Dairy Sci.* **2008**, *91*, 2307–2318. [CrossRef] [PubMed]
175. Ghaffari, M.H.; Bernhöft, K.; Etheve, S.; Immig, I.; Hölker, M.; Sauerwein, H.; Schweigert, F.J. Technical note: Rapid field test for the quantification of vitamin E, beta-carotene, and vitamin A in whole blood and plasma of dairy cattle. *J. Dairy Sci.* **2019**, *102*, 11744–11750. [CrossRef] [PubMed]

animals MDPI

Article

A Randomized Clinical Trial Evaluating the Effect of an Oral Calcium Bolus Supplementation Strategy in Postpartum Jersey Cows on Mastitis, Culling, Milk Production, and Reproductive Performance

Paulo R. Menta [1], Leticia Fernandes [1], Diego Poit [1], Maria Luiza Celestino [1], Vinicius S. Machado [1] and Rafael C. Neves [2,*]

[1] Department of Veterinary Sciences, College of Agricultural Sciences and Natural Resources, Texas Tech University, Lubbock, TX 79409, USA; paulo.menta@ttu.edu (P.R.M.); lfernandes@lelly.com (L.F.); diego.poit@usp.br (D.P.); mkcelestino@gmail.com (M.L.C.); Vinicius.Machado@ttu.edu (V.S.M.)

[2] Department of Veterinary Clinical Sciences, College of Veterinary Medicine, Purdue University, West Lafayette, IN 47907, USA

* Correspondence: rcneves@purdue.edu; Tel.: +1-765-496-1877

Citation: Menta, P.R.; Fernandes, L.; Poit, D.; Celestino, M.L.; Machado, V.S.; Neves, R.C. A Randomized Clinical Trial Evaluating the Effect of an Oral Calcium Bolus Supplementation Strategy in Postpartum Jersey Cows on Mastitis, Culling, Milk Production, and Reproductive Performance. *Animals* **2021**, *11*, 3361. https://doi.org/10.3390/ani11123361

Academic Editor: João Carlos Caetano Simões

Received: 30 October 2021
Accepted: 22 November 2021
Published: 24 November 2021

Simple Summary: The time around parturition is a challenging period in the lactation cycle of high-yielding dairy cows as it is characterized by several endocrine, metabolic, and physiological changes. Among those challenges, calcium demands are rapidly increased to support colostrogenesis and lactogenesis during a time of reduced dry matter intake; invariably, some cows may suffer from clinical and subclinical hypocalcemia. Oral calcium supplementation is a common preventative strategy adopted in the postpartum of high-yielding dairy cows to minimize the negative impact of suboptimal blood calcium concentration during this period. Despite a great number of studies evaluating the effects of oral calcium supplementation in postpartum Holstein cows, very limited information is available for the Jersey breed. This study aimed to evaluate the effect of an oral Ca supplementation strategy in the first 24 h after parturition on health and production outcomes in multiparous Jersey cows. Overall, treatment did not improve milk production or reproductive performance compared to control cows. Additionally, treatment had no effect on early lactation culling. A tendency for a minor impact on the odds of mastitis was evident though it depended on the parity of the cows.

Abstract: The objectives of this study were to evaluate the effects of a postpartum oral calcium supplementation strategy in multiparous Jersey cows on (1) the odds of clinical mastitis in the first 60 days in milk (DIM); (2) the odds of culling up to 60 DIM; (3) the risk of pregnancy in the first 150 DIM; (4) milk production in the first 15 weeks of lactation. A randomized clinical trial was performed in a dairy herd located in west Texas, United States. A total of 809 cows were used in the final analyses. Overall, postpartum oral calcium supplementation did not influence milk production, reproductive performance, or culling. Among second parity cows, oral calcium supplementation tended to decrease the odds of clinical mastitis in the first 60 DIM compared to controls; however, no differences were observed for cows in parities three and greater. To date, data evaluating the effect of postpartum oral calcium supplementation in multiparous Jersey cows are limited. In our study, oral calcium supplementation tended to reduce clinical mastitis in second parity cows. No positive benefits based on the reduction of culling, and improvement of milk production and reproductive performance were evident for the herd included in this study.

Keywords: dairy cow; Jersey; oral calcium bolus; calcium; hypocalcemia; mastitis; culling; milk production; reproduction

1. Introduction

The time around parturition is a challenging period in the lactation cycle of high-yielding dairy cows. Maladaptive responses to the increased nutrient requirements and stressors in the transition to lactation can impact lactation performance. Daily calcium (Ca) requirements increase approximately 1.6 times from late prepartum to early postpartum [1,2], highlighting the importance of Ca in the transition of the non-lactating to lactating state; those changes are due to the onset of colostrogenesis and lactogenesis. Invariably, suboptimum Ca concentration is common within 12 to 24 h after calving and can extend to a few days in lactation [3–5]. In order to meet the increased Ca requirements, the activation of homeostatic and homeorhetic mechanisms such as renal reabsorption, bone resorption, and intestinal absorption are essential [1]. Despite that, some cows fail in this process and may experience clinical or subclinical hypocalcemia in the postpartum period.

Clinical hypocalcemia (CH) is characterized by an acute reduction in blood Ca concentration below 1.4 mmol/L, and it is usually detected within 72 h after parturition [6]. Improvement of nutritional strategies applied in the prepartum period has helped to reduce CH over the years [7,8]. As an example, an acidogenic diet applied in the prepartum as a preventative dietary strategy against CH became popular and is proven to be an effective method [9–11].

Subclinical hypocalcemia (SCH) has caught more attention over the last decade because of its difficulty to diagnose and high prevalence among U.S. and European herds [7,12,13]. Even though it does not present evident signs, it has an important economic impact on dairy farms due to negative effects on dry matter intake at the beginning of lactation [14]. In addition, SCH has been associated with negative health outcomes such as retained placenta, metritis, impaired reproductive performance [15], displaced abomasum [16], increased culling rate [17], and impaired immune function [18]. Unfortunately, acidogenic diets have not been demonstrated to be as effective for SCH prevention as for CH [7]. Therefore, strategies to mitigate the potential effects of SCH via postpartum oral Ca supplementation are still widely adopted. In the U.S. for instance, 80% of the large farms used some combination of injectable, drench, or oral Ca as a preventative strategy to postpartum diseases [19]. The use of oral Ca supplementation increases blood Ca concentration [20], though the data evaluating the benefits of oral Ca are inconsistent across cows; some groups can benefit from it (e.g., high milk producers, lame cows) while others do not (e.g., first parity cows) [8,21]. Data evaluating the effect of postpartum oral Ca supplementation in health and production outcomes are limited for the Jersey breed.

The current literature is limited when evaluating the effect of postpartum oral Ca supplementation in Jersey cows, and results obtained in studies performed in Holstein cows may not directly apply to the Jersey breed. Jersey cows are well known for having an increased susceptibility to CH [6,22]; they mobilize more Ca into milk [23] and have a lower abundance of 1,25-dihidroxyvitamin D3 receptors in their intestine [22] which can contribute to a reduced dietary intestinal Ca absorption capacity. A study evaluating the effect of postpartum oral Ca supplementation for Jersey and Jersey–Holstein crossbreeds demonstrated an increase in blood Ca levels and a lower prevalence of SCH in treated cows [24]. The effect of postpartum oral Ca supplementation on milk yield was conditional to cow-level factors such as previous lactation length and calving locomotion score [25]. More studies evaluating the benefits of postpartum oral Ca supplementation in health, production, and reproductive outcomes for purebred Jersey cows are warranted.

The objective of our study was to determine the effect of an oral Ca supplementation strategy applied to multiparous Jersey cows (one bolus given soon after parturition followed by a second bolus 12–24 h after) on health outcomes, reproductive performance, and milk production. We hypothesized that postpartum oral Ca supplementation would decrease the odds of clinical diseases and improve milk production and reproductive performance. Our study was designed to mimic how postpartum Ca supplementation is commonly performed in U.S. dairy farms (blanket treatment) to better inform veterinarians and producers on the potential benefits of this strategy for the Jersey breed.

2. Materials and Methods

2.1. Study Design and Data Collection

A randomized clinical trial was conducted from July 2018 to April 2019 at a commercial dairy farm in west Texas milking 3800 Jersey and Jersey–Holstein crossbred cows. A sample size calculation determined that at least 788 cows were necessary for a study with 80% power and significant differences declared at $\alpha = 0.05$ assuming an expected difference of at least 1.0 kg of milk per day between treated and control cows, with a daily milk production SD of each group being of 5.0 kg, and an equal number of cows among groups (k = 1).

A more detailed description of herd demographics, average milk production, reproductive program used, and total mixed ration offered to the close-up and fresh cows can be found in Menta et al. (2021) [5]. Of note, prepartum cows were fed a negative dietary cation-anion diet. Cow enrollment occurred from July to December 2018; for this study, only purebred multiparous Jersey cows entering their 2nd or greater lactation were randomly assigned to the treatment or control group. The farm was provided with randomized enrollment sheets blocked by parity groups (2nd vs. 3rd and greater) prepared by the research group beforehand, and cows were sequentially enrolled by calving date and time by farm personnel. The treated cows received two doses of a commercial oral Ca bolus (Bovikalc®, Boehringer Ingelheim Vetmedica, Inc., St. Joseph, MO, USA) containing calcium chloride and calcium sulfate (43 g of Ca per bolus); the first dose was administered by farm personnel within 1 h after calving in a chute at the maternity area. The 2nd dose of the Ca bolus was administered by the research personnel the day after parturition, which occurred approximately 21 h and 20 min after calving (SD = 6 h) while cows were restrained in headlocks after the morning milking. In total, 852 multiparous cows were enrolled. Cows that had twins or that aborted (gestation length < 260 days) were not included in the study.

Milk yield was recorded by the research group weekly. Milk records were logged using a manually portable device (Pocket CowCard, Valley Agricultural Software, Tulare, CA, USA) which was automatically downloaded into the herd management software (DairyComp 305, Valley Agricultural Software, Tulare, CA, USA). Disease event data were recorded by farm personnel in DairyComp 305 using disease definition protocols accorded before the start of the study. Retained placenta (RP) was defined as the failure to expel fetal membranes within 12 h of parturition [26]. Clinical hypocalcemia (CH) was defined as a cow that was recumbent within 72 h of parturition accompanied by cool extremities and reduced ruminal contractions [27]. Displaced abomasum (DA) was defined as a classical resonant sound (indicating the presence of gas) during concurrent auscultation and percussion in an imaginary line from the coxae to the olecranon of the left/right flank region [27]. Mastitis was defined as any abnormalities during milk secretion, such as in color and clots, with or without local visible signs of inflammation in the udder [27]. Dystocia was defined as any human intervention during parturition. Body condition score (BCS) was evaluated at 4 DIM by the first author, who was blinded to treatment groups, using a 5-point scale [28].

2.2. Statistical Analyses

Cow-level data including health events, reproduction, and milk production were extracted from DairyComp 305 into Microsoft Excel (Microsoft Corp., Redmond, WA, USA) before statistical analyses in SAS v9.4 (SAS Institute Inc., Cary, NC, USA). Descriptive statistics were performed using the UNIVARIATE, FREQ, and MEANS procedures. For the binary and continuous outcomes, the GLIMMIX and MIXED procedures were used, respectively. Continuous data were evaluated for distribution of the residuals and homogeneity of variance after model fitting.

A binary variable representing calving-related problem(s) was created to characterize cows that suffered from dystocia (i.e., any cow requiring human intervention for calf extraction) and(or) had a stillbirth. Body condition score was categorized to represent thin (BCS $\leq$ 2.75), normal (BCS between 3.0 and up to 3.5), and over-conditioned cows (BCS $\geq$ 3.75). Parity was dichotomized to represent 2nd versus $\geq$3rd parities. Calving season was considered as a dichotomous variable (warmer months: cows calving from July to September 22, 2018; cooler months: calvings occurring from September 23 to December 2018). Average temperature for the warmer and cooler months were 26.4 °C (SD: 3.4 °C) and 11.7 °C (SD: 6.8 °C), respectively, according to temperature records retrieved from the National Weather Service database (National Oceanic and Administrative Association, United States Department of Commerce, Silver Spring, MD, USA) and based on the herd's ZIP code.

Univariable analyses were performed to screen variables associated with the outcomes at $p \leq 0.20$ before inclusion to the multivariable models. A manual backward stepwise selection procedure was used, and variables at $p \leq 0.05$ were retained as main effects. If a variable caused more than 20% change in one or more estimates, it was maintained as a confounder. The effect of treatment was forced in all models as it was the main predictor of interest. Parity was considered a confounder a priori and, therefore, included in all models regardless of the significance value. Potential 2-way interactions were tested in the final models between significant variables and the predictor of interest. Potential predictors considered in the models were calving-related problems, calving season, BCS score (mastitis, culling, pregnancy to 150 DIM, and milk yield models), mastitis up to 60 DIM (culling model), weekly milk test number, and previous gestation length (milk production model). As the studied herd did not participate in a Dairy Herd Improvement program, previous lactation 305 mature equivalent data were not available to be used as a covariate in the milk production model.

For health parameters, the odds of mastitis in the first 60 DIM was the sole outcome evaluated. Clinical hypocalcemia, RP, and DA had very low incidences during the study. Metritis diagnosis was not consistently performed across all cows in the herd and was, therefore, not evaluated. To evaluate the effect of treatment on the odds of mastitis and culling, a multivariable logistic regression model was fitted to the data using the GLIMMIX procedure. The LSMEANS option was used to calculate the least square means and standard error of the mean to report proportions. Cox's proportional hazard model was fitted using the PHREG procedure for pregnancy up to 150 DIM. Cows were left-censored if not diagnosed as being pregnant before being culled or if they died within the survival time. A generalized linear mixed model was fitted for the milk production data using the MIXED procedure while accounting for the repeated measurements within cow. To appropriately account for within-cow correlation, the error term was modeled by imposing a Toeplitz heterogenous covariance structure as it yielded the smallest Akaike's information criterion from the ones tested. For all models described above, statistical significance was declared if $p \leq 0.05$, and a tendency was considered if $0.05 < p < 0.1$.

3. Results

A total of 852 cows (treatment: 418 and control: 434) were initially enrolled in the study. Of the total number of enrolled cows, 27 and 16 cows in the treatment and control groups, respectively, were excluded. Exclusion of cows in the treatment group were for the following reasons: 3 cows were first parity, 9 were Jersey–Holstein crosses, 3 cows were diagnosed with mastitis at 0 DIM, and 12 did not receive their second bolus. Exclusion of cows in the control group were for the following reasons: two cows were first parity, four were Jersey–Holstein crosses, 3 cows were diagnosed with mastitis at 0 DIM, one cow had uterine prolapse, and six were treated with a Ca bolus by mistake. Thus, 809 cows (treatment: 391 and control: 418) remained in the experiment for final analyses; 310 (38%) were of second parity and 499 (62%) of third and greater parities. Descriptive statistics data depicting incidences of early lactation disorders for the cows included in the final analyses can be found in Table 1.

Table 1. Early lactation disorders in a randomized clinical trial evaluating the effects of an oral postpartum calcium supplementation strategy in multiparous Jersey cows (n = 809) in a dairy in west TX.

Item	Control (n = 418)		Treatment (n = 391)	
	Parity 2	Parity ≥ 3	Parity2	Parity ≥ 3
Dystocia	4	8	6	15
Stillbirth	4	10	4	5
Clinical hypocalcemia [1]	1	4	0	8
Retained placenta	0	3	0	1
Left displaced abomasum	1	1	0	1

[1] Clinical hypocalcemia diagnosis occurred within 24 h after calving.

3.1. Mastitis

A parity by treatment group interaction was found in the final model evaluating the odds of clinical mastitis in the first 60 DIM (p = 0.02). Among second parity cows, treatment tended to reduce the odds of clinical mastitis (OR = 0.49; p = 0.07) compared to controls. The same was not true among third and greater parity cows; there was no difference in the odds of clinical mastitis when treated cows were compared to controls (OR = 1.45; p = 0.49). The other variable retained in the final model was the effect of a calving-related problem (p = 0.49). Table 2 presents the odds ratio for the parity by treatment group interaction for the mastitis model.

Table 2. Odds of clinical mastitis in the first 60 DIM by parity in a randomized clinical trial evaluating the effect of an oral postpartum calcium supplementation strategy in multiparous Jersey cows (n = 809) in a dairy in west TX.

Comparison	Estimate	Odds Ratio	95% CI [1]	p-Value
Parity 2: treatment versus control	−0.71	0.49	0.18–1.33	0.07
Parity ≥ 3: treatment versus control	0.37	1.45	0.74–2.83	0.49

[1] Confidence interval.

3.2. Culling

Treatment was not associated with the odds of culling in the first 60 DIM (p = 0.72). Variables remaining in the final model were parity (p = 0.001), and calving-related problems (p = 0.01). Culling incidence in the first 60 DIM for the treated and controls cows were 8.7 and 9.3%, respectively. Table 3 presents the final logistic regression model evaluating the effect of postpartum oral Ca supplementation on the odds of culling in the first 60 DIM.

Table 3. Final logistic regression model evaluating the effect of an oral postpartum calcium supplementation strategy with culling in the first 60 DIM in a randomized clinical trial in multiparous Jersey cows (n = 809) in a dairy in west TX.

Variable	Estimate	SE [1]	p-Value
Intercept	−3.07	0.30	<0.001
Postpartum Ca supplementation			
Control	Ref [2]	–	–
Treatment	−0.09	0.25	0.72
Parity			
2	Ref [2]	–	–
≥3	1.01	0.31	0.001
Calving problem [3]			
No	Ref [1]	–	–
Yes	0.94	0.38	0.01

[1] Standard error. [2] Reference category (i.e., the value to which the variable level is being compared to while controlling for the effect of the other predictors in the model). [3] Calving problem: variable representing cows that suffered from dystocia and(or) that had a stillbirth.

3.3. Reproductive Performance

The Cox proportional hazards model revealed no effect of treatment on time to pregnancy ($p = 0.67$). Other variables retained in the final model were the effect of parity ($p = 0.91$) and calving-related problems ($p = 0.16$). The effect of calving season violated the assumption of proportional hazards over time and was included in the STRATA statement. Table 4 presents the final Cox proportional hazards model.

Table 4. Cox proportional hazards model evaluating the effect of an oral postpartum calcium supplementation strategy with reproductive efficiency in a randomized clinical trial in multiparous Jersey cows ($n = 809$) in a dairy in west TX.

Variable	Estimate	SE [1]	*p*-Value	Hazard Ratio	Hazard Ratio CI
Postpartum Ca supplementation					
Control	Ref [2]	–	–	–	–
Treatment	0.04	0.10	0.67	1.04	0.86–1.27
Parity					
2	Ref [2]	–	–	–	–
≥3	−0.01	0.10	0.91	0.99	0.8–1.20
Calving problem [3]					
No	Ref [2]	–	–	–	–
Yes	−0.37	0.26	0.16	0.69	0.41–1

[1] Standard error. [2] Reference category (i.e., the value to which the variable level is being compared to while controlling for the effect of the other predictors in the model). [3] Calving problem: variable representing cows that suffered from dystocia and(or) that had a stillbirth.

3.4. Milk Production

Treatment had no effect on milk production for the first 15 weeks of lactation ($p = 0.73$). On average, treated and control cows produced 33.5 kg/d (SEM ± 0.59) and 33.3 kg/d (SEM ± 0.36), respectively. Other variables included in the model were the effect of parity ($p = 0.22$), BCS score ($p = 0.25$), calving season ($p = 0.02$), gestation length ($p = 0.0001$), and week of milk measurement ($p < 0.0001$). The final milk model is presented in Table 5.

Table 5. Final linear mixed model evaluating the effect of an oral postpartum calcium supplementation strategy with milk production within 15 weeks of lactation in a randomized clinical trial in multiparous Jersey cows ($n = 809$) in a dairy in west TX.

Variable	Estimate	SE [1]	*p*-Value
Intercept	−9.62	10.83	0.35
Postpartum Ca supplementation			
Control	Ref [2]	–	–
Treatment	0.24	0.69	0.73
Parity			
2	Ref [2]	–	–
≥3	0.50	0.41	0.22
Test number	–	–	<0.001
Calving season [3]			
Warm	Ref [2]	–	–
Cool	−0.97	0.40	0.02
Gestation length (days)	0.15	0.04	<0.001
Body condition score			
Thin	Ref [2]	–	–
Normal	0.76	0.72	0.29
Over-conditioned	1.64	0.99	0.10

[1] Standard error. [2] Reference category (i.e., the value to which the variable level is being compared to while controlling for the effect of the other predictors in the model). [3] Calving season (warm: cows calving from 19 July to 22 September 2018; cool: cows calving from 23 September to 9 December 2018).

4. Discussion

Currently, studies evaluating the effect of postpartum oral Ca supplementation focusing on health, production, and reproduction in purebred Jersey cows are lacking. The objectives of this study were to evaluate the effect of an oral postpartum Ca supplementation strategy comprised of 43 g of calcium salts administered within 1 h after parturition, followed by a second dose the day after parturition, on the odds of clinical mastitis and culling within 60 DIM, as well as on milk production and reproductive efficiency in multiparous Jersey cows. Subclinical hypocalcemia is a risk factor for early lactation diseases, impaired reproduction, and culling in dairy herds [21,29,30]. It is a common belief among dairy farmers that oral Ca supplementation postpartum decreases the incidence of some postpartum diseases, and can potentially reduce culling and improve milk production and reproductive efficiency. Our results showed no effects of the oral Ca supplementation strategy in multiparous postpartum Jersey cows on early-lactation culling, milk production, or reproductive efficiency. There was a tendency for oral Ca supplementation in second parity cows to have reduced odds of mastitis in the first 60 DIM; the same was not true for third and greater parity cows.

The use of postpartum oral Ca supplementation in Holstein cows has been associated with improved cow performance. A study done comparing the effect of two different doses (43 and 86 g) of oral Ca supplementation showed that both treatments were able to increase the concentration of ionized Ca in the blood [31]. Additionally, the same study showed that oral Ca supplementation improved reproductive performance in multiparous cows only, while having a detrimental effect in primiparous cows. When oral Ca supplementation is administrated to specific groups of cows, it seems to have a more relevant impact on health [8,32]. Leno et al. (2018) demonstrated in Holstein cows that a single dose of Ca oral supplementation within 24 h postpartum improved health status for cows of greater parity and BCS, and lame cows [33]. Additionally, a recent study done in Jersey and Jersey–Holstein crosses showed that the effects of the treatment were dependent on cow-level factors and had minimal impacts on group-level assessments [25]. A limitation of our study is the absence of lameness information (data not collected by our group) and previous lactation milk yield (herd not participating in Dairy Herd Improvement program) which limited our ability to explore some cow-level dependent effects.

We evaluated the effect of oral Ca supplementation on the odds of clinical mastitis within 60 days after parturition and found that the results depended on parity; among second parity cows, treatment tended to decrease the odds of mastitis. Suboptimum Ca concentration in the blood is associated with a reduced concentration of Ca in the cytosol of leukocytes and a reduced immune response [34]. A study done on Jersey–Holstein crosses demonstrated improved neutrophil phagocytosis and oxidative burst when cows were supplemented with two oral Ca (50g) boluses in the first 24 h after parturition [35]. Domino et al. (2017) found that the postpartum administration of Ca boluses decreased the risk of mastitis in multiparous Holstein cows within the group of cows with a high relative herd rank only (a metric that was based on previous lactation milk production) [36]. However, other studies evaluating the effect of oral Ca boluses on health outcomes reported no effect of treatment on the risk of mastitis [8,31].

Treatment had no impact on the odds of culling in the first 60 DIM. Milk production and reproduction are two main factors affecting culling in dairy herds [37,38]. No treatment effect was found for milk production in the first 15 weeks of lactation or reproductive efficiency up to 150 DIM; therefore, it is not surprising that oral Ca supplementation also did not have any impact on the odds of culling. We are unaware of any studies reporting an association of postpartum oral Ca supplementation and decreased culling.

No effect of treatment on time to pregnancy was found. The current literature does not report any beneficial effect of postpartum prophylactic Ca administration on reproductive performance in Holstein and Jersey cows as a blanket support therapy [25,39,40]; Martinez et al. (2016) administered Ca boluses to Holsteins cows considered to be of high and low risk to develop metritis in a randomized block design and found a benefic impact on reproduction outcomes for cows in the high-risk metritis block [31]. No effect of oral Ca treatment was found in pregnancy at first service in a previous Jersey study [25]. The literature is inconsistent when evaluating low concentrations of Ca in the blood as a risk factor for impaired reproduction [5,30,37,41]. While we did not measure serum Ca dynamics in early postpartum, Valldecabres et al. (2021) demonstrated that Jersey cows with Ca $\leq$ 1.94 mmol/L within the first hours postpartum had decreased risk of pregnancy [41]. On the other hand, Menta et al. (2021) demonstrated no association of blood Ca levels in the first 3 d postpartum and pregnancy to first service [5]. Although we did not find positive effects of oral Ca boluses in Jersey cows with most outcomes evaluated, we cannot rule out that some cow subpopulations could benefit from this strategy. One potential limitation of our study is that we did not measure blood Ca concentrations following oral Ca supplementation to confirm the rise in systemic Ca levels post-treatment. However, we used a commercial product that has been used in previous studies and proven to increase blood Ca concentrations [31,36].

No effect of postpartum oral Ca was found on milk yield in the first 15 weeks after lactation. This result agrees with previous research conducted in Holstein cows [33,42,43]. Oetzel and Miller (2012) found a positive effect of Ca boluses for lame and high-producing cows only [8]. Our results are consistent with those of Valldecabres and Silva-del-Rio (2021), which was conducted in a population of Jersey and Jersey–Holstein crossbred cows [25]. A recent study done by our research group evaluated the association of Ca concentration in the first 3 DIM with milk production and found that Jersey cows with reduced Ca concentrations at 1 ($\leq$1.84 mmol/L) and 2 DIM ($\leq$2.04 mmol/L) had increased milk yield in the first 9 weeks after calving [5]. Therefore, further research evaluating how different oral Ca supplementation protocols that vary the timing of bolus administration relative to calving are warranted.

5. Conclusions

In summary, our randomized clinical trial demonstrated that prophylactic postpartum Ca supplementation to multiparous Jersey cows had no effects on culling, milk yield, and reproduction. Second parity cows that were supplemented with oral Ca boluses tended to have reduced odds of mastitis compared to non-supplemented cows; however, no treatment effect was evident for third and greater parity cows. This is one of the few large randomized clinical trials evaluating the effects of postpartum Ca supplementation in Jersey cows. Our data do not support blanket oral Ca supplementation in Jersey cows as the effects were minimal to none; however, targeted oral Ca supplementation for subpopulations of cows and at different times relative to parturition remain to be investigated.

Author Contributions: Conceptualization: V.S.M. and R.C.N.; methodology, R.C.N.; software, P.R.M. and R.C.N.; validation, P.R.M., L.F., D.P., M.L.C., V.S.M., and R.C.N.; data curation, P.R.M. and R.C.N.; formal analysis, P.R.M. and R.C.N.; investigation, P.R.M., L.F., D.P., and M.L.C.; resources, V.S.M. and R.C.N.; data curation, P.R.M. and R.C.N.; writing—original draft, P.R.M.; writing—review and editing, L.F., D.P., M.L.C., V.S.M., and R.C.N.; visualization, P.R.M. and R.C.N.; supervision, R.C.N.; project administration, P.R.M., V.S.M., and R.C.N.; funding acquisition, R.C.N. All authors have read and agreed to the published version of the manuscript.

Funding: This research received no external funding.

Institutional Review Board Statement: All animal procedures were reviewed and approved by the Texas Tech University Institutional Animal Care and Use Committee (protocol 18035-04).

Data Availability Statement: Data not available due to participating farm restrictions.

Acknowledgments: The authors thank the participating farm and employees for their support during the study.

Conflicts of Interest: The authors declare no conflict of interest.

References

1. DeGaris, P.J.; Lean, I.J. Milk fever in dairy cows: A review of pathophysiology and control principles. *Vet. J.* **2008**, *176*, 58–69. [CrossRef]
2. Goff, J.P.; Kimura, K.; Horst, R.L. Effect of Mastectomy on Milk Fever, Energy, and Vitamins A, E, and β-Carotene Status at Parturition 1. *J. Dairy Sci.* **2002**, *85*, 1427–1436. [CrossRef]
3. Goff, J.P. The monitoring, prevention, and treatment of milk fever and subclinical hypocalcemia in dairy cows. *Vet. J.* **2008**, *176*, 50–57. [CrossRef] [PubMed]
4. McArt, J.A.A.; Neves, R.C. Association of transient, persistent, or delayed subclinical hypocalcemia with early lactation disease, removal, and milk yield in Holstein cows. *J. Dairy Sci.* **2020**, *103*, 690–701. [CrossRef] [PubMed]
5. Menta, P.R.; Fernandes, L.; Poit, D.; Celestino, M.; Machado, V.S.; Ballou, M.A.; Neves, R.C. Association of calcium and energy balance metabolites in the first 3 days after parturition with disease and production outcomes in multiparous Jersey cows. *J. Dairy Sci.* **2021**, *103*, 17. [CrossRef]
6. Roche, J.R.; Berry, D.P. Periparturient climatic, animal, and management factors influencing the incidence of milk fever in grazing systems. *J. Dairy Sci.* **2006**, *89*, 2775–2783. [CrossRef]
7. Reinhardt, T.A.; Lippolis, J.D.; McCluskey, B.J.; Goff, J.P.; Horst, R.L. Prevalence of subclinical hypocalcemia in dairy herds. *Vet. J.* **2011**, *188*, 122–124. [CrossRef] [PubMed]
8. Oetzel, G.R.; Miller, B.E. Effect of oral calcium bolus supplementation on early-lactation health and milk yield in commercial dairy herds. *J. Dairy Sci.* **2012**, *95*, 7051–7065. [CrossRef] [PubMed]
9. Moore, S.J.; VandeHaar, M.J.; Sharma, B.K.; Pilbeam, T.E.; Beede, D.K.; Bucholtz, H.F.; Liesman, J.S.; Horst, R.L.; Goff, J.P. Effects of altering dietary cation-anion difference on calcium and energy metabolism in peripartum cows 1. *J. Dairy Sci.* **2000**, *83*, 2095–2104. [CrossRef]
10. Charbonneau, E.; Pellerin, D.; Oetzel, G.R. Impact of lowering dietary cation-anion difference in nonlactating dairy cows: A meta-analysis. *J. Dairy Sci.* **2006**, *89*, 537–548. [CrossRef]
11. Santos, J.E.P.; Lean, I.J.; Golder, H.; Block, E. Meta-analysis of the effects of prepartum dietary cation-anion difference on performance and health of dairy cows. *J.Dairy Sci.* **2019**, *102*, 2134–2154. [CrossRef] [PubMed]
12. Caixeta, L.S.; Ospina, P.A.; Capel, M.B.; Nydam, D.V. The association of subclinical hypocalcemia, negative energy balance and disease with bodyweight change during the first 30 days post-partum in dairy cows milked with automatic milking systems. *Vet. J.* **2015**, *204*, 150–156. [CrossRef]
13. Venjakob, P.L.; Borchardt, S.; Heuwieser, W. Hypocalcemia—Cow-level prevalence and preventive strategies in German dairy herds. *J. Dairy Sci.* **2017**, *100*, 9258–9266. [CrossRef] [PubMed]
14. Goff, J.P. Major advances in our understanding of nutritional influences on bovine health. *J. Dairy Sci.* **2006**, *89*, 1292–1301. [CrossRef]
15. Martinez, N.; Risco, C.A.; Lima, F.S.; Bisinotto, R.S.; Greco, L.F.; Ribeiro, E.S.; Maunsell, F.; Galvão, K.; Santos, J.E.P. Evaluation of peripartal calcium status, energetic profile, and neutrophil function in dairy cows at low or high risk of developing uterine disease. *J. Dairy Sci.* **2012**, *95*, 7158–7172. [CrossRef] [PubMed]
16. Chapinal, N.; Carson, M.; Duffield, T.F.; Capel, M.; Godden, S.; Overton, M.; Santos, J.E.P.; LeBlanc, S.J. The association of serum metabolites with clinical disease during the transition period. *J. Dairy Sci.* **2011**, *94*, 4897–4903. [CrossRef]
17. Roberts, T.; Chapinal, N.; LeBlanc, S.J.; Kelton, D.F.; Dubuc, J.; Duffield, T.F. Metabolic parameters in transition cows as indicators for early-lactation culling risk. *J. Dairy Sci.* **2012**, *95*, 3057–3063. [CrossRef]
18. Martinez, N.; Sinedino, L.D.P.; Bisinotto, R.S.; Ribeiro, E.S.; Gomes, G.C.; Lima, F.S.; Greco, L.F.; Risco, C.A.; Galvão, K.N.; Taylor-Rodriguez, D.; et al. Effect of induced subclinical hypocalcemia on physiological responses and neutrophil function in dairy cows. *J. Dairy Sci.* **2014**, *97*, 874–887. [CrossRef]
19. U.S. Department of Agriculture. Available online: https://www.aphis.usda.gov/animal_health/nahms/dairy/downloads/dairy96/Dairy96_dr_PartI.pdf (accessed on 23 August 2021).
20. Blanc, C.D.; Van der List, M.; Aly, S.S.; Rossow, H.A.; Silva-del-Río, N. Blood calcium dynamics after prophylactic treatment of subclinical hypocalcemia with oral or intravenous calcium. *J. Dairy Sci.* **2014**, *97*, 6901–6906. [CrossRef]
21. Martinez, N.; Sinedino, L.D.P.; Bisinotto, R.S.; Daetz, R.; Risco, C.A.; Galvão, K.N.; Thatcher, W.W.; Santos, J.E.P. Effects of oral calcium supplementation on productive and reproductive performance in Holstein cows. *J. Dairy Sci.* **2016**, *99*, 8417–8430. [CrossRef]
22. Goff, J.P.; Reinhardt, T.A.; Horst, R.L. Milk fever and dietary cation-anion balance effects on concentration of vitamin D receptor in tissue of periparturient dairy cows. *J. Dairy Sci.* **1995**, *78*, 2388–2394. [CrossRef]
23. Cerbulis, J.; Farrell, H.M. Composition of the Milks of Dairy Cattle. II. Ash, Calcium, Magnesium, and Phosphorus. *J. Dairy Sci.* **1976**, *59*, 589–593. [CrossRef]
24. Valldecabres, A.; Pires, J.A.A.; Silva-del-Río, N. Effect of prophylactic oral calcium supplementation on postpartum mineral status and markers of energy balance of multiparous Jersey cows. *J. Dairy Sci.* **2018**, *101*, 4460–4472. [CrossRef] [PubMed]
25. Valldecabres, A.; Silva-del-Río, N. Effects of postpartum oral calcium supplementation on milk yield, milk composition, and reproduction in multiparous Jersey and Jersey × Holstein crossbreed cows. *J. Dairy Sci.* **2020**, *104*, 795–805. [CrossRef] [PubMed]
26. Sheldon, I.M.; Lewis, G.S.; LeBlanc, S.; Gilbert, R.O. Defining postpartum uterine disease in cattle. *Theriogenology* **2006**, *65*, 1516–1530. [CrossRef]

27. Peek, S.F.; Divers, T.J. *Rebhun's Diseases of Dairy Cattle*, 3rd ed.; Elsevier: St. Louis, MO, USA, 2017; pp. 713–733, ISBN 9780323390552.
28. Ferguson, J.D.; Galligan, D.T.; Thomsen, N. Principal Descriptors of Body Condition Score in Holstein Cows. *J. Dairy Sci.* **1994**, *77*, 2695–2703. [CrossRef]
29. Ribeiro, E.S.; Lima, F.S.; Greco, L.F.; Bisinotto, R.S.; Monteiro, A.P.A.; Favoreto, M.; Ayres, H.; Marsola, R.S.; Martinez, N.; Thatcher, W.W.; et al. Prevalence of periparturient diseases and effects on fertility of seasonally calving grazing dairy cows supplemented with concentrates. *J. Dairy Sci.* **2013**, *96*, 5682–5697. [CrossRef]
30. Caixeta, L.S.; Ospina, P.A.; Capel, M.B.; Nydam, D.V. Association between subclinical hypocalcemia in the first 3 days of lactation and reproductive performance of dairy cows. *Theriogenology* **2017**, *94*, 1–7. [CrossRef]
31. Martinez, N.; Sinedino, L.D.P.; Bisinotto, R.S.; Daetz, R.; Lopera, C.; Risco, C.A.; Galvão, K.N.; Thatcher, W.W.; Santos, J.E.P. Effects of oral calcium supplementation on mineral and acid-base status, energy metabolites, and health of postpartum dairy cows. *J. Dairy Sci.* **2016**, *99*, 8397–8416. [CrossRef]
32. Mcart, J.A.A.; Oetzel, G.R. A stochastic estimate of the economic impact of oral calcium supplementation in postparturient dairy cows. *J. Dairy Sci.* **2015**, *98*, 7408–7418. [CrossRef]
33. Leno, B.M.; Neves, R.C.; Louge, I.M.; Curler, M.D.; Thomas, M.J.; Overton, T.R.; McArt, J.A.A. Differential effects of a single dose of oral calcium based on postpartum plasma calcium concentration in Holstein cows. *J. Dairy Sci.* **2018**, *101*, 3285–3302. [CrossRef] [PubMed]
34. Kimura, K.; Reinhardt, T.A.; Goff, J.P. Parturition and hypocalcemia blunts calcium signals in immune cells of dairy cattle. *J. Dairy Sci.* **2006**, *89*, 2588–2595. [CrossRef]
35. Reitsma, L.; Batchelder, T.; Davis, E.; Machado, V.; Neves, R.; Ballou, M. Effects of oral calcium bolus supplementation on intracellular polymorphonuclear leukocyte calcium levels and functionality in primiparous and multiparous dairy cows. *J. Dairy Sci.* **2020**, *103*, 103,11876–11888. [CrossRef] [PubMed]
36. Domino, A.R.; Korzec, H.C.; McArt, J.A.A. Field trial of 2 calcium supplements on early lactation health and production in multiparous Holstein cows. *J. Dairy Sci.* **2017**, *100*, 9681–9690. [CrossRef] [PubMed]
37. Neves, R.C.; Leno, B.M.; Curler, M.D.; Thomas, M.J.; Overton, T.R.; McArt, J.A.A. Association of immediate postpartum plasma calcium concentration with early-lactation clinical diseases, culling, reproduction, and milk production in Holstein cows. *J. Dairy Sci.* **2018**, *101*, 547–555. [CrossRef] [PubMed]
38. Bascom, S.S.; Young, A.J. A Summary of the Reasons Why Farmers Cull Cows. *J. Dairy Sci.* **1998**, *81*, 2299–2305. [CrossRef]
39. Pinedo, P.; Velez, J.; Solano, G.; Rodriguez, N.; Naves, J.; Schuenemann, G.M.; Risco, C. Effect of oral calcium administration on the cure and reproductive performance of Holstein cows diagnosed with puerperal metritis. *J. Dairy Sci.* **2017**, *100*, 2917–2927. [CrossRef]
40. Miltenburg, C.L.; Duffield, T.F.; Bienzle, D.; Scholtz, E.L.; LeBlanc, S.J. Randomized clinical trial of a calcium supplement for improvement of health in dairy cows in early lactation. *J. Dairy Sci.* **2016**, *99*, 6550–6562. [CrossRef]
41. Valldecabres, A.; Silva-del-RÃ, N. Association of low serum calcium concentration after calving with productive and reproductive performance in multiparous Jersey cows. *J. Dairy Sci.* **2021**, *104*, 11983–11994. [CrossRef]
42. Goff, J.P.; Horst, R.L.; Jardon, P.W.; Borelli, C.; Wedam, J. Field Trials of an Oral Calcium Propionate Paste as an Aid to Prevent Milk Fever in Periparturient Dairy Cows. *J. Dairy Sci.* **1996**, *79*, 378–383. [CrossRef]
43. Melendez, P.; Donovan, A.; Risco, C.A.; Hall, M.B.; Littell, R.; Goff, J. Metabolic responses of transition holstein cows fed anionic salts and supplemented at calving with calcium and energy. *J. Dairy Sci.* **2002**, *85*, 1085–1092. [CrossRef]

animals

MDPI

Article

Prediction of Calving to Conception Interval Length Using Algorithmic Analysis of Endometrial mRNA Expression in Bovine

Dawid Tobolski [1,*], Karolina Łukasik [2], Agnieszka Bacławska [2], Dariusz Jan Skarżyński [2], Miel Hostens [3] and Wojciech Barański [4]

1 Department of Internal Diseases with Clinic, Faculty of Veterinary Medicine, University of Warmia and Mazury, Oczapowskiego 14, 10-957 Olsztyn, Poland
2 Department of Reproductive Immunology and Pathology, Institute of Animal Reproduction and Food Research, Polish Academy of Sciences, Tuwima 10, 10-748 Olsztyn, Poland; k.lukasik@pan.olsztyn.pl (K.Ł.); a.baclawska@pan.olsztyn.pl (A.B.); d.skarzynski@pan.olsztyn.pl (D.J.S.)
3 Department of Farm Animal Health, University of Utrecht, Yalelaan 7, 3584 CL Utrecht, The Netherlands; m.m.hostens@uu.nl
4 Department of Animal Reproduction with Clinic, Faculty of Veterinary Medicine, University of Warmia and Mazury, Oczapowskiego 14, 10-957 Olsztyn, Poland; wojbar@uwm.edu.pl
* Correspondence: dawid.tobolski@uwm.edu.pl; Tel.: +48-50-245-0264

Simple Summary: Our study aimed to develop the unsupervised clustering model based solely on selected markers to investigate the association between calving conception interval length, subclinical endometritis, and endometrial gene expression. An algorithmic analysis of endometrial gene expression showed a higher predictive ability to identify cows exhibiting excellent fertility than previously used methods, highlighting the correlation between *INHBA/INHA* and calving–conception interval length.

Abstract: After parturition, the uterus undergoes significant reconstruction, allows the endometrium to create an environment for subsequent embryo development. Here, we used an unsupervised algorithmic approach to select characteristic endometrial mRNA expression patterns of proposed markers and investigate each marker's role as an individual indicator of reproductive success. Clinically healthy cows at a sixth week postpartum were examined, the percentage of neutrophils (PMNs%) in the cytological smear was calculated, and an endometrial biopsy was taken for qPCR. Based on pregnancy examination, cows were divided into three groups: Pregnant before 100 days postpartum (P100, n = 11), pregnant between 100–200-day (P200, n = 14), and culled (C, n = 10). Animals were also classified based on two PMNs% thresholds > 5% PMNs and > 10% PMNs. The expression of *IL1B*, *IL6*, *CXCL8*, and *IL17A* was higher in >10%PMNs. The expression of PTGS1 was higher in the P200 compared to P100. Upregulation of inhibin A subunit (*INHA*) and downregulation of inhibin β A subunit (*INHBA*) were observed in the P100. *INHBA/INHA* ratio was the most accurate linear predictor of the calving-to-conception interval. The application of the k-means algorithm allowed the identification of five unique expression patterns. The sensitivity and specificity of predicting allocation to P100 were 81% and 79%. We also documented the low efficiency of genes associated with subclinical endometritis and PMNs% in determining reproductive capability. These results suggested the presence of distinctive expression patterns in 6 weeks postpartum, correlated with cows' reproductive capacity. Furthermore, we proposed the *INHBA/INHA* ratio as an indicator of calving-to-conception interval length.

Keywords: postpartum diseases; activin; inhibin; cytokines; endometrium; subclinical endometritis; cow

Citation: Tobolski, D.; Łukasik, K.; Bacławska, A.; Skarżyński, D.J.; Hostens, M.; Barański, W. Prediction of Calving to Conception Interval Length Using Algorithmic Analysis of Endometrial mRNA Expression in Bovine. *Animals* **2021**, *11*, 236. https://doi.org/10.3390/ani11010236

Received: 17 December 2020
Accepted: 15 January 2021
Published: 19 January 2021

1. Introduction

The uterus undergoes involution after parturition that finishes around day 40–50 postpartum (PP). Endometrial return to the condition before pregnancy may be delayed due to uterine inflammatory processes such as metritis, clinical (CE), and subclinical (SE) endometritis or other causes related to calving ease, breed, metabolic status, and age of the cows [1]. Thus, the development of diseases in the postpartum period is considered the primary risk factor for an increased period between parturition and following conception (CCI) [2]. Minimizing the length of CCI could increase cow's milk yield relative to parturition and feed cost, increase calves' number, reduce culling and cost of recurrent insemination [3]. Therefore, the development of a forecasting tool for CCI's length is highly anticipated by veterinarians and farmers.

Subclinical endometritis is described as an inflammation of the endometrium manifested by an elevated level of neutrophils in the absence of purulent vaginal discharge. Generally, SE's effect on reproduction performance was evaluated based on the days diagnosed relative to parturition, increasing with time and neutrophils' threshold. Subclinical endometritis diagnosed at day 20–33 PP increased median days open from 112 in healthy cows to 141, from 100 to 162 when diagnosed at day 34–47 PP [4], and from 118 to 206 when diagnosed at day 40–60 PP [5]. A recent study predicting the pregnancy status using the percentage of polymorphonuclear leukocytes (PMNs%) evaluated between 42- and 49-days PP as a diagnostic method revealed a sensitivity of 36.9, 39, and 53.6% for day 100, 150, and 200, respectively. The percentage of PMNs showed the highest sums of sensitivity and specificity along with the lowest hazard ratio and odds ratio as a predictor of pregnancy status up to 100-day PP (compared to transrectal palpation measurement of the cervical diameter, ultrasonographic measurement of the fluid in uterus score, vaginoscopic detection of external uterine orifice hyperemia, vaginal discharge score). The PMNs percent, together with other tested predictors, shows a low predictive value to detect cows able to implant embryo before 100 days PP [6].

The diagnosis of metritis or CE is straightforward due to the visible discharge from the vagina and fluid presence in the uterus lumen during an ultrasound examination [2]. However, in the absence of clinical symptoms, evaluation of endometrial conditions is difficult, time-consuming, and almost impossible at a farm site. A commonly-used method to evaluate the status of the endometrium is to count the number of neutrophils infiltrating the mucosa layer done by a cytobrush [4], low-volume uterine lavage [5], cytotape [7], or uterine secretions [8]. Nevertheless, these listed diagnostic tools allow to collect and evaluate only superficial cells and mucus. Thus, a better method to investigate ongoing molecular changes during the endometrium's late postpartum recovery is a biopsy [9–11], as well as further research is needed to determine the association between biopsy findings and reproductive performance of the cows [12]. Currently used diagnostic methods focus on finding indicators describing the ongoing inflammatory process and correlations between PMNs and inflammatory mediators. Determination of the tissue markers of the cow's likelihood to become pregnant would allow the creation of tools for monitoring both the health of the individual cow as well as herd level.

Expression of inflammatory mediators, including cytokines, chemokines, is altered in the endometrium after parturition and during SE and CE. Previous studies showed that a high number of PMNs is associated with increased expression of interleukin 1 beta (*IL1B*), Interleukin 6 (*IL6*), Interleukin 8 (*CXCL8*), and Tumor Necrosis Factor between the fifth and eighth-week PP [13,14]. Increased expression of cytokines and chemokines in this period is not associated with SE or CE at week 4 PP but with an ongoing inflammatory process. The same authors also suggested that an excessive immune response in the late puerperium may cause poor reproduction performance in dairy cows [15]. Therefore, it is crucial to investigate the association of increased interleukins expression in the late postpartum period with the probability of cows becoming pregnant.

Endometrial production of prostaglandins (PGs) plays an essential role in pregnancy, parturition, and subsequent uterine involution. Production of PGs is controlled mostly by prostaglandin-endoperoxide synthase 2 (*PTGS2*) and prostaglandin-endoperoxide synthase 1 (*PTGS1*), which provide substrates for the downstream reactions. Prostaglandin F and E synthases (*PRXL2B, PTGES*) are responsible for prostaglandins F (PGF) and E (PGE) synthesis. They are essential for the correct progression of the estrus cycle and return to physiological cyclicity after parturition [16]. Cows with increased PMNs in endometrial smear had lower PGF concentrations and higher PGE than healthy cows in the fourth week postpartum, and the ratio of PGE to PGF production may be one of the causes of delayed involution or endometrial restoration [17].

Another group of factors potentially associated with involution and endometrial tissue remodeling is the activin and periostin (*POSTN*) pathways [18,19]. Two Inhibin βA subunits (*INHBA*) form activin A, while the *INHBA* subunit and the inhibin α subunit (*INHA*) form inhibin A, which, together with follistatin (*FST*) are antagonists of activin A [20]. Until now, studies confirmed a regulatory role of activins and inhibins in ovarian function, folliculogenesis, and crucial role in the proper function of the female reproductive tract [21,22]. Studies on other species underlined the importance of activin A and its subunit as a monomer in the inflammatory response, wound healing, scar formation, and organ fibrosis [23,24]. There are limited data relating to the endometrial expression levels and their local function. Salilew-Wondim [18] suggested *INHBA* among 28 other genes as a potential marker of endometrial inflammation, including change of *INHBA* expression as a potential mechanism of molecular dysregulation of uterine receptivity and homeostasis [18].

Hence, we hypothesized that both subclinical endometritis as a disease and markers proposed for its diagnosis are useful predictors of future reproductive performance. In this study, we aimed to use an unsupervised algorithmic approach to select characteristic endometrial mRNA expression patterns of proposed markers in 6 weeks postpartum and investigate the role of each marker as an indicator of reproductive success.

2. Materials and Methods

2.1. Ethic Statement

The Local Ethics Committee for animal experiments in Olsztyn approved the research by resolution No. 49/2016.

2.2. Animals and Study Design

The study was performed on one dairy farm with Polish Holstein-Friesian cows. The cows ranged from the first to the fifth lactation with an average 305 d milk production over 9000 kg/cow. The cows were housed in a free-stall barn and fed with a partial mixed ration. Animals were milked by means of a voluntary milking system and artificially inseminated by a single AI technician after 60 days of voluntary waiting period at every visible heat until pregnancy detection or culling. Heat detection was performed three times a day for 30 min by one observer. The outcome of each insemination was noted for the estimation of bulls' fertility. The quality of bulls' semen did not differ and met standards for the production of frozen bovine semen. Daily milk yield for the first 60 days of lactation was obtained from the DeLaval ALPRO database (Supplementary Materials Table S1).

The first vaginoscopy was performed between 21 to 29-day PP. Cows that showed any signs of clinical endometritis (CE) or received treatment for mastitis, lameness, and pneumovagina were excluded from the study. Forty-three remaining animals were re-examined between day 35 and 42 postpartum. Based on inspection of the tail, vulva, and vaginoscopy, five cows were diagnosed as CE and treated with intrauterine antibiotics; therefore, they were excluded from the study (3/5—clear mucus containing flecks of white pus, 1/5—<50% pus in mucus, 1/5—>50% pus in mucus). Ultrasound examination of the genital tract was done to exclude cows with ovarian cysts (n = 1), estrus at the time of examination (n = 2). The diameter of the cervix and uterine horns was measured (Table S1). Finally, thirty-five cows were included in the study cohort. Blood was collected from the coccygeal vein to determine the cow's metabolic condition at the time of sampling

(Table S1). Progesterone concentration in serum was measured to assess the influence of active corpus luteum on endometrial gene expression.

2.3. Sample Collection

Blood samples were collected from the coccygeal vein using the VACUETTE blood collection system (VACUETTE® TUBE 9 mL CAT Serum Clot Activator, Greiner Bio-One, Kremsmünster, Austria). Collected tubes were placed on ice and transported to the laboratory. Blood serum was separated by centrifugation at 4000 rpm at 4 °C for 20 min (Beckman Coulter, J-6 MC, Brea, CA, USA). The serum was transferred to 2 mL Eppendorf tubes, placed in the ultra-freezer, and stored at −80 °C until further analysis. The concentration of total cholesterol (TC), triglyceride (TG), non-esterified fatty acids (NEFA), β-Hydroxybutyrate (BHB) was measured using a biochemical analyzer (Cormay Group, ACCENT-200, Łomianki, Poland). Progesterone levels in the samples were measured using the Radio Immuno Assay method in the β-radiation counter (Pharmacia, Wallace 1410, Finland) (Table S1).

Uterine cytology samples were obtained at the second examination by the cytobrush method (Cervical Brush, Zarys International Group, Zabrze, Poland). Briefly, the brush was mounted on the mandrel and inserted into a sterile metal catheter. For the protection of the brush from vaginal contamination, the sterile gloves for rectal examination were used. The entire setup was inserted into the cow's genital tract and passed through the cervical canal under the second hand's control placed in the rectum. At the uterine end of the cervix, the glove protecting catheter was punctured. The catheter was carefully inserted into the uterus' right horn, and the sample was taken by clockwise rotation of the rodded brush. After sample collection, the brush was pulled into the catheter, and the complete setup was gently drawn back from the cow's genital tract. The brush was pushed from the catheter, and cytological material was transferred on the glass slide by rolling on it.

Uterine biopsy sample collection was obtained with biopsy forceps six weeks post-partum (Kevorkian's uterine biopsy forceps, Hauptner Herberholz, Solingen, Germany). Sterile forceps were placed into a rectal glove and passed in the same manner as a catheter into the uterine horn. Forceps were opened, and the uterine wall was gently pressed into the jaw that was closed. Endometrial tissue samples were obtained by quick retraction of the forceps from the uterus and immediately placed into tubes for low-temperature storage. Subsequently, at the farm, tubes were dipped into liquid nitrogen and stored. Next, the samples were placed in the ultra-freezer and stored at −80 °C until further analysis. Samples weighed around 50 mg. Only one tissue fragment was taken to minimize the influence of biopsy on the reproductive performance of cows. Ultrasound pregnancy diagnosis (Honda HS-1500 Ultrasound, Toyohashi, Japan) was performed every fourth week by a single veterinarian after sample collection until 200 days postpartum or the cow was culled.

2.4. Cytological Examination

The smear was fixed by air dry and dyed with Romanowski type staining (Hemavet, Kolchem, Łódź, Poland). The percentage of neutrophils was calculated based on the examination of 300 cells at 400× magnification under the light microscope. One observer evaluated all samples.

2.5. RNA Isolation

Endometrial samples were taken from the storage in tubes kept at −80 °C and homogenized using a ceramic mortar in liquid nitrogen to a form of fine tissue powder. Subsequently, samples were moved into a 1.5 mL Eppendorf tube with 400 microliters of phenozol, stored for 30 min, vortexed for 1 min, and centrifugated. According to the producer's protocol, total mRNA from endometrial samples was isolated with the Total RNA Mini Plus kit (A&A Biotechnology, Gdynia, Poland, #036-100). The isolated product was stored at −80 °C. Quantity and quality of the isolated total mRNA were measured with the NanoDrop spectrometer. The same amount (1000 ng) of mRNA from each sample was reverse transcribed using Maxima First Strand cDNA Synthesis Kit (Thermo Fisher, Waltham, MA, USA, #K1641). Reverse transcription was done according to the manufacturer's protocol. cDNA was frozen at −20 °C until further analysis.

2.6. Real-Time PCR

For Real-time PCR analysis, ABI Prism 7900 sequence detection system (Applied Biosystems, Life Technologies, Foster City, CA, USA) was used with Maxima SYBR Green/ROX qPCR Master Mix (2X) (Thermo Scientific, Waltham, MA, USA, #K0222). The samples' preparation was done according to the manufacturer's protocol using 15 ng cDNA per well in a total of 10 microliters of the reaction mix. PCR was performed in duplicates for every sample using 384-well plates (MicroAmp™ Optical 384-Well Reaction Plate with Barcode, #4309849). Two reference genes (*C20RF29, SLC30A6*) were used to normalize and calculate arbitrary gene relation units [25]. The primers were designed with web-based software Primer BLAST (http://www.ncbi.nlm.nih.gov/tools/primer-blast). The size of the amplified fragments and sequences of primers are presented in Supplementary Material Table S2. Miner software (http://miner.ewindup.info) was used for the relative quantification of mRNA.

2.7. Statistical and Machine Learning Analysis of Reproduction Success Based on mRNA Expression

Collected data were analyzed using Python 3.7 programming language (Python Software Foundation, https://www.python.org/) and R 3.6.0 (R Core Team, 2017, https://www.r-project.org/).

For the supervised data analysis approach, two thresholds of the neutrophil percentage of 5% [26] (n = 35, <5% PMNs = 20 and >5% PMNs = 15) and 10% [4] (n = 35, <10% PMNs = 26 and >10% PMNs = 9) were selected to check the effect of subclinical endometritis (SE) on mRNA expression. The normality assumption of studied parameters was tested using the Shapiro–Wilk test, and Levene's test was used to evaluate the equality of variances. A non-parametric Mann–Whitney U test was used to compare mRNA expression differences in the thresholds mentioned above.

Correlation between gene expression data, calving-to-conception interval (CCI), and PMNs percent in smears were calculated using the SciPy library spearman r function. Matplotlib and seaborn libraries were used to create a graphical representation of the data. Differences between groups were presented as fold difference between medians of normalized mRNA expression.

In order to check the differences in mRNA expression related to pregnancy outcomes, animals were divided: Pregnant animals up to 100 days (P100, n = 11), pregnant between 100- and 200-days PP (P200, n = 14) and other animals as culled (C, n = 10, including two cows sold to another farm at day 70 and 92 postpartum, two cows culled at 145 and 173 because of lameness and six not pregnant after 200 days).

Multivariable cox proportional hazards regression (package survival, function coxph) was used to determine the association of mRNA expression of selected genes with pregnancy risk to 200 days in milk. First, the model containing all selected genes was established, and variables with $p < 0.2$ were retained in the analysis. Next, using manual stepwise backward elimination variables with a $p < 0.05$ established model characterized by the lowest Akaike information criterion. The Schoenfeld residuals were used to check assumptions for cox proportional hazards regression (R package survival, function cox.zph).

The linear regression model (Python package SciPy) was performed using the log2 transformed mRNA expression for P100 and P200 groups predicting CCI. The minimalistic and highly accurate linear regression model was established based on the Akaike information criterion, correlations between mRNA expression, and studied genes' biological function. A simple model including a low number of genes increases the applicability of current study in veterinary practice by reducing the analysis's expense and duration.

Logistic regression (R function glm) was performed to predict pregnant animals up to 100 days after parturition. The backward elimination method (R function step) was used to select a model representing the lowest Akaike information criterion (R function AIC) and has the highest sensitivity and specificity. The forward elimination method was used to choose a model characterized by a small number of variables while maintaining high sensitivity and specificity.

Unsupervised data analysis was used to identify specific mRNA expression patterns of selected genes and link them to future reproductive performance. The analysis was carried out in two stages with machine learning algorithms (K-nearest Neighbor Clustering + Hierarchical Clustering and Random Forest Clustering + Hierarchical Clustering).

A K-nearest neighbor clustering (K-means, R function kmeans) algorithm was performed in 10,000 iterations on a matrix of all studied cows and log2 mRNA genes in a range of 2–10 clusters. The optimal number of clusters was determined using the elbow method and the Silhouette method. Initial clustering validation allows as to narrow k ranged to 3–6. Then, hierarchical clustering (R function hclust) was done in 1000 iterations until consensus was achieved to create dendrograms. 1-Pearson correlation distance (R function distanceMatrix) was used to calculate the distance matrix followed by the Complete-linkage clustering algorithm. Clusters were automatically divided into those that predict reproductive success in more than or equal to 50% of animals and those below 50% to determine the effectiveness of predicting cows' pregnancy up to 100 days after parturition. This approach allowed calculating the sensitivity and specificity of the models. The Least Significant Difference (LSD) test was used to determine which genes were particularly crucial in assigning the animal to the cluster for the model with the highest sensitivity, specificity, and low akaike information criterion (AIC). The clustering results for the best-suited model were presented in the form of an mRNA expression heatmap (R package ComplexHeatmap). Belonging to the pregnancy group and the 5% and 10% SE thresholds were added after clustering analysis as descriptive data and presented next to the heatmap for each cow taking part in the experiment.

Random Forest algorithm (R function randomForest) was run in unsupervised mode and 10,000 trees to generate the proximity matrix. The optimal number of cuts was determined using the "elbow" method and the Silhouette method. The hierarchical tree was cut in the range of 3–6 branches, narrowed based on initial clustering validation. Next, the proximity matrix was converted to the distance matrix. Hierarchical clustering (R function hclust) was done in 1000 iterations until consensus was achieved using the wards method. Clusters were divided into those that predict pregnancy in the same method presented in K-means, and sensitivity and specificity were calculated. The importance of each mRNA gene expression in Random Forest was investigated by a mean decrease in the Gini index.

3. Results

Descriptive data including a diameter of the cervix, a diameter of left and right uterine horns, mean daily milk yield for the first 60 days of lactation, and concentrations of total cholesterol, triglycerides, non-esterified fatty acids, β-Hydroxybutyrate across all studied groups are presented in Table S1. There were no significant differences between the studied groups in the presented biochemical parameters. There was no significant difference in the fertility of used bulls measured in the whole-farm population and the study population.

3.1. *Expression of Selected Genes*

3.1.1. mRNA Expression of IL1B, IL6, CXCL8, and IL17A

The mRNA expression of *IL1B*, *IL6*, *CXCL8*, and *IL17A* was significantly different between <10% PMNs and >10% PMNs groups ($p < 0.05$). *IL1B* and *IL6* expression were 2-fold higher in >10% PMNs group, and transcription of *CXCL8* and *IL17A* were 3.5-fold and 3-fold higher for >10% PMNs group.

There were no significant differences between groups using the threshold of 5% of PMNs ($p > 0.05$) as well as pregnant and not pregnant ($p > 0.05$) (Figure 1B).

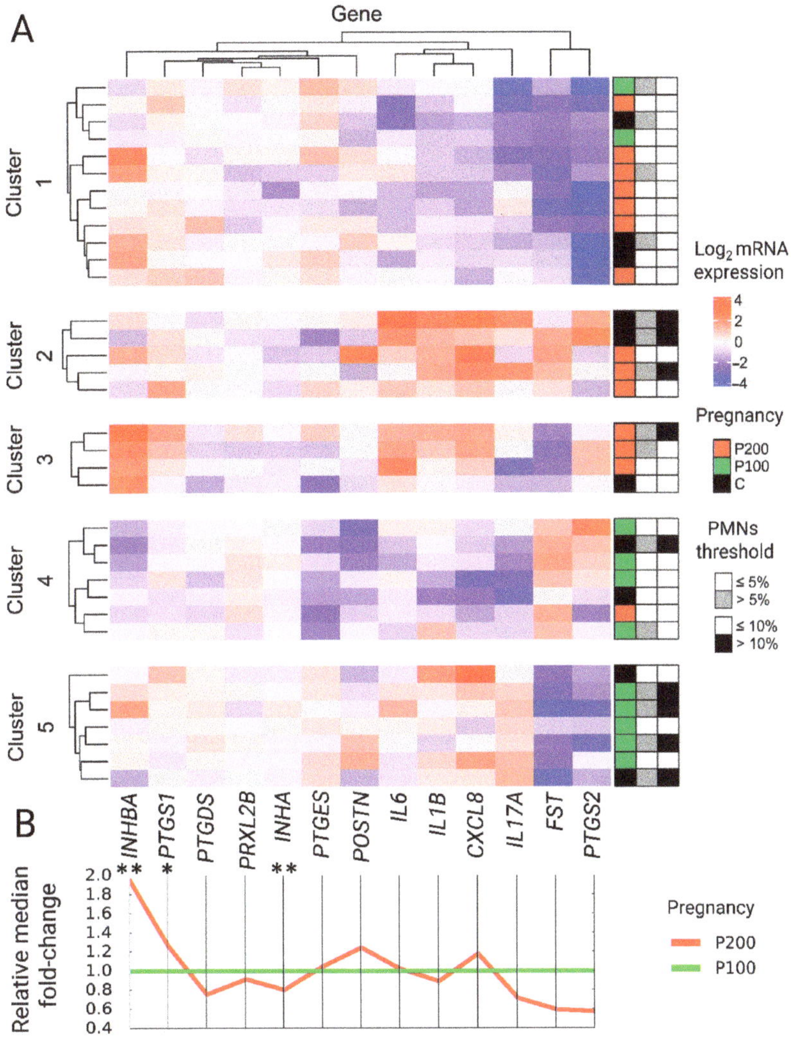

Figure 1. (**A**) Heat map of mRNA expression divided into 5 clusters. Cows with a similar expression pattern of marker genes were cluster together. Belonging to the pregnancy group, the 5% and 10% SE thresholds are presented next to the heatmap for each cow (row). (**B**) Median mRNA expression fold-change relative to the P100 group. * $p < 0.05$, ** $p < 0.01$.

3.1.2. mRNA Expression of INHA, INHBA, FST, and POSTN

Comparing the expression of *INHA* and *INHBA* in the P100 and P200 groups, significant differences were observed ($p < 0.05$). Transcription of *INHA* in group P100 was 1.3-fold higher compared to group P200 ($p < 0.05$), while a 2-fold lower expression was observed in the P100 group for *INHBA* compared to P200 ($p < 0.05$). The expression of *FST* and *POSTN* did not differ between groups. *INHBA*/*INHA* ratio was 3.5-fold lower in group P100 compared to P200 ($p < 0.05$) (Figure 1B).

INHBA, *INHA*, *FST*, and *POSTN* expressions were similar for 5% and 10% of PMN's thresholds ($p > 0.05$).

3.1.3. mRNA Expression of PRXL2B, PTGDS, PTGES, PTGS1, and PTGS2

The expression of *PTGS1* was 1.3-fold higher in the P200 group compering to P100 ($p < 0.05$). The mRNA expression of *PRXL2B*, *PTGDS*, *PTGES*, and *PTGS2* was similar, either using pregnancy or both cytological thresholds for group comparison ($p > 0.05$) (Figure 1B).

3.1.4. Correlation Analysis

The percentage of PMNs was correlated with the expression of *CXCL8*, *IL17A*, and *IL1B*. The rest of the analyzed genes were not significantly correlated with PMN's in the sixth week postpartum. CCI was negatively correlated with *INHA* while positively correlated with *INHBA* and *PTGS1* expression. Expression of *IL1B* and *CXCL8* was positively correlated with all interleukins' genes, while *IL6* and *IL17A* were correlated with other interleukins but not with each other (Table 1).

Table 1. Spearman rank correlation (upper triangle) and *p*-values (lower triangle) for the mRNA expression of studied markers.

	IL1B	*IL6*	*CXCL8*	*IL17A*	*INHA*	*INHBA*	*FST*	*POSTN*	*PRXL2B*	*PTGDS*	*PTGES*	*PTGS1*	*PTGS2*	PMNs%	CCI
IL1B	*IL1B*	0.500	0.834	0.590									0.504	0.380	
IL6	0.002	*IL6*	0.437			0.383							0.457		
CXCL8	0.001	0.009	*CXCL8*	0.484									0.374	0.522	
IL17A	0.001		0.003	*IL17A*										0.430	
INHA					*INHA*						0.356				−0.491
INHBA		0.023				*INHBA*	−0.449	0.510							0.654
FST						0.007	*FST*		0.400		−0.396		0.434		
POSTN						0.002		*POSTN*							
PRXL2B							0.017		*PRXL2B*	−0.395					
PTGDS									0.019	*PTGDS*					
PTGES					0.036		0.019				*PTGES*		−0.417		
PTGS1												*PTGS1*			0.480
PTGS2	0.002	0.006	0.027				0.009				0.013		*PTGS2*		
PMNs%	0.024		0.001	0.010										PMNs%	
CCI					0.013	0.001						0.015			CCI

3.2. *Conventional Models and Unsupervised Clustering*

3.2.1. Multivariable Cox Hazard Proportional Regression and Linear Regression

Cox hazard proportional regression indicates that higher endometrial mRNA expression of *INHBA* and *PTGS1* decreases the risk of being pregnant given day PP, while *INHA* and *POSTN* increase the risk of being pregnant (Table 2). Using *INHBA*/*INHA* in linear regression, we could predict CCI with $R^2 = 0.575$ (Figure 2, Table 3).

Table 2. Results of the multivariable Cox hazard proportional model in the studied population (n = 35) showing the effects of mRNA gene expression on the time to pregnancy to 200 DIM (backward elimination method).

Model	β	HR	SE	95% CI		*p*-Value
				0.025	0.975	
INHA	1.2528	3.5001	0.5778	1.1280	10.8609	0.030
INHBA	−1.1534	0.3156	0.321	0.1676	0.5941	<0.001
POSTN	0.7849	2.1923	0.2880	0.4561	1.2467	0.006
PTGS1	−0.7929	0.4525	0.4039	0.2051	0.9988	0.049

Concordance = 0.778, Likelihood ratio test = 26.76, Wald test = 17.45, Score (logrank) test = 21.33, Akaike information criterion = 97.25, Bayesian information criterion = 102.13.

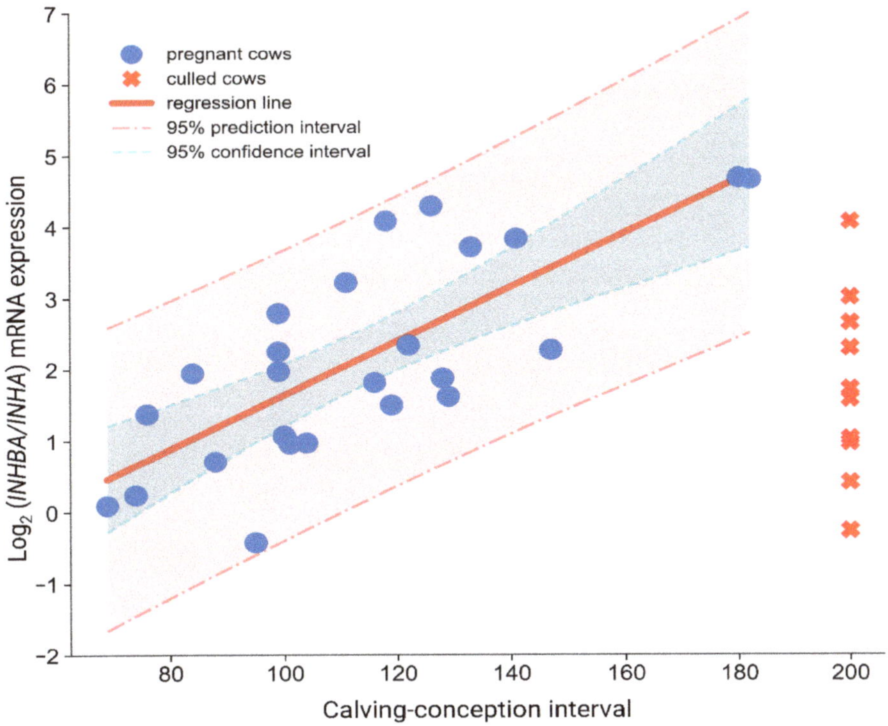

Figure 2. Relationship between calving conception interval and mRNA expression ratio of INHBA/INHA established using linear regression analysis.

Table 3. Linear regression model for calving conception interval prediction (INHBA/INHA ratio).

Model	Stand. β	Unstand. β	SE.	95% CI		*p*-Value
				0.025	0.975	
Intercept		80.9842	6.9968	66.5102	95.4582	<0.001
INHBA/INHA	0.7583	15.1439	2.7147	9.5281	20.7598	<0.001

R: 0.76, R^2: 0.57, Adjusted R^2: 0.56, RMSE: 19.22, Model *p*-value < 0.001.

3.2.2. Logistic Regression and Clustering

The most accurate model in predicting pregnancy up to 100 days PP used mRNA expression of *INHA, INHBA, FST, POSTN, IL1B, IL6, CXCL8,* and *PTGES* achieved sensitivity and specificity equal 100%. Whereas the model using the forward elimination method consisting of *INHA, INHBA,* and *POSTN* achieved a 73% sensitivity and a specificity of 87% (Table 4).

K-means and random forest analysis gave similar predicting accuracies. The optimal mRNA expression pattern clusters estimated by elbow and Silhouette methods for both algorithms was 3, while the highest sensitivity and specificity were achieved by dividing into 4 clusters for random forest and 5 clusters for K-means. An additional increase in cluster numbers over five significantly increases the Akaike information criterion (Table 4). Mean decrease in Gini index for 4 clusters random forest was *IL1B*—2.81, *CXCL8*—2.79, *PTGS2*—2.78, *INHBA*—2.70, *IL6*—2.69, *FST*—2.64, *PTGES*—2.64, *IL17A*—2.62, *POSTN*—2.58, *PRXL2B*—2.57, *PTGDS*—2.56, *PTGS1*—2.55, and *INHA*—2.55. The highest sensitivity—81%, and specificity—79%, were noted by the K-means algorithm, presented in the form of a heatmap (Figure 1A).

Table 4. Comparison of sensitivity, specificity, and precision of logistic regression, K-means clustering (k = 3–6), and random forest clustering (k = 3–6) in predicting allocation to P100.

Model	AIC	BIC	Sensitivity	Specificity	Precision
Logistic regression					
INHA	37.46	40.57	0.45	0.87	0.63
INHA + INHBA	35.00	39.67	0.55	0.87	0.66
INHA + INHBA + POSTN	31.94	38.17	0.73	0.87	0.73
INHA+ INHBA+ FST+ POSTN+ IL1B+ IL6+ CXCL8+ PTGES	18.00	31.99	1	1	1
K-Means Clustering					
K3	360.17	420.17	0.36	0.87	0.57
K4	356.89	437.77	0.36	0.87	0.57
K5	360.70	461.80	0.81	0.79	0.64
K6	365.19	486.51	0.81	0.79	0.64
Random Forest Clustering					
K3	380.27	440.93	0.27	0.87	0.5
K4	363.66	444.54	0.82	0.71	0.66
K5	364.87	465.97	0.82	0.75	0.53
K6	374.94	496.26	0.73	0.79	0.73

AIC: Akaike information criterion, BIC: Bayesian information criterion, AUC: Area under the curve.

Cluster 1 had the lowest expression of *PTGS2, IL1B, IL6,* low expression of *CXCL8,* and *IL17A,* no cow had >10% PMNs SE (0/12), and 4/12 had >5% PMNs SE. It was also characterized by high expression of *INHBA* and *PTGES*. Nevertheless, the low PMNs% did not turn into high fertility up to 100 days PP (2 pregnant/12).

Cluster 2 and 3 were characterized by the highest *IL1B, IL6, CXCL8,* and *PTGS1,* lowest expression of *INHA,* and the small number of cows pregnant up to 100 days PP (0 pregnant/9). Cluster 2 includes animals with the highest *IL17A* and a higher proportion of cows at risk of >10% PMNs SE (3/5). Cluster 3 consisted of animals with the highest expression of *INHBA* and *PTGS1*.

Cluster 4 and 5 were characterized by the highest *INHA* expression, lowest *INHBA* and *PTGS1* expression, and highest number of cows pregnant up to 100 days PP (9 pregnant/14). Cluster 4 includes animals that had low *IL1B*, *IL6*, and lowest *IL17A*, *CXCL8* expression, and a low likelihood of >10% PMNs SE (1/7). Cluster 5 consisted of animals with high *IL1B*, *IL17A*, and *CXCL8* expression but lower IL6 expression, then cluster 2 and 3 and an elevated risk of having >10% PMNs SE (4/7).

4. Discussion

Expression of selected genes as an indicator of future reproduction performance and subclinical endometritis.

Markers that were evaluated in the present study were combined in three groups: (1) Interleukins, (2) enzymes responsible for prostaglandins metabolism, and (3) factors involved in tissue restoration—taking into consideration two levels of PMNs (5% and 10%) for the diagnosis of SE and prediction of future reproduction performance.

(1) It was found that all examined interleukins were significantly higher in cows with a threshold of 10%, indicating a more robust inflammatory response. The present study agreed with previous findings showing increased expression of interleukins and its significant correlation with the number of immune cells in the endometrial biopsy or the evaluated cytological smear [15,27,28]. Correlation between PMNs% and interleukins expression was also confirmed when K-means clustering data were analyzed. Cows characterized by higher interleukins expression and >10% PMNs tend to cluster together. Formerly the difference in expression of *IL1B* between healthy and SE cows was not always manifested and depended highly on the experimental design or postpartum time [12,13,26,27]. While in agreement with the previous studies, we have found that a higher PMNs% in endometrial smear during SE is correlated with increasing endometrial expression of *CXCL8*. These data prove the usefulness of *CXCL8* mRNA expression measurement as a marker of linear change in tissue infiltration by neutrophils [13–15,29]. Interleukin 17A also proved to be a promising marker of SE and increased PMNs%. Johnson [30] showed the increased expression of interleukin 17A in animals with SE compared to the healthy, and the same significant relationship was observed in the present study for animals with >10% PMNs. Elevated expression of Interleukin 6 in the group of >10% PMNs was also observed in the current study what agrees with previous results analyzing variations between healthy and SE cows [14,30,31]. Our goal was to select specific interleukins that could serve as potent and certain markers validating the condition of subclinical endometritis, which undoubtedly succeeded. On the other hand, we were surprised by the absence of an apparent link between subclinical endometritis at the level of interleukin expression and future reproductive performance. This discrepancy could be easily noticed by clustering where cows with low interleukin expression were grouped at the same time, showing no signs of increased fertility (cluster 1), and on the contrary, a group of cows with high interleukin expression and PMNs% could have been pregnant before 100 days PP (cluster 5). The presented data indicate that interleukins are excellent markers of the ongoing inflammatory process or PMNs%. Nevertheless, low expression of interleukins or increase during disease and subsequent decrease due to the healing could not be used as a marker of future reproduction success. Therefore, our results could justify at mRNA expression level why cytobrush PMNs% used as a predictor of pregnancy status up to 100-day PP had low diagnostic performance (sensitivity 36.9%) [6] and indicates the need to select new and more accurate forecasting methods.

(2) There is a lack of agreement in the literature about whether the expression of enzymes involved in prostaglandin metabolism could be used as a valid marker of inflammation in subclinical endometritis. *PTGS1* is rarely examined in the endometrium because its role in physiological regulations of the bovine estrous cycle and pregnancy is limited, and its expression is almost not changed [16]. Studies have not confirmed significant differences in *PTGS1* expression between healthy and cows with SE at 4, 5, and 6 weeks PP [15], which agrees with our findings. While the results obtained in our

research show that animals pregnant before day 100 postpartum had lower *PTGS1* mRNA expression in the 6th-week postpartum. A change in the prostaglandin E synthesis pathway may explain the positive correlation between CCI and PTGS1 expression, initiating an imbalance in prostaglandin E to F production [17]. Under physiological status, endometrial *PTGS2* expression is positively correlated with *PTGES* expression in the endometrium during the estrous cycle and early pregnancy [16]. In the present study, the correlation was negative. The observed result might indicate that in the late stage of the uterus involution or under pathological status (i.e., subclinical endometritis), the endometrial expression pattern of *PTGS2* is different than during the physiological estrus cycle. It has been found previously that mRNA *PTGS2* is highly expressed only in the clinical form of endometritis [15]. In contrast, Ledgard [32] showed that higher *PTGS2* expression had an outcome of a greater pregnancy rate after embryo transfer. Similarly, higher expression of *PTGS2* was observed at 20-day of pregnancy comparing to 20-day of the estrus cycle in the caruncular endometrium [33]. In the present study, we found numerically higher levels of this gene expression in animals pregnant before 100-day postpartum and significantly lower expression in the first k-means cluster associated with subfertility. This variation was not correlated with the presence of SE but might contribute to infertility. A similar relationship was found in repeat breeder cows where control cows had higher *PTGS2* mRNA expression [34].

(3) Studies conducted by Salilew-Wondim [18] indicated the possibility of *POSTN* being a candidate gene that allows distinguishing healthy animals from cows with SE due to its role in tissue remodeling and fibrosis tissue formation. *POSTN* is also suggested as a potential mediator of conceptus elongation and recognition of the embryo in small ruminants by a dam [35]. *POSTN* expression remains high during epithelial inflammation or prolonged wound healing [36]. On the other hand, the level of *POSTN* is rapidly growing and subsequently falls equally quickly to a low level in quickly healing wounds or non-inflammatory tissue [37]. We considered this gene as a possible indicator of normal uterine involution. We did not find significant changes in the expression level between P100 and P200 groups, only numerically higher expression in P200 animals. Using logistic regression to predict animals pregnant before day 100 postpartum and linear regression predicting CCI *POSTN* showed a significant contribution to models. However, the highest-efficiency models' main components, both linear and binomial, in predicting future reproductive performance were INHBA and INHA expression. So far, there has been no research on the possibility of using *INHA* and *INHBA* as an indicator of normal or delayed uterine involution. Being aware that calving can damage uterine tissue and specific bacteria enter uterine lumen at that time. We hypothesized that the uterus during involution might use similar molecular mechanisms to wound healing. The role of activin in tissue remodeling during wound healing is extensively studied. Inhibition of activin action by over-expression of follistatin causes a delay in wound repair while reducing scar tissue formation [38]. Therefore, higher expression of *INHBA* may indicate the ongoing active remodeling of the tissue. Microarrays revealed *INHBA* as one of the most profoundly altered genes in the endometrial inflammation caused by lipopolysaccharides (LPS) [18]. The present study showed that animals ready to get pregnant before day 100 PP had lower endometrial mRNA expression of *INHBA* than the subfertile group, which may signal the termination of histological involution and return of endometrium to the status before pregnancy. Increased expression of *INHA* occurring in animals with higher reproduction performance can cause competition for the *INHBA* subunit. Thus, the switching mechanism between these two expression patterns may limit the formation of activin A in favor of Inhibin A working as a "molecular clock" of endometrial restoration. Highlighting the gap in knowledge and possible limitations of the study. More research is needed to determine the exact function of analyzed genes, the level of their protein products, the role of inhibin A and activin A in the endometrium after delivery and throughout the whole involution period, as well as validation of the findings on a large number of animals and farms. Using the proposed hypothesis at the mRNA expression level, we suggest that the *INHBA*/*INHA* ratio (Figure 2) is a far better indicator of the present status of the endometrium than the

use of *INHBA* and *INHA* separately. Obtained results show that this parameter can be applied as an indicator of endometrial restoration, and we found a linear relationship between *INHBA*/*INHA* ratio and future reproductive performance, namely lower ratio characterized cows exhibiting excellent fertility and shorter CCI.

Prediction of gene expression patterns associated with future reproduction performance using Machine Learning Algorithms.

In this study, the logistic regression model (*INHA, INHBA, FST, POSTN, IL1B, IL6, CXCL8,* and *PTGES*) identified by the backward elimination method had sensitivity and specificity equal to 100%. This highly accurate prediction is burdened with high bias. Conventional statistics, especially binomial logistic regression, has many limitations when analyzing correlated data. When applied to the new dataset, models created in that way frequently show far lower predictive abilities. The selection of machine learning algorithms has been proposed to interpret correlated data [39]. Until now, models take into consideration qualitative features in the form of mixed [40] or binomial models, bringing additional effort and cost of data collection. Our model is based solely on gene expression. The decision to use unsupervised algorithms was directed to find characteristic mRNA gene expression patterns of selected markers occurring six weeks after delivery and identify which of them are associated with improved and reduced fertility. K-means and random forest clustering approaches had similar sensitivity and specificity with the superiority of the first one. K-means method in the present study resulted in 5 distinct gene expression groups in sixth-week PP. The unsupervised division of cows confirmed the usefulness of endometrial gene expression as a prospective reproductive indicator. Two clusters were associated with high reproduction performance, a 3 cluster with reduced fertility. Extended validation of our result can offer a new way to assess information about the endometrial condition and the ability for embryo reception.

5. Conclusions

We provided evidence that lower expression of *INHBA* and higher INHA expression in the sixth week postpartum characterized cows showing better reproduction performance, and usefulness of *INHBA*/*INHA* ratio to predict CCI length. The k-means algorithm application allowed identifying five unique expression patterns of the studied genes at 6th-week postpartum, of which two were characterized by high reproduction performance. We also documented the low efficiency of gene markers associated with SE and PMNs% in determining reproductive capability.

Supplementary Materials: The following are available online at https://www.mdpi.com/2076-2615/11/1/236/s1, Table S1: The diameter of the cervix and diameter of left and right uterine horns, mean daily milk yield for the first 60 days of lactation and blood concentrations of total cholesterol, triglycerides, non-esterified fatty acids, β-Hydroxybutyrate across all studied groups (Mean ± SD); Table S2: Selected genes, primer forward, and reverse sequences used for RT-qPCR.

Author Contributions: Conceptualization, D.T. and W.B.; methodology, D.T., M.H., K.Ł., A.B., and D.J.S.; software, D.T. and M.H.; validation, D.T., W.B., and K.Ł.; formal analysis, D.T. and D.J.S.; investigation, D.T.; resources, D.T. and W.B.; data curation, D.T.; writing—original draft preparation, D.T.; writing—review and editing, D.T., W.B.; M.H., K.Ł., A.B., and D.J.S.; visualization, D.T.; supervision, D.T., W.B., M.H., and D.J.S.; project administration, D.T. and W.B.; funding acquisition, W.B. All authors have read and agreed to the published version of the manuscript.

Funding: Grant Support: Funded by KNOW (Leading National Research Centre) Scientific Consortium "Healthy Animal-Safe Food" decision of the Ministry of Science and Higher Education No. 05-1/KNOW2/2015. Project financially supported by Minister of Science and Higher Education in the range of the program entitled "Regional Initiative of Excellence" for the years 2019–2022, Project No. 010/RID/2018/19.

Institutional Review Board Statement: The Local Ethics Committee for animal experiments in Olsztyn approved the research by resolution No. 49/2016.

Acknowledgments: The authors would like to thank the staff of the farm.

Conflicts of Interest: The authors declare no conflict of interest.

References

1. Sheldon, I.M.; Lewis, G.S.; LeBlanc, S.; Gilbert, R.O. Defining postpartum uterine disease in cattle. *Theriogenology* **2006**, *65*, 1516–1530. [CrossRef] [PubMed]
2. Sheldon, I.M.; Cronin, J.G.; Bromfield, J.J. Tolerance and Innate Immunity Shape the Development of Postpartum Uterine Disease and the Impact of Endometritis in Dairy Cattle. *Annu. Rev. Anim. Biosci.* **2019**, *7*, 361–384. [CrossRef] [PubMed]
3. Hay, M.J.; Gunn, A.J.; Abuelo, A.; Brookes, V.J. The Effect of Abnormal Reproductive Tract Discharge on the Calving to Conception Interval of Dairy Cows. *Front. Vet. Sci.* **2019**, 6. [CrossRef] [PubMed]
4. Kasimanickam, R.; Duffield, T.F.; Foster, R.A.; Gartley, C.J.; Leslie, K.E.; Walton, J.S.; Johnson, W.H. Endometrial cytology and ultrasonography for the detection of subclinical endometritis in postpartum dairy cows. *Theriogenology* **2004**, *62*, 9–23. [CrossRef] [PubMed]
5. Gilbert, R.O.; Shin, S.T.; Guard, C.L.; Erb, H.N.; Frajblat, M. Prevalence of endometritis and its effects on reproductive performance of dairy cows. *Theriogenology* **2005**, *64*, 1879–1888. [CrossRef]
6. Koyama, T.; Omori, R.; Koyama, K.; Matsui, Y.; Sugimoto, M. Optimization of diagnostic methods and criteria of endometritis for various postpartum days to evaluate infertility in dairy cows. *Theriogenology* **2018**, *119*, 225–232. [CrossRef]
7. Pascottini, O.B.; Dini, P.; Hostens, M.; Ducatelle, R.; Opsomer, G. A novel cytologic sampling technique to diagnose subclinical endometritis and comparison of staining methods for endometrial cytology samples in dairy cows. *Theriogenology* **2015**, *84*, 1438–1446. [CrossRef]
8. Helfrich, A.L.; Reichenbach, H.D.; Meyerholz, M.M.; Schoon, H.A.; Arnold, G.J.; Frohlich, T.; Weber, F.; Zerbe, H. Novel sampling procedure to characterize bovine subclinical endometritis by uterine secretions and tissue. *Theriogenology* **2020**, *141*, 186–196. [CrossRef] [PubMed]
9. Bonnett, B.N.; Miller, R.B.; Martin, S.W.; Etherington, W.G.; Buckrell, B.C. Endometrial biopsy in Holstein-Friesian dairy cows. II. Correlations between histological criteria. *Can. J. Vet. Res.* **1991**, *55*, 162–167.
10. Meira, E.B., Jr.; Henriques, L.C.; Sa, L.R.; Gregory, L. Comparison of ultrasonography and histopathology for the diagnosis of endometritis in Holstein-Friesian cows. *J. Dairy Sci.* **2012**, *95*, 6969–6973. [CrossRef]
11. Madoz, L.V.; Giuliodori, M.J.; Migliorisi, A.L.; Jaureguiberry, M.; de la Sota, R.L. Endometrial cytology, biopsy, and bacteriology for the diagnosis of subclinical endometritis in grazing dairy cows. *J. Dairy Sci.* **2014**, *97*, 195–201. [CrossRef] [PubMed]
12. Wagener, K.; Gabler, C.; Drillich, M. A review of the ongoing discussion about definition, diagnosis and pathomechanism of subclinical endometritis in dairy cows. *Theriogenology* **2017**, *94*, 21–30. [CrossRef] [PubMed]
13. Fischer, C.; Drillich, M.; Odau, S.; Heuwieser, W.; Einspanier, R.; Gabler, C. Selected pro-inflammatory factor transcripts in bovine endometrial epithelial cells are regulated during the oestrous cycle and elevated in case of subclinical or clinical endometritis. *Reprod. Fertil Dev.* **2010**, *22*, 818–829. [CrossRef] [PubMed]
14. Galvao, K.N.; Santos, N.R.; Galvao, J.S.; Gilbert, R.O. Association between endometritis and endometrial cytokine expression in postpartum Holstein cows. *Theriogenology* **2011**, *76*, 290–299. [CrossRef]
15. Peter, S.; Michel, G.; Hahn, A.; Ibrahim, M.; Lubke-Becker, A.; Jung, M.; Einspanier, R.; Gabler, C. Puerperal influence of bovine uterine health status on the mRNA expression of pro-inflammatory factors. *J. Physiol. Pharm.* **2015**, *66*, 449–462.
16. Arosh, J.A.; Parent, J.; Chapdelaine, P.; Sirois, J.; Fortier, M.A. Expression of cyclooxygenases 1 and 2 and prostaglandin E synthase in bovine endometrial tissue during the estrous cycle. *Biol. Reprod.* **2002**, *67*, 161–169. [CrossRef] [PubMed]
17. Baranski, W.; Lukasik, K.; Skarzynski, D.; Sztachanska, M.; Zdunczyk, S.; Janowski, T. Secretion of prostaglandins and leukotrienes by endometrial cells in cows with subclinical and clinical endometritis. *Theriogenology* **2013**, *80*, 766–772. [CrossRef] [PubMed]
18. Salilew-Wondim, D.; Ibrahim, S.; Gebremedhn, S.; Tesfaye, D.; Heppelmann, M.; Bollwein, H.; Pfarrer, C.; Tholen, E.; Neuhoff, C.; Schellander, K.; et al. Clinical and subclinical endometritis induced alterations in bovine endometrial transcriptome and miRNome profile. *BMC Genom.* **2016**, *17*, 218. [CrossRef]
19. Foley, C.; Chapwanya, A.; Callanan, J.J.; Whiston, R.; Miranda-CasoLuengo, R.; Lu, J.; Meijer, W.G.; Lynn, D.J.; Cliona, O.F.; Meade, K.G. Integrated analysis of the local and systemic changes preceding the development of post-partum cytological endometritis. *BMC Genom.* **2015**, *16*, 811. [CrossRef]
20. Knight, P.G.; Glister, C. Potential local regulatory functions of inhibins, activins and follistatin in the ovary. *Reproduction* **2001**, *121*, 503–512. [CrossRef]
21. Findlay, J.K. An update on the roles of inhibin, activin, and follistatin as local regulators of folliculogenesis. *Biol. Reprod.* **1993**, *48*, 15–23. [CrossRef] [PubMed]
22. Woodruff, T.K.; Lyon, R.J.; Hansen, S.E.; Rice, G.C.; Mather, J.P. Inhibin and Activin Locally Regulate Rat Ovarian Folliculogenesis. *Endocrinology* **1990**, *127*, 3196–3205. [CrossRef] [PubMed]
23. Werner, S.; Alzheimer, C. Roles of activin in tissue repair, fibrosis, and inflammatory disease. *Cytokine Growth Factor Rev.* **2006**, *17*, 157–171. [CrossRef]
24. Phillips, D.J. Activins, inhibins and follistatins in the large domestic species. *Domest. Anim. Endocrinol.* **2005**, *28*, 1–16. [CrossRef] [PubMed]

25. Walker, C.G.; Meier, S.; Mitchell, M.D.; Roche, J.R.; Littlejohn, M. Evaluation of real-time PCR endogenous control genes for analysis of gene expression in bovine endometrium. *BMC Mol. Biol.* **2009**, *10*, 100. [CrossRef] [PubMed]
26. Baranski, W.; Podhalicz-Dziegielewska, M.; Zdunczyk, S.; Janowski, T. The diagnosis and prevalence of subclinical endometritis in cows evaluated by different cytologic thresholds. *Theriogenology* **2012**, *78*, 1939–1947. [CrossRef] [PubMed]
27. Herath, S.; Lilly, S.T.; Santos, N.R.; Gilbert, R.O.; Goetze, L.; Bryant, C.E.; White, J.O.; Cronin, J.; Sheldon, I.M. Expression of genes associated with immunity in the endometrium of cattle with disparate postpartum uterine disease and fertility. *Reprod. Biol. Endocrinol.* **2009**, *7*, 55. [CrossRef] [PubMed]
28. Chapwanya, A.; Meade, K.G.; Doherty, M.L.; Callanan, J.J.; Mee, J.F.; O'Farrelly, C. Histopathological and molecular evaluation of Holstein-Friesian cows postpartum: Toward an improved understanding of uterine innate immunity. *Theriogenology* **2009**, *71*, 1396–1407. [CrossRef]
29. Kasimanickam, R.; Kasimanickam, V.; Kastelic, J.P. Mucin 1 and cytokines mRNA in endometrium of dairy cows with postpartum uterine disease or repeat breeding. *Theriogenology* **2014**, *81*, 952–958.e952. [CrossRef]
30. Johnson, H.; Torres, C.G.; Carvallo, F.; Duchens, M.; Peralta, O.A. Endometrial expression of selected transcripts in postpartum of primiparous Holstein cows with clinical and subclinical endometritis. *Anim. Reprod. Sci.* **2015**, *156*, 34–39. [CrossRef]
31. Ghasemi, F.; Gonzalez-Cano, P.; Griebel, P.J.; Palmer, C. Proinflammatory cytokine gene expression in endometrial cytobrush samples harvested from cows with and without subclinical endometritis. *Theriogenology* **2012**, *78*, 1538–1547. [CrossRef] [PubMed]
32. Ledgard, A.M.; Meier, S.; Peterson, A.J. Evaluation of the uterine environment early in pregnancy establishment to characterise cows with a potentially superior ability to support conceptus survival. *Reprod. Fertil. Dev.* **2011**, *23*, 737–747. [CrossRef] [PubMed]
33. Mansouri-Attia, N.; Aubert, J.; Reinaud, P.; Giraud-Delville, C.; Taghouti, G.; Galio, L.; Everts, R.E.; Degrelle, S.; Richard, C.; Hue, I.; et al. Gene expression profiles of bovine caruncular and intercaruncular endometrium at implantation. *Physiol. Genom.* **2009**, *39*, 14–27. [CrossRef]
34. Wagener, K.; Pothmann, H.; Prunner, I.; Peter, S.; Erber, R.; Aurich, C.; Drillich, M.; Gabler, C. Endometrial mRNA expression of selected pro-inflammatory factors and mucins in repeat breeder cows with and without subclinical endometritis. *Theriogenology* **2017**, *90*, 237–244. [CrossRef] [PubMed]
35. Ahn, H.W.; Farmer, J.L.; Bazer, F.W.; Spencer, T.E. Progesterone and interferon tau-regulated genes in the ovine uterine endometrium: Identification of periostin as a potential mediator of conceptus elongation. *Reproduction* **2009**, *138*, 813–825. [CrossRef] [PubMed]
36. Hamilton, D.W. Functional role of periostin in development and wound repair: Implications for connective tissue disease. *J. Cell Commun. Signal.* **2008**, *2*, 9–17. [CrossRef]
37. Conway, S.J.; Izuhara, K.; Kudo, Y.; Litvin, J.; Markwald, R.; Ouyang, G.; Arron, J.R.; Holweg, C.T.; Kudo, A. The role of periostin in tissue remodeling across health and disease. *Cell Mol. Life Sci.* **2014**, *71*, 1279–1288. [CrossRef]
38. Sulyok, S.; Wankell, M.; Alzheimer, C.; Werner, S. Activin: An important regulator of wound repair, fibrosis, and neuroprotection. *Mol. Cell. Endocrinol.* **2004**, *225*, 127–132. [CrossRef]
39. Xu, W.; van Knegsel, A.T.M.; Vervoort, J.J.M.; Bruckmaier, R.M.; van Hoeij, R.J.; Kemp, B.; Saccenti, E. Prediction of metabolic status of dairy cows in early lactation with on-farm cow data and machine learning algorithms. *J. Dairy Sci.* **2019**, *102*, 10186–10201. [CrossRef]
40. Caruso, G.; Gattone, S.A.; Fortuna, F.; Di Battista, T. Cluster Analysis for mixed data: An application to credit risk evaluation. *Socio-Econ. Plan. Sci.* **2020**, 100850. [CrossRef]

animals

MDPI

Article

Pregnancy Loss (28–110 Days of Pregnancy) in Holstein Cows: A Retrospective Study

Aitor Fernandez-Novo [1], Octavi Fargas [2], Juan Manuel Loste [3], Francisco Sebastian [4], Natividad Perez-Villalobos [5], Jose Luis Pesantez-Pacheco [6], Raquel Patron-Collantes [7] and Susana Astiz [8,*]

1 Dpto. Producción y Sanidad Animal, Salud Pública Veterinaria y Ciencia y Tecnología de los Alimentos, Facultad de Veterinaria, Universidad CEU Cardenal Herrera, C/Tirant lo Blanc, 7, 46115 Alfara del Patriarca, Valencia, Spain; aitor.fernandeznovo@uchceu.es
2 VAPL S.L., C/Antoni Figueras 20, 08551 Tona, Barcelona, Spain; octavi.vaplslp@gmail.com
3 Akaborro s/n., 31860 Navarra, Spain; loste@albaitaritza.com
4 Cowvet SL, Avda. País Valenciano 6, 5, 46117 Betera, Valencia, Spain; cowvetsl@gmail.com
5 Facultad de Ciencias Biomédicas, Universidad Europea de Madrid, C/Tajo s/n, 28670 Villaviciosa de Odón, Madrid, Spain; natividad.perez@universidadeuropea.es
6 School of Veterinary Medicine and Zootechnics, Faculty of Agricultural Sciences, University of Cuenca, Avda. Doce de Octubre, 010150 Cuenca, Ecuador; jose.pesantez@ucuenca.edu.ec
7 TRIALVET S.L., C/Encina 22, Cabanillas de la Sierra, 28721 Madrid, Spain; rpatron@trialvet.com
8 Dpto. Reproducción Animal (INIA), Avda. Puerta de Hierro s/n., 28040 Madrid, Spain
* Correspondence: astiz.susana@inia.es; Tel.: +34-913473769; Fax: +34-3474011

Received: 22 April 2020; Accepted: 25 May 2020; Published: 26 May 2020

Simple Summary: High-yield dairy cow farms have implemented high technified management for the last few decades, aiming at optimizing productions with the best animal welfare canons. A key point to achieve this is the reproductive performance. Around 12% of cattle suffer pregnancy loss during the late embryonic/early foetal period (between 28 and 110 day of pregnancy). Thus, our objective was to study the pregnancy losses occurring in eight different Spanish high-yielding Holstein dairy herds, in locations with severe heat stress during the summer, to examine the link between pregnancy loss and different management factors. Some factors, previously confirmed as significant ones, such as the technician who performed artificial insemination (AI), fixed-time or after observed oestrus AI, the bull used, type of semen or season, did not affect pregnancy loss in our study. Moreover, older cows (compared to heifers), first artificial inseminations (compared to ≥2nd ones) and pregnancies after fixed-time-AI (compared to AI after observed oestrus and natural breedings) were definitively associated to higher pregnancy loss. Therefore, farmers and consultants should adapt their prevention strategies relating to pregnancy loss, particularly, to the parity of the cattle and to type and rank of AI.

Abstract: The objective of this retrospective study was to investigate the prevalence of pregnancy loss (PL; between 28–110 pregnancy days) and its relationship with factors: farm, year (2015–2018), season, artificial insemination (AI)-rank, parity, AI-type (fixed-time vs. oestrus-AI), previous PL, days in milk (DIM), fixed-time-AI protocol, AI-technician, bull, and semen-type (sexed vs. conventional). Data of 19,437 Holstein cattle AIs from eight Spanish farms were studied. Overall conception rate was 34.3% (6696/19,437) and PL 12.3% (822/6696). The PL was more likely to occur in primiparous (10.8%, odds ratio (OR) = 1.35; $p = 0.04$) and multiparous (15.3%; OR = 2.02, $p < 0.01$) than in heifers (PL = 6.9%, reference). Pregnancies achieved with AI after observed oestrus and natural breedings were associated with less PL than pregnancies after fixed-time-AI (12.7 vs. 11.9%; OR = 0.12, $p = 0.01$). First AIs related to higher PL than ≥2nd AIs (PL = 13.8% vs. 11.2; OR = 0.73, $p < 0.01$). The factors season, fixed-time-AI protocol, DIM, bull, AI-technician, or type of semen were not significantly associated with PL. Therefore, farmers and consultants should adapt their preventive strategies relating to PL, particularly, to the parity of the cattle.

Keywords: reproductive strategy; parity; season; rank of AI; type of AI

1. Introduction

Pregnancy loss (PL) during the late embryonic and early fetal phases of pregnancy is a major cause of infertility in dairy cows and a crucial factor for dairy cattle farms' economy [1,2]. Embryonic losses are categorized as early or late before and after day 25, respectively [1], the average rate of late embryonic loss being 10–12%, with this rate ranging broadly among farms from 3.5% to 26.3% [3].

Some studies have focused on possible predictors to forecast pregnancy loss [4], while other studies have concentrated on contributing factors. The following PL risk factors have been observed in epidemiological studies: parity, with multiparous showing higher PL than younger cows [5,6]; individual bulls [1,7–9]. Progesterone level at AI and during the first days of pregnancy has also been related to PL [10–12]. Similarly, the serum concentration of pregnancy-associated glycorproteins (PAGs) [13,14] is correlated to PL probability; the intensity of oestrus at AI [12,15], season at AI [16,17], the protocol for timed AI (FTAI) [18,19], twin pregnancy [9], and ovulation number [12]. Some other factors were not associated with pregnancy loss, such as the rank of AI [16], extended lactation [20], calving-AI interval [16], corpus luteum, and uterine blood flow at first pregnancy diagnosis observed with ultrasonography [10]. However, information is contradictory, and there are other studies that did not outline significant links between these factors and pregnancy loss [21,22].

In light of these observations, it is generally accepted that pregnancy loss in high-producing dairy cows is multifactorial, yet the exact underlying mechanisms remain unknown [3,23,24]. Previous reports have described the incidence and pattern of pregnancy loss in dairy cattle [25], but several of these studies assessed a limited number of cows or included the phases of pregnancy loss (from fertilization to pregnancy term) without clearly separating them [6,12,26]. Therefore, further information on this topic—clarifying factors and pregnancy phases, including a large number of pregnancies and exploring new risk issues—would be welcome, and even crucial, in helping us to find answers to practical questions, and guide management strategies towards improving reproductive efficiency in dairy farms. The hypothesis of the authors is that having lost a previous pregnancy is not a risk factor for future PL, while other factors such as farm, season, or AI rank play a key role. Therefore, the objective of this observational, retrospective study was to examine, in an extensive sample of pregnancies in high-yielding dairy cows, the relationship between pregnancy loss prevalence and the different factors described.

2. Materials and Methods

2.1. Ethics Statement

This is an observational, retrospective study based on data provided by collaborating veterinarian consultants relating to AIs and natural breedings performed at eight high-yielding, intensively managed dairy cattle farms located in different Spanish regions. No experimental intervention was performed, and farms carried out all their activities according to the EU Directive about the protection of animals in animal production [27].

2.2. Herds and Management

The study included a total of 19,437 AIs from dairy cattle and heifers (Holstein breed), performed during the period 2015–2018, from eight different intensively managed Spanish farms located in central and eastern Spain, where heat stress during the summer (June to August) is severe. Data taken from farm computers was provided by veterinary consultants. Data on heifers was provided only by Farms 4, 7, and 8.

Average milk yield and farm size were 34.8 ± 2.3 L/cow/day (range of 31.3 to 37.9 L/cow/day) and 746 ± 571 cows/herd (range of 253 to 1950 cows/herd). All herds were managed under similar conditions, being milked three times per day, with ad libitum access to water, and fed twice daily with a total mixed ration that was balanced to meet or exceed nutrient recommendations for lactating dairy cows. All farms are located in areas with severe heat stress during the summer, and cooling methods, such as fans with sprinkler systems on feeding and resting barns, were counted.

Reproductive management varied across farms, but in general, the farms used one long synchronization protocol for first AIs (Presynch-Ovsynch, Double Ovsynch, or G6G), and AI after observed oestrus for subsequent AIs if possible. If not detected in oestrus, all farms applied shorter synchronization protocols for 2nd or subsequent AIs (Ovsynch, 5d-Ovsynch, 5d-Cosynch, and 5d-Cosynch with intravaginal progesterone device). Pregnancy diagnoses were performed in all farms with ultrasound at 28–35 days for first pregnancy diagnosis, and between 100 and 110 days for second pregnancy diagnoses by the same experienced veterinarians at each farm. A positive diagnosis of pregnancy was made if the allantochorion and embryo in the uterine lumen were visualized and a heartbeat was detected, with ovaries being scanned only if problems arose when imaging the conceptus. On Farm 8, natural breeding was used for "repeating cows" with more than five AIs. In the case of heifers, farms implemented oestrus synchronization protocols with intravaginal progesterone device, maintained during 7 d with GnRH administration at the day of IPD insertion and prostaglandin at the day of IPD removal, or with prostaglandin injections 7 d apart, and AI after observed oestrus. Farm 8 included a natural service for repeating heifers after two failing AIs. Detailed information regarding farm, protocol, and AIs is summarized in Tables 1 and 2.

Table 1. Descriptive statistics for the type of reproductive strategy used in 19,437 breedings across eight Spanish Holstein farms.

Reproductive Strategy	Farms																Breedings by Strategy	
	1		2		3		4		5		6		7		8			
	AIs	%	AIs	%	AIs	%	AIs	%	AIs	%	AIs	%	AIs	%	AIs	%	AIs	%
PRES													364	10.8	180	2.8	544	2.8
DOV	132	16.3					10	0.2	190	12.4			121	3.6			453	2.3
G6G	84	10.3	31	4.4	628	33.7	1046	25.5	367	24.0	423	71.8			753	11.7	3332	17.1
OVS	155	19.1	28	3.9			978	23.8					401	11.9	928	14.4	2490	12.8
OVS-IPD	92	11.3					10	0.2							353	5.5	455	2.3
5dOVS									540	35.3	47	8.0					587	3.0
5dCO					401	21.5											401	2.1
5dCO-IPD					560	30.0											560	2.9
OE	349	43.0	651	91.7	276	14.8	2059	50.2	433	28.3	118	20.1	2482	73.7	4002	61.9	10,370	53.4
NS															245	3.8	245	1.3
Breedings by farm	812	4.2	710	3.7	1865	9.6	4103	21.1	1530	7.9	588	3.0	3368	17.3	6461	33.2	19,437	

AIs: artificial inseminations; PRES: timed AI after Presynch; DOV: timed AI after Double Ovsynch; OVS: timed AI after Ovsynch; OVS-IPD: timed AI after Ovsynch with intravaginal progesterone device; 5dOVS: timed AI after 5 days Ovsynch; 5dCO: timed AI after 5 days Cosynch; 5dCO-IPD: timed AI after 5 days Cosynch with intravaginal progesterone device; OE: AI after observed oestrus; NS: natural service.

Table 2. Breeding descriptive statistics for the type of reproductive strategy used in heifer, primiparous, and multiparous Holstein cattle across eight Spanish farms.

Reproductive Strategy	Cattle Type									Breedings by Strategy	
	Heifer			Primiparous			Multiparous				
	AIs	(n/N) %	% *	AIs	(n/N) %	% *	AIs	(n/N) %	% *	AIs	%
PRES	3	(3/544) 0.5	0.1	160	(160/544) 29.4	2.7	382	(382/544) 70.1	3.5	544	2.8
DOV				296	(296/453) 65.3	5.0	157	(157/453) 34.7	1.4	453	2.3
G6G				985	(985/3332) 29.5	16.5	2,347	(2347/3332) 70.4	21.5	3332	17.1
OVS	17	(17/2490) 0.6	0.6	802	(802/2490) 32.2	13.5	1,667	(1667/2490) 66.9	15.3	2490	12.8
OVS-IPD	267	(267/455) 58.7	10.4	70	(70/455) 15.4	1.2	118	(118/455) 25.9	1.1	455	2.3
5dOVS				159	(159/587) 27.1	2.7	428	(428/587) 72.6	3.9	587	3.0
5dCO				189	(189/401) 47.1	3.1	212	(212/401) 52.9	2.0	401	2.1
5dCO-IPD				262	(262/560) 46.8	4.4	298	(298/560) 53.2	2.7	560	2.9
OE	2074	(2074/10,370) 20.0	81.1	3024	(3024/10,370) 29.2	50.7	5,272	(5272/10,370) 50.8	48.3	10,370	53.4
NS	198	(198/245) 80.8	7.8	13	(13/245) 5.3	0.2	34	(34/245) 13.9	0.3	245	1.3
Breedings by cattle type	2557	(2557/19,437) 13.1		5965	(5965/19,437) 30.7		10,915	(10,915/19,437) 56.2		19,437	

% *: indicates the overall percentage of breedings after each reproductive strategy used within each cattle type (heifers, primiparous, and multiparous). For abbreviations, see Table 1.

2.3. Recorded Data

The following information was recorded for each AI or service: Farm (1 to 8); animal identification; year of AI; AI date (seasonality at service); rank of AI (1st to 11th); lactation order of the cow from 0 to 7th lactation and categorized as heifer, primiparous, or multiparous; reproductive strategy; days in milk; type of AI (fixed time AI with hormonally controlled follicle development vs. AI after observed oestrus and natural service, with natural follicle development (FTAI vs. OE + NS)), AI-technician, type of semen (conventional or sexed shorted seminal doses) and bull. Finally, in all AIs included, the factor if a pregnancy loss had occurred in the same lactation prior to the AI under consideration (previous PL in the same lactation, yes or not), and if the animal had suffered PL at least once before when considering all subsequent AIs, during the same and following lactations (previous PL during its lifetime, yes or no) was studied. Days in milk at service was logically only available in cows (40 to 848 d for first and further AIs); AI technician and type of semen (if sexed or conventional semen was used) was available only in farms 7 and 8; 9829 AIs); age at service in heifers was available only for heifers' AIs from farm 4 (716 services; 399 to 668 d of age);

The number of AIs included in the study was 812/19,437 from Farm 1 (4.2%); 710/19,437 from Farm 2 (3.7%); 1865/19,437 from Farm 3 (9.6%); 4103/19,437 from Farm 4 (21.1%); 1530/19,437 from Farm 5 (7.9%); 588/19,437 from Farm 6 (3.0%); 3368/19,437 from Farm 7 (17.3%); and 6461/19,437 (33.2%) from Farm 8.

Synchronization protocols recorded for timed AI (FTAI) were as follows:

PRES or Presynch-Ovsynch protocol, as described by Moreira et al. [28]. In brief, cows received 2 injections of PGF2α administered 14 d apart beginning 26 d before initiation of Ovsynch-FTAI protocol, which consists of GnRH1 on day 0, PGF2α on Day 7, GnRH2 on day 9 and FTAI 16 h after last GnRH. The authors included the modification of two prostaglandins 24 h apart before the last gonadotropin-releasing hormone or GnRH2; (total of FTAIs included n = 544; 2.8%); DOV: Double Ovsynch described by Souza et al. [29]. Briefly, treatment with GnRH, followed 7 d later by PGF2α, then 3 d later a GnRH treatment, and finally by the Ovsynch-FTAI protocol. Authors included the modification of two prostaglandins 24 h apart before the last GnRH (total of FTAIs included n = 4534; 2.3%); G6G described by Bello et al. [30]. In short, this protocol included presynchronization with PGF2α, followed 2 d later with GnRH, and then 6 d later with the first GnRH injection of the Ovsynch-FTAI protocol. Again, authors included the modification of two prostaglandins 24 h apart before the last GnRH [31] (total of FTAIs included n = 3332; 17.1%); OVS: classical Ovsynch [32], described before, with the modification of two prostaglandins 24 h apart before the last GnRH [33]; (total of FTAIs included n = 2490; 12.8%); OVS-IPD: classical Ovsynch with insertion of an intravaginal progesterone device between days 0 and 7 of the synchronization [34]; (total of FTAIs included n = 455; 2.3%); 5dCO or 5 d Cosynch, following Santos et al. [35]. Briefly, d 0 GnRH, d 5 and 6 PGF2α, d 9 GnRH and FTAI; (total of FTAIs included n = 401; 2.1%); 5dCO-IPD or 5 d Cosynch, identical to the previous protocol, with insertion of an intravaginal progesterone device between d 0 and 5 of the synchronization [36]; (total of FTAIs included n = 560; 2.9%); 5dOVS: 5 d Ovysnch or short Ovsynch. In brief, d 0 GnRH, d 5 and 6 PGF2α, d 7 GnRH and TAI 16 h after the last GnRH [35,37] (total of FTAIs included n = 587; 3%).

Natural service (NS; total services included 245; 1.3%) and AIs after observed oestrus (OE; total AIs included 10,370; 53.4%) were included as further reproductive strategies. Therefore, ten reproductive strategies were included in the analysis, the eight firstly described for FTAI, the natural service, and AI after OE.

Average Days in Milk (DIM) at first FTAI in cows was 86.13 ± 39.33 d, and the average age of heifers at first service was 438 ± 27.9 d. Average DIM at any service was 149.9 ± 87.75 d, ranging among farms from 114 ± 51.4 in Farm 2 to a maximum of 166 ± 103.7 in Farm 8, and the average age of heifers at service (716, all from farm 4, the only one to provide this data) was 459 ± 44.7 d.

A total of 7094 services/AI were first AIs (36.5%), and 12,346 were 2nd or subsequent AIs, of which 3277 (16.9%) were 5th or subsequent AIs. A total of 8822/19,437 were FTAI, and 10,615/19,437 were AI after observed oestrus and natural services. The AIs were performed in animals with different lactation orders, with 2557 services/AIs having been in heifers (13.1%), 5965 in primiparous cows (30.7%), and the rest (10,936 AIs; 56.2%) in multiparous cows (with lactation order up to 7th). From the total of 2557 services performed on heifers, 867 (34%) came from Farm 4, 716 (28%) from Farm 7, and 974 (38%) from Farm 8. All farms provided data on services on primiparous and multiparous animals, including data on lactation order 0 (heifers) to the 7th lactation.

A total of 14,629 breedings/AIs (75.3%) were performed during the cool season (September to May) and 4808 (24.7%) during the hot season (June, July, and August). The breakdown during the period of study was 2282/19,437 (11.7%) in 2015; 10,442/19,437 (53.7%) in 2016; 5443/19,437 (28%) in 2017; 1270/19,437 (6.5%) during 2018.

Frozen semen from 216 different bulls was used in the different farms, plus the four bulls used for natural mounting (Farm 8). From the total of 9829 AIs with information regarding if sexed or conventional semen was used, 1026 AIs were performed using sexed, frozen, commercial semen. A total of 16 experienced AI technicians/veterinarians performed all AIs during the time of data recording. Tables 1 and 2 show the breakdown of the AIs and services provided by farm, reproductive strategy followed, and animal type (heifer, primiparous, or multiparous).

Wiltbank et al. [3] described four pivotal periods of pregnancy when studying pregnancy loss from fertilization failure to fetal loss. In brief, the first period occurs during the first week after breeding with the absence of fertilization or death of the early embryo, with 20–50% of dairy cows experiencing pregnancy loss during this period. Key factors here are oocyte quality and progesterone concentrations during the preovulatory phase (with issues such as heat stress, inflammatory diseases, body condition being essential). The second pivotal period (days 8 to 27), covers embryo elongation and the maternal recognition of pregnancy, with losses around 30%, with a huge variation between farms. Maintenance of the corpus luteum (CL) of pregnancy is the key factor here. The third pivotal period (days 28 to 60), with losses of 12%, seems to depend on delays or defects in placentomes or embryo. The fourth period covers the third month of pregnancy and show a reduced pregnancy loss incidence (2%), that can be high in cows carrying twins in the same uterine horn. The current study aims to include pregnancy losses occurring in pivotal periods 3 and 4, corresponding to the phases of "late embryonic and early fetal losses" described in previous studies [5]. Consequently, we included all PL cases documented after a positive pregnancy diagnosis (>28 and <35 d of pregnancy) and with a further negative pregnancy diagnosis at days >70 and <110 after AI/service.

In order to determine whether a cow which had previously lost its pregnancy showed a different probability of getting pregnant and losing pregnancy during the rest of its life in services following that first PL, animals were classified according to whether they had lost a pregnancy at least once in their life, compared with those cows with no PL history (including only animals whose full history was known, and including the whole reproductive history of all these cows; n = 5614 services). An additional analysis sought to investigate whether an animal that had experienced PL changed its future probability of getting pregnant and losing pregnancy after subsequent AI/services during the same lactation after the previous PL.

2.4. Conception Rate: Descriptive Results

Conception rate (CR) was not the aim of the present study. Therefore, only descriptive results are outlined. The overall conception rate observed was 34.3% (6696/19,437). Descriptive data on CR values classified by reproductive strategy, farm, and other factors is summarized in Table 3.

Table 3. Conception rate (CR) descriptive statistics for farm, reproductive strategy, season, AI-rank, cattle type, type of AI, and previous pregnancy loss across eight Spanish Holstein farms.

Farm *n*	**Pregnancies/AI**	**CR (%)**	**Reproductive Strategy**	**Pregnancies/AI**	**CR (%)**
1	226/812	27.8	PRES	152/544	27.9
2	294/710	41.4	DOV	167/453	36.9
3	853/1865	45.7	G6G	1353/3332	40.6
4	1290/4103	31.4	OVS	687/2490	27.5
5	616/1530	40.3	OVS-IPD	143/455	31.4
6	276/588	46.9	5dOVS	224/587	38.2
7	1229/3368	36.5	5dCO	168/401	41.9
8	1912/6461	29.6	5dCO-IPD	237/560	42.3
			OE	3364/10,370	32.4
			NS	201/245	82.0
Season	Pregnancies/AI	CR (%)	AI-rank	Pregnancies/AI	CR (%)
Cool	5360/14,629	36.6	First AI	2692/7094	37.9
Warm	1336/4808	27.8	Second AI	3184/9066	35.1
			≥3rd AI	820/3277	25.0
Cattle type	Pregnancies/AI	CR (%)	Type of AI	Pregnancies/AI	CR (%)
Heifers	1288/2557	50.4	FTAI	3131/8822	35.5
Primiparous	2066/5965	34.6	OE and NS	3565/10,615	33.6
Multiparous	3342/10,915	30.6			
Previous PL/lactation	Pregnancies/AI	CR (%)	Previous PL/life	Pregnancies/AI	CR (%)
No previous PL	6253/18,055	34.6	No previous PL	1675/3947	42.4
Previous PL	443/1382	32.1	Previous PL	517/1667	31.0

PL: Pregnancy loss. FTAI: fixed-time artificial insemination. OE: artificial insemination after observed oestrus; NS: natural service. For remaining abbreviations, see Table 1.

2.5. Statistical Analyses

All data were analyzed using SPSS® 25 (IBM, Armonk, NY, USA). Probability values less than or equal to 0.05 were considered significant, and those between 0.05 and 0.1 were considered trends. All data are reported as a mean percentage (categorical variables) or as mean ± SD (numeric variables). Pregnancy loss probability was analyzed using logistic regression, odds ratios (OR) were determined, and a stepwise forward method based on the Wald statistics criterion of $p > 0.10$. The statistical model included farm, animal, and year as random effects to control the effect for repeated AIs.

Different subsets of data were analyzed separately in different regression models in order to control the reduction of the size of the total population analyzed. A first, main, analysis was performed including all pregnancies analyzed (n = 6696 and a total of pregnancies lost 822) and the following factors (available in all pregnancies): farm, animal, year, season, AI-rank, lactation order (categorized in heifers, primiparous, or multiparous), type of AI (fixed-time AI with hormonally controlled follicular development vs. AI after observed oestrus and natural service, with natural development of the follicle (FTAI vs. OE + NS)), reproductive strategy, and previous pregnancy loss. Information on type of semen used (if sexed or conventional semen) and AI-technician was only available in data coming from Farms 7 and 8. Therefore, an additional analysis was performed on these 3141 pregnancies, including all factors previously detailed. A third regression model was conducted on the subset of pregnancies coming from cows (excluding heifers) with a total of 5408 pregnancies, including the additional factor of average Days in Milk (DIM) at each AI. A fourth analysis was performed, including only heifers (n = 1288 pregnancies) with the factors: farm, animal, year, season, AI-rank, type of AI, reproductive strategy, and previous pregnancy loss. The last regression model conducted was on a reduced subset of data of heifers' pregnancies, including the additional factor and age at AI, only provided by Farm 4 (716 AIs). In all the models the following interactions were explored: reproductive strategy × farm; reproductive strategy × season; reproductive strategy × parity; reproductive strategy × type of semen, reproductive strategy × rank of AI, and type of AI × farm. Once demonstrated that interactions were not significant, the second phase in each analysis was performed without those not significant interactions.

3. Results

From a total of 6696 diagnosed pregnancies after a first pregnancy diagnosis and <day 110 of pregnancy (between 28–110 days after AI), a general pregnancy loss rate (PL) of 12.3% was found (822/6696). A PL percentage of 11.9% was observed during the cool season (636/5360) and 13.9% during the hot season (186/1336; $p = 0.32$). Regarding the order of AIs or AI-rank, the PL detected was 13.8% for pregnancies induced by first AIs (372/2692; $p < 0.01$), 11.1% for second services (353/3184), and 11.8% for third and subsequent services (97/820). Heifers achieved the lowest PL (6.9%; 89/1288), followed by primiparous cows (10.8%; 223/2066; $p = 0.04$), while multiparous cows showed the highest PL rate (15.3%; 510/3342; $p < 0.01$; difference primiparous vs. multiparous with $p > 0.05$).

PL results distributed for each type of reproductive strategy across the eight farms are summarized in Table 4. When segmenting the data and observing only heifers, the pregnancy loss rate by protocol was 0.5% (1/194) for natural service, 7.7% (76/982) for observed oestrus, 33.3% (2/6) for OVS, and 9.6% (10/104) for OVS-IPD.

Table 4. Pregnancy loss descriptive statistics for the type of reproductive strategy across eight Spanish Holstein farms.

Reproductive Strategy	Farms																PL by Strategy	
	1		2		3		4		5		6		7		8			
	n/N	%	n/N	%	n/N	%	n/N	%	n/N	%	n/N	%	n/N	%	n/N	%	n/N	%
PRES													21/109	19.3	14/43	32.6	35/152	23.0
DOV	5/50	16.3					1/2	50.0	12/89	14.5			4/26	15.4			22/167	13.2
G6G	3/24	12.5	1/16	6.3	33/346	9.5	69/413	16.7	24/130	18.5	21/197	10.7			23/227	10.1	174/ 1353	12.9
OVS	3/38	7.9	1/7	14.3			27/298	9.1					4/26	15.4	40/239	16.7	93/687	9.4
OVS-IPD	1/13	7.7					10	0.2							10/130	7.7	11/143	7.7
5dOVS									21/213	9.9	0/11	0.0					21/224	9.4
5dCO					19/168	11.3											19/168	11.3
5dCO-IPD					24/237	10.1											24/237	10.1
OE	9/101	8.9	23/271	8.5	11/102	10.8	66/577	11.4	15/184	8.2	9/68	13.2	131/989	13.2	157/1072	14.6	421/ 3365	12.5
NS															2/201	1.0	2/201	1.0
PL by farm n/N %	21/ 226	9.3	25/ 294	8.5	87/ 853	10.2	163/ 1290	12.6	72/ 616	11.7	30/ 276	10.9	178/ 1229	14.5	246/ 1912	12.9	822/ 6696	12.3

n/N indicates the number of pregnancy loss events/number of pregnancies. For the remaining abbreviations, see Table 1.

Logistic Regression Models on Pregnancy Loss Probability, Including all Recorded Factors and Different Data Subsets

The logistic regression analysis results on pregnancy loss, including all pregnancies studied (n = 6696), are summarized in Table 5. As described previously, individual animal, year and farm were introduced in the model as random effects. The analyses determined that the *P*-values were $p = 0.47$ for farm; for the factor animal: $p = 0.06$ and for the interactions reproductive strategy × parity, reproductive strategy × AI rank and AI type × farm, $p = 0.58$, $p = 0.87$ and $p = 0.30$, respectively. Therefore, only the factor year was kept in the final model (Table 5).

Table 5. Logistic regression statistics for pregnancy loss across eight Spanish Holstein farms.

Factors	Class	n/N	PL (%)	Odds Ratio	95% Confidence Interval			*p*-Value
Year	2015, reference	110/583	18.9					
	2016	500/3318	15.1	0.73	0.569	-	0.889	0.01
	2017	176/2402	7.9	0.32	0.246	-	0.428	<0.01
	2018	36/393	9.2	0.43	0.285	-	0.651	<0.01
Season	Cool, reference	636/5360	11.9					
	Warm	186/1362	13.9	1.10	0.912	-	1.319	0.32
AI. rank	First AI, reference	372/2692	13.8					
	≥2nd AI	450/4004	11.2	0.73	0.605	-	0.876	<0.01
Parity	Heifer, reference	89/1288	6.9					
	Primiparous	223/2066	10.8	1.35	1.019	-	1.803	0.04
	Multiparous	510/3342	15.3	2.02	1.546	-	2.654	<0.01
AI type	FTAI, reference	399/3131	12.7			-		
	OE + NS	423/3565	11.9	0.12	0.026	-	0.532	0.01

n/N indicates number of pregnancy loss events/number of pregnancies; AI: artificial insemination; PL: pregnancy Loss; FTAI: fixed-time AI; OE: AI after observed oestrus; NS: natural breeding. Reference factors within each class are those placed in first instance.

Second or more AIs, being a heifer and AI after observed oestrus or natural breeding, were associated with a lower PL rate, while not having a PL history during the same lactation tended to be associated with higher PL. The factors season, previously having lost a pregnancy in life and reproductive strategy, were not associated with PL ($p > 0.05$).

Having had a previous PL in the lactation, tended to have a reduced probability of losing pregnancy again during the same lactation (780/6253, 12.5% vs. 42/443, 9.5% of PL for cows without and with PL history, respectively; OR = 0.77; 95% CI 0.551–1.086; $p = 0.13$).

No relationship to PL was detected for the different hormonal protocols for fixed time AI and AI after observed oestrus ($p > 0.05$).

The sub-model observing only cows (total pregnancies: 5408), in order to analyse the same model adding days in milk, demonstrated that DIM did not influence PL ($p = 0.76$). Subsequently, factors not provided by all farms were added, such as AI-technitian and type of semen used. The model observing AI-technician and type of semen, provided only by Farms 7 and 8, included 9829 AIs and 3141 pregnancies (408 pregnancies after using sexed semen and 2733 after using conventional semen). In this model, AI-technician was not associated with pregnancy loss ($p = 0.90$), and neither was type of semen ($p = 0.62$). Pregnancy loss after the 3141 pregnancies wth sexed semen was 8.1% (33/408), while PL was 14.3% (391/2733) in pregnancies achieved with conventional semen. In this data subset, similarly to the subset of all included AIs, second or subsequent AIs were associated with lower PL compared with first services ($p = 0.04$; OR = 0.79; 95% CI: 0.63–0.99). Being a heifer as opposed to an adult cow was associated also with lower PL (OR = 2.16; 95% CI: 1.54–3.03 for primiparous and OR = 3.11; 95% CI: 2.27–4.26 for multiparous; $p < 0.01$). Having lost pregnancy related to a lower PL value in future pregnancies during the same lactation (OR = 0.62; 95% CI: 0.40–0.98; $p = 0.04$). No explored interaction and neither type of AI or season were significant factors for PL in this subset of AIs in cows.

Finally, the logistic regression model on the subset of heifers (total of AIs included 2557 and 1288 pregnancies) detected the factors Natural Service (PL = 0.5%; 1/194) and Ovsynch reproductive strategy (PL = 40%; 2/5), as associated with PL (OR = 0.004, 95% CI: 0.00–0.27, $p = 0.010$, and OR = 8.32, 95% CI: 1.30–53.23, $p = 0.03$, respectively).

4. Discussion

This study indicates a 12.3% overall rate of pregnancy loss between days 28 and 110 of pregnancy across all AIs and services included. The global PL rate observed (12.3%) in the present study is consistent with the 11.95% based on analysis of 24,391 pregnancies by Wiltbank et al. [3]. The hypothesis of the authors was that having lost previously pregnancy was not a risk factor for future PL. However, farm did note play a key role, while AI-rank and AI-type were significantly associated with PL.

Although in the present study the "farm" factor itself was not significantly associated with different PL results, we could consider Farm 4 to be an intrinsic factor worsening overall reproductive efficiency. In fact, Farm 4 was the farm with the second worst average conception rate (31.4%; Table 3), and it showed the worse PL rates after three different strategies (observed oestrus, G6G, and Ovsynch; Table 4), which would support a negative influence of the farm itself on its overall reproductive efficiency. The "farm" factor includes different features together, such as health status, management practices, and environmental conditions, which worsens the average farm level of PL as stated by other authors [3,8]. The factor animal was included as random factor, to control the effect of repeated AIs in one same animal. However, it was not significantly associated and could be removed from the final model.

On the contrary, the factor year of study, was strongly associated with PL with a decreasing PL rate being observed with time. We could, therefore, argue a general improvement of the reproductive management in general, which seems to be associated also to this reduction of PL with time.

The hot season at AI is a known risk factor for lower fertility [9,16,17]. However, heat stress is reported to mainly influence failure in very early pregnancy phases, such as fertilization and "Pregnancy Pivotal Periods" n 1 and 2 after Wiltbank et al.'s [3] classification (days of pregnancy from fertilization to day 28). Accordingly, in our study, season was not significantly associated with pregnancy loss during Pivotal Periods 3 and 4. García-Ispierto et al. [5] demonstrated a strong association between PL and heat stress, with the likelihood of pregnancy loss being associated with a PL increase by a factor of 1.05 for each additional unit of the mean maximum temperature–humidity index from days 21 to 30 of gestation. This factor may be a complex one, and is probably directly related to an additional factor connected with PL, the unilateral twinning pregnancy [37], which is the main risk factor for PL during the fourth pivotal period of pregnancy (60 to 90 days of pregnancy) [3] and which is linked to season [38]. Unfortunately, there was no information available on twin pregnancies or laterality of double CL in our dataset. The interaction reproductive strategy x season did neither result significant in this study. In farms with a severe heat stress (as it was the case of the farms in the study), farmers avoid investing resources in inseminating at fixed time during the hot season. Due to this reason, the number of pregancies achieved during the summer with hormonal protocols was limited.

Being a heifer as opposed to cows related to lower PL values. These results partly agree with other earlier observations, which showed that parity is positively associated with PL (more PL in high parity cows) [5,6]. Our study shows an eight-point PL difference between multiparous cows and heifers (15.3 vs. 6.9% of PL; $p < 0.01$), which is similar to the results of Starbuck et al. [39] who reported that older cows maintained 9% fewer pregnancies than younger ones. While embryo survival rates are assumed to be similar in heifers and low- to moderate-producing dairy cows, this influence could be related to a parity-related effect and could also be a consequence of milk yield level [25]. Accordingly, an increase in milk yield during the first 60 days in milk can increase early embryonic loss by almost 5 points [12]. Unfortunately, production data were not available in our data set.

In this study the type of AI was associated with PL. Pregnancies produced after natural follicle development (OE + NS) were associated to less PL (12.7% vs. 11.9%). In other studies, the occurrence of oestrus at AI, using tail chalk and tail head patch, was associated with a reduction in pregnancy loss in dairy cows [40,41]. Furthermore, Pereira et al. [41] reported that the reduction in pregnancy loss in cows that expressed oestrus at AI occurred regardless of the diameter of the pre-ovulatory follicle and the hormonal protocol. Madureira et al. [15] elucidated that the actual factor related to increased CR and decreased PL could be the intensity of oestrus expression, irrespective of the reproductive strategy implemented. In the present work, the type of AI with less PL pregnancies included those pregnancies after natural breeding, which may be the deffinitive factor, supporting the results of Madureira et al. [15] because animals inducing bulls to mount and tolerating natural breeding express estrous adequately. Moreover, the factor reproductive strategy which included all hormonal protocols and AI after observed oestrus, was not significantly associated with PL in the current study. The strategy for detecting oestrus in our farms, was to use activity devices where a strong oestrus expression is not required for insemination, which could explain this result. It seems that progesterone is the final modulator linking the better expression of oestrus with a better fertility and reduced pregnancy loss afterwards. Progesteron leads oestrous expression of higher intensity and greater fertility, and, consistently, increased progesterone concentration post-AI sustains embryo and fetal development [42], probably due to changes in the progesterone receptor profile within the endometrium [43].

The result that ≥2nd AIs were associated with less PL could reflect the less stressing situation with the cattle having overcome the lactation peak (more days in milk). However, this may also reflect just the low probability of occurring PL, together with the reduced number of AIs of higher rank, which numerically results in a overall reduced prevalence of PL in further AIs. The same fact may be behind the result obtained in the current study that animals having had a previous PL in the lactation tended to have a reduced probability of losing pregnancy again during the same lactation (12.5% vs. 9.5% of PL for cows without and with PL history, respectively). Necessarily, when a cow has experienced PL previously, further AIs in the same lactation are of a higher rank. Certain pathological events, although less likely to occur, might suggest the involvement of a possible infectious etiology in some cases of pregnancy loss [44]. However, this hypothesis is not supported by the fact that this association with PL of having had previous PL was not significant when observing AIs during the rest of the animal's life. The retrospective design of this study does not allow detection of protective factors, but based on these results, we can state that further AIs after a pregnancy loss event and higher rank of AIs are not associated with an enhanced probability of PL.

No relationship to PL was detected for the different hormonal protocols for fixed time AI and AI after observed oestrus. Similarly, the meta-analysis performed by Borchardt et al. [21] showed no PL difference before day 60 of pregnancy in cows inseminated after Presynch AI. Similarly, several authors have failed to link determined protocols to PL rates [21,45–48], although they specifically pointed out the limited number of herds with similar management practices and of pregnancies included in some of the previous studies, and, consequently, on the limited number of pregnancy losses analyzed. These circumstances, could be similar in the current work. Moreover, although the reproductive strategy for AI is directly linked to the AI-rank, with long protocols (PRES, DOV, and G6G) mostly used for first AIs [49], no significance was detected for the "AI rank × reproductive strategy" interaction on PL.

Regarding the relationship protocol and PL, observing only heifers, as opposed to all cows, gave rise to different results, with the Ovsynch protocol being associated with higher PL ($p = 0.03$). Heifers show higher individual fertility than cows [5], and it is well known that the Ovsynch protocol does not correctly synchronize the ovulation in dairy heifers [32], this also probably being the reason for a higher PL subsequently when pregnant, due to less functional CLs [23], as observed in lactating dairy cows [40]. Other factors included in the study, such as AI-technician, bull semen, sexed vs. conventional semen, or DIM, were not significantly associated with PL. However, based on the retrospective, epidemiological, observational design of the study, and the reduced sample size in these sub-subsets that included these factors, we cannot rule out the lack of statistical power to detect such effects.

5. Conclusions

Our results demonstrate that being a heifer, second or more breedings and AI after observed oestrus and natural breeding have a notable impact, being associated to less PL. Having lost pregnancy previously, was detected as a slighter influencing factor (statistical tendency), relating to less PI in future pregnancies in the same lactation. However, factors previously demonstrated as significant, such as season, bull, AI technician or type of semen, might not be associated with PL. Therefore, farmers can decide inseminating again cows which suffered previous PL and farmers and consultants should adapt their preventive strategies against PL to cattle parity. These results highlight again, the complex reproductive characteristics of the pregnancy loss event and the relevance of further studies on this issue.

Author Contributions: Conceptualization, S.A; methods, all authors; software: A.F.-N., O.F., J.M.L., F.S., J.L.P.-P., and S.A.; validation, A.F.-N., N.P.-V., and S.A.; data analysis: A.F.-N., O.F., J.M.L., F.S., J.L.P.-P., R.P.-C., and S.A.; investigation: A.F.-N. and S.A.; resources: A.F.-N., O.F., J.M.L., F.S., J.L.P.-P., R.P.-C., and S.A.; data curation: A.F.-N., O.F., J.M.L., F.S., J.L.P.-P., R.P.-C., and S.A.; drafting of the manuscript: A.F.-N., N.P.-V., and S.A.; review and editing of the manuscript, all authors; supervision: A.F.-N., N.P.-V., and S.A.; project administration: S.A. All authors have read and agreed to the published version of the manuscript.

Funding: This research received no external funding.

Acknowledgments: We thank Pedro Cuesta and Iagoba Cano (Department of Research Support, Complutense University of Madrid) for their help with statistical analyses; we thank Peter Bonney (European University of Madrid) for his thorough advise with the native, scientific English, and all the farmers and farm workers for contributing data.

Conflicts of Interest: The authors confirm that there are no conflict of interest.

References

1. Santos, J.E.P.; Thatcher, W.W.; Chebel, R.C.; Cerri, R.L.A.; Galvão, K.N. The effect of embryonic death rates in cattle on the efficacy of estrus synchronization programs. *Anim. Reprod. Sci.* **2004**, *82–83*, 513–535. [CrossRef]
2. Cabrera, V.E. A simple formulation and solution to the replacement problem: A practical tool to assess the economic cow value, the value of a new pregnancy, and the cost of a pregnancy loss. *J. Dairy Sci.* **2012**, *95*, 4683–4698. [CrossRef]
3. Wiltbank, M.C.; Baez, G.M.; Garcia-Guerra, A.; Toledo, M.Z.; Monteiro, P.L.J.; Melo, L.F.; Ochoa, J.C.; Santos, J.E.P.; Sartori, R. Pivotal periods for pregnancy loss during the first trimester of gestation in lactating dairy cows. *Theriogenology* **2016**, *86*, 239–253. [CrossRef]
4. Ealy, A.D.; Seekford, Z.K. Symposium review: Predicting pregnancy loss in dairy cattle. *J. Dairy Sci.* **2019**, *102*, 11798–11804. [CrossRef]
5. García-Ispierto, I.; López-Gatius, F.; Santolaria, P.; Yániz, J.L.; Nogareda, C.; López-Béjar, M.; De Rensis, F. Relationship between heat stress during the peri-implantation period and early fetal loss in dairy cattle. *Theriogenology* **2006**, *65*, 799–807. [CrossRef] [PubMed]
6. Lee, J.-I.; Kim, I.-H. Pregnancy loss in dairy cows: The contributing factors, the effects on reproductive performance and the economic impact. *J. Vet. Sci.* **2007**, *8*, 283. [CrossRef] [PubMed]
7. Labèrnia, J.; López-Gatius, F.; Santolaria, P.; López-Béjar, M.; Rutllant, J. Influence of management factors on pregnancy attrition in dairy cattle. *Theriogenology* **1996**, *45*, 1247–1253. [CrossRef]
8. Grimard, B.; Freret, S.; Chevallier, A.; Pinto, A.; Ponsart, C.; Humblot, P. Genetic and environmental factors influencing first service conception rate and late embryonic/foetal mortality in low fertility dairy herds. *Anim. Reprod. Sci.* **2006**, *91*, 31–44. [CrossRef] [PubMed]
9. López-Gatius, F. Factors of a noninfectious nature affecting fertility after artificial insemination in lactating dairy cows. A review. *Theriogenology* **2012**, *77*, 1029–1041. [CrossRef]
10. Kelley, D.E.; Galvão, K.N.; Mortensen, C.J.; Risco, C.A.; Ealy, A.D. Using Doppler ultrasonography on day 34 of pregnancy to predict pregnancy loss in lactating dairy cattle. *J. Dairy Sci.* **2017**, *100*, 3266–3271. [CrossRef]
11. Martins, J.P.N.; Wang, D.; Mu, N.; Rossi, G.F.; Martini, A.P.; Martins, V.R.; Pursley, J.R. Level of circulating concentrations of progesterone during ovulatory follicle development affects timing of pregnancy loss in lactating dairy cows. *J. Dairy Sci.* **2018**, *101*, 10505–10525. [CrossRef] [PubMed]

12. Nyman, S.; Gustafsson, H.; Berglund, B. Extent and pattern of pregnancy losses and progesterone levels during gestation in Swedish Red and Swedish Holstein dairy cows. *Acta Vet. Scand.* **2018**, *60*. [CrossRef] [PubMed]
13. Gatea, A.O.; Smith, M.F.; Pohler, K.G.; Egen, T.; Pereira, M.H.C.; Vasconselos, J.L.M.; Lawrence, J.C.; Green, J.A. The ability to predict pregnancy loss in cattle with ELISAs that detect pregnancy associated glycoproteins is antibody dependent. *Theriogenology* **2018**, *108*, 269–276. [CrossRef] [PubMed]
14. Northrop, E.J.; Rich, J.J.J.; Rhoades, J.R.; Perry, G.A. Comparison of two bovine serum pregnancy tests in detection of artificial insemination pregnancies and pregnancy loss in beef cattle. *PLoS ONE* **2019**, *14*, e0211179. [CrossRef] [PubMed]
15. Madureira, A.M.L.; Polsky, L.B.; Burnett, T.A.; Silper, B.F.; Soriano, S.; Sica, A.F.; Pohler, K.G.; Vasconcelos, J.L.M.; Cerri, R.L.A. Intensity of estrus following an estradiol-progesterone-based ovulation synchronization protocol influences fertility outcomes. *J. Dairy Sci.* **2019**, *102*, 3598–3608. [CrossRef]
16. Gehrke, M.; Zbylut, J. Factors connected with pregnancy loss in dairy cows. *Bull. Vet. Inst. Pulawy* **2011**, *55*, 457–464.
17. Souza, F.; Carneiro, L.C.; Cesar, J.; dos Santos, R.M. Non-infectious causes that increase early and mid-to-late pregnancy loss rates in a crossbreed dairy herd. *Trop. Anim. Health Prod.* **2019**, *51*, 759–765. [CrossRef]
18. Heidari, F.; Dirandeh, E.; Ansari Pirsaraei, Z.; Colazo, M.G. Modifications of the G6G timed-AI protocol improved pregnancy per AI and reduced pregnancy loss in lactating dairy cows. *Animal* **2017**, *11*, 2002–2009. [CrossRef]
19. Wijma, R.; Stangaferro, M.L.; Masello, M.; Granados, G.E.; Giordano, J.O. Resynchronization of ovulation protocols for dairy cows including or not including gonadotropin-releasing hormone to induce a new follicular wave: Effects on re-insemination pattern, ovarian responses, and pregnancy outcomes. *J. Dairy Sci.* **2017**, *100*, 7613–7625. [CrossRef]
20. Niozas, G.; Tsousis, G.; Steinhöfel, I.; Brozos, C.; Römer, A.; Wiedemann, S.; Bollwein, H.; Kaske, M. Extended lactation in high-yielding dairy cows. I. Effects on reproductive measurements. *J. Dairy Sci.* **2019**, *102*, 799–810. [CrossRef]
21. Borchardt, S.; Haimerl, P.; Heuwieser, W. Effect of insemination after estrous detection on pregnancy per artificial insemination and pregnancy loss in a Presynch-Ovsynch protocol: A meta-analysis. *J. Dairy Sci.* **2016**, *99*, 2248–2256. [CrossRef] [PubMed]
22. Martins, J.P.N.; Acevedo, M.J.T.; Cunha, T.O.; Piterini, C.; Pursley, J.R. The effect of presynchronization with prostaglandin F2α and gonadotropin-releasing hormone simultaneously, 7 d before Ovsynch, compared with Presynch-10/Ovsynch on luteal function and first-service pregnancies per artificial insemination. *J. Dairy Sci.* **2017**, *100*, 5107–5116. [CrossRef]
23. Santos, J.E.P.; Rutigliano, H.M.; Sá Filho, M.F. Risk factors for resumption of postpartum estrous cycles and embryonic survival in lactating dairy cows. *Anim. Reprod. Sci.* **2009**, *110*, 207–221. [CrossRef] [PubMed]
24. Lima, F.S.; Bisinotto, R.S.; Ribeiro, E.S.; Ayres, H.; Greco, L.F.; Galvão, K.N.; Risco, C.A.; Thatcher, W.W.; Santos, J.E.P. Effect of one or three timed artificial inseminations before natural service on reproductive performance of lactating dairy cows not observed for detection of estrus. *Theriogenology* **2012**, *77*, 1918–1927. [CrossRef] [PubMed]
25. Diskin, M.G.; Parr, M.H.; Morris, D.G. Embryo death in cattle: An update. *Reprod. Fertil. Dev.* **2011**, *24*, 244–251. [CrossRef] [PubMed]
26. Diskin, M.G.; Waters, S.M.; Parr, M.H.; Kenny, D.A. Pregnancy losses in cattle: Potential for improvement. *Reprod. Fertil. Dev.* **2016**, *28*, 83–93. [CrossRef]
27. COUNCIL DIRECTIVE 98/58/EC of 20 July 1998 Concerning the Protection of Animals Kept for Farming Purposes 1998. Available online: https://eur-lex.europa.eu/legal-content/EN/TXT/?uri=CELEX%3A31998L0058 (accessed on 20 January 2020).
28. Moreira, F.; Orlandi, C.; Risco, C.A.; Mattos, R.; Lopes, F.; Thatcher, W.W. Effects of presynchronization and bovine somatotropin on pregnancy rates to a timed artificial insemination protocol in lactating dairy cows. *J. Dairy Sci.* **2001**, *84*, 1646–1659. [CrossRef]
29. Souza, A.H.; Ayres, H.; Ferreira, R.M.; Wiltbank, M.C. A new presynchronization system (Double-Ovsynch) increases fertility at first postpartum timed AI in lactating dairy cows. *Theriogenology* **2008**, *70*, 208–215. [CrossRef]

30. Bello, N.M.; Steibel, J.P.; Pursley, J.R. Optimizing ovulation to first GnRH improved outcomes to each hormonal injection of ovsynch in lactating dairy cows. *J. Dairy Sci.* **2006**, *89*, 3413–3424. [CrossRef]
31. Ribeiro, E.S.; Bisinotto, R.S.; Favoreto, M.G.; Martins, L.T.; Cerri, R.L.A.; Silvestre, F.T.; Greco, L.F.; Thatcher, W.W.; Santos, J.E.P. Fertility in dairy cows following presynchronization and administering twice the luteolytic dose of prostaglandin F2α as one or two injections in the 5-day timed artificial insemination protocol. *Theriogenology* **2012**, *78*, 273–284. [CrossRef]
32. Pursley, J.R.; Mee, M.O.; Wiltbank, M.C. Synchronization of ovulation in dairy cows using PGF2α and GnRH. *Theriogenology* **1995**, *44*, 915–923. [CrossRef]
33. Brusveen, D.J.; Souza, A.H.; Wiltbank, M.C. Effects of additional prostaglandin F2alpha and estradiol-17beta during Ovsynch in lactating dairy cows. *J. Dairy Sci.* **2009**, *92*, 1412–1422. [CrossRef] [PubMed]
34. Lamb, G.C.; Stevenson, J.S.; Kesler, D.J.; Garverick, H.A.; Brown, D.R.; Salfen, B.E. Inclusion of an intravaginal progesterone insert plus GnRH and prostaglandin F2alpha for ovulation control in postpartum suckled beef cows. *J. Anim. Sci.* **2001**, *79*, 2253–2259. [CrossRef] [PubMed]
35. Santos, J.E.P.; Narciso, C.D.; Rivera, F.; Thatcher, W.W.; Chebel, R.C. Effect of reducing the period of follicle dominance in a timed artificial insemination protocol on reproduction of dairy cows. *J. Dairy Sci.* **2010**, *93*, 2976–2988. [CrossRef]
36. Bridges, G.A.; Helser, L.A.; Grum, D.E.; Mussard, M.L.; Gasser, C.L.; Day, M.L. Decreasing the interval between GnRH and PGF2alpha from 7 to 5 days and lengthening proestrus increases timed-AI pregnancy rates in beef cows. *Theriogenology* **2008**, *69*, 843–851. [CrossRef]
37. López-Gatius, F.; Andreu-Vázquez, C.; Mur-Novales, R.; Cabrera, V.E.; Hunter, R.H.F. The dilemma of twin pregnancies in dairy cattle. A review of practical prospects. *Livest. Sci.* **2017**, *197*, 12–16. [CrossRef]
38. López-Gatius, F.; García-Ispierto, I.; Hunter, R.H.F. Factors affecting spontaneous reduction of corpora lutea and twin embryos during the late embryonic/early fetal period in multiple-ovulating dairy cows. *Theriogenology* **2010**, *73*, 293–299. [CrossRef]
39. Starbuck, M.J.; Dailey, R.A.; Inskeep, E.K. Factors affecting retention of early pregnancy in dairy cattle. *Anim. Reprod. Sci.* **2004**, *84*, 27–39. [CrossRef]
40. Pereira, R.V.; Siler, J.D.; Ng, J.C.; Davis, M.A.; Grohn, Y.T.; Warnick, L.D. Effect of on-farm use of antimicrobial drugs on resistance in fecal *Escherichia coli* of preweaned dairy calves. *J. Dairy Sci.* **2014**, *97*, 7644–7654. [CrossRef]
41. Pereira, M.H.C.; Wiltbank, M.C.; Barbosa, L.F.S.P.; Costa, W.M.; Carvalho, M.a.P.; Vasconcelos, J.L.M. Effect of adding a gonadotropin-releasing-hormone treatment at the beginning and a second prostaglandin F2α treatment at the end of an estradiol-based protocol for timed artificial insemination in lactating dairy cows during cool or hot seasons of the year. *J. Dairy Sci.* **2015**, *98*, 947–959. [CrossRef]
42. Bisinotto, R.S.; Ribeiro, E.S.; Martins, L.T.; Marsola, R.S.; Greco, L.F.; Favoreto, M.G.; Risco, C.A.; Thatcher, W.W.; Santos, J.E.P. Effect of interval between induction of ovulation and artificial insemination (AI) and supplemental progesterone for resynchronization on fertility of dairy cows subjected to a 5-d timed AI program. *J. Dairy Sci.* **2010**, *93*, 5798–5808. [CrossRef] [PubMed]
43. Lonergan, P. Influence of progesterone on oocyte quality and embryo development in cows. *Theriogenology* **2011**, *76*, 1594–1601. [CrossRef] [PubMed]
44. Barkallah, M.; Gharbi, Y.; Hassena, A.B.; Slima, A.B.; Mallek, Z.; Gautier, M.; Greub, G.; Gdoura, R.; Fendri, I. Survey of infectious etiologies of bovine abortion during mid- to late gestation in dairy herds. *PLoS ONE* **2014**, *9*, e91549. [CrossRef] [PubMed]
45. Chebel, R.C.; Santos, J.E.P.; Cerri, R.L.A.; Rutigliano, H.M.; Bruno, R.G.S. Reproduction in dairy cows following progesterone insert presynchronization and resynchronization protocols. *J. Dairy Sci.* **2006**, *89*, 4205–4219. [CrossRef]
46. Chebel, R.C.; Santos, J.E.P. Effect of inseminating cows in estrus following a presynchronization protocol on reproductive and lactation performances. *J. Dairy Sci.* **2010**, *93*, 4632–4643. [CrossRef] [PubMed]
47. Bisinotto, R.S.; Castro, L.O.; Pansani, M.B.; Narciso, C.D.; Martinez, N.; Sinedino, L.D.P.; Pinto, T.L.C.; Van de Burgwal, N.S.; Bosman, H.M.; Surjus, R.S.; et al. Progesterone supplementation to lactating dairy cows without a corpus luteum at initiation of the Ovsynch protocol. *J. Dairy Sci.* **2015**, *98*, 2515–2528. [CrossRef] [PubMed]

48. Borchardt, S.; Haimerl, P.; Pohl, A.; Heuwieser, W. Evaluation of prostaglandin F2α versus prostaglandin F2α plus gonadotropin-releasing hormone as Presynch methods preceding an Ovsynch in lactating dairy cows: A meta-analysis. *J. Dairy Sci.* **2017**, *100*, 4065–4077. [CrossRef]
49. Carvalho, P.D.; Santos, V.G.; Giordano, J.O.; Wiltbank, M.C.; Fricke, P.M. Development of fertility programs to achieve high 21-day pregnancy rates in high-producing dairy cows. *Theriogenology* **2018**, *114*, 165–172. [CrossRef]

animals

MDPI

Review

Recent Possibilities for the Diagnosis of Early Pregnancy and Embryonic Mortality in Dairy Cows

Ottó Szenci

Department of Obstetrics and Food Animal Medicine Clinic, University of Veterinary Medicine Budapest, H-2225 Üllő-Dóra Major, Hungary; szenci.otto@univet.hu

Simple Summary: Pregnancy diagnosis plays an essential role in decreasing days open in dairy farms; therefore, it is very important to select an accurate method for diagnosing early pregnancy. Besides traditional pregnancy diagnoses made by rectal palpation of the uterus from 40 to 60 days after AI and measuring the serum or milk progesterone concentration between 18 to 24 days after AI, there are several new possibilities to diagnose early pregnancy in dairy farms. However, it is very important to emphasize that before introducing any new diagnostic test, we need to make sure the accuracy of that particular test to be able to decrease the rate of iatrogenic pregnancy losses caused by prostaglandin or resynchronization treatments. This review focuses on the diagnostic possibilities and limitations of early pregnancy diagnosis in the field.

Abstract: One of the most recent techniques for the on-farm diagnosis of early pregnancy (EP) in cattle is B-mode ultrasonography. Under field conditions, acceptable results may be achieved with ultrasonography from Days 25 to 30 post-AI. The reliability of the test greatly depends on the frequency of the transducer used, the skill of the examiner, the criterion used for a positive pregnancy diagnosis (PD), and the position of the uterus in the pelvic inlet. Non-pregnant animals can be selected accurately by evaluating blood flow in the corpus luteum around Day 20 after AI, meaning we can substantially improve the reproductive efficiency of our herd. Pregnancy protein assays (PSPB, PAG-1, and PSP60 RIA, commercial ELISA or rapid visual ELISA tests) may provide an alternative method to ultrasonography for determining early pregnancy or late embryonic/early fetal mortality (LEM/EFM) in dairy cows. Although the early pregnancy factor is the earliest specific indicator of fertilization, at present, its detection is entirely dependent on the use of the rosette inhibition test; therefore, its use in the field needs further developments. Recently found biomarkers like interferon-tau stimulated genes or microRNAs may help us diagnose early pregnancy in dairy cows; however, these tests need further developments before their general use in the farms becomes possible.

Keywords: dairy cow; ultrasonography; pregnancy proteins; embryonic mortality; fetal mortality

Citation: Szenci, O. Recent Possibilities for the Diagnosis of Early Pregnancy and Embryonic Mortality in Dairy Cows. *Animals* **2021**, *11*, 1666. https://doi.org/10.3390/ani11061666

Academic Editors: João Simões and Fiammetta Berlinguer

Received: 31 March 2021
Accepted: 27 May 2021
Published: 3 June 2021

1. Introduction

The extent of pregnancy loss (PL) in dairy cows can be estimated from the difference between the fertilization rate (FR) and the subsequent calving rate (CR). It is generally accepted that the FR in healthy dairy cattle is between 85% and 90% [1] or, more recently, around 80% [2]. The CR has decreased from 66% since 1951, to about 50% until 1975 [3], and furthermore to about 33.1% in Spain [4], 33.4% in Israel [5], 37% in Canada [6], 41% in Japan [7], and 40% to 44% in the USA [3,8], while it did not change (>60%) in heifers [9,10]. Therefore, 41% to 57% of pregnancies can be lost during gestation. Most of the losses (early embryonic mortality: EEM) may occur before the maternal recognition of gestation (<Day 16) when the life of the corpus luteum is not extended, and cows return to estrus [1,11]. After the maternal recognition of pregnancy, the life of the corpus luteum will be extended, and embryonic mortality (further 5% to 10%) occurring between Days 16 and 42 after AI is called late embryonic mortality (LEM), while occurring between Days 42

to 90 after AI is named early fetal mortality (EFM). Late fetal mortality between Day 90 and term is rare [12]. It has recently been reported that PL after the first month of gestation may range between 3.2% and 42.7% [13].

Recently, Pohler et al. [14] have suggested using the terms 'early embryonic development' until Day 30 and 'late embryonic development' between Days 31 and 45 of gestation. Due to the fact that different technical terms are used for this period, perhaps it would be easier to divide the former late embryonic period between Days 16 to 42 into two stages, namely, stage 1 between Days 16 and 30 and stage 2 between Days 31 and 42/45 of gestation.

The PL represents a considerable biological and economic waste for the farmer, therefore, the objective of the present review is to discuss recent methods suitable for diagnosing early pregnancy and pregnancy losses in dairy cattle.

2. Early Pregnancy Diagnosis (PD)

2.1. Real-Time B-Mode Ultrasonography

During ultrasonographic examination (UE), a cow was initially considered to be pregnant when an irregularly shaped, non-echogenic black spot (or spots) were recognized within the uterine lumen, representing the chorioallantoic fluid [15]. The demonstration of an embryo (later with a beating heart) and/or embryonic membranes provided additional confirmation of pregnancy [16,17]. A normal amount of chorioallantoic fluid in the uterine horn ipsilateral to the ovary with a corpus luteum provided additional confirmation of a normal gestation [17,18]. Where no such signs (Table 1) were found, the possibility of pregnancy was ruled out, giving a non-pregnancy diagnosis. The confirmation of ultrasonographic diagnoses was usually based on palpation per rectum of the uterus at 2 to 3 months post-AI, or upon spontaneous return to estrus after AI. A cow was also considered pregnant if an embryo proper with a beating heart was recognized at a final UE on Days 50 to 60 post-AI. Cows diagnosed as non-pregnant by palpation per rectum or by ultrasonography between Days 50 and 90 were usually designated as non-pregnant [19–21].

Table 1. Identifiable characteristics of the bovine conceptus during ultrasonographic examinations for pregnancy diagnosis, indicating the first day of detection.

Characteristics	First day of Detection in Heifers Curran et al. [21] Range	First Day of Detection in Heifers Curran et al. [21] Mean ± SEM	First Day of Detection after ET Totey et al. [22] Mean ± SD
Embryo proper	19 to 24	20.3 ± 0.3	19.5 ± 0.7
Heartbeat	19 to 24	20.9 ± 0.3	22.6 ± 0.9
Allantois	22 to 25	23.2 ± 0.3	23.1 ± 0.8
Spinal cord	26 to 33	29.1 ± 0.5	33.0 ± 1.5
Forelimb buds	28 to 31	29.1 ± 0.3	32.7 ± 1.3 *
Amnion	28 to 33	29.5 ± 0.5	25.1 ± 1.4
Hindlimb buds	30 to 33	31.2 ±0.3	32.9 ± 1.3 **
Placentomes	33 to 38	35.2 ± 1.0	-
Split hooves	42 to 49	44.6 ± 0.7	-
Fetal movement	42 to 50	44.8 ± 0.8	50.7 ± 1.0
Ribs	51 to 55	52.3 ± 0.5	60.9 ± 1.7

ET: embryo transfer, * forelimb, ** limb buds.

Under experimental conditions, embryonic vesicles in dairy cows can be detected as early as at 9 [23], 10 [20], or 12 days [24] of gestation; however, an accurate pregnancy diagnosis (100%) can be reached only on Days 20 and 22 of gestation [25], when the embryo (20.3 days) and its heartbeat (20.9 days) can be detected [21]. It is important to mention that fluid accumulation in the chorioallantois during early pregnancy can be easily confused with the uterine fluid within the uterus during pro-estrus and estrus [25].

Under on-farm conditions, between Days 22 to 49 after AI, the sensitivity, specificity, positive predictive value, and negative predictive value of ultrasonographic diagnoses made with a 5 MHz linear-array or sector transducer varied between 95% and 100%, 67% and 100%, 85.7% and 100%, and 84% and 100%, respectively (Table 2).

Table 2. Accuracy of ultrasonographic examinations for diagnosing early pregnancy in dairy cattle by using a 5.0 MHz linear-array or sector transducer.

Days after AI	N	Sensitivity (%)	Specificity (%)	Positive Predictive Value (%)	Negative Predictive value (%)	References
22–40	435	96.2	71.1	89.6	87.8	Filteau and DesCoteaux, [26]
25–29	101	98	91.3	93	97.6	Szenci et al. [27] [a]
30–39	143	100	100	100	100	
26 [b]	48	100	96.7	94.4	100	Romano et al. [17]
27 [b]	53	100	92.8	92.6	100	
26–33	85	97.7	87.8	89.6	97.2	Pieterse et al. [15]
29 [c]	151	100	96.3	91.3	100	Romano et al. [17]
30 [c]	151	100	97.4	91.9	100	
30–39	722	95	67	89	84	Hanzen and Laurent [28]
40–49	620	98	77	92	94	
31–35	323	98	80	85.7	97	Munoz del Real et al. [29]
36–40		97	93	93.6	97	

[a] Sector transducer, [b] Heifers, [c] Cows, NG: not given. Sensitivity = correct positive/correct positive + false negative × 100. Specificity = correct negative/correct negative + false positive × 100. Positive predictive value = correct positive/correct positive + false positive × 100. Negative predictive value = correct negative/correct negative + false negative × 100.

In contrast, Badtram et al. [30] reported that the sensitivity and specificity of the ultrasound test between Days 23 and 31 post-AI were only 68.8% and 71.7%, respectively. In a recent study, maximum sensitivity and negative predictive value were reached at Day 26 in dairy heifers and at Day 29 in dairy cows, while the specificity and positive predictive values were 96.7% and 94.4% for dairy heifers and 96.3% and 91.3% for dairy cows, respectively [17]. In contrast, according to Nation et al. [16], due to pregnancy loss, Days 28 to 35 after AI are too early for reliable detection of pregnancy in dairy cows.

Between Days 27 to 34 after AI, the sensitivity, specificity, positive predictive value, and negative predictive value of ultrasonographic diagnoses made by the use of 5–10 MHz transducer varied between 90% and 96.6%, 91.5% and 100%, 88.4% and 100%, and 92.3% and 97.8%, respectively (Table 3). When the recognition of an embryo proper with a beating heart was used as the criterion for a positive ultrasound PD, significantly ($p < 0.001$) more false negative and less false positive ultrasound diagnoses were made, in comparison with the recognition of chorioallantoic fluid [31]. In contrast, Silva et al. [18] reported that the presence of chorioallantoic fluid in the uterine lumen and a corpus luteum alone might lead to more false positive results than when an embryo was visualized by using a 5–10 MHz linear-array transducer on Day 27 after timed AI.

Table 3. Accuracy of ultrasonographic examinations for diagnosing early pregnancy in dairy cattle by using a high frequency linear-array transducer.

Days after AI	N	Sensitivity (%)	Specificity (%)	Positive Predictive Value (%)	Negative Predictive Value (%)	References
27	1673	96.5	93.4	89.7	97.8	Silva et al. [18] (5–10 MHz transducer)
28	100	92.7	91.5	88.4	94.7	Karen et al. [32] (6–10 MHz transducer)
28–35	497	96 [a]	83	91	92	Nation et al. [16] (7.5 MHz transducer)
		97 [b]	86	92	93	
29–30	138	90.4 [c]	96.0	95.0	92.3	Szenci et al. [31] (7.5 MHz transducer)
33–34	135	96.6 [c]	98.6	98.3	97.3	
33–34	135	90.0 [d]	100	100	92.5	
30	47	92.3	97.1	92.9	97.1	Abdullach et al. [33] (6.5 MHz transducer)

[a] Observation of ≥15 mm fluid in the uterine lumen and embryonic membranes; [b] Observation of an embryo with beating heart; [c] Recognition of allantoic fluid was used as the criterion for a positive pregnancy diagnosis; [d] Recognition of an embryo proper with a beating heart was used as the criterion for a positive pregnancy diagnosis.

The reliability of the test greatly depends on the frequency of the transducer used, the skill of the operator [30,34], the criterion used for a positive pregnancy diagnosis (uterine fluid or embryo [15,27,32]; uterine fluid, embryonic membranes, or embryo [16]; amniotic fluid and/or embryo surrounded by an amniotic sac [29]; normal amount of chorioallantoic fluid, embryo, ipsilateral CL [18]; embryo with a heartbeat and corpus luteum [35]; embryo with a heartbeat, fluid-filled uterine horn, ipsilateral corpus luteum [36,37]; amniotic fluid, embryo, or embryonic heartbeat [38]; accumulation of intrauterine gestural fluid in non-pregnant cows [25], and the position of the uterus in the pelvic inlet [39]. More incorrect non-pregnancy diagnoses were made between Days 24 to 38 in cows in which the uterus was located far cranial to the pelvic inlet, in comparison with cows in which the uterus was located within or close to the pelvic inlet [39]. It is important to mention if our ultrasonographic pregnancy diagnoses are based on detection of uterine fluid in the uterus on Day 29 after timed AI, at that time, these cows can be classified 3.8 times more likely as not pregnant 74 days after timed AI than those cows diagnosed pregnant based on visualization of an embryo with a heartbeat [36].

Special attention must be paid to the diagnosis of twin pregnancy by clearly locating the two embryos because twin pregnancy loss and spontaneous twin reduction have been reported to occur up to Day 90 of gestation [40,41]. Possibilities for diagnosing twin pregnancy in the field have been reviewed recently by Szelényi et al. [42].

2.2. *Color Flow Doppler Ultrasonography*

Color flow Doppler ultrasonography (CFDU) can be used to monitor blood flow in the corpus luteum (CL) around Day 20 after AI because luteal vascularization plays a decisive role in the functional evaluation of the corpus luteum [43–46]. On the other hand, there is a good correlation between decreasing blood flow during CL regression and progesterone concentration [47,48]. One of the main advantages of CFDU is its high sensitivity (99%) and negative predictive value (98.5%) of diagnosing non-pregnant dairy cattle on Day 20 after AI, which results in few false negative diagnoses [49]. According to Dubuc et al. [50], by monitoring the blood flow in the CL on Day 21 after AI in contrast with measuring the progesterone concentration, non-pregnancy in dairy cows on Day 32 after AI can also be predicted with high accuracy (sensitivity: 99.8%, negative predictive value: 99.3%). In a recent study, Siqueira et al. [51] have found that the reduction of blood flow in the CL takes place some days before any detectable changes in CL morphology and echotexture, and therefore, determination of the blood flow and the adjusted blood flow calculated from the ratio of luteal tissue area and blood flow were the best early predictors of non-pregnancy. In contrast, luteal tissue area and echotexture were found to be inconsistent early indicators of luteolysis.

2.3. Conceptus Proteins

Trophoblastic mono- and binucleate cells from the early bovine conceptus synthesize substantial amounts of proteins. Among these, one has been described as bovine pregnancy-specific protein B (bPSPB), which enters into the maternal circulation [52]. In addition, a bPSPB-related protein, designated bovine pregnancy-associated glycoprotein (bPAG; [53]) or bPAG-1 [54], as well as pregnancy serum protein 60 kDa (PSP60; [55]) have been described. Pregnancy proteins are inactive members of the aspartic proteinase family (pepsins, cathepsins, and renins also belong to this family), and are identical in genetic nucleotide sequence [56–58]. The isolated preparations of pregnancy proteins may differ in carbohydrate and sialic acid content, which may explain their minor differences in profile and disappearance from the maternal circulation after calving or EM [32,59,60]. Because the concentrations of these proteins gradually increase during pregnancy, they are good indicators of the presence of a live embryo [61]. The pregnancy proteins have been detected in the serum of some pregnant cows as early as Days 15 to 22 [61,62] or Day 22 [55,63] after AI. Due to the delayed appearance of these proteins in the blood in some cows, their use for PD provides more accurate results from Days 28 to 30 onwards [32,57,64]. Both bPSPB and bPAG-1 have been detected in the peripheral circulation during the postpartum period 70 to 100 days after calving [61,65]. Likewise, Mialon et al. [55] reported similar residual protein concentrations for PSP60 after calving in the Charolais (mean: 107 days/91–119/), Normande (mean: 84 days/56–105/), and Holstein (mean: 88 days/63–126/) breeds. In a recent study, 56.7% and 44.9% of the false positive diagnoses based on bPSPB and bPAG-1 tests, respectively, originated from cows that were inseminated within 70 days after calving [32]. These findings indicate that the presence of bPSPB and bPAG-1 in the plasma of cows during the early stages of the postpartum period may limit their use under field conditions. If only those cows are selected for the protein tests, which are inseminated after Day 50 [66] or Day 70 after calving [67–69], post-calving interference with the residual bPSPB or bPAG-1 in the peripheral circulation during the postpartum period can be minimal.

A further limitation after late embryonic mortality (LEM) is that protein levels may remain above the threshold level, although the concentration of both proteins decreases steadily [70,71]. This is probably related to the relatively long half-life (7–8 days for bPSPB and 3–4 days for bPAG-1) in the maternal circulation after EM [70,72].

Serrano et al. [73] reported that the herd, fetal sex, milk production, lactation number, and plasma progesterone concentrations did not significantly influence the plasma PAG-1 concentration, while twin pregnancy, the use of Limousin semen and conception during the cool period significantly increased its concentrations throughout gestation. Clone pregnancies comparing with control cows may also significantly increase pregnancy protein concentrations during the whole gestation, regardless of pregnancy outcome [74].

2.3.1. In-House PAGs ELISA Tests

Originally PSPB, PAG-1, and PSP60 radioimmunoassay tests were used to detect pregnancy proteins; however, in the meantime, it turned out that there are 21 PAG family members, and PAG-1 used for pregnancy diagnosis was not the earliest pregnancy protein produced by the trophoblast [75]. After recognizing this, the first in-house sandwich ELISA protocol was developed for measuring the circulating concentration of PAGs by Green et al. [76] and Friedrich and Holtz [77]. Between Days 26 and 58 after AI, the sensitivity, specificity, positive predictive value, and negative predictive value of the in-house PAGs ELISA tests varied between 94% and 100%, 77% and 94.2%, 90.7% and 97.8%, and 91.2% and 97.1%, respectively (Table 4).

Table 4. Accuracy of in-house early-pregnancy associated glycoproteins (PAGs) ELISA tests for diagnosing early pregnancy in dairy cattle.

Days after AI	N	Type of Sample	Sensitivity (%)	Specificity (%)	Positive Predictive Value (%)	Negative Predictive Value (%)	References
26–58	169	Serum	97.8	91.2	97.8	91.2	Piechotta et al. [78] [a]
27–35	2129	Plasma	95.1	89.0	90.1	94.5	Sinedino et al. [79] [b]
31–35	209		98.7	88.1	83.7	99.1	
26–30	106	Serum *	94	77	NG	NG	Friedrich and Holtz [77]
31–35	88		95	93	NG	NG	
36–40	57		95	93	NG	NG	
>40	128		100	92	NG	NG	
27	1673	Plasma	95.4	94.2	90.7	97.1	Silva et al. [18] [b]
28	100	Plasma	95.3	88.3	NG	NG	Thompson et al. [80] [b]
30	100	Plasma	100	90.9	NG	NG	

* using a threshold of 2 ng/mL; [a] Used an in-house ELISA assay described by Friedrich and Holtz [77]; [b] Used an in-house ELISA assay described by Green et al. [76] (2005); NG: not given.

Mercadante et al. [81] found higher PAGs concentrations (in-house ELISA) in primiparous cows compared with multiparous pregnant cows (in agreement with Ricci et al. [36], while Kaya et al. [82] found similar differences between heifers and lactating cows), during the second and later breeding compared with the first breeding postpartum, in cows experiencing clinical metritis, metabolic problems and left displaced abomasum after calving compared with cows not experiencing those clinical diseases and in cases of greater milk yield, while the body condition score did not influence it. According to Dufour et al. [38], the accuracy of the PAG test (commercial milk ELISA) was not influenced by parity, the number of days, since the last breeding, and the level of milk production, while Ricci et al. [37] found a negative correlation between plasma and milk PAG concentration (commercial ELISA) and milk production.

2.3.2. Commercial ELISA Tests

There are currently several ELISA tests are available on the market, as discussed below.

BioPRYN ELISA test for the detection of pregnancy-specific protein B in the serum as early as 28 days after breeding in cattle with no interference from a previous pregnancy as early as 73 days after calving.

Between Days 26 to 58 after AI, the sensitivity, specificity, positive predictive value, and negative predictive value of the BioPRYN ELISA test varied between 93.9% and 100%, 87% and 97.1%, 92% and 99.3%, and 91.7% and 97.8%, respectively (Table 5).

Martins et al. [83] compared the basal serum PSPB concentrations determined at Day 17 after AI with Day 23 results. They found that if the difference was more than 28% at that time the sensitivity, and the specificity of the ELISA test were higher (98% and 97%), compared to 92.8% and 97% obtained when only Day 23 values were evaluated. It is also important to mention that the concentrations of PSPB on Days 23 and 28 were related to pregnancy losses between Days 28 and 35 after AI, but not to those occurring in later periods (Days 35 to 56 and Days >56 to calving).

Middleton and Pursley [84] suggested comparing the results of serum PSPB samples withdrawn on Days 17 and 24 after AI to diagnose non-pregnant cows with 100% accuracy at Day 24 after AI and to predict early pregnancy loss.

Table 5. Accuracy of commercially available ByoPRYN pregnancy ELISA tests for diagnosing early pregnancy in dairy cattle.

Days after AI	N	Type of Sample	Sensitivity (%)	Specificity (%)	Positive Predictive Value (%)	Negative Predictive Value (%)	References
23	544	Serum	92.8	97.0	NG	NG	Martins et al. [83]
23 [a]			98.0	97.0			
28			100	91.6			
24	206	Serum	100 [b]	93.6 [b]	93.3 [b]	100 [b]	Middleton and Pursley [84]
26–58	185	Serum	98.0	97.1	99.3	91.7	Piechotta et al. [78]
28	246	Plasma	93.9	95.5	94.7	94.7	Romano and Larson [85]
30	229		96.0	93.9	92.2	96.8	
35	246		97.2	93.6	92.0	97.8	
28–41	507	Serum	97.0	95.9	96.2	96.7	Breed et al. [86]
35–60	976		96.6	92.2	95.0	94.7	
30–36	1742	Serum	95.1	68.6	NG	NG	Gábor et al. [87]
30–36	1336	Serum	100	87.8	NG	NG	Howard et al. [88]

[a] A cow was considered pregnant when there was an increase in serum PSPB from basal (Day 17) to Day 23 after AI of more than 28%; [b] Only non-pregnant animals were evaluated.

DG29® Bovine Blood Pregnancy ELISA Test for the detection of specific pregnancy-related protein in serum as early as 29 days after breeding in cattle with no interference from a previous pregnancy as early as 90 days after calving. Between Days 28 and 40, after AI the sensitivity, specificity, positive predictive value, and negative predictive value of the D29 test varied between 90.2% and 100%, 66.7% and 98.3%, 91% and 97.4%, and 93.7% and 100%, respectively (Table 6).

Table 6. Accuracy of commercially available D29® bovine blood pregnancy ELISA tests for diagnosing early pregnancy in dairy cattle.

Days after AI	N	Type of Sample	Sensitivity (%)	Specificity (%)	Positive Predictive Value (%)	Negative Predictive Value (%)	References
28	100	serum	90.2	98.3	97.4	93.7	Karen et al. [31]
29–36	202	blood	99.4	66.7 *	92.6	96.3	Paré et al. [89]
30–40	212	plasma	100	81.3	91	100	Moussafir et al. [90]

* 100% in non-inseminated cows.

The IDEXX Bovine Pregnancy ELISA Test for the detection of early pregnancy-associated glycoproteins in the serum or plasma of cattle as early as 28 days after breeding in cows with no interference from a previous pregnancy as early as 60 days after calving. Between Days 25 and ~41, after AI the sensitivity, specificity, positive predictive value, and negative predictive value of the test varied between 92% and 100%, 87% and 100%, 84% and 100%, and 94.2% and 100%, respectively (Table 7).

The IDEXX Milk Pregnancy ELISA Test for the detection of pregnancy-associated glycoproteins in bovine milk from 28 days after breeding in cows with no interference from a previous pregnancy as early as 60 days after calving. Between Days 28 to ~53, after AI the sensitivity, specificity, positive predictive value, and negative predictive value of the test varied between 96% and 100%, 83% and 97.9%, 79% and 98.5%, and 96% and 100%, respectively (Table 8). While on Day ≥60, the sensitivity and the specificity of the test varied between 98.5–99.2% and 95.5–96.7%, respectively, the positive predictive value and the negative predictive value were 99.8% and 80.8%, respectively (Table 8). It is important to mention that the plasma PAG levels turned out to be approximately twice higher than the milk PAG levels [36].

Table 7. Accuracy of commercially available IDEXX bovine pregnancy ELISA tests for diagnosing early pregnancy in dairy cattle.

Days after AI	N	Type of Sample	Sensitivity (%)	Specificity (%)	Positive Predictive Value (%)	Negative Predictive Value (%)	References
25	61	plasma	92.0	91.6	88.4	94.2	Kaya et al. [82]
28	84		97.1	100	100	98.0	
32	86		97.4	91.4	90.4	97.7	
25–32	231		95.9	94.7	93.1	96.9	
28	210	serum	100	94.9	94.4	100	Akkose et al. [91]
28	320	plasma	97 *	94 **	95 **	97 *	Mayo et al. [92]
30 ± 1	116	plasma	100	88.9	93.8	100	Commun et al. [93]
30 ± 1	116	serum	100	88.6	93.8	100	
32	141	plasma	100	87	84	100	Ricci et al. [37]
41 ± 2	116	plasma	100	100	100	100	Commun et al. [93]
41 ± 2	116	serum	98.4	96.8	98.4	96.8	

* ±1, ** ±2.

Table 8. Accuracy of commercially available milk pregnancy ELISA tests for diagnosing early pregnancy in dairy cattle.

Days after AI	N	Sensitivity (%)	Specificity (%)	Positive Predictive Value (%)	Negative Predictive Value (%)	References
28	320	96 *	94 *	94 *	96 *	Mayo et al. [92]
28–45	497	99	95	NG	NG	Dufour et al. [38]
28–35	1006	99.2	93.4	NG	NG	Fosgate et al. [34]
30 ± 1	116	98.1	90.3	94.5	96.6	Commun et al. [92]
32	135	98	83	79	99	Ricci et al. [37]
33–52	119	100	97.9	98.5	100	Lawson et al. [94]
41 ± 2	116	100	92.3	96.4	100	Commun et al. [93]
53 ± 2	116	98.1	96.2	98.1	96.2	
≥60	683	99.2	95.5	99.8	80.8	LeBlanc [95]
≥60	602	98.5	96.7	NG	NG	Byrem et al. [96]

NG: not given, * ±2.

The Rapid Visual Pregnancy ELISA test has been recently developed to detect early pregnancy-associated glycoproteins in bovine whole blood, plasma, or serum as early as 28 days after breeding with no interference from a previous pregnancy as early as 60 days after calving.

This test can be run without ELISA instrumentation and read visually [97]. Between Days 25 and 45 after AI, the sensitivity, specificity, positive predictive value, and negative predictive value of the visual blood test for dairy cattle were 99.8%, 91.7%, 92.7%, and 99.7%, respectively [97]. When the IDEXX visual ELISA test was used, the sensitivity, specificity, positive predictive value, and negative predictive value of the whole blood test were 98 ± 1%, 85 ± 3%, 87 ± 3%, and 98 ± 1%, respectively [92]. The high sensitivity and negative predictive values mean that very few truly pregnant cows were misdiagnosed as not pregnant. The overall accuracy of the test was 92 ± 2% [92]. When a microtiter plate reader was used to measure the optical density for individual wells in the ELISA plate, the overall accuracy of the test became 94 ± 1% [92]. According to the agreement analysis, a very good agreement between visual and milk ELISA tests (kappa: 0.92) and between visual and serum ELISA tests (kappa: 0.97), respectively, were reported [34].

The high overall accuracy of the test (98.9%) was reported when the BioPRYN Rapid Visual Pregnancy Test® was used in 92 dairy cattle on Day 28 in heifers and on Day 30 in cows. However, it is important to mention that there were three samples evaluated as 'to be rechecked'; however, later, it was not possible to evaluate them, and therefore, these three samples were removed from the dataset [98]. A somewhat lower overall accuracy (90%) was reported when the Ubio quickVET rapid visual test was used for plasma samples between Days 30 and 40 after AI [90], and a much lower accuracy (70%) was obtained when the Fassisi® BoviPreg visual test kit was used for serum samples on Day 30 after AI [99].

2.4. Early Pregnancy Factor (EPF)

The earliest specific indicator of fertilization and the continuing presence of a viable conceptus is a serum constituent that had originally been detected in mice [100]. This substance is known as the early pregnancy factor (EPF) and has also been described in women [101], sheep [102], cattle [103], and pigs [104].

The reported and extraordinary properties of EPF include:

Early appearance (within hours) after mating or insemination

Rapid disappearance following induced death or removal of the embryos [105,106].

These factors suggest that EPF may be the most useful tool for investigating early embryonic survival or failure [106–108]. According to Laleh et al. [109], the rosette inhibition test (RIT) has the potential to distinguish pregnant from non-pregnant dairy cows in the first week of pregnancy. However, the detection of EPF is entirely dependent on a bioassay that is not practical. After identifying immunosuppressive EPF with a molecular weight of approximately 200,000 and raising polyclonal antibodies against it [110], a new diagnostic test, the early conceptus factor (ECF) test, was developed in the USA for field use; however, it cannot accurately identify conception within days or at any time before Day 21 of gestation [111–115].

After EPF, a 10.84 kDa protein, chaperonin 10 [116] having immunosuppressive and growth factor properties [117], was identified. Chaperonin 10 belongs to the family of heat shock proteins but, unlike other members of this family, EPF is detected extracellularly [117]. Further experimental work is needed for the development of an accurate on-farm diagnostic test.

2.5. Current Developments in Early Pregnancy Diagnosis

During elongation of the blastocyst, trophectoderm cells secret interferon-tau (IFNT) into the uterine cavity. With its very low levels in extrauterine tissues and in the peripheral circulation, IFNT is regarded as a signal for the maternal recognition of bovine pregnancy. INFT contributes to the maintenance of the CL by blocking prostaglandin $F_{2\alpha}$ secretion of the endometrium. Currently, there is no accurate assay for diagnosing early pregnancy based on measuring IFNT concentrations. At the same time, the determination of interferon-tau stimulated genes (ISG) has been recently suggested as an alternative method for the indirect detection of the conceptus itself. According to a recent review [14], the relative abundance of ISG in total leukocytes, peripheral blood mononuclear cells and polymorphonuclear cells in pregnant cows from Days 18 to 20 after AI may be significantly higher than in non-pregnant cows.

Another biomarker for diagnosing early pregnancy is the measurement of circulating microRNAs; however, their use is currently limited to research investigations because standardized laboratory techniques are needed to isolate and measure them [14].

Proteomics analysis of the milk identified three possible biomarkers (lactoferrin, lactotransferrin, and alpha1G) for diagnosing early pregnancy [118], while proteomics analysis of the blood identified another three genes (Myxovirus resistance: MX1 and MX2 and oligoadenylate synthetase-1: OAS1), which can be used for early pregnancy diagnosis after validation on a large number of dairy cows [119]. Glycans may also play some critical roles in both the normal function of cells and in disease; therefore, bovine pregnancy can be predicted from a glycan biomarker present in a cow's milk some 2–4 weeks earlier than by the standard method of pregnancy detection using ultrasonography [120]. Circulating nucleic acids (CNAs) or preimplantation factor (PIF) can be another biomarker for diagnosing early pregnancy in dairy cows [121,122], while Barbato et al. [123] suggested measuring PAG-2 mRNA in maternal blood cells, which can be detected earlier than the PAG-1 placental proteins in water buffalo and in other ruminants, as well.

These new branches of diagnostic sciences may contribute to finding molecules that may be exclusively related to maternal metabolic alterations during early embryonic development and to signaling for maternal recognition of pregnancy and continued survival [124], and may contribute to the development of an accurate early pregnancy diagnostic test in dairy cows.

A new technology (in-line milk analysis system, Herd Navigator) has already made possible the automatic collection of milk samples at milking robots or in the milking parlor to analyze progesterone, lactate dehydrogenase, and beta-hydroxybutyrate to detect estrus, tissue damage, and metabolic disorders, respectively [125]. According to Bruinjé and Ambrose [126], by using this new technology for early pregnancy diagnosis, a high sensitivity (>95%) could be reached from Day 27 after AI, while the specificity was somewhat lower before Day 40 after AI. After finding an accurate biomarker for early pregnancy diagnosis, its continuous measurements during milking will make it possible to diagnose pregnancy loss much earlier, meaning we can greatly contribute to increasing reproductive efficiency in our dairy herds. The importance of this technology would also be emphasized by its ability to identify pregnant and non-pregnant animals in a timely manner with no animal handling, because even a simple transrectal examination of dairy cows can lead to increased plasma and salivary cortisol concentrations and changes in heart rate, heart rate variability, and behavior that are indicative of pain [127].

3. Diagnosis of Pregnancy Losses (PL)

3.1. *Ultrasonography*

One of the advantages of UEs is that PL can be recognized by the absence of a heartbeat, the detachment of the fetal membranes, the appearance of particles in the fetal fluids, or the lack of the embryo proper [25,128]. UEs have revealed that LEM may occur in up to 23% of pregnancies [28,129]. PL (8%) diagnosed by ultrasonography in cows between Days 26 and 58 post-AI occurred at approximately Day 29 (n = 1), Day 33 (n = 3), Day 37 (n = 3), Day 40 (n = 2), Day 44 (n = 1), and Day 56 (n = 1) after AI. The exact day of occurrence of LEM/EFM could not be determined because UE was performed at intervals of 3–4 days [71].

According to Kelly et al. [130], decreased crown-rump length and progesterone concentration measured on Day 34 of gestation tended to be associated with an increased odds ratio for pregnancy loss, whereas CL perfusion and reduced blood flow of the uterine arteries evaluated by Doppler ultrasonography could not be used for predicting pregnancy loss in lactating dairy cattle.

3.2. *Pregnancy Proteins*

After diagnosing spontaneous cases of LEM by ultrasonography, both plasma bPSPB and bPAG-1 levels began to decline in most cases, while the CL continued to produce progesterone [63,71,72]. This confirms the previous observations [70,131], and demonstrates that lower progesterone concentrations are not the cause of conceptus death.

Although the concentrations of both proteins decrease steadily [70,71] after spontaneous or induced LEM/EFM, they reach the threshold level only after a relatively long half-life, namely, about 7 to 8 days for the bPSPB RIA test [70], and 3 to 4 days for the bPAG-1 RIA test [72]. Thus, they can contribute to the elevation of false positive pregnancy diagnoses on the farm. Similar results were reported by Giordano et al. [63] when inducing LEM on Day 39 of gestation and using the PSPB commercial and the PAG in-house ELISA tests. Although the threshold levels for these tests were not determined, they reached the basal levels in both tests approximately 5 to 7 days after inducing LEM. Based on commercial blood and milk ELISA tests, the threshold levels are reached approximately 7 to 14 days after pregnancy loss [37].

According to Mercadante et al. [81], reduced PAG concentrations (in-house ELISA) at Day 32 after AI may predict pregnancy loss between Days 46 and 74 of gestation. Based on positive and negative predictive value analysis, a circulating concentration of PAG (in-house ELISA) below 1.4 ng/mL in lactating dairy cattle following timed AI and 1.85 ng/mL following timed embryo transfer on Day 7 was 95% accurate in predicting LEM/EFM (between Days 31 and 59) at Day 31 of gestation [13]. It has been recently reported that cows being pregnant at Day 31 of gestation and maintaining the pregnancy until Day 59 had significantly higher circulating concentrations of PAG (commercial ELISA test) at Day 31 of gestation compared with cows that experienced LEM/EFM between Days 31 and 59 of gestation [37]. In contrast, although there was a significant difference in the PAG concentrations measured on Day 24 after AI between pregnant and non-pregnant multiparous cows while in heifers only a tendency was detected at Day 31 after AI, the circulating concentrations of PAG at Day 24 of gestation in animals that maintained pregnancy until Day 60 compared to animals that lost pregnancy between Days 31 and 60 of gestation, were not significantly different [132].

López-Gatius et al. [133] reported that low or very high plasma pregnancy-associated glycoprotein-1 (PAG-RIA) levels on Day 35 of gestation in cows were related to a subsequent pregnancy loss. Similarly, Gábor et al. [134] also found pregnancy losses in cows with high PSPB (commercial ELISA) concentrations (>1.1ng/mL) and in cows with low concentrations of both PSPB and progesterone (0.6–1.1 and <2ng/mL, respectively), however the prevalence of pregnancy loss was significantly lower in cows with high PSPB concentrations (15%) between Days 29 and 35 of gestation than in those with low concentrations (76.3%).

It has been recently reported that different PAG ELISA assays may accurately detect pregnancy; however, the ability to predict embryo survival vs. mortality during early gestation appears to be antibody-/assay-dependent [135].

In order to be able to decrease the effect of false positive diagnoses, due to pregnancy loss on the farm, it is necessary to repeat the pregnancy tests [136]. By using commercial plasma or milk ELISA tests, the optimal time for the first pregnancy diagnosis is around Day 32 after AI, when plasma and milk PAG levels are at an early peak. After this period, all pregnant cows should be retested on Day 74 after AI or later, when plasma and milk PAG levels rebound from their nadirs [37]. In contrast to the gradual increase in PAG-1 concentration (RIA) throughout gestation [61], plasma and milk PAG levels (commercial ELISA) reached a peak at Day 32 of gestation and then started to decrease to a nadir from Days 53 to 60 for the plasma PAG level and from Days 46 to 67 for the milk PAG level, followed by a gradual increase in PAG levels from Days 74 to 102 after AI [37].

The potential clinical significance of diagnosing pregnancy loss using ultrasonography or pregnancy protein tests and treating the cows with prostaglandin as soon as possible is that these measures may reduce the number of days before re-insemination [137].

4. Future Perspectives

Pregnancy diagnosis plays an essential role in decreasing days open in dairy farms; therefore, it is very important to select an accurate method for diagnosing early pregnancy, because the cost of each day open past 100 DIM may reach $4.00 [138] or €2.5 to 6.5 [139], respectively. Besides traditional pregnancy diagnoses made by rectal palpation of the uterus from 40 to 60 days after AI and measuring the serum or milk progesterone concentration between 18 to 24 days after AI [14,140,141], there are several new possibilities to diagnose early pregnancy in dairy farms; however, it is very important to emphasize that before introducing any new diagnostic test we need to make sure the accuracy of that particular test to be able to decrease the rate of iatrogenic pregnancy losses caused by prostaglandin treatment to reduce the interval to the next AI service [140] or resynchronization of the cows [142,143]. Furthermore, the new pregnancy diagnostic results must be confirmed by the old diagnostic method to decrease the negative effects of false negative diagnoses [144]. Linear-array/sector B-mode [145] and Doppler ultrasonography [14] may exceed the other diagnostic methods in the amount of information collecting from each animal during scanning, however, their uses greatly depend on the operator proficiency and availability [145].

5. Conclusions

The successful genetic selection for higher milk production caused a dramatic decline in the reproductive performance of dairy cows all over the world during the last decades. Achievement of optimum herd reproductive performance (calving interval of 12 or 13 months with the first calf born at 24 months of age) requires concentrated management activities especially during calving and during the first 100 days following calving. One of the most important management activities needed to pursue during the early postpartum period to reach or approach the optimal reproductive performance is diagnosis of early pregnancy diagnosis and embryonic mortality in dairy cows. There are several diagnostic methods available for the dairy farms such as rectal palpation, transrectal ultrasonography, chemical and hormone assays, however, transrectal B-mode and Doppler ultrasonography may exceed the other diagnostic methods in the amount of information collecting from each animal during scanning in the farm. The advantages and disadvantages of the different diagnostic methods were discussed in order to be able to select the most accurate method for the dairy.

Funding: This research received no external funding.

Conflicts of Interest: The authors declare no conflict of interest.

References

1. Roche, J.F.; Boland, M.P.; McGeady, T.A. Reproductive wastage following artificial insemination in cattle. *Vet. Rec.* **1981**, *109*, 95–97. [CrossRef]
2. Saacke, R.G.; Dalton, C.; Nadir, S.; Nebel, R.L.; Bame, H. Relationship of seminal traits and insemination time to fertilization rate and embryo quality. *Anim. Reprod. Sci.* **2000**, *60–61*, 663–677. [CrossRef]
3. Butler, W.R. Review: Effect of protein nutrition on ovarian and uterine physiology in dairy cattle. *J. Dairy Sci.* **1998**, *81*, 2533–2539. [CrossRef]
4. López-Gatius, F. Is fertility declining in dairy cattle? A retrospective study in northeastern Spain. *Theriogenology* **2003**, *60*, 89–99. [CrossRef]
5. Galon, N. Factors affecting reproductive performance in Israeli dairy herds. In *Factors Affecting Reproductive Performance in the Cow*; Szenci, O., Bajcsy, Á.C., Eds.; Hungarian Association for Buiatrics: Budapest, Hungary, 2008; pp. 28–36.
6. Bouchard, E.; Du Tremblay, D. Dairy herd production and reproduction in Quebec and Canada. In *Factors Affecting Reproductive Performance in the Cow*; Szenci, O., Bajcsy, Á.C., Eds.; Hungarian Association for Buiatrics: Budapest, Hungary, 2008; pp. 60–66.
7. Nakao, T. Declining fertility in dairy cows in Japan and efforts to improve the fertility. In *Factors Affecting Reproductive Performance in the Cow*; Szenci, O., Bajcsy, Á.C., Eds.; Hungarian Association for Buiatrics: Budapest, Hungary, 2008; pp. 38–48.
8. Wiltbank, M.C.; Baez, G.M.; Garcia-Guerra, A.; Toledo, M.Z.; Monteiro, P.L.; Melo, L.F.; Ochoa, J.C.; Santos, J.E.; Sartori, R. Pivotal periods for pregnancy loss during the first trimester of gestation in lactating dairy cows. *Theriogenology* **2016**, *86*, 239–253. [CrossRef] [PubMed]
9. Butler, W.R.; Smith, R.D. Interrelationships between energy balance on postpartum reproductive function in dairy cattle. *J. Dairy Sci.* **1989**, *72*, 767–783. [CrossRef]
10. Ujvári, V.; Niles, D.; Dinnyés, A.; Nagy, K.; Szenci, O. Retrospective evaluation of the results of embryo transfer, or insemination with sexed and traditional semen in heifers at an American Holstein-Friesian dairy farm. *Magy. Állatorv. Lapja* **2020**, *142*, 599–605.
11. Long, S. Abnormal development of the conceptus and its consequences. In *Veterinary Reproduction and Obstetrics*, 9th ed.; Noakes, D.E., Parkinson, T.J., England, G.C.W., Eds.; Saunders Elsevier: Edinburgh, UK, 2009; pp. 123–145.
12. López-Gatius, F.; Szenci, O.; Bech-Sàbat, G.; García-Ispierto, I.; Serrano, B.; Santolaria, P.; Yániz, J.L. Factors of non-infectious nature affecting late embryonic and early foetal loss in high producing dairy herds in north-eastern Spain. Literature review. *Magy. Állatorv. Lapja* **2009**, *131*, 515–531.
13. Pohler, K.G.; Pereira, M.H.C.; Lopes, F.R.; Lawrence, J.C.; Keisler, D.H.; Smith, M.F.; Vasconcelos, J.L.M.; Green, J.A. Circulating concentrations of bovine pregnancy-associated glycoproteins and late embryonic mortality in lactating dairy herds. *J. Dairy Sci.* **2016**, *99*, 1584–1594. [CrossRef]

14. Pohler, K.G.; Reese, S.T.; Franco, G.A.; Oliveira Filho, R.V.; Paiva, R.; Fernandez, L.; Melo, G.; Vasconcelos, J.L.M.; Cooke, R.; Poole, R.K. New approaches to diagnose and target reproductive failure in cattle. *Anim. Reprod.* **2020**, *17*, e20200057. [CrossRef]
15. Pieterse, M.C.; Szenci, O.; Willemse, A.H.; Bajcsy, A.C.; Dieleman, S.J.; Taverne, M.A.M. Early pregnancy diagnosis in cattle by means of linear-array real-time ultrasound scanning of the uterus and a quantitative and qualitative milk progesterone test. *Theriogenology* **1990**, *33*, 697–707. [CrossRef]
16. Nation, D.P.; Malmo, J.; Davis, G.M.; Macmillan, K.L. Accuracy of bovine pregnancy detection using transrectal ultrasonography at 28 to 35 days after insemination. *Aust. Vet. J.* **2003**, *81*, 63–65. [CrossRef]
17. Romano, J.E.; Thompson, J.A.; Forrest, D.W.; Westhusin, M.E.; Tomaszweski, M.A.; Kraemer, D.C. Early pregnancy diagnosis by transrectal ultrasonography in dairy cattle. *Theriogenology* **2006**, *66*, 1034–1041. [CrossRef] [PubMed]
18. Silva, E.; Sterry, R.A.; Kolb, D.; Mathialagan, N.; McGrath, M.F.; Ballam, J.M.; Fricke, P.M. Accuracy of a pregnancy-associated glycoprotein (PAG) ELISA to determine pregnancy status of lactating dairy cows twenty-seven days after timed AI. *J. Dairy Sci.* **2007**, *90*, 4612–4622. [CrossRef] [PubMed]
19. Taverne, M.A.M.; Szenci, O.; Szétag, J.; Piros, A. Pregnancy diagnosis in cows with linear-array real-time ultrasound technique: A preliminary note. *Vet. Q.* **1985**, *7*, 264–270. [CrossRef] [PubMed]
20. Curran, S.; Pierson, R.A.; Ginther, O.J. Ultrasonographic appearance of the bovine conceptus from days 10 through 20. *J. Am. Vet. Med. Assoc.* **1986**, *189*, 1289–1294.
21. Curran, S.; Pierson, R.A.; Ginther, O.J. Ultrasonographic appearance of the bovine conceptus from days 20 through 60. *J. Am. Vet. Med. Assoc.* **1986**, *189*, 1295–1302. [PubMed]
22. Totey, S.M.; Singh, G.; Taneja, M.; Talwar, G.P. Ultrasonography for detection of early pregnancy following embryo transfer in unknown breed of Bos indicus cows. *Theriogenology* **1991**, *35*, 487–497. [CrossRef]
23. Boyd, J.S.; Omran, S.N.; Ayliffe, T.R. Use of a high frequency transducer with real time B-mode ultrasound scanning to identify early pregnancy in cows. *Vet. Rec.* **1988**, *123*, 8–11. [CrossRef]
24. Pierson, R.; Ginther, O. Ultrasonography for detection of pregnancy and study of embryonic development in heifers. *Theriogenology* **1984**, *22*, 225–233. [CrossRef]
25. Kastelic, J.P.; Bergfelt, D.R.; Ginther, O.J. Ultrasonic detection of the conceptus and characterisation of intrauterine fluid on days 10 to 22 in heifers. *Theriogenology* **1991**, *35*, 569–581. [CrossRef]
26. Filteau, V.; DesCoteaux, L. Valeur predictive de l'utilisation de l'appareil echographique pour le diagnostic precoce de la gestation chez la vache laitiere. *Médecin Vétérinaire Québec* **1998**, *28*, 81–85.
27. Szenci, O.; Piros, A.; Kovács, L. Early bovine pregnancy diagnosis by a battery operated portable ultrasonic scanner the Ultra-Scan. In Proceedings of the 16th World Buiatrics Congress, Salvador, Brazil, 13–17 August 1990; pp. 219–223.
28. Hanzen, C.; Laurent, Y. Application de l'echographie bidimensionnelle au diagnostic de gestation et l'evaluation de l'incidence de la mortalit, embryonnaire dans l'espece bovine. *Ann. Med. Vet.* **1991**, *134*, 481–487.
29. Munoz del Real, L.M.; Valencia, G.L.; Corral, A.; Renteria, T. Evaluation of ultrasound for pregnancy diagnosis between 20 and 40 days post insemination in dairy cows. *J. Anim. Vet. Adv.* **2006**, *5*, 949–957.
30. Badtram, G.A.; Gaines, J.D.; Thomas, C.B.; Bosu, W.T.K. Factors influencing the accuracy of early pregnancy detection in cattle by real-time ultrasound scanning of the uterus. *Theriogenology* **1991**, *35*, 1153–1167. [CrossRef]
31. Szenci, O.; Beckers, J.F.; Humblot, P.; Sulon, J.; Sasser, G.; Taverne, M.A.M.; Varga, J.; Baltusen, R.; Schekk, G. Comparison of ultrasonography bovine pregnancy-specific protein B, and bovine pregnancy-associated glycoprotein 1 tests for pregnancy detection in dairy cows. *Theriogenology* **1998**, *50*, 77–88. [CrossRef]
32. Karen, A.; De Sousa, N.M.; Beckers, J.F.; Bajcsy, Á.C.; Tibold, J.; Mádl, I.; Szenci, O. Comparison of a commercial bovine pregnancy-associated glycoprotein ELISA test and a pregnancy-associated glycoprotein radiomimmunoassay test for early pregnancy diagnosis in dairy cattle. *Anim. Reprod. Sci.* **2015**, *159*, 31–37. [CrossRef] [PubMed]
33. Abdullach, M.; Mohanty, T.K.; Kumaresan, A.; Mohanty, A.K.; Madkar, A.R.; Baithalu, R.K.; Bhakat, M. Early pregnancy diagnosis in dairy cattle: Economic importance and accuracy of ultrasonography. *Adv. Anim. Vet. Sci.* **2014**, *2*, 464–467. [CrossRef]
34. Fosgate, G.T.; Motimele, B.; Ganswindt, A.; Irons, P.C. Bayesian latent class model to estimate the accuracy of pregnancy diagnosis by transrectal ultrasonography and laboratory detection of pregnancy-associated glycoproteins in dairy cows. *Prev. Vet. Med.* **2017**, *145*, 100–109. [CrossRef]
35. Szenci, O.; Varga, J.; Bajcsy, A.C. Role of early pregnancy diagnosis be means of ultrasonography in improving reproductive efficiency in a dairy herd: A retrospective study. *Bov. Pract.* **1999**, *33*, 67–69.
36. Giordano, J.O.; Fricke, P.M. Accuracy of pregnancy diagnosis outcomes using transrectal ultrasonography 29 days after artificial insemination in lactating dairy cows. *J. Dairy Sci.* **2012**, *95* (Suppl. 2), 75.
37. Ricci, A.; Carvalho, P.D.; Amundson, M.C.; Fourdraine, R.H.; Vincenti, L.; Fricke, P.M. Factors associated with pregnancy-associated glycoprotein (PAG) levels in plasma and milk of Holstein cows during early pregnancy and their effect on the accuracy of pregnancy diagnosis. *J. Dairy Sci.* **2015**, *98*, 2502–2514. [CrossRef] [PubMed]
38. Dufour, S.; Durocher, J.; Dubuc, J.; Dendukuri, N.; Hassan, S.; Buczinski, S. Bayesian estimation of sensitivity and specificity of a milk pregnancy-associated glycoprotein-based ELISA and of transrectal ultrasonographic exam for diagnosis of pregnancy at 28-45 days following breeding in dairy cows. *Prev. Vet. Med.* **2017**, *140*, 122–133. [CrossRef] [PubMed]
39. Szenci, O.; Gyulai, G.; Nagy, P.; Kovács, L.; Varga, J.; Taverne, M.A.M. Effect of uterus position relative to the pelvic inlet on the accuracy of early bovine pregnancy diagnosis by means of ultrasonography. *Vet. Q.* **1995**, *17*, 37–39. [CrossRef] [PubMed]

40. López-Gatius, F.; Hunter, R.H.F. Spontaneous reduction of advanced twin embryos: Its occurrence and clinical relevance in dairy cattle. *Theriogenology* **2005**, *63*, 118–125. [CrossRef]
41. Szelényi, Z.; Szenci, O.; Répási, A.; de Sousa, M.N.; Beckers, J.F. Accuracy of diagnosing double corpora lutea and twin pregnancy by measuring serum progesterone and bovine pregnancy-associated glycoprotein 1 in the first trimester of gestation in dairy cows. *Theriogenology* **2015**, *84*, 76–81. [CrossRef]
42. Szelényi, Z.; Szenci, O.; Kovács, L.; García-Ispierto, I. Practical aspects of twin pregnancy diagnosis in cattle. *Animals* **2021**, *11*, 1061. [CrossRef]
43. Herzog, K.; Bollwein, H. Application of Doppler ultrasonography in cattle reproduction. *Reprod. Domest. Anim.* **2007**, *42* (Suppl. 2), 51–58. [CrossRef]
44. Matsui, M.; Miyamoto, A. Evaluation of ovarian blood flow by colour Doppler ultrasound: Practical use for reproductive management in the cow. *Vet. J.* **2009**, *181*, 232–240. [CrossRef] [PubMed]
45. Lüttgenau, J.; Bollwein, H. Evaluation of bovine luteal blood flow by using color Doppler ultrasonography. *Reprod. Biol.* **2014**, *14*, 103–109. [CrossRef]
46. Bollwein, H.; Heppelmann, M.; Lüttgenau, J. Ultrasonographic Doppler use for female reproduction management. *Vet. Clin. Food Anim.* **2016**, *32*, 149–164. [CrossRef] [PubMed]
47. Herzog, K.; Brockhan-Ludemann, M.; Kasle, M.; Beindorff, N.; Paul, V.; Niemann, H.; Bollwein, H. Luteal blood flow is a more appropriate indicator for luteal function during the bovine estrous cycle than luteal size. *Theriogenology* **2010**, *73*, 691–697. [CrossRef] [PubMed]
48. Herzog, K.; Voss, C.; Kastelic, J.P.; Beindorff, N.; Paul, V.; Niemann, H.; Bollwein, H. Luteal blood flow increases during the first three weeks of pregnancy in lactating dairy cows. *Theriogenology* **2011**, *75*, 549–554. [CrossRef] [PubMed]
49. Siqueira, L.G.; Areas, V.S.; Ghetti, A.M.; Fonseca, J.F.; Palhao, M.P.; Fernandes, C.A.; Viana, J.H. Color Doppler flow imaging for the early detection of nonpregnant cattle at 20 days after timed artificial insemination. *J. Dairy Sci.* **2013**, *96*, 6461–6472. [CrossRef]
50. Dubuc, J.; Houle, J.; Rousseau, M.; Roy, J.-P.; Buczinski, S. Short communication: Accuracy of corpus luteum color flow Doppler ultrasonography to diagnose nonpregnancy in dairy cows on day 21 after insemination. *J. Dairy Sci.* **2020**, *103*, 2019–2023. [CrossRef]
51. Siqueira, L.G.; Arashiro, E.K.; Ghetti, A.M.; Souza, E.D.; Feres, L.F.; Pfeifer, L.F.; Fonseca, J.F.; Viana, J.H. Vascular and morphological features of the corpus luteum 12 to 20 days after timed artificial insemination in dairy cattle. *J. Dairy Sci.* **2019**, *102*, 5612–5622. [CrossRef] [PubMed]
52. Butler, J.E.; Hamilton, W.C.; Sasser, R.G.; Ruder, C.A.; Hass, G.M.; Williams, R.J. Detection and partial characterization of two bovine pregnancy-specific proteins. *Biol. Reprod.* **1982**, *26*, 925–933. [CrossRef]
53. Zoli, A.P.; Beckers, J.F.; Wouters-Ballman, P.; Closset, J.; Falmagne, P.; Ectors, F. Purification and characterization of a bovine pregnancy associated glycoprotein. *Biol. Reprod.* **1991**, *45*, 1–10. [CrossRef]
54. Xie, S.; Low, B.G.; Nagel, R.J.; Kramer, K.K.; Anthony, R.V.; Zoli, A.P.; Beckers, J.F.; Roberts, R.M. Identification of the major pregnancy-specific antigens of cattle and sheep as inactive members of the aspartic proteinase family. *Proc. Natl. Acad. Sci. USA* **1991**, *88*, 10247–10251. [CrossRef]
55. Mialon, M.M.; Camous, S.; Renand, G.; Martal, J.; Ménissier, F. Peripheral concentrations of a 60-kDa pregnancy serum protein during gestation and after calving and in relationship to embryonic mortality in cattle. *Reprod. Nutr. Dev.* **1993**, *33*, 269–282. [CrossRef]
56. Lynch, R.A.; Alexander, B.M.; Sasser, R.G. The cloning and expression of the bovine pregnancy-specific protein B (bPSPB) gene. *Biol. Reprod.* **1992**, *46* (Suppl. 1), 72.
57. Vasquez, M.I.; Horta, A.E.M.; Marques, C.C.; Sasser, R.G.; Humblot, P. Levels of bPSPB throughout single and twin pregnancies after AI or transfer of IVM/IVF cattle embryos. *Anim. Reprod. Sci.* **1995**, *38*, 279–289. [CrossRef]
58. Xie, S.; Green, J.; Beckers, J.F.; Roberts, R.M. The gene encoding bovine pregnancy-associated glycoprotein-1, an inactive member of the aspartic proteinase family. *Gene* **1995**, *159*, 193–197. [CrossRef]
59. Beckers, J.F.; Zarrouk, A.; Batalha, E.S.; Garbayo, J.M.; Mester, L.; Szenci, O. Endocrinology of pregnancy: Chorionic somatomammotropins and pregnancy-associated glycoproteins: Review. *Acta Vet. Hung.* **1998**, *46*, 175–189.
60. Beckers, J.F.; Drion, P.V.; Garbayo, J.M.; Perényi, Z.; Zarrouk, A.; Sulon, J.; Remy, B.; Szenci, O. Pregnancy associated glycoproteins in ruminants: Inactive members of the aspartic proteinase family. *Acta Vet. Hung.* **1999**, *47*, 461–469.
61. Zoli, A.P.; Guilbault, L.A.; Delahaut, P.; Ortiz, W.B.; Beckers, J.F. Radioimmunoassay of a bovine pregnancy-associated glycoprotein in serum: Its application for pregnancy diagnosis. *Biol. Reprod.* **1992**, *46*, 83–92. [CrossRef]
62. Sasser, R.G.; Ruder, C.A.; Ivani, K.A.; Butler, J.E.; Hamilton, W.C. Detection of pregnancy by radioimmunoassay of a novel pregnancy-specific protein in serum of cows and a profile of serum concentrations during gestation. *Biol. Reprod.* **1986**, *35*, 936–942. [CrossRef] [PubMed]
63. Giordano, J.O.; Guenther, J.N.; Lopes, G.; Fricke, P.M. Changes in serum pregnancy-associated glycoprotein, pregnancy-specific protein B, and progesterone concentrations before and after induction of pregnancy loss in lactating dairy cows. *J. Dairy Sci.* **2012**, *95*, 683–697. [CrossRef]
64. Humblot, P.; Jeanguyot, N.; Ruder, C.A.; Leriche, I.; Thibier, M.; Sasser, R.G. Accuracy of pregnancy diagnosis by PSPB RIA in the plasma of dairy cows 28 days after AI. In Proceedings of the 11th International Congress on Animal Reproduction and Artificial Insemination, Dublin, Ireland, 26–30 June 1988; Volume 2, p. 3.

65. Kirakofe, G.H.; Wright, J.M.; Schalles, R.R.; Ruder, C.A.; Paris, S.; Sasser, R.G. Pregnancy-specific protein B in serum of postpartum beef cows. *J. Anim. Sci.* **1993**, *71*, 2199–2205. [CrossRef]
66. Sasser, R.G.; Alexander, B.M.; Ruder, C.A. Pregnancy detection in postpartum cows by measurement of pregnancy specific protein B (PSPB). *J. Anim. Sci.* **1991**, *69*, 660.
67. Ruder, C.A.; Sasser, R.G. Source of bovine pregnancy-specific protein B (bPSPB) during the postpartum period and estimation of half-life of bPSPB. *J. Anim. Sci.* **1986**, *63*, 335.
68. Humblot, P.; Camous, S.; Martal, J.; Charlery, J.; Jeanguyot, N.; Thibier, M.; Sasser, R.G. Pregnancy specific protein B, progesterone concentrations and embryonic mortality during early pregnancy in dairy cows. *J. Reprod. Fertil.* **1988**, *83*, 215–223. [CrossRef]
69. Humblot, P.; Camous, S.; Martal, J.; Charlery, J.; Jeanguyot, N.; Thibier, M.; Sasser, R.G. Diagnosis of pregnancy by radioimmunoassay of a pregnancy-specific protein in the plasma of dairy cows. *Theriogenology* **1988**, *30*, 257–267. [CrossRef]
70. Semambo, D.K.N.; Eckersall, P.D.; Sasser, R.G.; Ayliffe, T.R. Pregnancy-specific protein B and progesterone in monitoring viability of the embryo in early pregnancy in the cow after experimental infection with Actinomyces pyogenes. *Theriogenology* **1992**, *37*, 741–748. [CrossRef]
71. Szenci, O.; Humblot, P.; Beckers, J.F.; Sasser, G.; Sulon, J.; Baltusen, R.; Varga, J.; Bajcsy, Á.C.; Taverne, M.A.M. Plasma profiles of progesterone and conceptus proteins in cows with spontaneous embryonic/foetal mortality as diagnosed by ultrasonography. *Vet. J.* **2000**, *159*, 287–290. [CrossRef]
72. Szenci, O.; Beckers, J.F.; Sulon, J.; Bevers, M.M.; Börzsönyi, L.; Fodor, L.; Kovács, F.; Taverne, M.A.M. Effect of induction of late embryonic mortality on plasma profiles of pregnancy associated glycoproteins in heifers. *Vet. J.* **2003**, *165*, 307–313. [CrossRef]
73. Serrano, B.; López-Gatius, F.; Santolaria, P.; Almería, S.; García-Ispierto, I.; Bech-Sabat, G.; Sulon, J.; de Sousa, N.M.; Beckers, J.F.; Yániz, J.L. Factors affecting plasma pregnancy-associated glycoprotein 1 concentrations throughout gestation in high-producing dairy cows. *Reprod. Domest. Anim.* **2009**, *44*, 600–605. [CrossRef]
74. Chavatte-Palmer, P.; de Sous, N.; Laigre, P.; Camous, S.; Ponter, A.A.; Beckers, J.F.; Heyman, Y. Ultrasound fetal measurements and pregnancy associated glycoprotein secretion in early pregnancy in cattle recipients carrying somatic clones. *Theriogenology* **2006**, *66*, 829–840. [CrossRef]
75. Green, J.A.; Xie, S.; Quan, X.; Bao, B.; Gan, X.; Mathialagan, N.; Beckers, J.F.; Roberts, R.M. Pregnancy-associated bovine and ovine glycoproteins exhibit spatially and temporally distinct expression patterns during pregnancy. *Biol. Reprod.* **2000**, *62*, 1624–1631. [CrossRef] [PubMed]
76. Green, J.A.; Parks, T.E.; Avalle, M.P.; Telugu, B.P.; McLain, A.L.; Peterson, A.J.; McMillan, W.; Mathialagan, N.; Hook, R.R.; Xie, S.; et al. The establishment of an ELISA for the detection of pregnancy-associated glycoproteins (PAGs) in the serum of pregnant cows and heifers. *Theriogenology* **2005**, *63*, 1481–1503. [CrossRef] [PubMed]
77. Friedrich, M.; Holtz, W. Establishment of an ELISA for measuring bovine pregnancy-associated glycoprotein in serum or milk and its application for early pregnancy detection. *Reprod. Domest. Anim.* **2010**, *45*, 142–146. [CrossRef] [PubMed]
78. Piechotta, M.; Bollwein, J.; Friedrich, M.; Heilkenbrinker, T.; Passavant, C.; Branen, J.; Sasser, G.; Hoedemaker, M.; Bollwein, H. Comparison of commercial ELISA blood tests for early pregnancy detection in dairy cows. *J. Reprod. Dev.* **2011**, *57*, 72–75. [CrossRef]
79. Sinedino, L.D.P.; Lima, F.S.; Bisinotto, R.S.; Cerri, R.L.A.; Santos, J.E.P. Effect of early or late resynchronization based on different methods of pregnancy diagnosis on reproductive performance of dairy cows. *J. Dairy. Sci.* **2014**, *97*, 4932–4941. [CrossRef]
80. Thompson, I.; Cerri, R.; Kim, I.; Green, J.; Santos, J.; Thatcher, W. Effects of resynchronization programs on pregnancy per artificial insemination, progesterone, and pregnancy-associated glycoproteins in plasma of lactating dairy cows. *J. Dairy Sci.* **2010**, *93*, 4006–4018. [CrossRef] [PubMed]
81. Mercadante, P.M.; Ribeiro, E.S.; Risco, C.; Ealy, A.D. Associations between pregnancy-associated glycoproteins and pregnancy outcomes, milk yield, parity, and clinical diseases in high-producing dairy cows. *J. Dairy Sci.* **2016**, *99*, 3031–3040. [CrossRef]
82. Kaya, M.S.; Köse, M.; Bozkaya, F.; Mutlu, H.; Uçar, E.H.; Atli, M.O. Early pregnancy diagnosis using a commercial ELISA test based on pregnancy-associated glycoproteins in Holstein-Friesian heifers and lactating cows. *Turk. J. Vet. Anim. Sci.* **2016**, *40*, 694–699. [CrossRef]
83. Martins, J.; Wang, D.; Mu, N.; Rossi, G.; Martini, A.; Martins, V.; Pursley, J. Level of circulating concentrations of progesterone during ovulatory follicle development affects timing of pregnancy loss in lactating dairy cows. *J. Dairy Sci.* **2018**, *101*, 10505–10525. [CrossRef]
84. Middleton, E.; Pursley, J. Blood samples before and after embryonic attachment accurately determine non-pregnant lactating dairy cows at 24 d post-artificial insemination using a commercially available assay for pregnancy-specific protein B. *J. Dairy Sci.* **2019**, *102*, 7570–7575. [CrossRef] [PubMed]
85. Romano, J.E.; Larson, J.E. Accuracy of pregnancy specific protein-B test for early pregnancy diagnosis in dairy cattle. *Theriogenology* **2010**, *74*, 932–939. [CrossRef]
86. Breed, M.W.; Guard, C.L.; White, M.E.; Smith, M.C.; Warnick, L.D. Comparison of pregnancy diagnosis in dairy cattle by use of a commercial ELISA and palpation per rectum. *J. Am. Vet. Med. Assoc.* **2009**, *235*, 292–298. [CrossRef]
87. Gábor, G.; Tóth, F.; Sasser, G.; Szász, F.; Bárány, I.; Wölfling, A.; Völgyi-Csík, J. Ways of decrease the period between calvings in dairy cows. Early pregnancy detection by Biopryn ELISA-test. *Magy. Állatorv. Lapja* **2004**, *126*, 459–464.
88. Howard, J.; Gábor, G.; Gray, T.; Passavant, C.; Ahmadzadeh, A.; Sasser, N.; Pals, D.; Sasser, G. BioPRYN, a blood-based pregnancy test for managing breeding and pregnancy in cattle. *J. Anim. Sci.* **2007**, *58*, 295–298.

89. Paré, J.; Audet-Grenier, M.H.; Rouillier, P.; Sirard, M.A. Evaluation of the DG29 test for early detection of pregnancy in cattle. *Can. Vet. J.* **2008**, *49*, 1119–1121. [PubMed]
90. Moussafir, Z.; Allaia, L.; El Khalila, K.; Essamadib, A.; El Amiri, B. Could a bovine pregnancy rapid test be an alternative to a commercial pregnancy-associated glycoprotein ELISA test in dairy cattle? *Anim. Reprod. Sci.* **2018**, *192*, 78–83. [CrossRef]
91. Akkose, M.; Cebi-Sen, C.; Kirmit, A.; Korkmaz, O.; Yazlik, M.O.; Gocen, M.; Kuru, M.; Akçay, E. Pregnancy-associated glycoproteins (PAGs) and progesterone concentrations in Holstein heifers following two methods of estrus syncronization. *Veterinaria México OA* **2019**, *2019*, 646. [CrossRef]
92. Mayo, L.M.; Moore, S.G.; Poock, S.E.; Silvia, W.J.; Lucy, M.C. Technical note: Validation of a chemical pregnancy test in dairy cows that uses whole blood, shortened incubation times, and visual readout. *J. Dairy Sci.* **2016**, *99*, 7634–7641. [CrossRef]
93. Commun, L.; Velek, K.; Barbry, J.-B.; Pun, S.; Rice, A.; Mestek, A.; Egli, C.; Leterme, S. Detection of pregnancy-associated glycoproteins in milk and blood as a test for early pregnancy in dairy cows. *J. Vet. Diagn. Investig.* **2016**, *28*, 207–213. [CrossRef] [PubMed]
94. Lawson, B.C.; Shahzad, A.H.; Dolecheck, K.A.; Martel, E.L.; Velek, K.A.; Ray, D.L.; Lawrence, J.C.; Silvia, W.J. A pregnancy detection assay using milk samples: Evaluation and considerations. *J. Dairy Sci.* **2014**, *97*, 6316–6325. [CrossRef]
95. LeBlanc, S.J. Short communication: Field evaluation of a pregnancy confirmation test using milk samples in dairy cows. *J. Dairy Sci.* **2013**, *96*, 2345–2348. [CrossRef]
96. Byrem, T.M.; Velek, K.; Pearse, H.L. The Detection of Pregnancy Associated Glycoproteins (PAG) in Routine Milk Recording Samples as an Indicator of Pregnancy in Dairy Cattle. 2012. Available online: https://www.icar.org/wp-content/uploads/2015/09/Byrem.pdf (accessed on 15 September 2015).
97. Green, J.C.; Volkmann, D.H.; Poock, S.E.; McGrath, M.F.; Ehrhardt, M.; Moseley, A.E.; Lucy, M.C. Technical note: A rapid enzyme-linked immunosorbent assay blood test for pregnancy in dairy and beef cattle. *J. Dairy Sci.* **2009**, *92*, 3819–3824. [CrossRef]
98. Hameed, O.A.; Mustafa, M.; Madi, N.; Tate, W. Evaluation of a rapid visual pregnancy (Bio-RPD) test for detection of pregnancy specific protein B (PSPB) in cow serum. *J. Anim. Sci. Livest. Prod.* **2020**, *4*, No.3:3.
99. Karakus, O.; Kadar, C.; Kuru, M.; Kaya, S.; Demir, M.C.; Arı, U.C.; Zonturlu, A.K.; Öztürkler, Y. Efficacy of a commercial test kit to determine early pregnancy in cows using whole blood and blood serum. *Trop. Anim. Health Prod.* **2021**, *53*, 28. [CrossRef]
100. Morton, H.; Hegh, V.; Clunie, G.J.A. Immunosuppression detected in pregnant mice by rosette inhibition. *Nature* **1974**, *249*, 459–460. [CrossRef] [PubMed]
101. Morton, H.; Rolfe, B.E.; Anderson, M.; Morrison, J. An early pregnancy factor detected in human serum by rosette inhibition test. *Lancet* **1977**, *1*, 394–397. [CrossRef]
102. Morton, H.; Clunie, G.J.A.; Shaw, F.D. A test for early pregnancy in sheep. *Res. Vet. Sci.* **1979**, *26*, 261–262. [CrossRef]
103. Nancarrow, C.D.; Wallace, A.L.L.; Grewal, A.S. The early pregnancy factor of sheep and cattle. *J. Reprod. Fertil.* **1981**, *30*, 191–197. [CrossRef]
104. Paisley, L.G.; Davis, W.C.; Anderson, P.B.; Mickelsen, W.D. Detection of early pregnancy factor in swine: A need for dialogue. *Theriogenology* **1982**, *18*, 393–401. [CrossRef]
105. Rolfe, B.; Cavanagh, A.; Forde, C.; Bastin, F.; Chen, C.; Morton, H. Modified rosette inhibition test with mouse lymphocytes for detection of early pregnancy factor in human pregnancy serum. *J. Immunol. Methods* **1984**, *70*, 1–11. [CrossRef]
106. Yamazaki, S.; Kawahata, K.; Goto, T.; Takahashi, J.; Yasuda, Y. EPF as a marker for early embryonic losses in repeat breeders. *J. Reprod. Dev.* **1995**, *41*, 129–132. [CrossRef]
107. Rolfe, B. Detection of foetal westage. *Fertil. Steril.* **1982**, *37*, 655–660. [CrossRef]
108. Koch, E. Early pregnancy factor: Its significance as an indicator of fertilization and embryonic mortality. In *Embryonic Mortality in Farm Animals*; Screenan, J.M., Diskin, M.G., Eds.; Martinus Nijhoff Publishers: Dordrecht, The Netherlands, 1986; pp. 74–92.
109. Laleh, V.G.; Laleh, R.G.; Pirany, N.; Ahrabi, M.M. Measurement of EPF for detection of cow pregnancy using rosette inhibition test. *Theriogenology* **2008**, *70*, 105–107. [CrossRef] [PubMed]
110. Threlfall, W.R. Immunosuppressive early pregnancy factor (ISEPF) determination for pregnancy diagnosis in dairy cows. *Theriogenology* **1994**, *41*, 317. [CrossRef]
111. Adams, C.S.; Jardon, P.W. Evaluation of the early conception factor tests in cows 3–7 days post-breeding. *Proc. Am. Assoc. Bov. Pract.* **1999**, *32*, 240–241. [CrossRef]
112. DesCoteaux, L.; Carrière, P.D.; Bigras-Poulin, M. Evaluation of the Early Conception Factor (ECF) dipstick test in dairy cows between days 11 and 15 post-breeding. *Bov. Pract.* **2000**, *34*, 87–91.
113. Cordoba, M.C.; Sartori, R.; Fricke, P.M. Assessment of a commercially available early conception factor (ECF) test for determining pregnancy status of dairy cattle. *J. Dairy Sci.* **2001**, *84*, 1884–1889. [CrossRef]
114. Gandy, B.; Tucker, W.; Ryan, P.; Williams, A.; Tucker, A.; Moore, A.; Godfrey, R.; Willard, S. Evaluation of the early conception factor (ECF™) test for the detection of nonpregnancy in dairy cattle. *Theriogenology* **2001**, *56*, 637–647. [CrossRef]
115. Whisnant, C.S.; Pagels, L.A.; Daves, M.G. Case study: Effectiveness of a commercial early conception factor test for use in cattle. *Prof. Anim. Sci.* **2001**, *17*, 51–53. [CrossRef]
116. Cavanagh, A.C. Identification of early pregnancy factor as chaperonin 10: Implications for understanding its role. *Rev. Reprod.* **1996**, *1*, 28–32. [CrossRef]
117. Morton, H. Early pregnancy factor: An extracellular chaperonin 10 homologue. *Immunol. Cell Biol.* **1998**, *76*, 483–496. [CrossRef]

118. Han, R.X.; Kim, H.R.; Diao, Y.F.; Lee, M.G.; il Jin, D. Detection of early pregnancy-specific proteins in Holstein milk. *J. Proteom.* **2012**, *75*, 3221–3229. [CrossRef] [PubMed]
119. Panda, B.S.K.; Mohapatrab, S.K.; Chaudharya, D.; Alhussienc, M.N.; Kapilab, R.; Danga, A.K. Proteomics and transcriptomics study reveals the utility of ISGs as novel molecules for early pregnancy diagnosis in dairy cows. *J. Reprod. Immunol.* **2020**, *140*, 103148. [CrossRef] [PubMed]
120. Taron, C.H.; Rudd, P.M. Glycomics: A Rapidly Evolving Field with a Sweet Future. Available online: https://www.researchgate.net/publication/311058088_Glycomics_a_rapidly_evolving_field_with_a_sweet_future/citations (accessed on 15 January 2016).
121. Mayer, J.; Soller, J.T.; Beck, J.; Purwins, V.; Wemheuer, W.; Schutz, E.; Brenig, B. Early pregnancy diagnosis in dairy cows using circulating nucleic acids. *Theriogenology* **2013**, *79*, 173–179. [CrossRef]
122. Ramu, S.; Stamatkin, C.; Timms, L.; Ruble, M.; Roussev, R.G.; Barnea, E.R. PreImplantation factor (PIF) detection in maternal circulation in early pregnancy correlates with live birth (bovine model). *Reprod. Biol. Endocrin.* **2013**, *11*, 105. [CrossRef] [PubMed]
123. Barbato, O.; Guelfi, G.; Barile, V.L.; Menchetti, L.; Tortiello, C.; Canali, C.; Brecchia, G.; Traina, G.; Beckers, J.F.; Sousa, N.M. Using real-time PCR to identify pregnancy-associated glycoprotein 2 (PAG-2) in water buffalo (Bubalus bubalis) blood in early pregnany. *Theriogenology* **2017**, *89*, 106–113. [CrossRef] [PubMed]
124. Balhara, A.K.; Gupta, M.; Singh, S.; Ashok, K.; Mohanty, A.K.; Singh, I. Early pregnancy diagnosis in bovines: Current status and future directions. *Sci. World J.* 2013. [CrossRef] [PubMed]
125. Saint-Dizier, M.; Chastant-Maillard, S. Potential of connected devices to optimize cattle reproduction. *Theriogenology* **2018**, *112*, 53–56. [CrossRef] [PubMed]
126. Bruinjé, T.C.; Ambrose, D.J. Technical note: Validation of an automated in-line milk progesterone analysis system to diagnose pregnancy in dairy cattle. *J. Dairy Sci.* **2019**, *102*, 3615–3621. [CrossRef]
127. Kovács, L.; Kézér, F.L.; Kulcsár-Huszenicza, M.; Ruff, F.; Szenci, O.; Jurkovich, V. Hypothalamic-pituitary-adrenal and cardiac autonomic responses to transrectal examination differ with behavioral reactivity in dairy cows. *J. Dairy Sci.* **2016**, *99*, 7444–7457. [CrossRef] [PubMed]
128. Semambo, D.K.N.; Boyd, J.S.; Taylor, D.J.; Ayliffe, T.R.; Omran, S.N. Ultrasonographic study of early embryonic loss induced by Actinomyces pyogenes in cattle. *Vet. Rec.* **1992**, *131*, 7–12. [CrossRef]
129. Chaffaux, S.; Reddy, G.N.S.; Valon, F.; Thibier, M. Transrectal real-time ultrasound scanning for diagnosing pregnancy and for monitoring embryonic mortality in dairy cattle. *Anim. Reprod. Sci.* **1986**, *10*, 193–200. [CrossRef]
130. Kelly, D.E.; Galvao, K.N.; Mortensen, C.J.; Risco, C.A.; Ealy, A.D. Using Doppler ultrasonography on day 34 of pregnancy to predict pregnancy loss in lactating dairy cattle. *J. Dairy Sci.* **2017**, *100*, 3266–3271. [CrossRef]
131. Kastelic, J.P.; Northey, D.L.; Ginther, O.J. Spontaneous embryonic death on days 20 to 40 in heifers. *Theriogenology* **1991**, *35*, 361–363. [CrossRef]
132. Reese, S.T.; Pereira, M.H.C.; Edwards, J.L.; Vasconcelos, J.L.M.; Pohler, K.G. Pregnancy diagnosis in cattle using pregnancy associated glycoprotein concentration in circulation at day 24 of gestation. *Theriogenology* **2018**, *106*, 178–185. [CrossRef]
133. López-Gatius, F.; Hunter, R.H.; Garbayo, J.M.; Santolaria, P.; Yaniz, J.; Serrano, B.; Ayad, A.; De Sousa, N.M.; Beckers, J.F. Plasma concentrations of pregnancy-associated glycoprotein-1 (PAG-1) in high producing dairy cows suffering early fetal loss during the warm season. *Theriogenology* **2007**, *67*, 1324–1330. [CrossRef] [PubMed]
134. Gábor, G.; Kastelic, J.P.; Abonyi-Tóth, Z.; Gábor, P.; Endrődi, T.; Balogh, O.G. Pregnancy loss in dairy cattle: Relationship of ultrasound, blood pregnancy-specific protein B, progesterone and production variables. *Reprod. Domest. Anim.* **2016**, *51*, 467–473. [CrossRef]
135. Gatea, A.O.; Smith, M.F.; Pohler, K.G.; Egen, T.; Pereira, M.H.C.; Vasconselos, J.L.M.; Lawrence, J.C.; Green, J.A. The ability to predict pregnancy loss in cattle with ELISAs that detect pregnancy associated glycoproteins is antibody dependent. *Theriogenology* **2018**, *108*, 269–276. [CrossRef] [PubMed]
136. López-Gatius, F.; García-Ispierto, I. Ultrasound and endocrine findings that help to assess the risk of late embryo/early foetal loss by non-infectious cause in dairy cattle. *Reprod. Domest. Anim.* **2010**, *45* (Suppl. 3), 15–24. [CrossRef]
137. Baxter, S.J.; Ward, W.R. Incidence of fetal loss in dairy cattle after pregnancy diagnosis using an ultrasound scanner. *Vet. Rec.* **1997**, *140*, 287–288. [CrossRef] [PubMed]
138. Des Coteaux, L.; Fetrow, J. Does it pay to use an ultrasound machine for early pregnancy diagnosis in dairy cows? In Proceedings of the 31st AABP Annual Meeting, Spokane, WA, USA, 24–27 September 1998; pp. 172–174.
139. Fodor, I.; Ózsvári, L. Early pregnancy diagnosis on large dairy farms and its role in improving profitability: Literature review. *AWETH* **2018**, *14*, 22–36.
140. Fricke, P.M.; Ricci, A.; Giordano, J.O.; Carvalho, P.D. Methods for and implementation of pregnancy diagnosis in dairy cows. *Vet. Clin. Food. Anim.* **2016**, *32*, 165–180. [CrossRef]
141. Reese, S.T.; Pereira, M.C.; Vasconcelos, J.L.M.; Smith, M.F.; Green, J.A.; Geary, T.W.; Peres, R.F.G.; Perry, G.A.; Pohler, K.G. Markers of pregnancy: How early can we detect pregnancies in cattle using pregnancy-associated glycoproteins (PAGs) and microRNAs? *Anim. Reprod.* **2016**, *13*, 200–208. [CrossRef]
142. Silva, E.; Sterry, R.A.; Kolb, D.; Mathialagan, N.; McGrath, M.F.; Ballam, J.M.; Fricke, P.M. Effect of interval to resynchronization of ovulation on fertility of lactating Holstein cows when using transrectal ultrasonography or a pregnancy-associated glycoptorein (PAG) ELISA to diagnose pregnancy status. *J. Dairy Sci.* **2009**, *92*, 3643–3650. [CrossRef]

143. Green, J.C.; Newsom, E.M.; Lucy, M.C. Incorporation of a rapid pregnancy-associated glycoprotein ELISA into a CIDR-Ovsynch resynchronization program for a 28 day re-insemination interval. *Theriogenology* **2011**, *75*, 320–328. [CrossRef] [PubMed]
144. Szenci, O.; Taverne, M.A.M.; Beckers, J.F.; Sulon, J.; Varga, J.; Börzsönyi, L.; Hanzen, C.; Schekk, G. Evaluation of false ultrasonographic diagnoses in cows measuring plasma levels of bovine pregnancy associated glycoprotein (bPAG). *Vet. Rec.* **1998**, *142*, 304–306. [CrossRef] [PubMed]
145. Fricke, P.M. Scanning the future—Ultrasonography as a reproductive management tool for dairy cattle. *J. Dairy Sci.* **2002**, *85*, 1918–1926. [CrossRef]

animals

MDPI

Article

Sizing Milking Groups in Small Cow Dairies of Mediterranean Countries

Angela Calvo * **and Gianfranco Airoldi**

Department of Agricultural, Forest and Food Sciences (DISAFA), University of Torino, Largo P. Braccini 2, 10095 Grugliasco, Turin, Italy; gianfranco.airoldi@unito.it

* Correspondence: angela.calvo@unito.it; Tel.: +39-116-708-592

Received: 31 March 2020; Accepted: 1 May 2020; Published: 4 May 2020

Simple Summary: The dimensioning of cow milking systems has been studied for many years by many authors, but nobody has addressed the studies on small cow dairies that are actually present in many Mediterranean countries (with a number milked cows up to 100–120). The number of cows, the financial resource, the skilled workforce and the availability of time are parameters influencing the selection of the milking system also in smaller herds. What is actually lacking in these dairy farms is milking unit dimensioning, whatever the used milking system. This study analyzed the milk routines performed by milkers with different milking systems in small dairies (tie-stall and little parlors). The work-flow analysis was the starting point to develop different models for predicting the optimal milking unit sizing.

Abstract: A dairy farmer chooses the number of milking groups in function of the herd size, stall type and milking system also in small cow dairies (number of animals lower than 100–120). In these dairies, there are different milking systems (bucket, trolley, pipeline, little autotandem, herringbone or parallel parlors) and each of them has a different work routine. The knowledge of the routine is the starting point for assessing the milking installation, because it determines the number of milked cows per hour. Different milking systems have common tasks (as pre-dipping, inspecting foremilk, udder preparation, attaching teat cups, post-dipping), but in the meantime there are different operations that characterize each specific routine (e.g., animal entry and exit if there is a parlor, bucket, trolley or milking group positioning if tie-stall). For this reason, we surveyed twenty small dairy farms located in the Piedmont Region (Italy) with different milking systems to correctly acquire the specific milking routines. Different models were therefore studied using the observed routines in in the examined farms. These models were then used to calculate the number of milked cows per hour and the number of milking groups. The main findings were simple equations, specific for each milking system, easily accessible by the farmer to correctly size his milking system.

Keywords: milking system; work routine; parlor; milking model; small dairy

1. Introduction

Mechanical milking consists of a logical sequence of straightforward repetitive tasks for the milkers (after preparing the milking equipment and, when necessary, the parlor): animal approaching, udder preparation, pre-dipping, attaching the milking unit, waiting for the milk flow, detaching the milking unit (if the automatic detacher is not present), post-dipping [1].

The main concern of the mechanical milking was at the beginning to increase the efficiency of the milking process to improve milk quality and quantity, and the labor efficiency [2–5]. Some studies introduced the time and motion method to establish the work routine [3,6] because this methodology provides a remarkable measurement of performance, and it is also the baseline for the milking mathematical models [7–9]. Computer-assisted simulation models of milking parlor performance were therefore accomplished since the seventies to improve the milking throughput and the labor performance [10–12], especially to reduce the daily routine labor and consequently the related milking costs [13,14]. In the following years, the researchers addressed their studies on mechanized milking systems also to the operators' and animals' welfare, considering the effects on the routine of the milking tasks [15–17]. Some studies, for example, found that the optimum time to attach teat cups had benefits on the teat condition as well as on a higher milk yield [18,19]. The main goal of more recent studies was also to forecast the optimum parlor [1,20], and to assist the farmer for the most efficient parlor design [21] and management [22–24], because the milking center is the most expensive item for a dairy farm [25].

In compliance with the herd size and with the stall type (free or tie), the farmer must select the milking system [17], and consequently the number of the milking units and the necessary labor force. The main aspect to work out for a rational decision is the work routine of the milking operation. The time occurred for both the pre-milking and post milking tasks is essential to determine the number of milking stalls and milking units that a milker can manage in the operation timeframe. Milking routine defines how one or more milkers perform a mechanic milking procedure over multiple cows (independently by the animal stalling). The preliminary knowledge of the usage of time for preparing and finishing the milking determines the number of cows that can be milked per hour and therefore the precondition for assessing milking installations [26].

Many authors assessed the need to increase the efficiency of the milking process through the optimization of milking routine and some of them applied their results mainly to the parlors (parallel, herringbone and rotary) [10,11,14,20,23].

For many decades, mechanical milking is spread everywhere, in a different type of dairy farms (cows, sheep, goats, buffaloes and others). It is nowadays common also in traditional small-scale dairy farms of the Mediterranean areas [27–29], where different (and sometimes outdated) milking systems are present (bucket, trolley, tethered, small parlors). As in large herd sizes, though also in smaller herds, the number of cows, the financial resource, the skilled workforce and the availability of time are parameters influencing the selection of the milking system. What is actually lacking in these dairy farms (with a number milked cows up to 100–120) is an indication for the sizing of the necessary milking groups, whatever is the used milking system. The main goal of this work was therefore to suggest different models for predicting the optimal milking unit dimensioning to address the farmers to the best choice. The modeling is an alternative of the stiff tables of manufacturers, which are not always flexible in the adaptation to specific small dairy needs.

2. Materials and Methods

2.1. Preliminary Assumptions

Achieving good and realistic results in milking operations in small cow dairies, both in labor productivity and animal welfare, only one experienced and qualified milker, familiar with the physiology of dairy cows, was used. The milking systems were moreover all timely revised and accurately washed after each milking.

2.2. Milking Extraction Time (Milk Flow Time)

Many studies reported the effect of milk flow time on milking performance [4,30,31] as the most influencing activity. Even though the milk extraction is not a real operation of the routine, its duration highly influences the work routine time. For example, the milking time could be the key element of the operator's idling time in case of cows with high milk flow time, or it may cause a dangerous over milking in the opposite situation, when a cow has a lower milk flow time than other animals and automatic cluster removals are not present.

The milking extraction time of each cow depends on multiple variables: animal milk yield, correct feeding and stimulation during the preparation tasks, the physiologic predisposition of the cow to release milk more or less quickly.

Since the seventies, many authors studied and simulated the milking extraction time, and they found average times around 5–5.5 min [2,6,10,32,33]. In the following decades, however, the milk yield per milked cow increased, thanks to the genetic improvement of the animals, and thereby the milking extraction time increased [30]. New relationships between the total milking time and the milk yield were assessed. As things stand, a milking extraction time between 6 and 8 min is reasonable [34].

2.3. Milking Routine: Definitions

The milking routine is the amount of the operator's tasks performed during the milking of each cow. Each task requires a certain amount of time, and the sum of these times in addition to the milking extraction time determines the number of cows that the operator can milk per hour. The work routine is defined as the function of the amount and type of the milking equipment, the operators' skills and the duration of the milking operations [3].

Even though each milking system has its work routine with specific sequences of operations, there are some common tasks, mostly manually performed by the milker: pre-dipping, inspecting foremilk, udder preparation, attaching teat cups, removing teat cups (if the milking cluster removal is absent), post-dipping (disinfecting teats). Some definitions are necessary before introducing the proposed models for sizing the milking system in the function of the tie-stall or parlor type (Table 1).

Table 1. Parameters used in the milking sizing model.

Simbol	Unit	Meaning	Further Information
Nc	n	Number of milked cows	
Tm	min	Milk extraction (or flow) time	6–8 min
Tri	s	Early (initial) routine	Animal entry (*Ten*); bucket, trolley, or cluster positioning (*Tpos*), pre-dipping (*Tpre*), foremilk inspecting (*Tfor*), udder preparation (*Tup*), teat cups attach (*Tatt*)
Trf	s	Ending (final) routine	Teat cups removal (*Trem*), post-dipping (*Tpost*), bucket or tank empting (*Tempt*), relocation of the mobile parts of the milking system (*Trr*), animal exit (*Tex*)
Tr	s	Milking routine	30–180 s
Td	s	Downtime	Unproductive time due to unpredictable events (always present and split among *Tri*, *Trf* and *Trr*)
Tc	s	Milker's time for milking each cow	*Tc* is always higher than, or equal to, the milking routine *Tr*
Ng	n	Number of stalls	Only in parlor
Ng	n	Number of milking groups	
Tmw	s	Milker waiting time	Unproductive routine time (for example due to the absence of the milking cluster removal

The milking routine *Tr* (the time used by the operator for milking each cow) is split into the following categories: early (initial) routine (*Tri*, Equation (1)), ending (final) routine (*Trf*, Equation (2)) and downtime (*Td*), always present whatever the milking system. The entry (*Ten*) and the exit (T*ex*) of the animals, in and out of the parlor, are common parameters in *Tri* and *Trf* in parlors. Bucket, trolley, or cluster positioning (*Tpos*) is a *Tri* parameter in tie-stall, as the emptying of the bucket or tank in the trolley (*Tempt*) in *Trf*. The relocation of the mobile parts of the milking system (*Trr*) must be instead included in the final routine *Trf* in the trolley system (Equation (2)).

$$Tri = Ten^{*} + Tpos^{**} + Tpre + Tfor + Tup + Tatt \text{ (min)} \quad (1)$$

* only parlor, ** only tie-stall.

$$Trf = Trem + Tpost + Tempt^{*} + Trr^{**} + Tex^{***} \text{ (min)} \tag{2}$$

* only bucket and trolley, ** only trolley, *** only parlor.

Downtimes are always present in each phase of the milking routine and must be considered in the calculation of the total routine time, *Tr* (Equation (3)).

$$Tr = Tri + Trf + Td \text{ (min)} \tag{3}$$

The output milker's time, *Tc*, for milking each cow depends on the milking system and includes the calculated routine time, *Tr* (Equation (3)), the milk flow time, *Tm*, and, when present, the possible unproductive times, *Tmw* (as the milker's waiting time when the automatic cluster removal is absent). Equations (1)–(3) are used whatever the milking system. The washing and the maintenance of the milking unit, as well as the transportation of the milk tanks outside the stall at the end of the milking operation in tie-stall, were not considered in the calculation of the routine time, *Tr*, since it is strictly related to the milking tasks with the cows.

2.4. Inspected Milking Systems: Tethering Cows

In tie-stalls, cows are constrained to their stall barns, where they are directly milked. Tethering cows are still present in little dairy farms of many Mediterranean countries [35], where the milking is often carried out with trolley or bucket milking systems. In both these systems, the milker's time for each cow is greater than the routine time (*Tc* > *Tr*), because he/she has to wait for the completion of the milk flow time before moving to the next animal. If a milk-line (stall barn with pipeline) is not present, the labor productivity is low, as the milker (besides the basic routine) must move the trolley or the bucket alongside the next cow and transport the bucket outside the barn when full.

2.5. Inspected Milking Systems: Little Parlors

There are many types of parlor, the choice of which depends on the herd size and characteristics, the economic impact on the dairy, the number and the ability of the workers, and the automation level [2,5,25]. Parlor type affects building size, cow traffic to and from parlor, milking routine and mechanization level. The number of the daily milking is also important, for example, in the small cow herds of Italy, two milking per days are planned and animals are milked every about 12 h (usually at 4:00 and at 16:00). Two milkings/day represents a standard situation in small cow herds, not only in Italy. As also observed by other authors [36,37], it is convenient to maintain the same milking intervals during the day (every 12 h) in the cows, because milking time is one of the components that modulates the milk lipolytic system.

Parlors may be individual and collective (batch milking). In the first type, cows stand nose-to-tail inside individual stalls (tandem and autotandem parlors). In the second, cows are milked together after their entry in the milking stalls. In collective parlors, the milking operations are sequentially carried out, and it is important that the milking extraction time is almost the same for each animal. The swing-over milking system permits to shift the milking unit from one stall to another positioned in front of it, speeding up the milking routine (Figure 1).

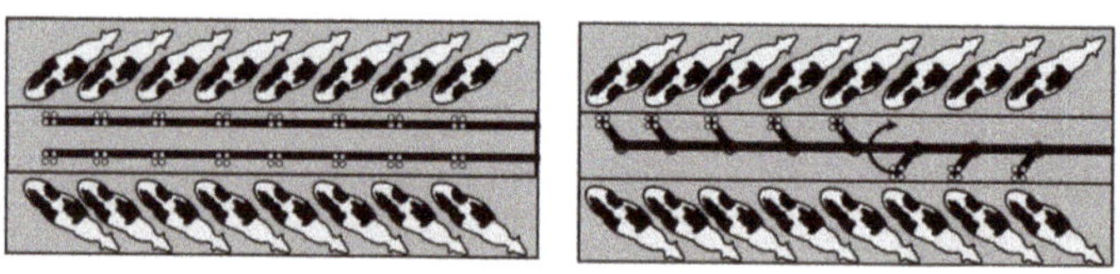

Figure 1. Herringbone parlor without (left) and with milking swing-over milking system (right).

A brief description of the milking systems is provided to understand the different modelled routines in the result chapter. Rotary parlors, which are more complex and overly expensive in small dairy herds of Mediterranean countries, were not considered in this work.

2.5.1. Tandem (Side-Opening) and Autotandem

In this type of parlor, the cows move individually to their milking stalls entering and exiting through gates manually (tandem) or automatically (autotandem) opened up and closed (Figure 2). The milking operation is performed individually, and cows may have a different milk flow time without affecting the milking routine of the other animals.

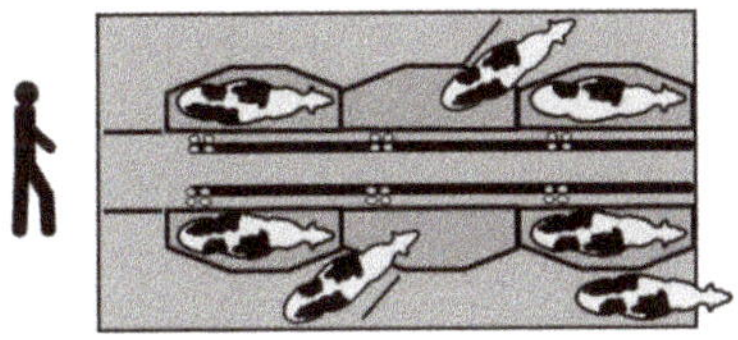

Figure 2. Autotandem parlor with 3 + 3 stalls.

2.5.2. Herringbone

In herringbone and parallel parlors, cows are handled in groups. The size is variable from 4 × 2 to 30 × 2 [25]. Many herringbone and parallel parlors are equipped with rapid-exit stalls (by freeing all cows at once from one side of the parlor in a direction perpendicular to the entry lane, Figure 3), which increase the milking parlor efficiency and improve the milking routine.

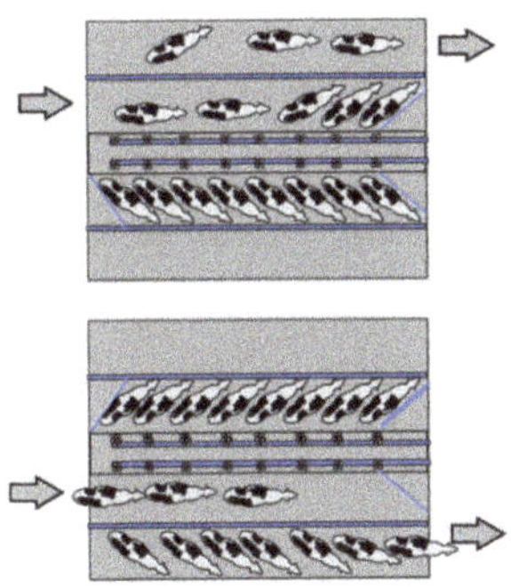

Figure 3. Herringbone parlor with rapid-exit system.

2.5.3. Parallel

Parallel parlor resembles the herringbone, but cows are arranged perpendicular to the edge of the milker's pit, where he/she works back to the animal (Figure 4). Advantages of this system for the lowering of the milking routine are: higher displacement of the cows (animals move faster) and fewer movements of the milker. As for herringbone, also in this case, a milking group for each stall or a milking group every two stalls (swing-over) may be provided.

2.6. Study of the Models

The studied models were based on the available literature [8,11,20,25,37–39], following the indication provided by Armstrong et al. [3], and analyzing the work flow during the milking operation in twenty small cow dairy farms located in north-west Italy. The dairy farms were located in the Stura mountain valley, in the Monferrato hills and in the Po valley in Piedmont Region. Piedmontese cows were present in the farms with trolley and bucket systems, while the Italian Friesian breed were in the other farms equipped with the milk-line and the parlors. The distribution of the surveyed milking systems was: three buckets, three trolleys and four milk-lines in tie-stalls, two autotandem, three herringbone and five parallel in parlors. The milking operations in each farm were surveyed for three days, while the number of lactating cows and milk yield per day per cow were furnished by the farmer.

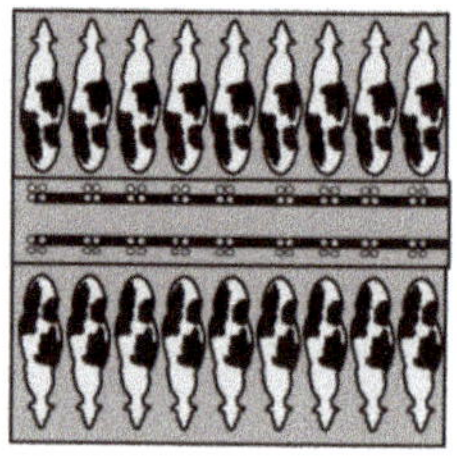

Figure 4. Parallel parlor.

2.7. Data Elaboration

Each routine task (pre-dipping, post-dipping, foremilk inspection, teat cups attach and so on) has a short duration, and, for this reason, they were acquired and studied using seconds. Initial and finale routines *Tri* and *Trf* were instead counted in minutes (with seconds expressed in hundreds), to have the same dimension unit of the milk extraction time (as used by other Authors) and to make comparisons with other researches. Collected data in the twenty dairy farms were analyzed using IBM SPSS Statistics (version 25, International Business Machines Corporation, Armonk, New York, NY, USA). The GLM (general linear model, to analyze quantitative data and to understand how the mean response relates to one or more independent predictors) was used to assess the effects of the variable milking system on *Tri* and *Trf* routines, with a confidence level $p = 0.01$. The homoscedastic condition (assumption of equal variance) was previously tested by the Levene's test.

3. Results

3.1. Tie-Stalls

3.1.1. Trolley

In trolley systems, all components (vacuum and milk system) are positioned on a mobile frame and transported inside the barn (Figure 5). One or two milking groups (maximum) are housed on the trolley, managed by one operator.

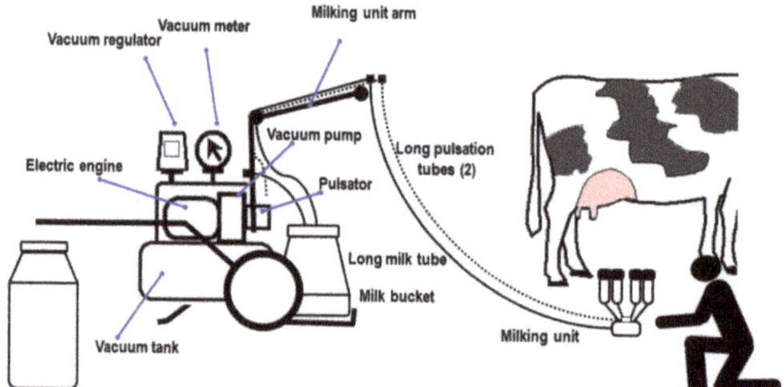

Figure 5. Trolley milking system.

The time *Tc* used by the milker for a single cow is calculated as in Equation (4), while the number of milked cows per hour is obtained by Equation (5).

$$Tc = Tr + Tm \tag{4}$$

$$Nc = \frac{60}{Tc} \tag{5}$$

With two milking units, the milking operation can be simultaneously performed by the same operator on two cows, but he/she must wait the ending of the milk flow regardless. In this case, an unproductive time is present ($Tmw = Tm - Tri$), to the detriment of the milking routine.

3.1.2. Bucket

In bucket systems, the motor unit and the vacuum system are located in a special room near the stable. The mobile milking buckets are connected to the vacuum pipeline (Figure 6), and the operator must transfer the milk from the buckets to the transport bulk tank and move it out of the stable. He/she is able to control up to three milking groups.

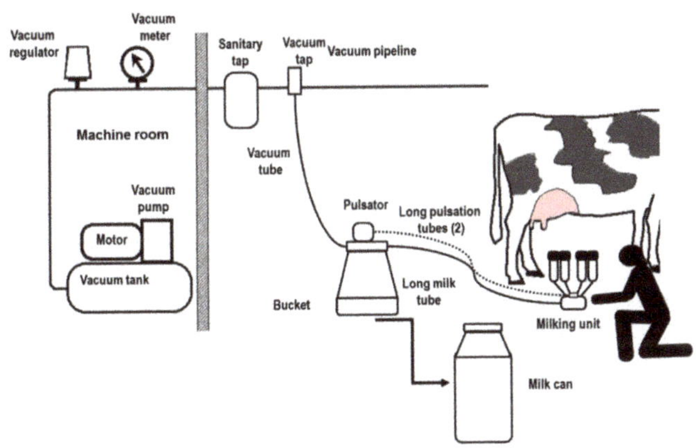

Figure 6. Bucket milking system.

The number of milked cows per hour *Nc* depends on the number of the milking groups *Ng* managed by the operator (Equation (6)). In this equation, *Tm/Tr* is usually a decimal number that must be rounded up or down. The rounding up (*ru*) may cause the serious problem of over-milking the cow, and, for this reason, in this milking system it is better to round down (rd) the fraction *Tm/Tr*.

$$Ng = rd\left(\frac{Tm}{Tr}\right) + 1 \tag{6}$$

Equation (7) gives the time *Tc* dedicated by the milker to each cow, while the number *Nc* of milked cows per hour per group is given by Equation (5).

$$Tc = Tr + \frac{Tm - Tr \times (Ng - 1)}{Ng} \tag{7}$$

3.1.3. Milk-Line

The milk-line has two rooms near the barn: a machine room and a milk room. There are two fixed pipes for milk and for vacuum (Figure 7).

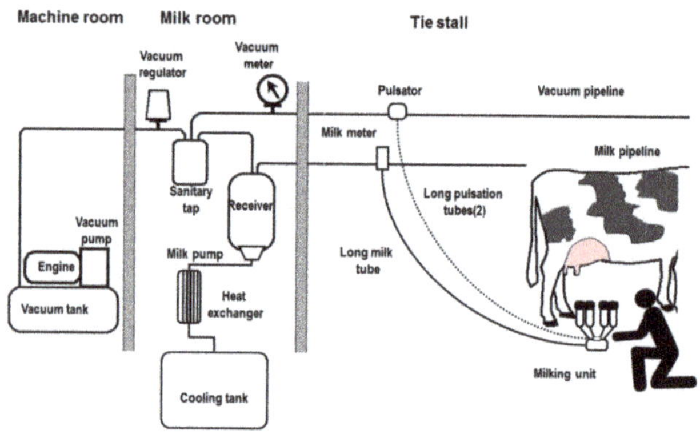

Figure 7. Milk-line.

The number of milking groups *Ng* and the milker's time per cow *Tc* are once again calculated by Equations (6) and (7) if the milking cluster removal is not present. With the milking cluster removal, the number of groups is calculated using Equation (8), rounding up the ratio *Tm/Tr*. The cow may wait for the final routine without any over-milking risk (Figure 8).

$$Ng = ru\left(\frac{Tm}{Tr}\right) + 1 \tag{8}$$

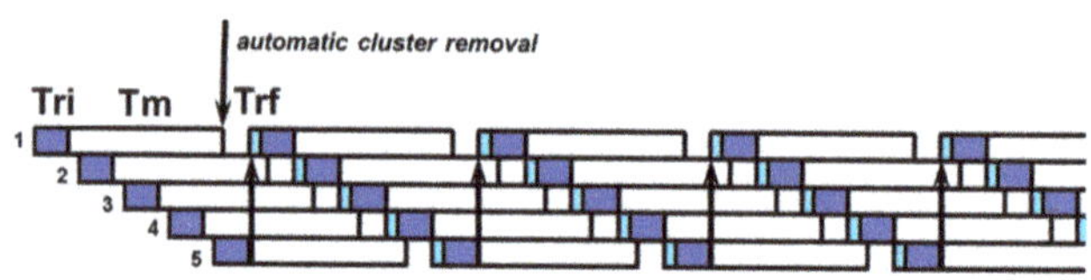

Figure 8. Milk-line routine in the presence of milking cluster removal (Ng = 5 and Tm = 6 min).

If the milking cluster removal is not present, the number of milked per cows per hour *Nc* is calculated by means of Equation (5), otherwise (with automatic milking removal) with Equation (9).

$$Nc = \frac{60}{Tr} \tag{9}$$

Table 2 gives the routine for bucket and trolley systems, while Table 3 shows the same routine for the milk-line without and with the automatic cluster removal (NOACR and YESACR respectively). The routine times were obtained by directly observing the milking operations in the examined diaries and then comparing them with the bibliography [38–42].

The initial routine *Tri* in the bucket system is close to the *Tri* in the milk-line without automatic cluster removal, while the final routine *Trf* is definitely shorter in all the milk-line systems (here, there are not buckets to empty). The automatic cluster removal, other than eliminating the over-milking risk, reduces the initial routine *Tri* by 20%, and the final routine *Trf* by 50% (Table 4).

Table 2. Routine of trolley and bucket systems.

Task	Time (s)			Routine Time (s)		Routine Time (min)	
	Input Data	Trolley	Bucket	Trolley	Bucket	Trolley	Bucket
Bucket or trolley positioning Pre-dipping Foremilk inspection Udder preparation Teat cups attach	Tpos Tpre Tfor Tup Tatt	10 15 8 20 14	28 15 8 20 14	*Tri* = 67 + 8 *	*Tri* = 85 + 10 *	*Tri* = 1.25	*Tri* = 1.58
Teat cups removal Post-dipping Bucket or container empting Relocation of trolley	Trem Tpost Tempt *Trr*	25 9 20 30	25 9 50 0	*Trf* = 84 + 16 *	*Trf* = 84 + 10 *	*Trf* = 1.67	*Trf* = 1.57
Downtimes *	Td	24	20				

* Divided among *Tri*, *Trf* and *Trr* (this last only for trolley).

Table 3. Routine of milk-lines.

Task	Time (s)			Routine Time (s)		Routine Time (min)	
	Input data	NOACR [1]	YESACR [2]	NOACR	YESACR	NOACR	YESACR
Milking group positioning Pre-dipping Foremilk inspecting Udder preparation Teat cups attach	Tpos Tpre Tfor Tup Tatt	23 15 8 20 14	15 15 8 20 14	*Tri* = 80 + 10 *	*Tri* = 72 + 6 *	*Tri* = 1.5	*Tri* = 1.3
Teat cups removal Post-dipping Group removal	Trem Tpost Tempt	25 9 15	- 9 12	*Trf* = 49 + 5 *	*Trf* = 21 + 6 *	*Trf* = 0.9	*Trf* = 0.45
Downtimes *	Td	15	12				

[1] NOACR: without automatic cluster removal. [2] YESACR: with automatic cluster removal. * Divided between *Tri* and *Trf*.

Table 4. Initial and final routine in trolley, bucket and milk-line systems.

Routine Time (min)	Trolley	Bucket	Milk-Line NOACR	Milk-Line YESACR
Tri	1.25	1.58	1.50	1.30
Trf	1.03	1.57	0.90	0.45

3.2. Parlor

In the parlor, the milking time per cow, *Tc*, is always equal to the routine time, *Tr*, while the number of milking groups (and, consequently, the number of milked cows) depends on the milk flow time *Tm* and on the routine time *Tr*. The milking cluster removal is supposed to always be present in the examined dairies to avoid over-milking, as it is necessary to work with many animals simultaneously. The initial routine, *Tri*, and the final routine, *Trf*, are shorter than in tie-stalls, as there is no need to move trolleys, buckets or milking groups. After having performed the initial routine on the first cow, the milker can move on to the second animal for the initial routine and so on, until the first cow has finished being milked. The milker then returns to the first cow and carries out the final routine.

3.2.1. Autotandem (Side-Open) with Automatic Milking Cluster Removal

In side-open parlors, the swing-over is not present, because animals are handled individually and not in groups. The number of milking units, *Ng*, is given by Equation (8), because the automatic milking removal is present. If *Ng* is an odd number, the shrewdness is to add a further milking unit, because the number of stalls is even (cows are positioned on two parallel rows). *Tc* is equal to *Tr* and *Nc* is given by Equation (8). Figure 9 portrays the routine in the examined autotandem parlor (six stalls and *Te* = 6 min).

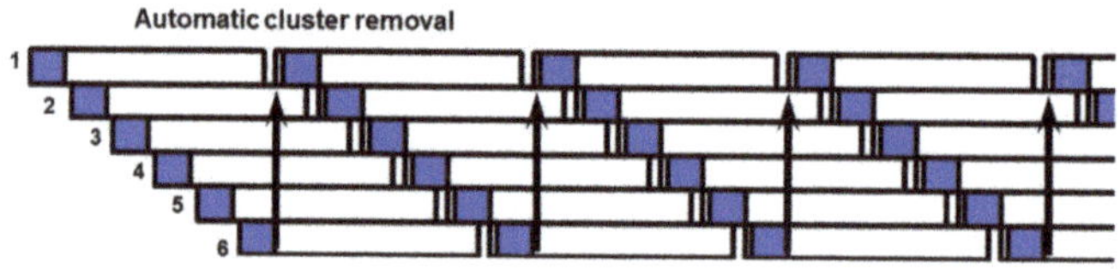

Figure 9. Example of a milking routine in an autotandem parlor (3 + 3 stalls and *Te* = 6 min).

3.2.2. Herringbone and Parallel Parlors

Initial and final routines are the same as in autotandem, but the entrance of the animals and the pre-dipping operations take less time, as the operations are sequentially performed (collective animal milking) on two parallel rows. Without swing-over, the number of milking groups *Ng* (equal to the number of stalls *Ns*) is calculated distinguishing if *Tm/Tr* is an integer number or not (Equation (10)).

$$\begin{array}{c} \frac{Tm}{Tr} = INT\left(\frac{Tm}{Tr}\right) \rightarrow Ng = \frac{Tm}{Tr} \times 2 \\ \frac{Tm}{Tr} \neq INT\left(\frac{Tm}{Tr}\right) \rightarrow Ng = \left(INT\left(\frac{Tm}{Tr}\right) + 1\right) \times 2 \end{array} \tag{10}$$

When the swing-over is present, the number of milking units *Ng* (half of the number of stalls *Ns*) is now half of the previous calculated by Equation (10), while Equation (11) gives the number of stalls *Ns*. The number of milked cows per hour *Nc* is in Equation (9).

$$Ns = 2 \times Ng \tag{11}$$

The milking routine in parallel parlors is the same as herringbone, but with a lower entry time of the animals (and consequently with a lower *Tr*). The milking routine observed in a parallel parlor with 16 stalls and *Te* = 8 min is depicted in Figure 10.

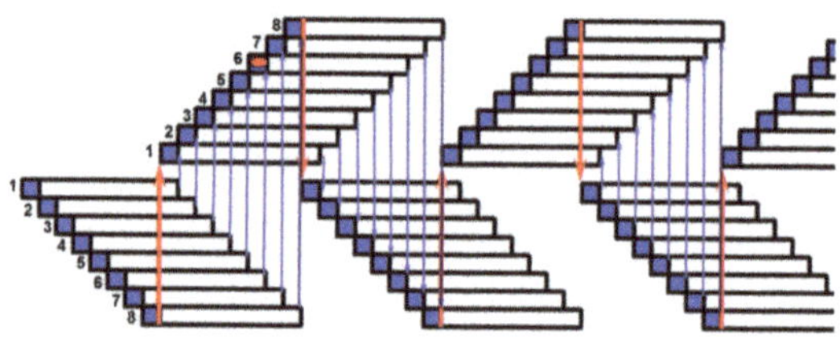

Figure 10. Milking routine observed in a parallel parlor (8+8 stalls and *Te* = 8 min).

Table 5 shows the initial and final routines in the examined parlors: autotandem, herringbone and parallel.

Table 5. Routine of autotandem, herringbone and parallel parlors.

Task	Input Data	Time (s) Auto Tandem	Time (s) Herring Bone	Time (s) Parallel	Routine Time (s) Auto Tandem	Routine Time (s) Herring Bone	Routine Time (s) Parallel
Animal entry	*Ten*	15	8	6			
Pre-dipping	*Tpre*	15	9	9			
Foremilk inspecting	*Tfor*	6	6	6	*Tri* = 53 + 10 *	*Tri* = 40 + 6 *	*Tri* = 38 + 4 *
Udder preparation	*Tup*	7	7	7			
Teat cups attach	*Tatt*	10	10	10			
Post-dipping	*Tpost*	7	7	7	*Trf* = 9 + 4 *	*Trf* = 9 + 5 *	*Trf* = 9 + 4 *
Animal exit	*Tex*	2	2	2			
Downtimes *	*Td*	14 *	11 *	8 *			

* Divided between *Tri* and *Trf*.

Final routines, *Trf*, are almost the same in all these parlors, while there is a difference between the individual (autotandem) and collective parlors in the initial routine, *Tri* (Table 6). In herringbone and parallel parlors, in fact, animal entry and pre-dipping are fastened by the group management of the cows.

Table 6. Initial and final routine in autotandem, herringbone and parallel parlors.

Routine Time (min)	Auto Tandem	Herring Bone	Parallel
Tri	1.05	0.77	0.7
Trf	0.22	0.23	0.22

3.3. The Examined Dairy Farms

We examined the milking tasks performed in different dairy farms with the same milking system (three trolleys, three buckets, four milk-lines, two autotandem, three herringbone and five parallel), and the most interesting result of the survey was that the milker executed the same initial and final routines in the farms with the same milking system.

3.3.1. Tie-Stall Dairy Farms

The number of lactation cows was between 22 and 45 in the farms with trolley, bucket and milk-line systems without automatic cluster removal, increasing to 50 animals in the farm with the milk-line equipped with the automatic cluster removal (the sole investigated, because this milking system is not spread in the Piedmont Region). The milk flow time was in the range 5.5–7.6 min, with higher values in the milk-line systems (Table 7).

Table 7. Descriptive statistic of the examined tie-stall dairy farms.

	Trolley		Bucket		Milk-Line NOACR		Milk-Line YESACR	
Dairy Farm Data	Average	SD	Average	SD	Average	SD	Average	SD
lactating cows (n)	24.7	2.52	34.3	6.0	35.0	8.89	50	-
milk flow time average (min)	6.4	0.12	6.7	0.17	6.9	0.10	7.1	-
milk flow time min (min)	5.5	0.12	5.4	0.30	5.8	0.17	6.8	-
milk flow time max (min)	7.1	0.12	7.2	0.06	7.6	0.06	7.6	-

Concerning the daily milk yield, some differences were observed (Figure 11).

In the farms with trolley and bucket systems, the milk yield was around 18.7 L per day, against an average of 32.5 L per day in the farms with the milk-line system. This difference is due to the breed present in these last farms (Italian Fresian cows), more productive than the Piedmontese breed cows [43].

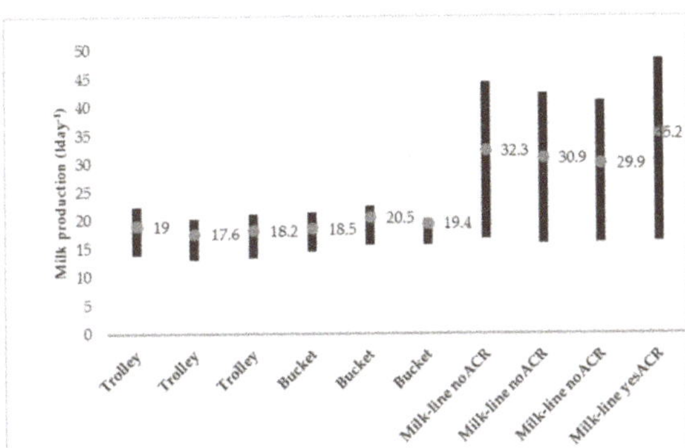

Figure 11. Milk production (min, max and mean) in the surveyed tie-stall dairy farms (lday^{-1}).

3.3.2. Dairy Farms with Parlors

From 95 to 130 lactating cows (Italian Fresian breed cows) were present in the dairy farms equipped with the parlor. All these farms, but three (located in the Stura valley), were located in the Monferrato hills and in the Po valley near Cuneo.

The observed milk flow time was between 5.9 and 8.1 min, with lower and more homogeneous values observed in herringbone and parallel parlors (Table 8).

Table 8. Descriptive statistic of the examined dairy farms with a parlor.

	Autotandem		Herringbone		Parallel	
Dairy Farm Data	**Mean**	**SD**	**Mean**	**SD**	**Mean**	**SD**
lactating cows (n)	98.5	4.95	111.7	10.41	121.4	7.40
milk flow time average (min)	7.2	0.07	6.8	0.06	7.0	0.03
milk flow time min (min)	5.9	0.10	6.3	0.04	6.5	0.02
milk flow time max (min)	8.1	0.07	7.4	0.06	7.5	0.20

The milk yield per cow (l day^{-1}) was always around 40 L per day, with peaks of 50 L in two farms with the parallel parlor and the highest number of lactating cows (more than 120). A slightly lower production (around 32 L) was observed in two farms with the autotandem milking system and with the lowest number of lactating cows (less than 100).

3.3.3. Differences in the Observed Routines

Milking systems statistically influenced both the initial (*Tri*) and final (*Trf*) routines (Table 9), due to the high differences between tie-stall and parlors routines (Tables 4 and 5).

Table 9. General linear model (GLM) statistic of *Tri* and *Trf* in the function of the milking system.

Origin		ss	df	Ms	F	*p*-Value
Correct model	*Tri*	7.169 [a]	6	1.195	131.464	0.000
	Trf	22.204 [b]	6	3.701	282.077	0.000
Intercept	*Tri*	64.386	1	64.386	7083.875	0.000
	Trf	28.773	1	28.773	2193.215	0.000
Milking system	*Tri*	7.169	6	1.1948	131.464	0.000
	Trf	22.204	6	3.701	282.076	0.000

ss: sum of squares; df: degrees of freedom; Ms: mean square; F: F-value. a: R-squared = 0.937 (R-squared corrected = 0.930). b: R-squared = 0.970 (R-squared corrected = 0.966).

On the other hand, the Tukey post-hoc test highlighted similarities among some milking systems (Table 10) for both *Tri* and *Trf*. The highest statistical significances were found in *Tri* of parallel and herringbone parlors (0.70 and 0.75 average minutes, respectively), and in trolley and milk-lines with automatic cluster removal (1.26 and 1.30 average minutes). *Trf* were significantly the same in parallel (average 0.23 min), herringbone (average 0.25 min) and autotandem parlors (average 0.27 min). Bucket and trolley milking systems had similar *Trf* times (1.57 and 1.68 min), but with a lower statistical significance (0.61).

Table 10. Tukey post-hoc statistic of *Tri* and *Trf*.

Tri						
Milk_System	**Test**	**Subset**				
	N	1	2	3	4	5
Parallel	15	0.70				
Herringbone	9	0.75				
Autotandem	6		1.01			
Trolley	9			1.26		
Milk-line YESACR	3			1.30	1.30	
Milk-line NOACR	9				1.46	1.46
Bucket	9					1.59
Sign.		0.97	1.00	0.98	0.05	0.12

Trf					
Milk_System	**Title**	**Subset**			
	N	1	2	3	4
Parallel	15	0.23			
Herringbone	9	0.25			
Autotandem	6	0.27	0.27		
Milk-line YESACR	3		0.45		
Milk-line NOACR	9			0.94	
Bucket	9				1.57
Trolley	9				1.68
Sign.		0.99	0.07	1.00	0.61

4. Discussion

The efficiency of a milking system is usually assessed on the basis of multiple parameters, which, in our situation (small herds with a number of animals not higher than 120), are reduced in cows milked per hour and the number of stalls to complete the milking in time (about 2 h). For this reason, the focus was to discuss the above-mentioned results, as obtained by the models and by the surveyed farms.

4.1. Tie-Stall

In Table 11, there are the calculated parameters *Tr*, *Tc*, *Ng* and *Nc* in two different scenarios: milk flow time *Tm* = 6 min and *Tm* = 8 min.

Table 11. Comparison of milking systems parameters in tie-stall.

Milking System	Tm = 6 min				Tm = 8 min			
	Tr	Tc	Ng	Nc	Tr	Tc	Ng	Nc
Trolley	2.91	8.91	2	13.46	2.91	10.91	2	5.50
Bucket	3.15	4.58	2	26.23	3.15	3.72	3	16.14
NOACR Milk-line	2.40	2.80	3	21.43	2.40	2.60	4	23.08
YESACR Milk-line	1.75	1.55	5	34.29	1.75	1.63	6	34.29

The trolley system is less convenient than the bucket system because it takes more time and the number of milked cows per hour is lower (due to the fact that the milker has to wait the complete the milking of two cows before moving the whole trolley to the following animals). In both cases, from the routine point of view, the milker has many unproductive times (*Tmw*), Figure 12.

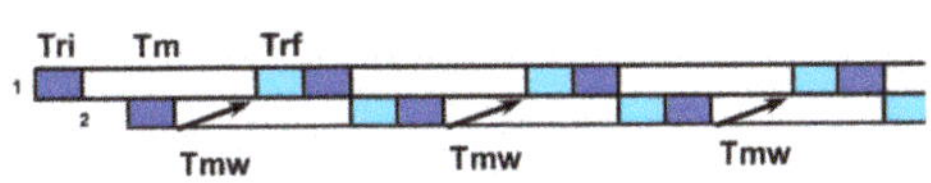

Figure 12. Example of milker's waiting time, *Tmw,* when using the bucket milking system.

The number of the milked cows per hour was 13.46 for the trolley system and 26.23 for the bucket (Table 11), allowing the milking of the lactating cows in the surveyed farms, 24.7 and 34.3 in average, also considering the high standard deviation (Table 7).

The milk-line provides, as expected, a more efficient milking, especially when the automatic cluster removal is present. In this case, the number *Nc* of milked cows per hour (34.29) does not change, whatever is the milk flow time (Table 11), because it is calculated using only the routine time *Tr*, independently from the milk flow time. Table 11 also shows a firm decrement of the time dedicated by the milker to each cow in the milk-line when the automatic cluster removal is present (-44.6% if *Tm* = 6 min, −37.5% if *Tm* = 8 min), as also found by Rasmussen [41]. The presence of the automatic cluster removal also significantly decreases the over-milking caused by a prolonged teat stimulation, with the risk of cause teat congestion [44]. The number of calculated milked cows per hour (Table 11) with the milk-lines were 21.43 (NOACR) and 34.29 (YESACR), which allowed for the milking of the observed 35 and 50 mean cows (respectively) in time.

4.2. Herringbone and Parallel Parlors

In this example, the routine time *Tr* is 1.27 min in autotandem, 1 min in herringbone and 0.92 min in parallel (Table 12). The preparation time (initial routine *Tri*) in herringbone and parallel parlors is similar to the values recommended by Peychev et al. [38] (between 45 and 90 s, Table 5). Without considering the animal entry, the *Tri* calculated in herringbone and parallel parlors (Table 5) are in line with the data found by Smith et al. [39] (32 s).

Table 12. Comparison of milking systems parameters in different parlor types.

	Tm = 6 min					*Tm* = 8 min				
Parlor Type	***Tr***	***Tc***	***Ng***	***Ns***	***Nc***	***Tr***	***Tc***	***Ng***	***Ns***	***Nc***
Autotandem	1.27	1.27	6	(3 + 3)	47.4	1.27	1.27	8	(4 + 4)	47.4
Herringbone	1.00	1.00	12	(6 + 6)	60.0	1.00	1.00	16	(8 + 8)	60.0
Herringbone with swing-over	1.00	1.00	6	(6 + 6)	60.0	1.00	1.00	8	(8 + 8)	60.0
Parallel	0.92	0.92	14	(7 + 7)	65.5	0.92	0.92	18	(9 + 9)	65.5
Parallel with swing-over	0.92	0.92	7	(7 + 7)	65.5	0.92	0.92	9	(9 + 9)	65.5

As *Nc* is directly obtained by *Tr*, regardless of the milk flow time, the number of milked cows does not differ in the two scenarios with a different milk flow time (6 and 8 min). The autotandem is less efficient than herringbone and parallel parlors, since the time of both animal entry and pre-dipping is greater (cows move individually and they are not milked together).

Higher is the number, *Nc*, of milked cows per hour, and lower is the routine time, *Tr*; this is a self-evident consequence of the calculation of *Nc*, but the numbers are confirmed also by other authors, independently by this equation [4,45]. The number of milked cows per hour in the double-8 herringbone (60) is slightly lower than 64, as found by Burks et al. in 1998 [1], but is in line with Krumm et al. [23], and the 65 cows per hour of double-9 parallel are compliant with Hatem et al. [20].

Carreira et al. [42] studied and developed some algorithms for the best sizing of herringbone and parallel parlors, starting from the routine time and the milk flow time, as we did in this work, but without distinguishing the parlor type. As also observed here, they got similar results (Figure 13), recording any change when the number of clusters per milker was greater than (or equal to) 12 (we indeed observed two different values in the number of milked cows per hour with 14 and 18 clusters, due to the slightly differences in the routines of herringbone and parallel parlors, assumed to be equal by the other authors).

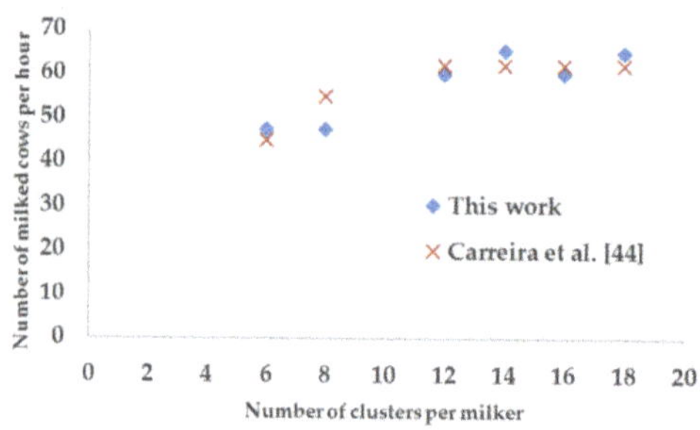

Figure 13. Number of milked cows per hour in function of the number of clusters per milker (comparison between this work and the study of Carreira et al. [42]).

The milker expertise and the type of milking facility affect labor and routine efficiencies: for this reason, the input values used in these models may change; nevertheless, the equations remain valid. In the farms equipped with parlors, the number of calculated milked cows per hour *Nc* (Table 12) allowed for the milking of all the lactating cows present in the surveyed farms (Table 8) in time.

5. Conclusions

Even though it is not easy to find literature concerning the milking systems in tie-stall and little parlors, there are still many dairy farms in Mediterranean countries with a low number of cows (less than 100–120) that cannot face high investment costs in expensive milking plants. It would appear apparently naïve modelling milking with low numbers of animals, but there is still a lack in the work organization of these small herds, with few people working a great amount of hours per day. For this reason, we surveyed twenty small dairy farms located in Piedmont Region (Italy) equipped with different milking systems to correctly acquire the milking routines. Different models were therefore studied using the observed routines in the examined farms. These models were developed independently by the routine times measured in each farm. The measured times were the method to validate the number of milked cows per hour and the number of milking groups in these small dairies. The simple models presented in this paper could therefore support the breeder to correctly dimension and choose the most effective milking system for his dairy farm.

Author Contributions: Conceptualization, G.A.; methodology, G.A.; validation, G.A. and A.C.; formal analysis, G.A. and A.C.; investigation, A.C.; resources, A.C.; data curation, G.A. and A.C.; writing—original draft preparation, A.C.; writing—review and editing, G.A. and A.C.; visualization, G.A. and A.C.; supervision, G.A. and A.C. All authors have read and agreed to the published version of the manuscript.

Funding: This research received no external funding.

Conflicts of Interest: The authors declare no conflict of interest.

References

1. Burks, T.F.; Turner, L.W.; Gates, R.S.; Bridges, T.C.; Crist, W.L. A stochastic simulation model for predicting cow throughputs in dairy parlors. *Trans. ASAE* **1998**, *41*, 1789–1797. [CrossRef]
2. Bickert, W.G.; Speicher, J.A.; Armstrong, D.V. Milking systems for large herds. *J. Dairy Sci.* **1974**, *57*, 369–377. [CrossRef]
3. Armstrong, D.V.; Quick, A.J. Time and motion to measure milking parlor performance. *J. Dairy Sci.* **1984**, *69*, 1169–1177. [CrossRef]
4. Blake, R.W.; Mcdaniel, B.T. Relationships among rates of milk flow, machine time, udder conformation, and management aspects of milking efficiency: A review. *J. Dairy Sci.* **1978**, *61*, 363–378. [CrossRef]
5. Smith, J.F.; Armstrong, D.V.; Gamroth, M.J. Labor management considerations in selecting milking parlor type & size. In Proceedings of the Western Dairy Management Conference, Las Vegas, NV, USA, 13–15 March 1997; pp. 43–49.
6. Thomas, C.V.; DeLorenzo, M.A.; Bray, D.R. Prediction of individual cow milking time for milking parlor simulation models. *J. Dairy Sci.* **1993**, *76*, 2184–2194. [CrossRef]
7. Chang, W.; Barry, M.C.; Jones, L.R.; Merrill, W.G. A computerized milking parlor evaluation system. *J. Dairy Sci.* **1992**, *75*, 2578–2586. [CrossRef]
8. Chang, W.; Streeter, D.H.; Jones, L.R. An object-oriented model for simulating milking parlor operations. *J. Dairy Sci.* **1994**, *77*, 84–93. [CrossRef]
9. Thomas, C.V.; DeLorenzo, M.A.; Bray, D.R. A network simulation model of large herringbone and parallel milking parlors. *J. Dairy Sci.* **1996**, *79*, 1960–1971. [CrossRef]
10. Price, D.R.; Heathington, K.W.; Peart, R.M. Computer simulation of dairy milking parlors. *Trans. Amer. Soc. Agr. Eng.* **1972**, *15*, 317–319. [CrossRef]
11. Micke, C.F.; Appleman, R.D. Simulating herringbone and side-opening milking parlor operations. *J. Dairy Sci.* **1973**, *56*, 1063–1069. [CrossRef]
12. Bickert, W.G.; Gerrish, J.B.; Hurt, J.H. Simulated milking rates in various semi-automatic milking parlors. *Trans. Am. Soc. Agric. Eng.* **1972**, *15*, 482–489. [CrossRef]

13. Wagner, A.; Palmer, R.W.; Bewley, J.; Jackson-Smith, D.B. Producer satisfaction, efficiency, and investment cost factors of different milking systems. *J. Dairy Sci.* **2001**, *84*, 1890–1898. [CrossRef]
14. Thomas, C.V.; DeLorenzo, M.A.; Bray, D.R.; Weldon, R.N.; Bucklin, R.A.; Martin, J.G., III. A stochastic economic analysis of large herringbone and parallel milking parlors. *J. Dairy Sci.* **1997**, *80*, 2418–2428. [CrossRef]
15. Blowey, R.; Edmondson, P. The milking routine and its effect on mastitis. In *Mastitis Control in Dairy Herds, Chapter 6*; Farming Press Books: Ipswich, UK, 1995; pp. 83–86.
16. Baines, J.R. The role of machine milking in quality milk production and ensuring dairy cow welfare. *Cattle Pract.* **2010**, *18*, 48–52.
17. Smith, J.F.; Armstrong, D.V.; Gamroth, M.J.; Harner, J., III. Factors affecting milking parlor efficiency and operator walking distance. *Appl. Eng. Agric.* **1998**, *14*, 643–650. [CrossRef]
18. Hamann, J.; Dodd, F.H. Milking routines. In *Machine Milking and Lactation, Chapter 3*; Bramley, A.J., Dodd, F.H., Mein, G.A., Bramley, J.A., Eds.; Insight Books: Berkshire, UK, 1992; pp. 69–96.
19. Reneau, J.K.; Farnsworth, R.J.; Johnson, D.G. Practical milking routines. In Proceedings of the National Mastitis Council Regional Meeting, East Lansing, MI, USA, 18 August 1994; pp. 22–32.
20. Nitzan, R.; Bruckental, I.; Bar Shira, Z.; Maltz, E.; Halachmi, I. Stochastic models for simulating parallel, rotary, and side-opening milking parlors. *J. Dairy Sci.* **2006**, *89*, 4462–4472. [CrossRef]
21. Hatem, M.; Samer, M.; Grimm, H.; Doluschitz, R.; Jungbluth, T. An expert system for planning and designing corral systems and their concrete constructions for dairy farms in hot climates. In Proceedings of the CIGR Workshop Section II: "Animal Housing in Hot Climates", CIGR Working Group 13 & China Agricultural University, Chongqing, China, 22–25 October 2009.
22. Armstrong, D.V.; Gamroth, M.J.; Smith, J.F. Milking Parlor Performance. In Proceedings of the Western Dairy Management Conference, Las Vegas, NV, USA, 4–6 April 2001; pp. 7–12.
23. Krumm, A.; Hartmut, G.; Ordloff, D. Working times and milking capacity in Swingover Parlours. *Landtechnik* **2004**, *6*, 42–343.
24. Herd, D.; Seufert, H. Work organization in large parallel or herringbone milking Parlours. *Landtechnik* **2007**, *2*, 110–111.
25. Smith, J.F.; Armstrong, D.V.; Gamroth, M.J.; Martin, J.G. Planning the milking center in expanding dairies. *J. Dairy Sci.* **1997**, *80*, 1866–1871. [CrossRef]
26. Graeme, A.M.; Reinemann, D.J. *Machine Milking*; CreateSpace: Scotts Valley, CA, USA, 2014; Volume 1, ISBN 1517603110.
27. Marnet, P.G. Milking procedures and facilities. In *Milk and Dairy Products in Human Nutrition: Production, Composition and Health*; Young, W., Park, G., Haenlein, F.W., Eds.; John Wiley and Sons: Hoboken, NJ, USA, 2013; pp. 46–64. [CrossRef]
28. Douphrate, D.I.; Hagevoort, G.R.; Nonnenmann, M.W.; Kolstrup, C.L.; Reynolds, S.J.; Jakob, M.; Kinsel, M. The dairy industry: A brief description of production practices, trends, and farm characteristics around the world. *J. Agromedicine* **2013**, *18*, 187–197. [CrossRef]
29. Sturaro, E.; Marchiori, E.; Cocca, G.; Penasa, M.; Ramanzin, M.; Bittante, G. Dairy systems in mountainous areas: Farm animal biodiversity, milk production and destination, and land use. *Livest. Sci.* **2013**, *158*, 157–168. [CrossRef]
30. Petersen, M.L.; Hansen, B.; Young, W.; Miller, K.P. Rates of milk flow and milking times resulting from selection for milk yield. *J. Dairy Sci.* **1986**, *69*, 556–563. [CrossRef]
31. Markos, H.G.; Touchberry, R.W. Heritability of rates of milk flow and time required to milk. *J. Dairy Sci.* **1970**, *53*, 188–194. [CrossRef]
32. Appleman, R.D.; Micke, C.F. Relationships of milking systems to adequacy of milking. *J. Dairy Sci.* **1973**, *56*, 1450–1458. [CrossRef]
33. Touchberry, R.W.; Markos, H. G Variations in the time required to milk cows. *J. Dairy Sci.* **1970**, *53*, 176–187. [CrossRef]
34. Tamburini, A.; Sandrucci, A.; Nicoletti, C.; Zanini, L. Milking procedures and milk ejection in Italian Brown cows. *Ital. J. Anim. Sci.* **2007**, *6* Suppl. 1, 478–480. [CrossRef]
35. Raussi, S. Human-cattle interactions in group housing. *Appl. Anim. Behav. Sci.* **2003**, *80*, 245–262. [CrossRef]
36. Vanbergue, E.; Peyraud, J.L.; Ferlay, A.; Miranda, G.; Martin, P.; Hurtaud, C. Effects of feeding level, type of forage and milking time on milk lipolytic system in dairy cows. *Livest. Sci.* **2018**, *217*, 116–126. [CrossRef]

37. Bachman, K.C.; Hayen, M.J.; Morse, D.; Wilcox, C.J. Effect of pregnancy, milk yield, and somatic cell count on bovine milk fat hydrolysis. *J. Dairy Sci.* **1988**, *71*, 925–931. [CrossRef]
38. Peychev, K.; Georgiev, D.; Dineva, G.; Dimova, V. Structure-time analysis and development of dairy cows machine milking models in "Herringbone" milking parlors. *Bulg. J. Agric. Sci.* **2019**, *25*, 196–200.
39. Smith, J.F.; Harner, J., III; Armstrong, D.V.; Fuhrmann, T.; Gamroth, M.J.; Brouk, M.J.; Reid, D.A.; Bray, D. Selecting and managing your milking facility. In Proceedings of the VI Western Dairy Management Conference, Reno, NV, USA, 12–14 March 2003; pp. 119–131.
40. CRPA (Centro Ricerche Produzioni Animali). *Strutture, Attrezzature e Impianti Per Vacche Da Latte*; L'Informatore Agrario: Verona, Italy, 2002; ISBN 8872201705.
41. Rasmussen, M.D. Influence of switch level of automatic cluster removers on milking performance and udder health. *J. Dairy Res.* **1993**, *60*, 287–297. [CrossRef] [PubMed]
42. Carreira, X.C.; Marino, R.A.; Perez-Cancio, X.; Fernandez, M.E. Design of algorithms for evaluating the efficiency of labour and variability of some group milking parlour parameters. *Livest. Sci.* **2013**, *151*, 252–263. [CrossRef]
43. Boettcher, P. Breeding for improvement of functional traits in dairy cattle. *Ital. J. Anim. Sci.* **2005**, *4*, 7–16. [CrossRef]
44. Hamann, J.; Osteras, O.; Mayntz, M.; Woyke, W. Functional parameters of milking units with regard to teat tissue treatment. In *Teat Tissue Reactions to Machine Milking and New Infection Risk*; Bulletin of the IDF (International Dairy Federation) No. 297/1994; IDF: Brussels, Belgium, 1994; pp. 23–34.
45. Chambliss, R.L., Jr. *Labor and Capital Requirements in Herringbone and other Elevated Stall Milking Parlors*; Virginia Agricultural Experiment Station, Virginia Polytechnic Institute: Blacksburg, VA, USA, 1962; p. 46.

animals

MDPI

Article

Production and Health Management from Grazing to Confinement Systems of Largest Dairy Bovine Farms in Azores: A Farmers' Perspective

Ivo Medeiros [1], Aitor Fernandez-Novo [2], Susana Astiz [3] and João Simões [1,*]

1 Veterinary and Animal Research Centre (CECAV), Department of Veterinary Sciences, School of Agricultural and Veterinary Sciences, University of Trás-os-Montes and Alto Douro (UTAD), 5000-801 Vila Real, Portugal; al62766@utad.eu
2 Department of Veterinary Medicine, School of Biomedical and Health Sciences, Universidad Europea de Madrid, C/Tajo s/n, Villaviciosa De Odón, 28670 Madrid, Spain; aitor.fernandez@universidadeuropea.es
3 Animal Reproduction Department, National Institute of Agronomic Research (INIA), Puerta De Hierro Avenue s/n, CP, 28040 Madrid, Spain; astiz.susana@inia.es
* Correspondence: jsimoes@utad.pt; Tel.: +351-259-350-666

Citation: Medeiros, I.; Fernandez-Novo, A.; Astiz, S.; Simões, J. Production and Health Management from Grazing to Confinement Systems of Largest Dairy Bovine Farms in Azores: A Farmers' Perspective. *Animals* **2021**, *11*, 3394. https://doi.org/10.3390/ani11123394

Academic Editor: Umberto Bernabucci

Received: 25 October 2021
Accepted: 25 November 2021
Published: 27 November 2021

Simple Summary: This study aimed to evaluate differences and critical factors in production and health management between dairy cattle farms with fixed milk parlours (FMP), and mobile milk systems (MMS) from Azorean grasslands. According to the farmers' perspective, calf diarrhea, calf pneumonia, infertility/poor reproductive management, and mastitis were the main problems that farms faced in 2020. FMP was associated with more advanced and mechanized production systems, with a higher adherence to preventive and biosecurity control programs, than traditional MMS farms. MMS farms also showed a greater vocation for dual-purpose farming (beef and milk), smaller herd sizes and more grazing time for cows. In conclusion, inherent and non-inherent differences in production and health management between FMP and MMS were quantified by authors. These results indicate that a greater adoption of preventive veterinary medicine and biosecurity measures should be taken, especially among MMS farms. The education of farmers should also be improved and stimulated.

Abstract: The intensification of bovine milk production in the Azores has led farmers to increase farm size and specialization in grasslands, implementing confined and semi-confined production systems. Fixed milking parlours (FMP) have progressively gained more popularity, at the expense of conventional mobile milking systems (MMS). The present study aimed to evaluate the associations between production and health management in dairy cattle farms, with FMP or MMS, in grasslands (São Miguel, Azores), according to the farmers' perspective. A total of 102 questions about production and health management were surveyed in 105 farms with >30 dairy cows each. Farms with FMP were associated ($p \leq 0.05$) with larger herd size, better facilities, and specialized management, however, the adoption of preventive and biosecurity measures should be improved by these farmers. MMS farms implemented a lower level of disease prevention or control programs, less frequent transhumance, and showed a wider vocation to dual-purpose (milk and cross beef) than FMP farms. In conclusion, MMS and FMP farms tried to optimize yield and economic viability in different ways using grasslands. Several biosecurity and health prevention constraints were identified for improvement.

Keywords: herd health; milking management; production systems

1. Introduction

The termination of milk quotas in the European Union has led to a more liberal and competitive market, with a more volatile and uncertain milk price [1]. As such, farms have changed their methods of production and operation. Dairy farms are increasing in size with growing efficiency, specialized work, and a higher adoption of preventive measures [2–4].

Areas with a mild temperature and sufficient humidity and rainfall, make pasture-based dairy production possible for bovine farms [5]. Thus, several regions worldwide, such as São Miguel Island (Azores) and New Zealand have adopted a "traditional" dairy milk production system, based on pastures with cows grazing during most of the year. Cows are naturally grazing animals, and pastures are their normal environment, where they can express normal behavior, and this theoretically creates the highest welfare level [6,7]. In fact, grazing herds achieve better animal welfare indicators, and a lower prevalence of certain diseases, such as lameness or even mastitis [8,9].

Despite this, cows fed exclusively or mostly on grazing sometimes cannot achieve their nutritional needs, and are thus limited in their productivity [10]. Moreover, one of the key limitations of a full grazing system has historically been hand milking. The first milking system was introduced in the 1800s, and has greatly evolved up to the present day [11], with the surge of robotic milking systems. Herringbone parlors were invented in New Zealand (1952), featuring advances in vacuum systems [12], to provide milking rooms for large-scale milking. Fixed milking parlors (FMP) are also used in grazing systems, but mobile milking systems (MMS) have also been developed, thus allowing the use of separate or rotational grasslands [13]. Other advantages are keeping cows in their natural environment, independently of distance from farming facilities, and reducing stress due to excessive animal movement.

The fall of the price of milk and an increasingly competitive market have led to the need to increase farm efficiency [14]. In the case of improvement of farms on pastures, farmers have chosen different production alternatives by using more intensified management [15]. Furthermore, some farms have reduced or removed all grazing, switching to an indoor-based system. Simultaneously, efforts have been made to optimize outdoor systems [16], including the use of grazing-based robotic milking systems, and improving the quality of grass, grass growth, grazing conditions, and grazing management [17–19].

There is an obvious relationship between veterinary health management and types of dairy systems, with health management progressively changing from curative to preventive programs with increasing farm size and individual milk yield [3]. The increases in milk production and farm size can lead to negative effects on cows, including the rising prevalence of certain diseases, or the emergence of new health issues that are directly linked to this kind of system [20], which are related to (reduced) cattle resilience [21]. These kinds of problems can be observed, in both FMP and MMS systems, with a certain level of intensification of their management systems [22].

Traditionally, MMS are the most often used milking systems in the Azores [5], based on 100% grazing time and dairy cattle transhumance between grasslands. In recent decades, together with general management and production system changes, these milking systems have been progressively replaced by FMP. Although these changes are occurring, there have been few formal studies on the level and impact of this shift in this region. However, it is known that the effects can be diverse [23], and that there are areas in the world, such as New Zealand and Ireland, where although the systems have evolved, the usage of pastures has prevailed [24,25]. Despite their relevance, the impacts of these changes on farm performance, management conditions, and animal health of these "changing" or "staying traditional" systems in the Azores, have been ignored.

Therefore, the aim of this study was to recover information on how Azorean farms are actually functioning in terms of production and health management. This was done by surveying the main actors and drivers of these changes, i.e., the farmers of Azorean dairy cattle farms on São Miguel Island, and differentiating between farms using MMS, which are "traditional systems", or using FMP, which are "evolved systems". Our hypothesis was that health management shows several differences depending on the production system, due to inherent differences that can be quantified and described by farmers. The ultimate goal of this study was to characterize herd health management of these dairy farms, thus identifying critical factors that should be improved, in order to increase Azorean dairy industry competitiveness.

2. Materials and Methods

2.1. Local, Sample Size and Selection of Farms

The survey addressed dairy farmers of São Miguel Island (Azores). Located in the middle of the Atlantic Ocean, the Azores is an archipelago with 9 islands, with mild temperatures (minimum and maximum temperature: 12.0–18.4 °C, respectively), humidity (minimum and maximal relative humidity: 89.0–97.4%) [26,27], and an abundant rainfall climate with precipitation of 960.6 ± 201 mm per year, with 75% of the precipitation falling between October and March [26,27].

Considering the mean size of Azorean farms (26 ± 3.9 cows; ±SD), only dairy and dual-purpose (from crossbreeding calves) farms with ≥30 adult cows were selected to obtain the largest ones due to their potential economies of scale. A total of 110 questionnaires were personally distributed by the first author from February to April 2021. Farmers were interviewed face to face (15 interviews were performed by the first author), or online via Google Forms [28] (90 respondents), after having personally contacted the farmer or having talked to them over the telephone.

Intensive or semi-intensive production systems were defined according to the time that the cows stayed indoors (confined), with availability of a milking parlor or fixed fixed milking parlors (FMP), or outdoors (grazing) using mobile milking systems (MMS), respectively, during the whole or greater part of the day. Cows from semi-intensive production systems grazed for at least 8 h per day [29]. All the farms used cattle with a genetic merit for milk-yield production.

2.2. Survey

The questionnaire addressed herd health management issues and major health problems observed during 2020 by farmers, as a modified version of the questionnaire used in a previous study [30]. Our questionnaire was completed by a preliminary assessment of management practices occurring in Azorean farms. The structured questionnaire consisted of eleven topics: characterization of the farm; biosecurity; calving and fertility; rearing management (up to 12 months of age); lameness; nutrition; reproduction; milking practices and mastitis; disease prevention; dry cow management; and major problems.

The questionnaires consisted of 102 questions, both closed and binary, with one additional table to express ordinal categories of intensity of the referred problem/issue. A total of 15 questions addressed management indicators and farm characteristics. A mastitis prevalence 5-point scale was proposed for affected cows: 1: 10%; 2: 10–20%; 3: 20–30%; 4: 30–40%; and, 5: >40% of prevalence. For health problem intensity estimation, the following scale was proposed: 1-Not problematic; 2-Less problematic; 3-Problematic; 4-Quite Problematic; 5-Serious/uncontrolled at the farm level.

2.3. Statistical Analysis

All data were recorded and statistically analyzed under the regulations of the General Data Protection Regulation (GDPR), in accordance with European regulation [31].

The minimum sample size of questionnaires (n = 99) was calculated according to Thrusfield, by considering the 95% confidence level, a 5% margin of error (Z = 1.96) and a 93% expected response rate, and adjusted for finite populations (3247 farms) [32].

All information was coded numerically, in order to assist analysis and guarantee anonymity. Uncategorized data were recoded into ordinal level data. Categorical data were entered into a database. Surveys not fully answered contributed partially to the responding topics, with a univariate model applied to maximize the number of answers per question. Two groups were formed to compare differences, depending on the type of milking systems used: FMP versus MMS farmers (Figure 1). Differences between percentages were evaluated with the Pearson chi-square test. A non-normal distribution of all continuous variables, including all five-point scales from categories of intensity/prevalence, was confirmed using the Shapiro–Wilk test. Therefore, a non-parametric one-way ANOVA model, followed by Van der Waerden post hoc analysis to test significance, was used [33]. The results are described as the mean percentage, or mean score and variation, as ±SEM for a significance level of 0.05. JMP® 14 software for Windows (SAS Institute, Cary, NC, USA) was used.

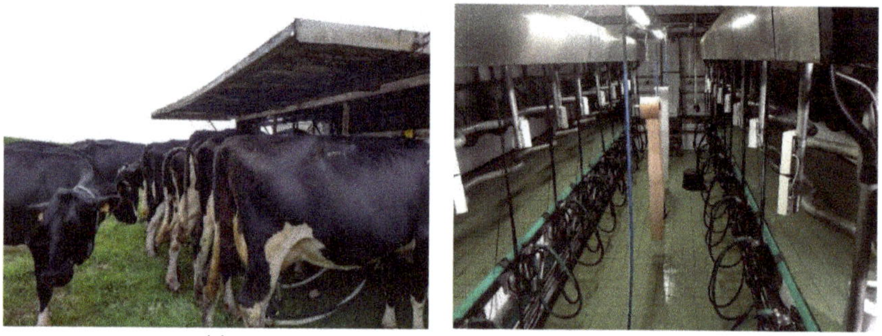

(a) (b)

Figure 1. Traditional mobile milking system (**a**) and fixed milking parlors (**b**).

3. Results

The overall response rate was 95.5% (105/110). In the respondents, we found 82.9% (87/105) with FMP, versus 17.1% (18/105) with MMS.

3.1. Characterization of the Farms, Biosecurity and Veterinary Advice

The estimated farm size (total cows) was 35% higher in FMP (213.1 ± 11.7; n = 87) than MMS (157.8 ± 21.9; n = 18; $p < 0.05$; Table S1) farms. A higher proportion of dual-purpose farms Artificial Insemination (AI) performed with beef breeds to have crossbreed calves), was observed in the MMS (50.0%; 9/18) as opposed to in the FMP group (24.1%; 21/87; $p < 0.05$). Semi-intensive dairy production was observed in 74.3% (78/105; $p < 0.001$) of the farms without differences between groups ($p = 0.33$). The remaining farms practiced intensive production.

Herringbone parlor (64.4%; 56/87), tandem parlor (34.5%; 30/87) and robotic milking machines (1.1%; 1/87) were observed in FMP farms. All MMS farms had Herringbone parlors. Refrigerated milk bulk tanks were more frequent in FMP (90.8%; 79/87) than in MMS farms (22.2%; 4/18; $p < 0.001$), as was the inclusion into an official animal welfare program (44.8%; 39/87 and 16.7%; 3/18, respectively, $p < 0.05$).

A higher proportion of FMP farms (58.6%; 50/87) had isolated sick pens/bays for sick animals than in MMS farms (27.8%; 5/18; $p < 0.05$). Animal transhumance is a common practice in the Azores, and it is defined as the movement of animals using public roads, so the animals can move from one pasture to another. This practice was less frequently practiced in FMP (42.5%; 37/87) compared with MMS (83.3%; 15/18; $p < 0.01$) farms.

The veterinary assistance provided by veterinarians working exclusively with farmer co-operatives, partially nationally subsidized, differed between types of farms (MMS: 94.4%; 17/18; FMP: 57.1%; 48/84; $p < 0.01$), with 21.8% (19/84) of FMP farms reporting veterinary assistance only from private veterinarians (21 farmers did not respond to this question).

3.2. Reproductive Management

No differences in breeding methods were found between farms (Table 1), with artificial insemination being mainly and almost exclusively implemented in adult cows, while in the heifers, the choice fell mostly to natural mating. The estimated mean number of services per pregnant (P) adult cow was very similar between FMP (2.3 ± 0.2 AI/P) and MMS (2.1 ± 0.1 AI/P; $p = 0.99$) farms.

Table 1. Reproduction management in dairy farms with fixed (FMP) and mobile (MMS) milking systems.

Factor	Farms		p Value
	FMP	MMS	
Breeding			
Adult females (number of animals)	139.1 ± 8.0 (32–400) [(1)]	100.8 ± 13.0 (20–200)	0.04
Heifers (number of animals)	31.5 ± 2.3 (8–130)	24.8 ± 3.9 (8–70)	0.17
Age at first breeding (months)	16.1 ± 0.3 (12–20)	15.3 ± 0.7 (12–18)	0.14
Breeding methods—Adult cows			
Artificial insemination (n = 46)	42.5% (37/87) [a (2)]	50.0% (9/18) [a]	0.82
Natural service (n = 8)	8.1% (7/87) [b]	5.6% (1/18) [b]	
Both (n = 51)	49.4% (43/87) [a]	44.4% (8/18) [a]	
Breeding methods Heifers:			
Artificial insemination (n = 18)	16.1%(14/87) [a]	22.2% (4/18) [a]	0.82
Natural service (n = 75)	72.4% (63/87) [b]	66.7% (12/18) [b]	
Both (n = 12)	11.5% (10/87) [a]	11.1% (2/18) [a]	
Artificial insemination performed by (n = 103):			
Technician (n = 94)	91.9% (79/86) [a]	88.2% (15/18) [a]	0.08
Farmer (n = 8)	8.1% (7/86) [b]	5.9% (1/18) [b]	
Both (n = 1)	0.0% (0/86)	5.9% (1/18) [b]	
Reproductive management:			
Beef sire semen (n = 103) [(4)]	92.9% (79/85)	94.4% (17/18)	0.82
Estimated mean number of services per pregnancy, cows (n = 103)	2.3 ± 0.2 (1–5)	2.1 ± 0.1 (1–4)	0.99
Reproductive examination during open days [(3)] (n = 104)	73.6% (64/87)	35.3% (6/17)	0.002
Ancillary oestrus detection devices	39.1% (34/87)	38.9% (7/18)	0.99
Protocols of oestrus or ovulation induction/synchronization	65.5% (57/87)	16.7% (3/18)	<0.001
Pregnancy diagnosis	79.3% (69/87)	50.0% (9/18)	0.01
Pregnancy diagnosis method (n = 78):			
Manual transrectal palpation (exclusively)	13.0% (9/69) [a]	55.6% (5/9) [a]	0.002
Ultrasonography	87.0% (60/69) [b]	44.4% (4/9) [b]	0.002
Abortion:			
Estimated total abortion number detected by farmer in 2020	3.5 ± 0.3 (0–10)	3.1 ± 0.5 (0–8)	0.41
Abortion timing (n = 101):			
Up to 3 months	10.7% (10/84) [a]	41.2%(7/17) [a]	0.005
3–6 months	69.1%(58/84) [b]	52.9% (9/17) [b]	
>6 months	17.8% (17/84)% [a]	5.9% (1/17) [a]	
Venereal disease diagnosis of sires (mating; n = 57)	2.3% (1/43)	8.3% (1/12)	0.33
Laboratory diagnosis, according to farm history, of (n = 42):			
IBRV (n = 16)	75.0% (12/16)	25.0% (4/16)	0.005
BVDV (n = 16)	81.3% (13/16)	18.8% (3/16)	<0.001
Neosporosis (n = 8)	75.0% (6/8)	25.0% (2/8)	-
Toxin/fungi (n = 2)	50.0% (1/2)	50.0% (1/2)	-

[a, b] Different superscript letters for the same column: $p < 0.01$. %: Percentage of farms with an affirmative response; n: Number of respondents. Omitted values means n = 105. [(1)] arithmetic mean ± standard error of mean (min–max). [(2)] (n/N): number of affirmative responses/number of total respondents. [(3)] Previous evaluation of uterine involution/content and ovarian examinations of breeding cows. Abbreviations: BVDV, bovine viral diarrhea virus; IBRV, infectious bovine rhinotracheitis virus. [(4)] Beef sire semen was used in selected dairy cows for crossbreeding purposes to obtain beef calves.

Reproductive management included reproductive examination up to pregnancy (p = 0.002), estrus or ovulation induction/synchronization protocols ($p < 0.001$) and routine pregnancy diagnosis (p = 0.002), which was more frequently implemented in FMP farms.

In FMP farms, most of the abortions were in the middle of gestation (3–6 months of pregnancy), while in MMS farms, the majority of abortion tended to occur up to six months into gestation ($p < 0.01$). The laboratory diagnosis of abortive agents at abortion occurrence during 2020 was low (10.5%; 15/105), with no significant differences (p = 0.49) between types of farms. Nevertheless, and according to the history of the farms, in 76.2% (32/42) of the cases, an infectious/toxic etiology was identified in FMP farms ($p < 0.001$).

3.3. *Rearing Management (Up to 12 Months)*

Seasonal calving distribution, according to grass availability, tended to be more frequent in MMS farms (22.2%; 4/18; p = 0.07; Table S2). Calving pens were more common in FMP farms (51.7%; 45/87; $p < 0.01$). Retained placenta was classified as a significant problem for 25.7% (27/105) of the farmers, with no significant differences between farms ($p = 0.83$). FMP farms mainly (64%; 55/86) buried placentas at pasture, while MMS farms did not discharge them out at all (44.4%; 8/18; $p < 0.05$).

Appropriate calf barns were also more commonly observed in FMP (89.7%; 78/87) farms compared with MMS (72.2%; 8/13; $p = 0.05$) farms. Colostrum tended to be administered for more days in MMS (4.7 $\pm$ 0.3 days) than in FMP (3.9 $\pm$ 0.5 days; $p = 0.10$) farms. Most farms, independent of milking system, did not store colostrum (15.2%; 16/105; $p = 0.45$) but considered diarrhea as their main problem at calving (69.5%; 73/105). Pneumonia in calves was considered a major problem in MMS farms (72.2%; 13/18; $p < 0.01$), which also reared males calves more frequently for fattening (61.1%;11/18; $p < 0.01$). Preventive measures, such as vaccination of pregnant cattle to prevent pneumonia and diarrhea in calves (25.3%; 22/87 vs. 5.6%; 1/18, respectively; $p = 0.07$) and vaccination of calves up to 12 weeks of age (13.8%; 12/87 vs. 0.0%; 0/18, respectively; $p = 0.09$), tended to be more frequently adopted by FMP than MMS farms.

3.4. Nutrition and Metabolic Disease Prevention

A higher level of nutritional assessment was performed in FMP farms. Nevertheless, both farm groups were regularly assessed by a nutritionist (88.6%; 93/105; $p = 0.96$), regardless of the type of farm. Body condition scoring was more commonly implemented in FMP farms (60.9%; 53/87; $p < 0.05$; Table 2).

Table 2. Nutritional management and metabolic disease prevention in dairy farms with fixed (FMP) or mobile (MMS) milking systems.

Factor	Farms		*p* Value
	FMP	MMS	
Nutritional assessment:			
Scoring body condition (peripartum; $n = 104$)	60.9% (53/87) [(1)]	29.4% (5/17)	0.02
Forage nutritional analyses (FNA, $n = 104$)	93.1% (81/87)	58.8% (10/17)	<0.001
Diet adjustment based on FNA results ($n = 94$)	90.5% (76/84)	60.0% (6/10)	0.006
Feeding management:			
Unifeed system ($n = 104$)	80.5% (70/87)	41.2% (7/17)	<0.001
Adding concentrate feed to unifeed ($n = 75$)	50.0% (34/68)	42.9% (3/7)	0.72
Own forages	93.1% (81/87)	100% (18/18)	0.25
Corn silage [(2)]	94.3% (82/87)	88.9% (16/18)	0.41
Grass silage ($n = 100$) [(3)]	8.3% (7/84)	6.3% (1/16)	0.78
Baled grass silage ($n = 103$) [(4)]	98.8% (83/84)	100% (17/17)	0.66
Hay rolls ($n = 101$) [(5)]	23.5% (20/85)	25.0% (4/16)	0.90
Straw ($n = 103$) [(6)]	32.6% (28/86)	43.8% (17/16)	0.39
Access to pasture (grassland)	73.6% (64/87)	94.4% (17/18)	0.06
Feed concentrate during milking	85.1% (74/87)	100% (18/18)	0.08
Dry cow diet [(7)] ($n = 104$)	23.0% (28/87)	58.8% (10/17)	0.003
Feed concentrate to dry cows	32.2% (20/87)	22.2% (4/18)	0.40
Water source of the farm:			
Pit water ($n = 2$)	1.1%^{a} (1/87) ^{a}	5.6% (1/18) ^{a}	0.32
Riverside ($n = 25$)	25.3% (22/87) ^{b}	16.7% (1/18) ^{a}	
Municipal water supply ($n = 76$)	72.4% (63/87) ^{c}	72.2% (13/18) ^{b}	
Wellspring ($n = 11$)	1.1% (1/87) ^{a}	5.6% (1/18) ^{a}	

n: Number of respondents. Omitted values means $n = 105$. %: Percentage of farms with an affirmative response. $^{a, b, c}$ Different letters for the same column: $p < 0.01$. [(1)] (n/N): number of affirmative responses/number of total respondents. [(2)] Corn silage is used by 81.4% (79/97) of the farms during the whole year. [(3)] Grass silage is used by 50% (4/8) of the farms during the whole year. [(4)] Baled grass silage is used by 12.5% (3/24) of the farms during the whole year. [(5)] Hay rolls are used by 22.9% (8/35) of the farms during the whole year. [(6)] Only 1.4% (1/73) of farmers also fed animals alfalfa. [(7)] Farmers who do not use a specific dry cow diet reported that they fed cows at pasturage ($n = 59$) and/or baled grass silage ($n = 48$), corn silage ($n = 15$), straw ($n = 6$) and/or grass silage ($n = 1$) segregated from lactating cows.

Forage nutritional analyses ($p < 0.001$), as well as diet adjustments ($p < 0.01$), were more frequent in FMP farms than in MMS farms. Additionally, the total mixed ration

(TMR) system, also named the "unifeed" system, was more frequently found in FMP farms ($p < 0.001$), with forages coming from their own production in all MMS farms ($p = 0.02$).

Water was administered ad libitum in all farms. Mostly, the water of both groups of farms came from the municipal water supply (72.4%; 76/105).

3.5. Milking and Mastitis

Differences were observed in the milking procedures of farms. Pre-dipping ($p < 0.001$) and paper towels ($p < 0.001$) but not post-dipping ($p = 0.30$) were more frequently implemented in FMP farms (Table 3).

Table 3. Milking procedures and mastitis scores, according to fixed (FMP) or mobile (MMS) milking system farms.

Factor	Farms		*p* Value
	FMP	MMS	
Milking procedures			
Pre-dipping	63.1% (53/84) [1]	16.7% (3/18)	<0.001
Post-dipping	98.8% (83/84)	94.4% (17/18)	0.3
Paper towels	78.6% (53/84)	11.1%(3/18)	<0.001
Gloves	47.2% (40/84)	33.3% (6/18)	0.27
Separate teatcups for mastitis cows	7.1% (6/84)	5.6% (1/18)	0.83
Teatcup disinfection after use by mastitis cows	4.8% (4/84)	0.0% (0/18)	0.35
Hot water cleaning machine	9.5% (40/84)	5.6% (1/18)	0.59
Mastitis			
Mastitis incidence (Score [3])	2.2 ± 0.1 (1–5) [2]	2.1 ± 0.2 (1–5)	0.54
Culling or death of mastitic cows	17.2% (15/87)	22.2% (4/18)	0.62
Estimated somatic cells count (log 10)	2.38 ± 0.30 (2.00–2.45)	2.45 ± 0.02 (2.00–2.60)	0.07

%: Percentage of farms with an affirmative response. *n*: Number of respondents. Omitted values means $n = 105$. [1] number of affirmative responses/number of total respondents. [2] arithmetic mean ± standard error of mean (min–max). (*n*/N):. [3] Scale 1 to 5, according to the percentage of affected cows with mastitis during 2020: 1:10%; 2:10–20%; 3:20–30%; 4:30–40%; 5: >40%.

Mastitis incidence and mortality did not differ among farms (18.1%; 19/105; $p = 0.62$), but the estimated somatic cell count tended to be higher in MMS farms ($p = 0.07$).

3.6. Lameness

Lameness was considered a major problem in 41% (43/105) of all farms. Nevertheless, only 34% (33/105) of farmers implemented a continuous lameness control program (Table S3). Approximately half (48.3%; 42/87) of the FMP farms had a functional footbath, contrary to MMS farms (5.6%; 1/18; $p = 0.001$). On most of the farms (77.2%; 78/101), trimming was only performed as treatment after the detection of lameness, without differences between groups.

3.7. Drying-Off and Prepartum Care

No significant differences in cow management between groups were observed at dry off, including the length of the dry period (45–60 d; Table S4). Drying-off anti-biotherapy was administered in 94.3% (99/105) of the farms. The mean percentages of the main procedures during the dry period and prepartum are reported in Figure 2.

The dry cows joined lactating cows in the prepartum period in 90.5% (95/105) of the farms without differences between groups ($p = 0.26$). FMP farms (95.3%; 82/86) tended to include a higher percentage of pregnant heifers in the lactating herd during the prepartum period, compared with MMS farms (83.3%; 15/18; $p = 0.06$).

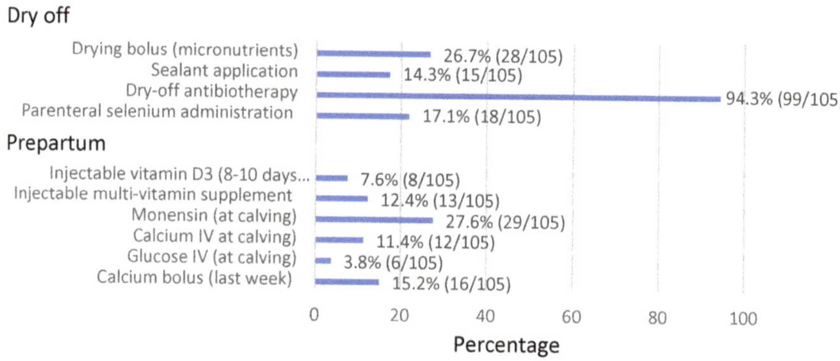

Figure 2. Main practices, micronutrient, and calcium administrations at dry-off and prepartum periods.

3.8. Disease Prevention and Major Problems

In general, a low to moderate level of preventive measures was implemented in farms, with some differences between FMP and MMS groups (Table 4). The frequency of blood sampling for disease diagnosis was higher in MMS (11.1%; 2/18) than in FMP (1.2%; 1/87; $p < 0.05$) farms. Conversely, the use of insecticide during the hot season/periods was more frequent in FMP farms (86.2%; 75/87; $p = 0.05$).

Table 4. Preventive health measures were adopted by fixed (FMP) and mobile (MMS) milking system farms.

Factor	Farms		p Value
	FMP	MMS	
Preventive measures			
Disease prevalence monitoring (serum samples)	1.1% (1/87) [(1)]	11.1% (2/18)	0.02
Parasitic disease monitoring (faecal samples)	0.0%	0.0%	-
Mineral monitoring (serum samples)	0.0%	0.0%	-
Mineral diet supplementation during dry period	62.1% (54/87)	50.0% (8/18)	0.17
Insecticide during hot season/periods	86.2% (75/87)	66.7% (12/18)	0.05
Regular deworming	56.3% (49/87)	50.0% (9/18)	0.42
Vaccination *:			
Clostridial diseases (n = 3)	1.1% (1/87) [a]	11.1% (2/18)	0.02
IBR/BVD (n = 37)	36.8% (32/87) [b]	27.8% (5/18)	0.47
Mastitis (n = 16)	13.8% (12/87) [c]	22.2% (4/18)	0.37
Respiratory complex disease (n = 16)	14.9% (13/87) [c]	16.7% (3/18)	0.90

%: Percentage of farms with an affirmative response. [(1)] (n/N): Number of affirmative responses/number of total respondents. [a, b, c] Different superscript letters for the same column: $p < 0.01$. n: number of farms with an affirmative response. Omitted values means n = 105. * total of farms using vacines = 61 (some farms used more than one vaccine type). Abbreviations: BVD, bovine viral diarrhea; IBRV, infectious bovine rhinotracheitis.

Vaccination for clostridial diseases, bovine rhinotracheitis virus/bovine viral diarrhea, mastitis and/or respiratory complex disease was implemented in 42.9% (45/105) of the farms, with no differences between FMP and MMS ($p = 0.50$) groups.

Blood sampling for mineral quantification and fecal sampling for parasitic disease diagnosis were not performed by any of the farms questioned.

According to the farmers surveyed, no natal differences were observed regarding the problems that affected FMP and MMS dairy farms in 2020 (Figure 3). However, neonatal diarrhea (score point = 4.0 ± 0.2 vs. 3.6 ± 0.1; $p = 0.04$) and infertility/reproductive problems (score point = 3.3 ± 0.2 vs. 2.9 ± 0.1; $p = 0.05$) were slightly more problematic for MMS farms, as opposed to FMP farms.

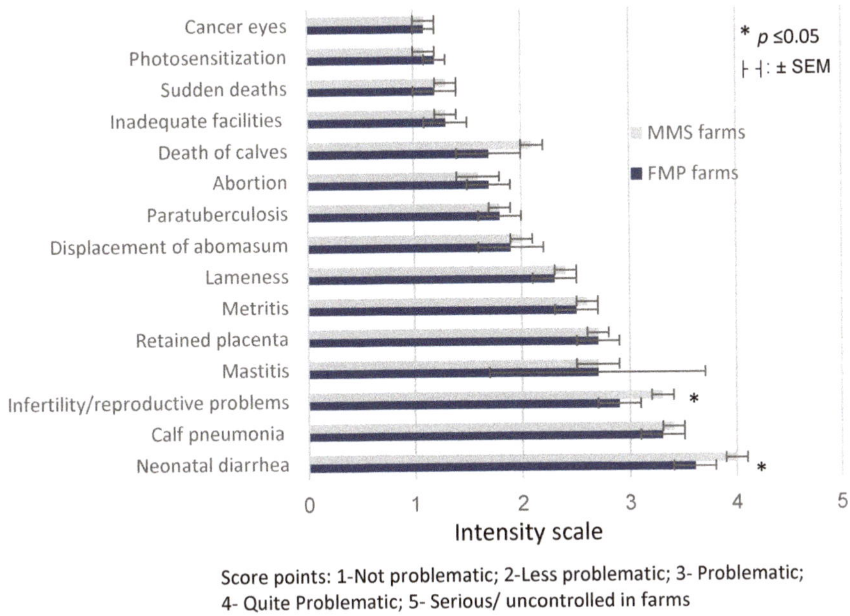

Figure 3. Score intensity scale of major problems suffered in 2020, according to the farmers' perceptions, in dairy farms with mobile (MMS) versus fixed (FMP) milking systems.

Neonatal diarrhea, calf pneumonia, mastitis, retained placenta, metritis, lameness and infertility/reproductive problems were the most problematic diseases/issues indicated by farmers from both groups of farms.

4. Discussion

Overall, larger herds, better facilities, greater focus on prevention, and constant nutritional and reproductive assessments were more frequently observed in FMP farms than in MMS farms. In contrast, a higher proportion of dual-purpose farms (milk and beef production) was found among the MMS farms trying to increase income. Therefore, the single purpose to produce milk, observed more often in Azorean FMP farms than in MMS farms, seems to be in line with the new reality of the European market, which has been without milk quotas since 2015, requiring a more efficient and specialized milk production.

The (semi)intensification level of dairy milk production systems (>70%), was similar in both farm groups in our study. In fact, supplementation to grazing dairy animals is required to maintain an adequate level of milk yield [34], together with an improvement in the stocking rate in the Azores, e.g., 3.2 cows/ha, as reported by Morais et al. (2018) [35]. Despite this last issue, important management differences were observed between the types of farms.

Regarding reproductive management, differences were observed between the types of farmers surveyed (see Table 1), with FMP farmers more frequently implementing complete reproductive management protocols and appropriate tools. This is to say, reproductive examination during open days, to select cows for breeding, and to treat diagnosed pathologies, induction and synchronization of estrus and ovulation, pregnancy diagnosis, and ultrasonography. All these interventions are crucial to optimize the reproductive output of dairy farms [4]. Further research is required to quantify differences between production systems.

In our study, a total of 39% of the farmers, independent of farm group, used at least one ancillary device to detect estrous. This result denotes the progressive adoption of technologies to improve fertility. In a similar survey of Canadian dairy farms [36], 89% of farmers used visual detection of estrous as their only method (3.5 observation times per day), i.e., only a low proportion of farms used ancillary devices. Nevertheless, in this last study, fixed-time artificial insemination was mainly implemented [36]. However, it is well known that several devices to detect estrus have been largely and efficiently implemented worldwide [37,38].

Failure in heat detection and low conception rates are major reproductive problems [39], with infertility and poor production being the main causes for culling dairy cows in the USA [40]. In fact, poor reproductive management represents losses of up to 231€ per cow per year, due to a decrease in milk yield and a high calving interval [39]. Therefore, the implementation of adequate reproductive strategies [41] is essential in dairy farms and should be enhanced in Azorean farms, especially among MMS farms. Pregnancy diagnosis ($p = 0.01$), protocols for estrus/ovulation induction ($p < 0.001$) and reproductive examination during open days ($p = 0.002$), were performed less often in MMS farms than in FMP farms. All these results were expected, since indoor systems allow better reproductive control and management [42].

In the present study, the number of calves born on the farms (2020) was, as expected, higher in FMP farms, since these farms had larger herd sizes (213.1 $\pm$ 11.7 and 157.8 $\pm$ 21.9 total animals for FMP and MMS, respectively; $p < 0.05$), and the trade of pregnant heifers or adult cows remained low. Appropriate calving pens were more commonly found in FMP farms, which are essential to control the vitality of the newborns, adequate immunity transference, and the mothers' health [43]. Nevertheless, the advantages and limitations of calving indoors or outdoors are still up for debate [44].

In MMS farms, we observed a higher tendency towards seasonal calving according to grass availability ($p = 0.07$). This is due to the importance of grass in the production system of these farms. This practice is in line with what is very common in Ireland and New Zealand, where seasonal calving is largely adopted, so local farms can take advantage of their animal production potential at the time of grass growing [45]. Additionally, in our study, only 30% of farms minimized calving during summer. This calving seasonality has the potential advantage of mitigating calving heat stress [46], and initiating new lactation in more developed grass periods, even if forages are stored. It has been observed that calving in late winter is most profitable in grazing systems, independent of milk premium price [47].

Rearing the animals is a very sensitive part of the production system, with correct hygiene, general management, and appropriate colostrum administration being essential to prevent the main health problems for young cattle: enteric and respiratory diseases being responsible for the highest mortality and morbidity rates [48–50].

It is essential to provide a sufficient volume of high-quality colostrum in the first hours of life [51]. Very few farms in our study (17.2% FMP and 11.1% MMS farms) had colostrum storage banks, and even fewer had an appropriate instrument to evaluate the quality of the colostrum prior to storage. All these aspects, as well as cleanliness of the pens and colostrum quality assessment, have been associated with pneumonia or diarrhea in calves [52]. Therefore, if these practices improved, the incidence of neonatal diseases would possibly decrease on São Miguel dairy farms (see Figure 2). Indeed, diarrhea and pneumonia in calves were the main problems faced by Azorean farmers, which is another issue to be addressed. In addition to direct losses [49], it is known that calves raised without diarrhea and/or pneumonia achieve greater longevity, yield, and profitability; furthermore, they are healthier cows [53]. In our study, FMP farms more frequently had calf sheds, administered colostrum for more days, and utilized more preventive measures in rearing than MMS farms did, evidencing poorer rearing management on these MMS farms.

Additionally, a tendency of FMP farms to use vaccination protocols more frequently, to prevent pneumonia and diarrhea in calves, was observed (see Table S2). Vaccination of dams and calves up to 12 months probably contributed to approximately half of the FPM farmers considering pneumonia as the main problem in their calves (34.5%; 30/87) when compared to that reported by MMS farmers (72.2%; 13/18; $p < 0.01$). Nevertheless, calf density can also contribute to pneumonia in MMS, and is higher in that type of farm due to its dual-purpose production. Moreover, calves exposed to extreme weather may not be able to regulate their body temperature with their own thermoregulation mechanisms, leading to significant losses in performance, and average daily gain [54]. This circumstance is more frequently observed in MMS farms than in FPM farms.

In terms of biosecurity, FMP farms adopted more measures, such as the existence of spaces to isolate sick animals, and less movement of animals using public roads. However, biosecurity measures were scarcely implemented by the dairy farms surveyed on São Miguel Island, which is an important observation of the study. In MMS farms, there was a tendency for more frequent movement of animals (transhumance), and of purchases and sales of living animals ($p = 0.10$). Despite this, only 38.5% of the respondent farmers (see Table S1) implemented quarantine measures when introducing animals. The movement of animals is the main cause of the spread of disease, and the introduction of new animals is a farm's highest risk of being infected with new diseases [55,56]. Moreover, important biosecurity practices, such as providing their own clothing to visitors or the use of footbaths, were not implemented at all in the farms surveyed on São Miguel Island. For example, in Belgium, between 66% and 61% carry out such methods [55]. The different kinds of production systems found in the Azores probably contribute to these differences. Finally, all Azorean farms bury dead animals on pastures, a practice almost abandoned in the rest of the world. In the USA, only 27.3% of farms bury their dead cows, while 29.2% add carcasses to compost, and 27.2% put them to render [40]. In fact, the burial of dead animals in the EU is forbidden, preventing contamination of soils and water by pathogenic and zoonotic microbials [57]. Nevertheless, some exceptions [58] are made in remote areas and specific conditions, such as Azores.

More frequent forage nutritional analysis, mainly when a new diet is available, and diet adjustment, according to lactational curve of dairy cattle, should be improved by farmers to optimize milk yield [59]. Cows on FMP farms spent more time indoors, making it easier to control the diets and intake rate, and to provide adequate feed quality, as demonstrated in previous studies [10]. Adequate nutritional management is essential for cattle health, welfare, reproductive performance, and milk yield [42], thus also being essential for farm productivity and economic sustainability [42,60]. In the present study, MMS farms, when compared to FMP farms, received nutritional assessments less frequently (see Table 2). As a result, fewer forage nutritional analyses were conducted ($p < 0.001$), and fewer scoring body conditions were performed ($p = 0.002$) on MMS farms. Consistent with this, the TMR system was more commonly observed in FMP farms ($p < 0.001$), with this technology allowing better nutritional management, and the possibility of giving animals homogeneous feed and a balanced diet [10,61,62].

Nutrition is also fundamental at the peri- and postpartum periods, during which cows undergo several metabolic and hormonal changes, which compromise their immune functions [63]. Negative energy balance is common in postpartum dairy cows, due to low feed intake and the inability to reach cow energetic requirements for milk yield [64]. Appropriate nutritional management of the peripartum cow is essential to avoid such problems [65], together with other practices [66], reducing the incidence of postpartum problems such as retained placenta, mastitis, metritis, and endometritis. These issues can all lead to several economic losses, and early culling rates [66,67]. Certain nutritional supplements, such as vitamin D, lysine, methionine, and others can be considered, in order to increase performance after calving [68,69]. In our study, monensin supplementation was the most common practice (27.6%).

In addition to the type of machine used to milk (used to classify the type of farms), relevant differences were found in the milking routines and hygiene practices during milking (see Table 3), with FMP farms implementing indicated routines more frequently [70], such as pre- and post-dipping and the use of paper towels. It is easier to implement adequate milking routines in fixed milking parlors than with mobile milking systems [71,72]. However, even when observing only FMP farms, we detected a huge difference in the possibilities of receiving specialized technical advice on the island, compared to that of other regions in Europe and Portugal; this clearly limits the implementation of new technologies, and the possibility of adequate evolution in the Azores [5], despite it being one of the main milk-producing regions in Portugal. Adequate milking practices, new technologies, and udder health control programs are key points to control mastitis. Although mastitis was not one of the top three most important diseases affecting São Miguel dairy farms, according to farmers' perceptions, it induces large economic losses which is in in agreement with the literature [73–75].

After rearing, reproductive problems, mastitis, and placenta retention were the challenges highlighted by our farmers as their major problems (see Figure 3). Similarly, infertility, clinical mastitis and lameness were the principal diseases that affected American dairy farms in 2014 [40].

In our study, 41% of the farmers considered lameness as a major problem, which can be considered a high prevalence for grazing systems. There are several causes that can justify this evidence, for example, the influence of (hard) floors found in public roads (transhumance); the high, steady humidity, mainly in winter season; scarcity of trimming programs and footbath use. Lameness negatively influences productive and reproductive traits [76], causing significant economic impact to farms [77].

There is a low to moderate adherence/application of preventive measures to combat these and other diseases on São Miguel Island. Only 42.9% of the surveyed farms carry out at least one vaccination. There was a large discrepancy when compared to data from previous studies, such as on Irish dairy farms, where only 13% did not apply any vaccinations [78]. The low vaccination rate observed in our study probably relates to the cost of vaccination, lack of immediate health improvement, previous experience of failure to control the disease by the farmers, and a general lack of education of farmers. However, vaccination by itself is not enough. Proactive co-operation between veterinarians and farmers is essential to optimize health and sustainability in dairy farms [79].

All these needs detected in the present study indicate an urgent necessity for improvement in farmer education, as previously seen in other regions [49]. Assuring that the farmers have the knowledge and equipment to record their own data, and are then able to calculate and interpret basic indicators, is essential to the early detection of diseases and problems [2]. In fact, the greater part of the health information reported in this study came from farmers' perception. A health data record is important to objectively evaluate the herd health and take appropriate decisions, and can contribute to a low degree of health program implementation. Nevertheless, farmer education is only a part of the equation to drive behavioral change. According to Michie et al. [80] individual behavior changes are related to capability (knowledge and skills, via education), motivation (brain stimulating process) and opportunity (outside events). The dairy industry is a business which provides incomes to farmers, and is the strongest motivation for farmers. The production efficiency and herd health management improvements, and new opportunities coming from social demands, including consumer perspectives on animal welfare and environmental impact, seem to be crucial keys for behavioral changes.

Consumers have a general opinion that grazing cattle are in a better condition in terms of health and welfare [81]. This is important and may change the markets. In fact, consumers have increasingly higher interest in animal production conditions, animal welfare, sustainability of the systems, and environmental protection [82], which are the main drivers of changes in legislation, especially in the European Union [83]. All of these drivers of change can be considered an opportunity for areas that produce mainly based on pasture.

This evolution also must be compatible with preserving the particular aspects of the Azorean Islands, and production systems closely linked to the natural environment and to agrarian populations [5,81,82]. The environmental impact of the dairy industry should be attenuated, ensuring economic viability of the farms with dairy added-value programs such as "happy cows" [5].

Our study reveals that both FMP and MMS have the potential for progress, improving their efficiency and preserving animal welfare, in extensive and intensive production systems.

5. Conclusions

There is a clear difference in health management between these two types of farms; MMS farms are associated with a more traditional production approach, while FMP farms have transitioned to more specialized dairy systems. However, all farms on São Miguel Island would benefit from animal welfare, productivity, resource efficiency, and sustainability. This would be further improved by the implementation of preventive and structured control programs, assessed by professional advisors, and thus enhance health, welfare efficiency and profitability. However, São Miguel Island dairy farms must continue taking advantage of their idiosyncrasy, benefiting from the natural resources available, stressing the production of "green milk".

Supplementary Materials: The following are available online at https://www.mdpi.com/article/10.3390/ani11123394/s1, Table S1. Characterization of farms and main biosecurity measures in dairy farms with fixed (FMP) versus mobile (MMS) milking machines. Table S2. Calving, rearing management and facilities in dairy farms with fixed (FMP) versus mobile (MMS) milking systems. Table S3. Lameness and lameness control in dairy farms with fixed (FMP) versus mobile (MMS) milking systems. Table S4. Dry-off time and prepartum measures adopted in dairy farms with fixed (FMP) versus mobile (MMS) milking systems.

Author Contributions: Conceptualization, methodology, formal analysis, data curation, writing—original draft preparation, and writing—review and editing were equally performed by I.M., A.F.-N., J.S. and S.A. All authors have read and agreed to the published version of the manuscript.

Funding: This research was funded by UIDB/CVT/00772/2020 (Fundação para a Ciência e Tecnologia).

Institutional Review Board Statement: The study was conducted according to the guidelines of the Declaration of Helsinki and was approved by the Institutional Review Board of the University of Trás-os-Montes and Alto Douro (UTAD) with protocol code request n. 2020111053000832 on 10 November 2020.

Informed Consent Statement: Informed consent was obtained from all respondents.

Data Availability Statement: The data that support the findings of this study are available on request from the corresponding author (J.S.).

Acknowledgments: This paper is obtained from the M.Sc. thesis in Veterinary Medicine of the first author at the University of Trás-os-Montes and Alto Douro (UTAD), Vila Real, Portugal.

Conflicts of Interest: The authors declare no conflict of interest.

References

1. Schulte, H.D.; Musshoff, O.; Meuwissen, M.P.M. Considering Milk Price Volatility for Investment Decisions on the Farm Level after European Milk Quota Abolition. *J. Dairy Sci.* **2018**, *101*, 7531–7539. [CrossRef]
2. LeBlanc, S.J.; Lissemore, K.D.; Kelton, D.F.; Duffield, T.F.; Leslie, K.E. Major Advances in Disease Prevention in Dairy Cattle. *J. Dairy Sci.* **2006**, *89*, 1267–1279. [CrossRef]
3. Derks, M.; van Werven, T.; Hogeveen, H.; Kremer, W.D.J. Associations between Farmer Participation in Veterinary Herd Health Management Programs and Farm Performance. *J. Dairy Sci.* **2014**, *97*, 1336–1347. [CrossRef] [PubMed]
4. Crowe, M.A.; Hostens, M.; Opsomer, G. Reproductive Management in Dairy Cows—The Future. *Ir. Vet. J.* **2018**, *71*, 1. [CrossRef] [PubMed]
5. De Almeida, A.M.; Alvarenga, P.; Fangueiro, D. The Dairy Sector in the Azores Islands: Possibilities and Main Constraints towards Increased Added Value. *Trop. Anim. Health Prod.* **2020**, *53*, 40. [CrossRef] [PubMed]
6. Arnott, G.; Ferris, C.P.; O'Connell, N.E. Review: Welfare of Dairy Cows in Continuously Housed and Pasture-Based Production Systems. *Anim. Int. J. Anim. Biosci.* **2017**, *11*, 261–273. [CrossRef] [PubMed]
7. Smid, A.-M.C.; Weary, D.M.; von Keyserlingk, M.A.G. The Influence of Different Types of Outdoor Access on Dairy Cattle Behavior. *Front. Vet. Sci.* **2020**, *7*, 257. [CrossRef] [PubMed]
8. Legrand, A.L.; von Keyserlingk, M.A.G.; Weary, D.M. Preference and Usage of Pasture versus Free-Stall Housing by Lactating Dairy Cattle. *J. Dairy Sci.* **2009**, *92*, 3651–3658. [CrossRef] [PubMed]
9. Crump, A.; Jenkins, K.; Bethell, E.J.; Ferris, C.P.; Kabboush, H.; Weller, J.; Arnott, G. Optimism and Pasture Access in Dairy Cows. *Sci. Rep.* **2021**, *11*, 4882. [CrossRef] [PubMed]

10. Charlton, G.L.; Rutter, S.M.; East, M.; Sinclair, L.A. Effects of Providing Total Mixed Rations Indoors and on Pasture on the Behavior of Lactating Dairy Cattle and Their Preference to Be Indoors or on Pasture. *J. Dairy Sci.* **2011**, *94*, 3875–3884. [CrossRef]
11. Bewley, J.M.; Robertson, L.M.; Eckelkamp, E.A. A 100-Year Review: Lactating Dairy Cattle Housing Management. *J. Dairy Sci.* **2017**, *100*, 10418–10431. [CrossRef]
12. Dood, F.H.; Hall, H.S. History and Development. In *Machine Milking and Lactation*; Insight Books: Huntington, VT, USA, 1992; Chapter 1.
13. Greenhalgh, J.; Domigan, I.; Greig, B.; Herud, G. *Pilot Study of a Mobile Milking System: An Investigation of Its Potential Use in New Zealand*; Centre of Excelence in Farm Bussiness Management: Wellington, New Zealand, 2012.
14. Sobczyński, T.; Klepacka, A.M.; Revoredo-Giha, C.; Florkowski, W.J. Dairy Farm Cost Efficiency in Leading Milk-Producing Regions in Poland. *J. Dairy Sci.* **2015**, *98*, 8294–8307. [CrossRef] [PubMed]
15. Charlton, G.L.; Rutter, S.M.; East, M.; Sinclair, L.A. The Motivation of Dairy Cows for Access to Pasture. *J. Dairy Sci.* **2013**, *96*, 4387–4396. [CrossRef] [PubMed]
16. Hennessy, D.; Delaby, L.; van den Pol-van Dasselaar, A.; Shalloo, L. Increasing Grazing in Dairy Cow Milk Production Systems in Europe. *Sustainability* **2020**, *12*, 2443. [CrossRef]
17. Daley, C.A.; Abbott, A.; Doyle, P.S.; Nader, G.A.; Larson, S. A Review of Fatty Acid Profiles and Antioxidant Content in Grass-Fed and Grain-Fed Beef. *Nutr. J.* **2010**, *9*, 10. [CrossRef]
18. Lessire, F.; Moula, N.; Hornick, J.-L.; Dufrasne, I. Systematic Review and Meta-Analysis: Identification of Factors Influencing Milking Frequency of Cows in Automatic Milking Systems Combined with Grazing. *Animals* **2020**, *10*, 913. [CrossRef] [PubMed]
19. Prache, S.; Martin, B.; Coppa, M. Review: Authentication of Grass-Fed Meat and Dairy Products from Cattle and Sheep. *Anim. Int. J. Anim. Biosci.* **2020**, *14*, 854–863. [CrossRef]
20. Walsh, S.W.; Williams, E.J.; Evans, A.C.O. A Review of the Causes of Poor Fertility in High Milk Producing Dairy Cows. *Anim. Reprod. Sci.* **2011**, *123*, 127–138. [CrossRef]
21. Berghof, T.V.L.; Poppe, M.; Mulder, H.A. Opportunities to Improve Resilience in Animal Breeding Programs. *Front. Genet.* **2019**, *9*, 692. [CrossRef]
22. Ma, W.; Bicknell, K.; Renwick, A. Production Intensification and Animal Health Expenditure on Dairy Farms in New Zealand. *J. Dairy Sci.* **2020**, *103*, 1598–1607. [CrossRef]
23. Azevedo, C.; Pacheco, D.; Soares, L.; Romão, R.; Moitoso, M.; Maldonado, J.; Guix, R.; Simões, J. Prevalence of Contagious and Environmental Mastitis-Causing Bacteria in Bulk Tank Milk and Its Relationships with Milking Practices of Dairy Cattle Herds in São Miguel Island (Azores). *Trop. Anim. Health Prod.* **2016**, *48*, 451–459. [CrossRef] [PubMed]
24. Macdonald, K.A.; Penno, J.W.; Lancaster, J.A.S.; Bryant, A.M.; Kidd, J.M.; Roche, J.R. Production and Economic Responses to Intensification of Pasture-Based Dairy Production Systems. *J. Dairy Sci.* **2017**, *100*, 6602–6619. [CrossRef] [PubMed]
25. Roche, J.R.; Berry, D.P.; Bryant, A.M.; Burke, C.R.; Butler, S.T.; Dillon, P.G.; Donaghy, D.J.; Horan, B.; Macdonald, K.A.; Macmillan, K.L. A 100-Year Review: A Century of Change in Temperate Grazing Dairy Systems. *J. Dairy Sci.* **2017**, *100*, 10189–10233. [CrossRef] [PubMed]
26. Dutra Silva, L.; Brito de Azevedo, E.; Vieira Reis, F.; Bento Elias, R.; Silva, L. Limitations of Species Distribution Models Based on Available Climate Change Data: A Case Study in the Azorean Forest. *Forests* **2019**, *10*, 575. [CrossRef]
27. Hernández, A.; Kutiel, H.; Trigo, R.M.; Valente, M.A.; Sigró, J.; Cropper, T.; Santo, F.E. New Azores Archipelago Daily Precipitation Dataset and Its Links with Large-Scale Modes of Climate Variability: Azores Archipelago Precipitation and Modes of Climate Variability. *Int. J. Climatol.* **2016**, *36*, 4439–4454. [CrossRef]
28. Medeiros, I. Herd Health Management of Dairy Cattle on the Island of São Miguel. Available online: https://docs.google.com/forms/d/e/1FAIpQLSdX8UCKp_RcbaT4ycDTp2XQGNRMvuSCrjWuaLDjfckX76Teiw/viewform?usp=embed_facebook (accessed on 18 November 2021).
29. Pugliese, M.; Biondi, V.; Passantino, A.; Licitra, F.; Alibrandi, A.; Zanghi, A.; Conte, F.; Marino, G. Welfare Assessment in Intensive and Semi-intensive Dairy Cattle Management System in Sicily. *Anim. Sci. J.* **2021**, *92*. [CrossRef]
30. O'Shaughnessy, J.; Mee, J.F.; Doherty, M.L.; Crosson, P.; Barrett, D.; O'Grady, L.; Earley, B. Herd Health Status and Management Practices on 16 Irish Suckler Beef Farms. *Ir. Vet. J.* **2013**, *66*, 21. [CrossRef]
31. Regulation (EU) 2016/679 of the European Parliament and of the Council of 27 April 2016 on the Protection of Natural Persons with Regard to the Processing of Personal Data and on the Free Movement of Such Data, and Repealing Directive 95/46/EC (General Data Protection Regulation). *Off. J. Eur. Union* **2016**, *59*, 1–88.
32. Thrusfield, M. *Veterinary Epidemiology*, 2nd ed.; Blackwell Science: Oxford, UK, 2005.
33. Luepsen, H. Comparison of Nonparametric Analysis of Variance Methods: A Vote for van Der Waerden. *Commun. Stat. Simul. Comput.* **2018**, *47*, 2547–2576. [CrossRef]
34. Tedeschi, L.O.; Molle, G.; Menendez, H.M.; Cannas, A.; Fonseca, M.A. The Assessment of Supplementation Requirements of Grazing Ruminants Using Nutrition Models. *Transl. Anim. Sci.* **2019**, *3*, 811–828. [CrossRef]
35. Morais, T.; Teixeira, R.; Rodrigues, N.; Domingos, T. Carbon Footprint of Milk from Pasture-Based Dairy Farms in Azores, Portugal. *Sustainability* **2018**, *10*, 3658. [CrossRef]
36. Denis-Robichaud, J.; Cerri, R.L.A.; Jones-Bitton, A.; LeBlanc, S.J. Survey of Reproduction Management on Canadian Dairy Farms. *J. Dairy Sci.* **2016**, *99*, 9339–9351. [CrossRef]

37. Saint-Dizier, M.; Chastant-Maillard, S. Towards an Automated Detection of Oestrus in Dairy Cattle: Automated Oestrus Detection in Dairy Cattle. *Reprod. Domest. Anim.* **2012**, *47*, 1056–1061. [CrossRef]
38. Mottram, T. Animal Board Invited Review: Precision Livestock Farming for Dairy Cows with a Focus on Oestrus Detection. *Animal* **2016**, *10*, 1575–1584. [CrossRef] [PubMed]
39. Inchaisri, C.; Jorritsma, R.; Vos, P.L.A.M.; van der Weijden, G.C.; Hogeveen, H. Economic Consequences of Reproductive Performance in Dairy Cattle. *Theriogenology* **2010**, *74*, 835–846. [CrossRef]
40. USDA Health and Management Practices on U.S. Dairy Operations. 2014. Available online: https://www.aphis.usda.gov/animal_health/nahms/dairy/downloads/dairy14/Dairy14_dr_PartIII.pdf (accessed on 22 April 2021).
41. Brooks, S.; House, J.K.; Ingenhoff, L. The Effect of Participation in a Regular Reproductive Herd Health Program on 80-Day Submission Rate and 100-Day in-Calf Rate in Year-Round Calving Dairy Farms. *Aust. Vet. J.* **2021**, *99*, 79–85. [CrossRef] [PubMed]
42. Schuenemann, G.M.; Eastridge, M.L.; Weiss, W.P.; Workman, J.D.; Bas, S.; Rajala-Schultz, P. Dairy Nutrition Management: Assessing a Comprehensive Continuing Education Program for Veterinary Practitioners. *J. Dairy Sci.* **2011**, *94*, 2648–2656. [CrossRef]
43. Kertz, A.F.; Hill, T.M.; Quigley, J.D.; Heinrichs, A.J.; Linn, J.G.; Drackley, J.K. A 100-Year Review: Calf Nutrition and Management. *J. Dairy Sci.* **2017**, *100*, 10151–10172. [CrossRef]
44. Matamala, F.; Strappini, A.; Sepúlveda-Varas, P. Dairy Cow Behaviour around Calving: Its Relationship with Management Practices and Environmental Conditions. *Austral J. Vet. Sci.* **2021**, *53*, 9–22. [CrossRef]
45. Timlin, M.; Tobin, J.T.; Brodkorb, A.; Murphy, E.G.; Dillon, P.; Hennessy, D.; O'Donovan, M.; Pierce, K.M.; O'Callaghan, T.F. The Impact of Seasonality in Pasture-Based Production Systems on Milk Composition and Functionality. *Foods* **2021**, *10*, 607. [CrossRef]
46. Tao, S.; Orellana Rivas, R.M.; Marins, T.N.; Chen, Y.-C.; Gao, J.; Bernard, J.K. Impact of Heat Stress on Lactational Performance of Dairy Cows. *Theriogenology* **2020**, *150*, 437–444. [CrossRef]
47. Spaans, O.K.; Macdonald, K.A.; Neal, M.; Auldist, M.J.; Lancaster, J.A.S.; Bryant, A.M.; Doole, G.J.; Roche, J.R. A Quantitative Case Study Assessment of Biophysical and Economic Effects from Altering Season of Calving in Temperate Pasture-Based Dairy Systems. *J. Dairy Sci.* **2019**, *102*, 11523–11535. [CrossRef] [PubMed]
48. Godden, S.M.; Lombard, J.E.; Woolums, A.R. Colostrum Management for Dairy Calves. *Vet. Clin. N. Am. Food Anim. Pract.* **2019**, *35*, 535–556. [CrossRef] [PubMed]
49. Relić, R.; Starič, J.; Ježek, J. Management Practices That Influence the Welfare of Calves on Small Family Farms. *J. Dairy Res.* **2020**, *87*, 93–98. [CrossRef] [PubMed]
50. Heinemann, C.; Leubner, C.D.; Hayer, J.J.; Steinhoff-Wagner, J. Hygiene Management in Newborn Individually Housed Dairy Calves Focusing on Housing and Feeding Practices. *J. Anim. Sci.* **2021**, *99*. [CrossRef]
51. Heinrichs, A.J.; Jones, C.M.; Erickson, P.S.; Chester-Jones, H.; Anderson, J.L. Symposium Review: Colostrum Management and Calf Nutrition for Profitable and Sustainable Dairy Farms. *J. Dairy Sci.* **2020**, *103*, 5694–5699. [CrossRef]
52. Calderón-Amor, J.; Gallo, C. Dairy Calf Welfare and Factors Associated with Diarrhea and Respiratory Disease among Chilean Dairy Farms. *Animals* **2020**, *10*, 1115. [CrossRef]
53. Svensson, C.; Alvåsen, K.; Eldh, A.C.; Frössling, J.; Lomander, H. Veterinary Herd Health Management-Experience among Farmers and Farm Managers in Swedish Dairy Production. *Prev. Vet. Med.* **2018**, *155*, 45–52. [CrossRef] [PubMed]
54. Roland, L.; Drillich, M.; Klein-Jöbstl, D.; Iwersen, M. Invited Review: Influence of Climatic Conditions on the Development, Performance, and Health of Calves. *J. Dairy Sci.* **2016**, *99*, 2438–2452. [CrossRef]
55. Sarrazin, S.; Cay, A.B.; Laureyns, J.; Dewulf, J. A Survey on Biosecurity and Management Practices in Selected Belgian Cattle Farms. *Prev. Vet. Med.* **2014**, *117*, 129–139. [CrossRef]
56. Denis-Robichaud, J.; Kelton, D.F.; Bauman, C.A.; Barkema, H.W.; Keefe, G.P.; Dubuc, J. Biosecurity and Herd Health Management Practices on Canadian Dairy Farms. *J. Dairy Sci.* **2019**, *102*, 9536–9547. [CrossRef]
57. Gwyther, C.L.; Williams, A.P.; Golyshin, P.N.; Edwards-Jones, G.; Jones, D.L. The Environmental and Biosecurity Characteristics of Livestock Carcass Disposal Methods: A Review. *Waste Manag.* **2011**, *31*, 767–778. [CrossRef]
58. Regulation (EC) No 1069/2009 of the European Parliament and of the Council of 21 October 2009 Laying down Health Rules as Regards Animal by-Products and Derived Products Not Intended for Human Consumption and Repealing Regulation (EC) No 1774/2002 (Animal by-Products Regulation). *Off. J. Eur. Union* **2009**, *300*, 1–33.
59. Erickson, P.S.; Kalscheur, K.F. Nutrition and Feeding of Dairy Cattle. In *Animal Agriculture*; Academic Press: London, UK, 2020; pp. 157–180. ISBN 978-0-12-817052-6.
60. Bach, L. Effects of Nutrition and Genetics on Fertility in Dairy Cows. *Reprod. Fertil. Dev.* **2018**, *31*, 40–54. [CrossRef] [PubMed]
61. Cardoso, F.C.; Kalscheur, K.F.; Drackley, J.K. Symposium Review: Nutrition Strategies for Improved Health, Production, and Fertility during the Transition Period. *J. Dairy Sci.* **2020**, *103*, 5684–5693. [CrossRef]
62. Casasús, I.; Villalba, D.; Gracia Chapullé, J.L. Los Sistemas Unifeed En La Alimentación de Rumiantes. *Ganad. Rev. Téc. Ganad.* **2012**, *79*, 56–60.
63. Drackley, J.K.; Cardoso, F.C. Prepartum and Postpartum Nutritional Management to Optimize Fertility in High-Yielding Dairy Cows in Confined TMR Systems. *Anim. Int. J. Anim. Biosci.* **2014**, *8* (Suppl. 1), 5–14. [CrossRef] [PubMed]
64. Xu, W.; van Knegsel, A.; Saccenti, E.; van Hoeij, R.; Kemp, B.; Vervoort, J. Metabolomics of Milk Reflects a Negative Energy Balance in Cows. *J. Proteome Res.* **2020**, *19*, 2942–2949. [CrossRef]

65. Taniguchi, A.; Nishikawa, T.; Morita, Y. Nutritional Condition in the Dry Period Is Related to the Incidence of Postpartum Subclinical Endometritis in Dairy Cattle. *Anim. Biosci.* **2021**, *34*, 539–545. [CrossRef] [PubMed]
66. Sheldon, I.M.; Molinari, P.C.C.; Ormsby, T.J.R.; Bromfield, J.J. Preventing Postpartum Uterine Disease in Dairy Cattle Depends on Avoiding, Tolerating and Resisting Pathogenic Bacteria. *Theriogenology* **2020**, *150*, 158–165. [CrossRef]
67. Piñeiro, J.M.; Menichetti, B.T.; Barragan, A.A.; Relling, A.E.; Weiss, W.P.; Bas, S.; Schuenemann, G.M. Associations of Postpartum Lying Time with Culling, Milk Yield, Cyclicity, and Reproductive Performance of Lactating Dairy Cows. *J. Dairy Sci.* **2019**, *102*, 3362–3375. [CrossRef]
68. Lee, C.; Lobos, N.E.; Weiss, W.P. Effects of Supplementing Rumen-Protected Lysine and Methionine during Prepartum and Postpartum Periods on Performance of Dairy Cows. *J. Dairy Sci.* **2019**, *102*, 11026–11039. [CrossRef]
69. Wisnieski, L.; Brown, J.L.; Holcombe, S.J.; Gandy, J.C.; Sordillo, L.M. Serum Vitamin D Concentrations at Dry-off and Close-up Predict Increased Postpartum Urine Ketone Concentrations in Dairy Cattle. *J. Dairy Sci.* **2020**, *103*, 1795–1806. [CrossRef]
70. Ruegg, P.L. A 100-Year Review: Mastitis Detection, Management, and Prevention. *J. Dairy Sci.* **2017**, *100*, 10381–10397. [CrossRef]
71. Akam, D.N. Technical Development of Milking Machines. *BSAP Occas. Publ.* **1980**, *2*, 55–64. [CrossRef]
72. Odorčić, M.; Rasmussen, M.D.; Paulrud, C.O.; Bruckmaier, R.M. Review: Milking Machine Settings, Teat Condition and Milking Efficiency in Dairy Cows. *Anim. Int. J. Anim. Biosci.* **2019**, *13*, s94–s99. [CrossRef] [PubMed]
73. Ismail, Z.B. Mastitis Vaccines in Dairy Cows: Recent Developments and Recommendations of Application. *Vet. World* **2017**, *10*, 1057–1062. [CrossRef] [PubMed]
74. Cobirka, M.; Tancin, V.; Slama, P. Epidemiology and Classification of Mastitis. *Anim. Open Access J.* **2020**, *10*, 2212. [CrossRef] [PubMed]
75. Zigo, F.; Vasil', M.; Ondrašovičová, S.; Výrostková, J.; Bujok, J.; Pecka-Kielb, E. Maintaining Optimal Mammary Gland Health and Prevention of Mastitis. *Front. Vet. Sci.* **2021**, *8*, 607311. [CrossRef] [PubMed]
76. Ranjbar, S.; Rabiee, A.R.; Gunn, A.; House, J.K. Identifying Risk Factors Associated with Lameness in Pasture-Based Dairy Herds. *J. Dairy Sci.* **2016**, *99*, 7495–7505. [CrossRef] [PubMed]
77. Puerto, M.A.; Shepley, E.; Cue, R.I.; Warner, D.; Dubuc, J.; Vasseur, E. The Hidden Cost of Disease: II. Impact of the First Incidence of Lameness on Production and Economic Indicators of Primiparous Dairy Cows. *J. Dairy Sci.* **2021**, *104*, 7944–7955. [CrossRef] [PubMed]
78. Sayers, R.G.; Sayers, G.P.; Mee, J.F.; Good, M.; Bermingham, M.L.; Grant, J.; Dillon, P.G. Implementing Biosecurity Measures on Dairy Farms in Ireland. *Vet. J.* **2013**, *197*, 259–267. [CrossRef] [PubMed]
79. Higgins, H.M.; Huxley, J.N.; Wapenaar, W.; Green, M.J. Proactive Dairy Cattle Disease Control in the UK: Veterinary Surgeons' Involvement and Associated Characteristics. *Vet. Rec.* **2013**, *173*, 246. [CrossRef]
80. Michie, S.; van Stralen, M.M.; West, R. The Behaviour Change Wheel: A New Method for Characterising and Designing Behaviour Change Interventions. *Implement. Sci.* **2011**, *6*, 42. [CrossRef]
81. Armbrecht, L.; Lambertz, C.; Albers, D.; Gauly, M. Assessment of Welfare Indicators in Dairy Farms Offering Pasture at Differing Levels. *Anim. Int. J. Anim. Biosci.* **2019**, *13*, 2336–2347. [CrossRef] [PubMed]
82. Cannas da Silva, J.; Noordhuizen, J.P.T.M.; Vagneur, M.; Bexiga, R.; Gelfert, C.C.; Baumgartner, W. Veterinary Dairy Herd Health Management in Europe: Constraints and Perspectives. *Vet. Q.* **2006**, *28*, 23–32. [CrossRef]
83. *European Comission Scoping Study Delivering on EU Food Safety and Nutrition in 2050—Scenarios of Future Change and Policy Responses*; Final Report; European Commision: Brussels, Belgium, 2013.

MDPI
St. Alban-Anlage 66
4052 Basel
Switzerland
Tel. +41 61 683 77 34
Fax +41 61 302 89 18
www.mdpi.com

Animals Editorial Office
E-mail: animals@mdpi.com
www.mdpi.com/journal/animals